Edited by
Ben L. Feringa and
Wesley R. Browne

Molecular Switches

Volume 1

Related Titles

Urban, M. W. (Ed.)

Handbook of Stimuli-Responsive Materials

2011
ISBN: 978-3-527-32700-3

Samori, P., Cacialli, F. (Eds.)

Functional Supramolecular Architectures

for Organic Electronics and Nanotechnology

2011
ISBN: 978-3-527-32611-2

Sauvage, J.-P., Gaspard, P. (Eds.)

From Non-Covalent Assemblies to Molecular Machines

2010
ISBN: 978-3-527-32277-0

Pignataro, B. (Ed.)

Ideas in Chemistry and Molecular Sciences

Advances in Nanotechnology, Materials and Devices

2010
ISBN: 978-3-527-32543-6

Albini, A., Fagnoni, M. (Eds.)

Handbook of Synthetic Photochemistry

2010
ISBN: 978-3-527-32391-3

Balzani, V., Credi, A., Venturi, M.

Molecular Devices and Machines

Concepts and Perspectives for the Nanoworld

Second edition, 2008
ISBN: 978-3-527-31800-1

Edited by Ben L. Feringa and Wesley R. Browne

Molecular Switches

Second, Completely Revised and Enlarged Edition

Volume 1

WILEY-VCH Verlag GmbH & Co. KGaA

The Editors

Prof. Dr. Ben L. Feringa
Stratingh Institute for Chemistry
& Zernike Institute for
Advanced Materials
Faculty of Mathematics and
Natural Sciences
University of Groningen
Nijenborgh 4
9747 AG Groningen
The Netherlands

Dr. Wesley R. Browne
Stratingh Institute for Chemistry
& Zernike Institute for
Advanced Materials
Faculty of Mathematics and
Natural Sciences
University of Groningen
Nijenborgh 4
9747 AG Groningen
The Netherlands

Cover
The graphic material used in the cover illustration was kindly provided by the editors Ben L. Feringa and Wesley R. Browne (University of Groningen)

All books published by **Wiley-VCH** are carefully produced. Nevertheless, authors, editors, and publisher do not warrant the information contained in these books, including this book, to be free of errors. Readers are advised to keep in mind that statements, data, illustrations, procedural details or other items may inadvertently be inaccurate.

Library of Congress Card No.: applied for

British Library Cataloguing-in-Publication Data
A catalogue record for this book is available from the British Library.

Bibliographic information published by the Deutsche Nationalbibliothek
The Deutsche Nationalbibliothek lists this publication in the Deutsche Nationalbibliografie; detailed bibliographic data are available on the Internet at <http://dnb.d-nb.de>.

© 2011 WILEY-VCH Verlag & Co. KGaA, Boschstr. 12, 69469 Weinheim, Germany

All rights reserved (including those of translation into other languages). No part of this book may be reproduced in any form – by photoprinting, microfilm, or any other means – nor transmitted or translated into a machine language without written permission from the publishers. Registered names, trademarks, etc. used in this book, even when not specifically marked as such, are not to be considered unprotected by law.

Cover Design   Formgeber, Eppelheim
Typesetting   Laserwords Private Limited, Chennai, India
Printing and Binding   Fabulous Printers Pte Ltd, Singapore

Printed in Singapore
Printed on acid-free paper

ISBN: 978-3-527-31365-5
ePDF ISBN: 978-3-527-63442-2
ePub ISBN: 978-3-527-63441-5
Mobi ISBN: 978-3-527-63443-9
oBook ISBN: 978-3-527-63440-8

Contents

Preface *XVII*
List of Contributors *XIX*
Abbreviations *XXVII*

Part I   **Molecular Switching** *1*

1   **Multifunctional Diarylethenes** *3*
C. Chad Warford, Vincent Lemieux, and Neil R. Branda
1.1   Introduction *3*
1.2   Electrochemical Ring-Closing and Ring-Opening of DTEs *4*
1.2.1   Electrochemical Behaviour of DTEs *4*
1.2.2   Fully Functional Photo- and Electrochromic DTEs *14*
1.3   Using Dithienylethenes to Modulate How Chemicals React or Interact with Others *14*
1.3.1   Control of Chemical Behaviour through Photoswitching *14*
1.3.2   Controlling Molecular Interactions and Reactions Using the Steric Differences in the DTE Photoisomers *16*
1.3.3   Controlling Molecular Reactions and Interactions Using the Electronic Differences in the DTE Photoisomers *19*
1.3.4   A Specific Approach to Using the Changes in Location of π-Bonds to Control Reactivity *21*
1.4   Gated Photochromism *24*
1.4.1   'Gated' Photochromism Based on Steric Effects *24*
1.4.2   Intramolecular 'Gating' *25*
1.4.3   Intermolecular 'Gating' *25*
1.4.4   Gating Based on Electronic Effects *27*
1.5   Reactivity-Gated Photochromism Using the Functional Group Effect *31*
1.6   Conclusion *32*
    References *32*

2	**Photoswitchable Molecular Systems Based on Spiropyrans and Spirooxazines** 37
	Vladimir I. Minkin
2.1	Introduction 37
2.2	Mechanism of the Photochromic Reaction 39
2.3	Switching of Physical Properties and Biological Activities via Photochromic Rearrangements of Functionalized Spiropyrans and Spirooxazines 47
2.3.1	Light-Switchable Fluorescence Modulation 47
2.3.2	Photocontrolled Magnetization 51
2.3.3	Photoswitching of Macroscopic Properties of Bulk Materials and Membranes 54
2.3.4	Photoswitchable Modulation of Biological Activities 58
2.4	Spiropyrans and Spirooxazines as Photodynamic Sensors for Metal Ions 61
2.5	Modulation of the Intramolecular Energy-Transfer Processes in SP/SPO-Containing Transition-Metal Complexes 67
2.6	Spiropyran-Containing Photoresponsive Polymers 69
2.7	Spiropyran/Spirooxazine-Containing Biphotochromic Systems 71
2.8	Concluding Remarks 73
	Acknowledgement 74
	References 74
3	**Fulgides and Related Compounds** 81
	Yasushi Yokoyama, Tsuyoshi Gushiken, and Takashi Ubukata
3.1	Introduction – Fulgides 81
3.2	Reviews Dealing with Fulgides 82
3.3	Introduction of New Fulgides towards Molecular Switches 82
3.4	Photophysics of Molecular Switches 84
3.4.1	Investigations into Reaction Pathways 84
3.4.2	Two-Photon-Absorption Excitation 85
3.4.3	Energy Transfer 86
3.5	Towards Optical Recording 87
3.5.1	Wavelength-Multiplied Recording 87
3.5.2	Incident-Angle-Multiplied Recording 87
3.5.3	Nondestructive Readout with Fluorescence 89
3.5.4	Recording with Optical Anisotropy 89
3.5.5	Formation of Nanostructures of Fulgides 90
3.6	Understanding of Molecular Structures from Calculations 91
3.7	Development of Photochromic Switches Closely Related to Fulgides 92
3.8	Perspectives of Research with Fulgides 93
	References 93

4	**Transition Metal-Complexed Catenanes and Rotaxanes as Molecular Machine Prototypes** 97
	Christian Tock, Julien Frey, and Jean Pierre Sauvage
4.1	Introduction 97
4.2	Copper-Complexed [2]Catenanes in Motion: the Archetypes 98
4.2.1	A Copper-Complexed [2]Catenane in Motion with Two Distinct Geometries 98
4.2.2	A Copper-Complexed [2]Catenane in Motion with Three Distinct Geometries 99
4.3	Fighting the Kinetic Inertness of the First Copper-Based Machines; Fast-Moving Pirouetting Rotaxanes 102
4.4	Molecular Motions Driven by Chemical Reactions – Use of a Chemical Reaction to Induce the Contraction/Stretching Process of a Muscle-Like Rotaxane Dimer 106
4.5	Electrochemically Controlled Intramolecular Motion within a Heterodinuclear Bismacrocycle Transition-Metal Complex 111
4.6	Ru(II)-Complexes as Light-Driven Molecular Machine Prototypes 112
4.7	Conclusion and Prospective 116
	References 116
5	**Chiroptical Molecular Switches** 121
	Wesley R. Browne and Ben L. Feringa
5.1	Introduction 121
5.2	Molecular Switching 122
5.2.1	Chiroptical Switches Based on Overcrowded Alkenes 126
5.2.1.1	Enantiomeric Photochromic Switches 127
5.2.1.2	Diastereomeric Photochromic Switches 128
5.2.2	Azobenzene-Based Chiroptical Switching 128
5.2.3	Diarylethene-Based Chiroptical Switches 134
5.3	Chiral Fulgides 138
5.3.1	Redox-Based Chiroptical Molecular Switching 139
5.3.2	Miscellaneous Chiroptical Switches 143
5.3.3	Chiroptical Switching of Luminescence 144
5.4	Light-Driven Molecular Rotary Motors 145
5.4.1	First- and Second-Generation Motors 146
5.4.2	Light-Driven Motors on Surfaces 159
5.4.3	Transmission of Molecular Chiroptical Switching from Bicomponent Molecules to Polymers 164
5.5	Liquid Crystals 167
5.6	Gels 171
5.7	Conclusions and Perspectives 172
	References 173

6	**Multistate/Multifunctional Molecular-Level Systems: Photochromic Flavylium Compounds** 181
	Fernando Pina, A. Jorge Parola, Raquel Gomes, Mauro Maestri, and Vincenzo Balzani
6.1	Introduction 181
6.2	Energy Stimulation 182
6.3	Photochromic Systems 182
6.4	Bistable and Multistable Systems 184
6.5	Nature of the Species Involved in the Chemistry of Flavylium Compounds 186
6.5.1	Thermodynamics of Flavylium Compounds 188
6.6	Thermal Reactions of the 4′-Methoxyflavylium Ion 189
6.7	Photochemical Behaviour of the 4′-Methoxyflavylium Ion 191
6.7.1	Continuous Irradiation 191
6.7.2	Pulsed Irradiation 192
6.8	Flavylium Ions with OH Substituents 193
6.9	Flavylium Ions with Other Substituents 195
6.10	Energy-Level Diagrams 198
6.11	Chemical Process Networks 200
6.11.1	Write-Lock-Read-Unlock-Erase Cycles 203
6.11.2	Reading without Writing in a *Write-Lock-Read-Unlock-Erase Cycle* 206
6.11.3	Micelle Effect on the Write-Lock-Read-Unlock-Erase Cycle 208
6.11.4	Permanent and Temporary Memories 210
6.11.5	Oscillating Absorbance Patterns 211
6.11.6	Colour-Tap Effect 211
6.11.7	Logic Operations 212
6.11.8	Multiple Reaction Patterns 215
6.11.9	Upper-Level Multistate Cycles 216
6.11.10	Multiswitchable System Operated by Proton, Electron and Photon Inputs 219
6.11.11	Nonaqueous Media and Steps towards Solid-State Devices 221
6.12	Conclusions 222
	Acknowledgements 222
	References 223
7	**Nucleic-Acid-Based Switches** 227
	Eike Friedrichs and Friedrich C. Simmel
7.1	Molecular Switches Made from DNA and RNA 227
7.2	Switchable Ribozymes 229
7.2.1	Ribozyme Switching by Antisense Interaction 230
7.2.2	Ribozyme Deactivation by Steric Hindrance 231
7.2.3	Ribozyme Activation by Complex Stabilization 231
7.2.4	Ligand-Induced Stabilization of the Ribozyme Domain 231
7.3	Regulatory RNA Molecules 232

7.3.1	Riboswitches 232
7.3.2	Synthetic RNA Regulatory Switches 234
7.4	Sensor Applications 236
7.4.1	Switches are Sensors 236
7.4.2	Sensor-Construction Requirements 236
7.4.3	Signal Amplification for Lowering Detection Limit 238
7.5	DNA Computing 240
7.6	DNA Machines 241
7.6.1	Prototype Machines Based on the i-Motif Transition 241
7.6.2	Tweezers – a Prototype System for Reversible Switching Devices 242
7.6.3	Switchable Aptamers 243
7.6.4	Devices Based on Double-Crossover Motifs 243
7.6.5	Walkers – towards DNA-Based Motors 245
7.7	Switchable Molecular Networks and Materials 247
7.8	Conclusion and Outlook 248
	Acknowledgements 249
	References 249

Part II	**Switching in Containers, Polymers and Channels** 257
8	**Switching Processes in Cavitands, Containers and Capsules** 259
	Vladimir A. Azov and François Diederich
8.1	Introduction 259
8.2	Switchable Covalently Constructed Cavitands and Container Molecules 261
8.2.1	Characterization of *Vase* and *Kite* Conformations in the Solid State and in Solution 262
8.2.2	Cavitand Immobilization on Surfaces and Switching at Interfaces 266
8.2.3	Synthetic Modifications of the Upper Rim 268
8.2.4	Modular Construction of Extended Switches with Giant Expansion–Contraction Cycles 272
8.2.5	Electrochemically Triggered Switching 274
8.2.6	Switching Molecular Containers 276
8.2.7	Cucurbit[n]urils 282
8.3	H-Bonded Molecular Capsules 283
8.3.1	Glycoluril-Derived H-Bonded Capsules 284
8.3.2	Calix[4]arene and Resorcin[4]arene Capsules 284
8.3.3	Multicomponent Self-Assembled Molecular Containers 290
8.4	Assembly and Disassembly of Metal-Ion-Coordination Cages 290
8.5	Conclusions 293
	Acknowledgements 293
	References 294

9	**Cyclodextrin-Based Switches** *301*
	He Tian and Qiao-Chun Wang
9.1	Introduction *301*
9.2	In and Out Switching *304*
9.3	Back and Forth Switching *306*
9.4	Displacement Switching *310*
9.5	Coordination Switching *313*
9.6	Rearrangement Switching *314*
9.7	Conclusion and Perspective *316*
	Acknowledgement *317*
	References *317*

10	**Photoswitchable Polypeptides** *321*
	Francesco Ciardelli, Simona Bronco, Osvaldo Pieroni, and Andrea Pucci
10.1	Photoresponsive Polypeptides *321*
10.2	Light-Induced Conformational Transitions *324*
10.2.1	Azobenzene-Containing Polypeptides *324*
10.2.2	Spiropyran-Containing Polypeptides *335*
10.2.3	Thioxopeptide Chromophore *339*
10.3	Photostimulated Aggregation–Disaggregation Effects *342*
10.4	Photoeffects in Molecular and Thin Films *344*
10.5	Photoresponsive Polypeptide Membranes *347*
10.6	Summary and Recent Developments *350*
10.7	Towards More Complex Biorelated Photoswitchable Polypeptides *354*
	References *356*

11	**Ion Translocation within Multisite Receptors** *361*
	Valeria Amendola, Marco Bonizzoni, and Luigi Fabbrizzi
11.1	Introduction *361*
11.2	Metal-Ion Translocation: Changing Metal's Oxidation State *362*
11.3	Metal-Ion Translocation: Changing through a pH Variation the Coordinating Properties of One Receptor's Compartment *366*
11.4	The Simultaneous Translocation of Two Metal Ions *381*
11.5	Redox-Driven Anion Translocation *386*
11.6	Anion Swapping in a Heteroditopic Receptor, Driven by a Concentration Gradient *392*
11.7	Conclusions and Perspectives: Further Types of Molecular Machines? *396*
	References *397*

12	**Optically Induced Processes in Azopolymers** *399*
	Cleber R. Mendonça, Débora T. Balogh, Leonardo De Boni, David S. dos Santos Jr., Valtencir Zucolotto, and Osvaldo N. Oliveira Jr.
12.1	Introduction *399*

12.2	Azoaromatic Compounds: Synthesis, Functionality and Film Fabrication *400*
12.3	Applications *401*
12.3.1	Optical Storage *401*
12.3.1.1	Optically Induced Birefringence *401*
12.3.1.2	Optical Storage Experimental Setup *402*
12.3.2	Nonlinear Optical Properties of Azochromophores *406*
12.3.2.1	Two-Photon-Induced Birefringence *408*
12.3.2.2	Coherent Control of the Optically Induced Birefringence *411*
12.3.3	Photoinscription of Surface-Relief Gratings *412*
12.4	Final Remarks and Prospects *417*
	Acknowledgements *417*
	References *417*
13	**Photoresponsive Polymers** *423*
	Zouheir Sekkat and Wolfgang Knoll
13.1	Introduction *423*
13.2	Photo-Orientation by Photoisomerization *423*
13.2.1	Introduction *423*
13.2.2	Photoisomerization of Azobenzenes *425*
13.2.3	Photo-Orientation by Photoisomerization *427*
13.2.3.1	Base Ground Work *427*
13.2.3.2	Theory of Photo-Orientation *429*
13.2.3.2.1	Purely Polarized Transitions Symmetry *430*
13.2.3.2.2	Phenomenological Theory and General Equations *431*
13.2.3.2.3	Dynamical Behaviour of Photo-Orientation *434*
13.2.3.2.4	Early Time Evolution of Photo-Orientation *436*
13.2.3.2.5	Steady State of A ↔ B Photo-Orientation *437*
13.2.4	Photo-Orientation of Azobenzenes: Individualizable Isomers *439*
13.2.4.1	Reorientation within the *trans* → *cis* Photoisomerization *440*
13.2.4.2	Reorientation within the *cis* → *trans* Thermal Isomerization *443*
13.2.5	Photo-Orientation of Azo Dyes: Spectrally Overlapping Isomers *444*
13.2.6	Photo-Orientation of Photochromic Spiropyrans and Diarylethenes *448*
13.2.6.1	Photoisomerization of Spiropyrans and Diarylethenes *449*
13.2.6.2	Spectral Features of Photo-Orientation *450*
13.2.6.3	Photo-Orientation Dynamics and Transitions Symmetry *450*
13.3	Photoisomerization and Photo-Orientation of Azo Dye in Films of Polymer: Molecular Interaction, Free Volume and Polymer Structural Effects *458*
13.3.1	Introduction *458*
13.3.2	Photoisomerization of Azobenzenes in Molecularly Thin Self-Assembled Monolayers: Photo-Orientation and Photomodulation of the Optical Thickness *460*
13.3.2.1	Photoisomerization of Azo-SAMs *460*

13.3.2.2	Photo-Orientation in Molecularly Thin Layers (Smart Monolayers) 460
13.3.2.3	Photomodulation of the Optical Thickness of Molecularly Thin Layers 463
13.3.3	Photoisomerization and Photo-Orientation of Azobenzenes in Supramolecular Assemblies: Photocontrol of the Structural and Optical Properties of Langmuir–Blodgett–Kuhn Multilayers of Hairy-Rod Azo-Polyglutamates 467
13.3.4	Polymer Structural Effects on Photo-Orientation 473
13.3.4.1	Photoisomerization and Photo-Orientation of High-Temperature Azo-Polyimides 473
13.3.5	Photoisomerization and Photo-Orientation of Flexible Azo-Polyurethanes 479
13.3.6	Pressure Effects on Photoisomerization and Photo-Orientation 486
13.4	Photoisomerization Effects in Organic Nonlinear Optics: Photoassisted Poling and Depoling and Polarizability Switching 491
13.4.1	Introduction 491
13.4.2	Photoassisted Poling 492
13.4.3	Photoinduced Depoling 498
13.4.4	Polarizability Switching by Photoisomerization 500
13.5	Conclusion 503
	Acknowledgements 504
	Appendix 13.A Quantum-Yield Determination 505
	Rau's Method 505
	Fischer's Method 506
	Appendix 13.B Derivation of Equations for Determination of Anisotropy 507
	Appendix 13.C From Molecular to Macroscopic Nonlinear Optical Properties 509
	References 511
14	**Responsive Molecular Gels** 517
	Jaap de Jong, Ben L. Feringa, and Jan van Esch
14.1	Introduction 517
14.1.1	Responsive Chemical Gels 517
14.1.2	Responsive Physical Gels 518
14.1.3	Triggering Signals and Anticipated Responses 519
14.2	Chemoresponsive Gels 520
14.2.1	Chemoresponsive Gels by Host–Guest Complexation 521
14.2.2	Metal-Ion and Anion-Responsive Gels 527
14.2.3	Gel-Sol Phase Transitions Triggered by pH Changes 531
14.2.4	Chemoresponsive Gel Systems 538
14.2.5	Enzyme-Responsive Gel Systems 540
14.3	Physicoresponsive Gels 544
14.3.1	An Unusual Temperature-Responsive LMOG Gel 545

14.3.2	Responses to Mechanical Stress 545
14.3.3	Light-Responsive Gels 549
14.4	Conclusions 558
	References 559

15 Switchable Proteins and Channels 563
Matthew Volgraf, Matthew Banghart, and Dirk Trauner

15.1	Introduction 563
15.2	Photoswitch Characteristics 564
15.2.1	Common Photoswitches 566
15.3	Photoswitch Incorporation 567
15.3.1	Bioconjugation Techniques 567
15.3.2	Unnatural Amino Acids 568
15.4	Designing Photoswitchable Proteins 569
15.5	Photoswitchable Enzymes 571
15.5.1	Random Modification of Enzyme Surfaces 571
15.5.2	Photochromic Amino Acids 572
15.5.3	Modification of Cysteine Mutants 575
15.5.4	Photoswitchable Affinity Labels (PALs) 578
15.6	Photoswitchable Ion Channels 579
15.6.1	The Nicotinic Acetylcholine Receptor (nAChR) 579
15.6.2	Gramicidin A 580
15.6.3	The Voltage-Gated K^+ Channel 581
15.6.4	The Ionotropic Glutamate Receptor (iGluR) 583
15.6.5	α-Hemolysin 585
15.6.6	The Mechanosensitive Channel of Large Conductance (MscL) 587
15.7	Future Challenges 588
15.8	Concluding Remarks 590
	References 591

Part III Molecular Switching in Logic Systems and Electronics 595

16 Reading and Powering Molecular Machines by Light 597
Vincenzo Balzani, Monica Semeraro, Margherita Venturi, and Alberto Credi

16.1	Introduction 597
16.2	Basic Concepts 598
16.2.1	Molecular Motions in Artificial Systems: Terms and Definitions 598
16.2.2	Energy Supply and Monitoring Signals 600
16.2.3	Other Features 602
16.3	Interlocked Molecular Species as Nanoscale Machines 602
16.4	Molecular Machines Monitored by Light 604
16.4.1	An Acid–Base Controllable Molecular Shuttle 604
16.4.2	Molecular Elevators 607
16.5	Molecular Machines Powered and Monitored by Light 611

16.5.1	Pseudorotaxane Threading–Dethreading Based on Photoisomerization Processes *611*
16.5.2	Pseudorotaxane Threading–Dethreading Based on Photoinduced Proton Transfer *614*
16.5.3	Molecular Shuttles Based on Photoinduced Electron Transfer *617*
16.6	Conclusion and Perspectives *622*
	Acknowledgements *623*
	References *624*

17	**Photoinduced Motion Associated with Monolayers** *629*
	Kunihiro Ichimura and Takahiro Seki
17.1	Introduction *629*
17.2	Background to Photoinduced Motion of Monolayers *630*
17.3	Photoswitchable Flat Monolayers *631*
17.3.1	LB Films *631*
17.3.2	SAMs Formed by Silylation *633*
17.3.3	SAMs by the Au-Thiol Method *635*
17.3.4	SAMs from Cyclic Amphiphiles *639*
17.4	Photoswitchable Surfaces with Controlled Roughness *641*
17.4.1	Background and Theory *641*
17.4.2	Rough Surfaces Covered with Thin Photochromic Films *642*
17.5	Light-Guided Liquid Motion *645*
17.6	Photoinduced Motion on Water Surface *651*
17.6.1	Photomechanical Effects in Monolayers *651*
17.6.2	Dynamic Pattern Propagation and Collective Reorientation by Light *652*
17.6.3	Photoresponse of Molecules with Unconventional Architecture *653*
17.6.3.1	Urea Derivatives *653*
17.6.3.2	Metal-Coordinated Macrocyclics *653*
17.6.3.3	Dendrimers and Dendrons *653*
17.7	Photoinduced Morphology and Switching at Nanometre Levels *656*
17.7.1	Azobenzene Derivatives *656*
17.7.2	Spiropyran Derivatives *657*
17.8	Photoinduced Morphologies in Two-Component Systems *658*
17.9	2D Block-Copolymer Systems *660*
17.9.1	Monolayers of Photoresponsive Block-Copolymers *660*
17.9.2	Thin Films of Block-Copolymers *661*
17.9.3	Incorporation of Hierarchical Structures in Relief Structures *662*
17.10	Summary *665*
	References *665*

18	**Molecular Logic Systems** *669*
	A. Prasanna de Silva, Thomas P. Vance, Boontana Wannalerse, and Matthew E.S. West
18.1	Introduction *669*

18.2	YES Logic	*670*
18.3	NOT Logic	*673*
18.4	AND Logic	*673*
18.5	OR Logic	*676*
18.6	NAND Logic	*677*
18.7	INH Logic	*678*
18.8	NOR Logic	*680*
18.9	XOR Logic	*681*
18.10	Three-Input AND Logic	*681*
18.11	Three-Input NOR Logic	*682*
18.12	EnNOR Logic	*683*
18.13	Arithmetic and Gaming	*683*
18.13.1	Half-Adders	*683*
18.13.2	Half-Subtractors	*688*
18.13.3	Combined Half-Adders and Half-Subtractors	*689*
18.13.4	Combined Full-Adders and Full-Subtractors	*690*
18.13.5	Tic-Tac-Toe	*691*
18.14	An Application of Molecular Logic: Molecular Computational Identification (MCID)	*692*
18.15	Conclusion	*693*
	Acknowledgements	*693*
	References	*694*

19 Electron- and Energy-Transfer Mechanisms for Fluorescence Modulation with Photochromic Switches *697*

Tiziana Benelli, Massimiliano Tomasulo, and Françisco M. Raymo

19.1	Fluorescence	*697*
19.2	Electron Transfer	*699*
19.3	Energy Transfer	*700*
19.4	Photochromism	*702*
19.5	Fluorescence Modulation in Fluorophore–Photochrome Conjugates	*704*
19.6	Fluorescence Modulation in Nanostructured Assemblies	*707*
19.7	Fluorescence Modulation in Multilayer Constructs	*711*
19.8	Conclusions	*713*
	References	*714*

20 Conductance Properties of Switchable Molecules *719*

Sense Jan van der Molen and Peter Liljeroth

20.1	Introduction	*719*
20.2	Intrinsic Switches and Extrinsic Switching	*721*
20.2.1	Functionality Loss	*721*
20.2.2	Stimuli	*722*
20.2.3	Stimuli and Directionality of Switching	*724*
20.3	Quantum Charge Transport through Molecular Junctions	*724*

20.4	Experimental Methods	734
20.4.1	Scanning Tunnelling Microscopy	734
20.4.2	Metal–Molecule–Metal Devices	737
20.4.2.1	Devices Based on Self-Assembled Monolayers	737
20.4.2.2	Devices Using Nano-Objects as Intermediates	738
20.4.3	Single-Molecule Junctions	739
20.5	Transport Studies on Switchable Molecules	742
20.5.1	Extrinsic Switching	742
20.5.2	Interlocked Molecular Switches	747
20.5.3	Tautomerization	752
20.5.4	Photochromic Switches	756
20.5.4.1	Diarylethenes	756
20.5.4.2	Azobenzenes	763
20.6	Conclusions and Outlook	766
	Acknowledgements	768
	References	768

Index 779

Preface

Nature has been particularly gracious to the molecular designers at the nanoscale by offering a myriad of examples of the most ingenious and complex dynamic systems. But at the same time it is fascinating to realise the elegance, effectiveness and apparent simplicity of several of its basic molecular concepts. When you read this sentence, the large collection of molecular switches that make this happen operate as a result of the simple photochemical *cis – trans* isomerisation of a tiny olefin unit in the protein rhodopsin in your eye. The process of vision is arguably the most fantastic among nature's numerous systems that can be triggered by a switching process at the molecular level.

This work is about the design, functioning and application of molecular switches, in particular illustrating progress made over the past decade. Research on molecular switches covers a wide range of frontiers in science from molecular computing to sensors, displays and smart materials and from drug delivery to control of biomolecular processes. In the ongoing quest for nano-devices and molecular machines, the design of molecular switching elements integrated with a variety of functions is a formidable challenge. Research on molecular switches has been greatly stimulated by prospects of memory elements as small as the single molecule and their potential for information technology. It is particularly rewarding to see how this field has been flourishing with the first electronic devices based on molecular switching elements now demonstrated. On the other hand these developments also make clear how long and windy a road it can be from molecular function to functional device. But as the saying goes, *it is a long road that has no turns*.

The present two-volume work builds on the 2001 book Molecular Switches but is not simply a revised edition. Several chapters have been updated covering both basic principles and recent developments for completeness however for further background and those topics not fully covered, the reader is referred to the previous edition. As the field has seen spectacular development in the past years it is evident that we have tried to cover also many recent topics related to molecular switches in this second edition.

The chapters cover the structural diversity of molecular switches including discussion on various switching principles and methodology to the study their dynamic behaviour. Particular emphasis is on the dynamic control of function and materials properties. Furthermore, the use of molecular switches as trigger elements to control assembly, organization and function at different hierarchical levels and in macromolecular, mesoscopic and supramolecular systems is illustrated.

In the first section the focus is on different types of molecular switches including multilevel switching, nucleic acid based switches and molecular machines. The second section covers switchable containers, gels and polymers while chapters on switchable receptors, proteins and channels illustrate the potential in biomolecular sciences. In the third section, progress and prospects for molecular switching in logic systems and electronics and to control motion is discussed. The book ends with a chapter discussing the state of affairs with respect to photoresponsive molecular wires and devices, arguably one of the most rapidly developing areas of molecular switching in recent years.

The combination of topics demonstrates the multidisciplinary nature of research on molecular switches. Several contributions in this work also illustrate two other key aspects of research on molecular switches; first, it brings a responsive element to molecules and systems that allows triggering and control on command and second, the switching element is frequently part of a more complex molecular system with several components acting in concert. The lessons learned from the approaches described in these volumes hopefully will be also beneficial to numerous young researchers entering into molecular nanoscience, systems chemistry and synthetic biology. It was not our intention to be comprehensive and unfortunately not all relevant topics could be covered. However, we feel that this handbook gives a good perspective on the potential of the emerging field of molecular switches.

This second edition was only possible by the great efforts of the numerous contributors. We are particularly grateful to all authors for their excellent chapters. Join us on a fascinating journey through the dynamic scientific landscape opened by the introduction of molecular switches.

We hope your interest is switched on and that this book serves as a source of inspiration.

Centre for Systems Chemistry,   *Wesley R. Browne,*
University of Groningen   *Ben L. Feringa*
Groningen, May 2011

List of Contributors

Valeria Amendola
Università di Pavia
Dipartimento di Chimica
Viale Taramelli, 24
27100 Pavia
Italy

Vladimir A. Azov
Universität Bremen
Institut für Organische Chemie
Leobener Str. NW 2C
28334 Bremen
Germany

Débora T. Balogh
Universidade de São Paulo
Instituto de Física de São Carlos
CP 369
13560-970 São Carlos, São Paulo
Brazil

Vincenzo Balzani
Università di Bologna
Dipartimento di Chimica
'G. Ciamician'
Via Selmi 2
40126 Bologna
Italy

Matthew Banghart
Genentech, Inc.
1 DNA WaySouth San Francisco
CA 94080
USA

Tiziana Benelli
University of Miami
Department of Chemistry
Center for Supramolecular
Science
1301 Memorial Drive
Coral Gables, FL 33146-0431
USA

Leonardo De Boni
Universidade de São Paulo
Instituto de Física de São Carlos
CP 369
13560-970 São Carlos, São Paulo
Brazil

Marco Bonizzoni
University of Alabama
Department of Chemistry
Tuscaloosa
AL 35487
USA

List of Contributors

Neil R. Branda
Simon Fraser University
4D LABS
8888 University Drive
Burnaby BC, V5A 1S6
Canada

Simona Bronco
Università di Pisa
PolyLab-CNR
c/o Dipartimento di Chimica e
Chimica Industriale
Via Risorgimento 35
56126 Pisa
Italy

Wesley R. Browne
University of Groningen
Stratingh Institute for Chemistry
& Zernike Institute for
Advanced Materials
Faculty of Mathematics and
Natural Sciences
Nijenborgh 4
9747 AG Groningen
The Netherlands

Francesco Ciardelli
Università di Pisa
Dipartimento di Chimica e
Chimica Industriale
Via Risorgimento 35
56126 Pisa
Italy

Alberto Credi
Università di Bologna
Dipartimento di Chimica
'G. Ciamician'
Via Selmi 2
40126 Bologna
Italy

François Diederich
Laboratorium für
Organische Chemie
ETH Zürich
Hönggerberg HCI
8093 Zürich
Switzerland

Jan van Esch
Delft University of Technology
Department of Chemical
Engineering
Self Assembling Systems
Julianalaan 136
2628 BL Delft
The Netherlands

Luigi Fabbrizzi
Università di Pavia
Dipartimento di Chimica
Viale Taramelli, 24
27100 Pavia
Italy

Ben L. Feringa
University of Groningen
Stratingh Institute for Chemistry
& Zernike Institute for
Advanced Materials
Faculty of Mathematics and
Natural Sciences
Nijenborgh 4
9747 AG Groningen
The Netherlands

Julien Frey
Swiss Federal Institute of
Technology (EPFL)
Laboratory of Photonics &
Interfaces
1015 Lausanne
Switzerland

Eike Friedrichs
Technische Universität München
Biomolecular Systems and
Bionanotechnology
Physics Department and
ZNN/WSI
Am Coulombwall 4a
85748 Garching
Germany

Raquel Gomes
Universidade Nova de Lisboa
REQUIMTE
Departamento de Química
Faculdade de Ciências e
Tecnologia
2829-516, Monte de Caparica
Portugal

and

University of Gent
Department of Inorganic and
Physical Chemistry
Physics and Chemistry of
Nanostructures Group
Krijgslaan 281 (S3)
9000 Gent
Belgium

Tsuyoshi Gushiken
Yokohama National University
Graduate School of Engineering
Department of Advanced
Materials Chemistry
Tokiwadai
Hodogaya
Yokohama 240-8501
Japan

Kunihiro Ichimura
Tokyo Institute of Technology
4259 Nagatsuta
Yokohama 226-8503
Japan

Jaap de Jong
University of Groningen
Stratingh Institute for Chemistry
Faculty of Mathematics and
Natural Sciences
Nijenborgh 4
9497 AG Groningen
The Netherlands

Wolfgang Knoll
AIT Austrian Institute of
Technology
Vienna
Austria

Vincent Lemieux
Simon Fraser University
4D LABS
8888 University Drive
Burnaby BC, V5A 1S6
Canada

Peter Liljeroth
University of Utrecht
Condensed Matter and Interfaces
Debye Institute for Nanomaterials
Science
3508 TA Utrecht
The Netherlands

and

Aalto University School of
Sciences
Department of Applied Physics
P.O. Box 15100
00076 Aalto
Finland

Mauro Maestri
Università di Bologna
Dipartimento di Chimica
'G. Ciamician'
Via Selmi 2
40126 Bologna
Italy

Cleber R. Mendonça
Universidade de São Paulo
Instituto de Física de São Carlos
CP 369
13560-970 São Carlos, São Paulo
Brazil

Vladimir I. Minkin
Southern Federal University
Institute of Physical and Organic
Chemistry
194/2 Stachka Ave
344090 Rostov on Don
Russian Federation

Sense Jan van der Molen
Leiden University
Niels Bohrweg 2
Kamerlingh Onnes Laboratorium
2333 CA Leiden
The Netherlands

Osvaldo N. Oliveira Jr.
Universidade de São Paulo
Instituto de Física de São Carlos
CP 369
13560-970 São Carlos, São Paulo
Brazil

A. Jorge Parola
Universidade Nova de Lisboa
REQUIMTE
Departamento de Química
Faculdade de Ciências e
Tecnologia
2829-516 Monte de Caparica
Portugal

Osvaldo Pieroni
Università di Pisa
Dipartimento di Chimica e
Chimica Industriale
Via Risorgimento 35
56126 Pisa
Italy

Fernando Pina
Universidade Nova de Lisboa
REQUIMTE
Departamento de Química
Faculdade de Ciências e
Tecnologia
2829-516 Monte de Caparica
Portugal

Andrea Pucci
Università di Pisa
Dipartimento di Chimica e
Chimica Industriale
Via Risorgimento 35
56126 Pisa
Italy

and

CNR NANO
Instituto Nanoscienze-CNR
piazza San Silverstro 12
56127 Pisa
Italy

Françisco M. Raymo
University of Miami
Department of Chemistry
Center for Supramolecular
Science
1301 Memorial Drive
Coral Gables, FL 33146-0431
USA

David S. dos Santos Jr.
Universidade de São Paulo
Instituto de Física de São Carlos
CP 369
13560-970 São Carlos, São Paulo
Brazil

Zouheir Sekkat
Alakhawayn University in Ifrane
School of Science and Engineering
Hassan II Avenue
Ifrane 53000
Morocco

and

Osaka University
Department of Applied Physics
Yamada-oka 2-1, Suita
Osaka 565-0871
Japan

Jean Pierre Sauvage
Université de Strasbourg
CNRS UMR 7177
Institut de Chimie
Laboratoire de Chimie
Organo Minérale
4, rue Blaise Pascal
67070 Strasbourg Cedex
France

Takahiro Seki
Nagoya University
Graduate School of Engineering
Department of Molecular Design
and Engineering
Furo-cho, Chikusa
Nagoya 464-8603
Japan

Monica Semeraro
Università di Bologna
Dipartimento di Chimica
'G. Ciamician'
Via Selmi 2
40126 Bologna
Italy

A. Prasanna de Silva
Queen's University
School of Chemistry and
Chemical Engineering
BT9 5AG Belfast
Northern Ireland

Friedrich C. Simmel
Technische Universität München
Biomolecular Systems and
Bionanotechnology
Physics Department and
ZNN/WSI
Am Coulombwall 4a
85748 Garching
Germany

Christian Tock
Luxinnovation
7, rue Alcide de Gasperi
1615 Luxembourg
Luxembourg

He Tian
East China University of
Science and Technology
Key Lab for Advanced Materials
and Institute of Fine Chemicals
130 Meilong Road
Shanghai 20037
P.R. China

Massimiliano Tomasulo
University of Miami
Department of Chemistry
Center for Supramolecular
Science
1301 Memorial Drive
Coral Gables, FL 33146-0431
USA

Dirk Trauner
University of Munich
Department of Chemistry and
Center of Integrated Protein
Science
Butenandtstr. 5-13, Haus F
81377 München
Germany

Takashi Ubukata
Yokohama National University
Graduate School of Engineering
Department of Advanced
Materials Chemistry
Tokiwadai
Hodogaya
Yokohama 240-8501
Japan

Thomas P. Vance
Queen's University
School of Chemistry and
Chemical Engineering
BT9 5AG Belfast
Northern Ireland

Margherita Venturi
Università di Bologna
Dipartimento di Chimica
'G. Ciamician'
Via Selmi 2
40126 Bologna
Italy

Matthew Volgraf
Department of Neurobiology
Harvard Medical School
Boston, MA 02115
USA

Qiao-Chun Wang
East China University of
Science and Technology
Key Lab for Advanced Materials
and Institute of Fine Chemicals
130 Meilong Road
Shanghai 20037
P.R. China

Boontana Wannalerse
Queen's University
School of Chemistry and
Chemical Engineering
BT9 5AG Belfast
Northern Ireland

and

Faculty of Science
Chulalongkorn University
Department of Chemistry
Bangkok 10330
Thailand

C. Chad Warford
Simon Fraser University
4D LABS
8888 University Drive
Burnaby BC, V5A 1S6
Canada

Matthew E.S. West
Queen's University
School of Chemistry and
Chemical Engineering
BT9 5AG Belfast
Northern Ireland

Yasushi Yokoyama
Yokohama National University
Graduate School of Engineering
Department of Advanced
Materials Chemistry
Tokiwadai
Hodogaya
Yokohama 240-8501
Japan

Valtencir Zucolotto
Universidade de São Paulo
Instituto de Física de São Carlos
CP 369
13560-970 São Carlos, São Paulo
Brazil

Abbreviations

αHL	α-Hemolysin
ABTS	2,2′-azino-bis3-ethylbenzthiazoline-6-sulfonic acid
ADA	1-adamantaneacetate
AFM	atomic force microscopy
ANI	4-amino-1,8-naphthalimide
ANS	8-anilinonaphthalene-1-sulfonic acid
ATR	attenuated total reflection
ATS	3-aminopropyltriethoxysilane
Az	azobenzene
AzOH	4-(phenylazo)phenetyl alcohol
Azo-PUR	azo-polyurethanes
BAM	Brewster-angle microscopy
BCAII	bovine carbonic anhydrase II
BN	binaphthyl
BODIPY	boron dipyromethene; 4,4-difluoro-4-bora-3a,4a-diaza-sindacene
BPB	bromophenol blue
BPDN	bipyridyl-dinitro oligophenylene-ethynylene dithiol
BSA	Bovine serum albumin
CAP	catabolite activator protein
cCMP	cytidine 2′,3′-cyclic monophosphate
CD	circular dichroism
CN	coordination number
CNDO/S	complete neglect of differential overlap/spectroscopy
ConA	concanavalin A
CPIMA	center on polymer interfaces and macromolecular assemblies
CPK	Corey, Pauling, Koltun
CRA	calix[4]resorcinarenes
crRNA	cis-repression RNA
CSTR	continuous-stirred-tank reactor
CT	charge transfer
CTAB	cetyltrimethylammonium bromide
CV	cyclic voltammetry
CyD	cyclodextrins

DAC	dodecyl ammonium chloride
DCE	1,2-dichloroethane
DE	diarylethene
DFT	density-functional theory
diMe-tpy	5,5″-dimethyl-2,2′:6′,2″-terpyridine
DMF	N,N-dimethylformamide
DNA	deoxyribonucleic acid
DNP	1,5-dioxynaphthalene
DON	dioxynaphthalene
dpp	2,9-diphenyl-1,10-phenanthroline
dppp	1,2-bis(diphenylphosphino)propane
DR1	disperse red one
DR19	disperse 19
DTE	dithienylethene
dto	dithiooxalate
ee	enantiomeric excess
EET	electronic energy transfer
EFIPE	electric-field-induced Pockels effect
EFISH	electric-field-induced second harmonic
en	ethylenediamine
EO	electro-optical
EPL	expressed protein ligation
ES-MS	electrospray mass spectroscopy
eT	electron-transfer
FCS	fluorescence correlation spectroscopy
FMN	flavin mononucleotide
FRET	fluorescence resonance energy transfer
FTIR	Fourier transform infrared
FU	functional unit
GDH	glucose dehydrogenase
GFP	green fluorescent protein
GIXR	grazing-angle X-ray reflectivity
hCAI	human carbonic anhydrase I
HCR	hybridization chain reaction
HEK	human embryonic kidney
HFP	hexafluoro-2-propanol
HOMO	highest occupied molecular orbital
HRP	horseradish peroxidase
HTP	helical twisting power
ICD	induced circular dichroism
ICT	internal charge transfer
IETS	inelastic electron tunnelling spectroscopy
iGluR	ionotropic glutamate receptor
imH	imidazole
IPS	3-isocyanatopropyltriethoxysilane

IR	infrared
LB	Langmuir–Blodgett
LBD	ligand-binding domain
LBK	Langmuir–Blodgett–Kuhn
LbL	layer-by-layer
LC	liquid-crystal
LD-LISC	ligand-driven light-induced spin change
LDOS	local density of states
LF	ligand field
LMOG	low molecular mass gelators
LMW	low molecular weight
LPL	linearly polarized light
LUMO	lowest unoccupied molecular orbital
MAQ	maleimide, azobenzene and quaternary ammonium
MCBJs	Mechanically controllable break-junctions
MEH-PPV	poly[2-methoxy,5-(2′-ethyl-hexyloxy)-1,4-phenylenevinylene]
MLCT	metal-to-ligand charge transfer
mRNA	messenger RNA
MscL	mechanosensitive channel of large conductance
nAChR	nicotinic acetylcholine receptor
NCL	native chemical ligation
NHS	N-hydroxy succinimide
NLO	nonlinear optical
NMTAA	N-Methylthioacetamide
nNOS	neuronal nitric oxide synthase
NOESY-NMR	nuclear Overhauser effect spectroscopy nuclear magnetic resonance
OHB	orientational hole burning
ONPC	β-D-cellobioside
OPE	oligo(phenylene ethynylene)
ORTEP	Oak ridge thermal ellipsoid plot program
OTf	triflate
PAH	poly(allylamine hydrochloride)
PAL	photoswitchable affinity labels
PAM	4-phenylazophenyl maleimide
PAP	photoassisted poling
PAP	phenylazophenylalanine
PCR	polymerase chain reaction
PCS	point-contact spectroscopy
2PE	two-photon excitation
PEO	poly(ethylene oxide)
PET	photoinduced electron transfer
PHEMA	poly(2-hydroxyethyl methacrylate)
phen	1,10-phenanthroline
PID	photoinduced depoling

PMMA	poly-methyl-methacrylate
PmPV	poly[(*m*-phenylenevinylene)-*co*-(dioctoxy-*p*-phenylenevinylene)]
PS	polystyrene
PSS	photostationary states
PTL	photoswitchable tethered ligands
PVA	poly(vinyl alcohol)
QY	quantum yield
RBS	ribosome binding site
RCA	rolling circle amplification
RCM	ring-closing metathesis
REMD	replica exchange molecular dynamics
RFID	radiofrequency identification
RGD	arginine-glycine-aspartate
SAM	S-adenosyl-methionine
SAM	self-assembled monolayer
SDS	sodium dodecyl sulfate
SELEX	systematic evolution of ligands by exponential enrichment
SEM	scanning electron microscopy
SERS	surface-enhanced Raman spectroscopy
SFVS	sum-frequency vibrational spectroscopy
SHG	second-harmonic generation
siRNA	short interfering RNA
SNOM	scanning near-field optical microscopy
SP	spiropyrans
SPO	spirooxazines
SRG	surface-relief gratings
STM	scanning tunnelling microscopy
STS	scanning tunnelling spectroscopy
Taq Pol	Thermus acquaticus polymerase
taRNA	trans-activating RNA
TBDS	*tert*-butyldiphenylchlorosilane
TE	transverse electric
terpy	2,2′,6′,2″-terpyridine
TFA	trifluoroacetic acid
TM	transverse magnetic
TMD	transmembrane domain
TMP	trimethylphosphate
TPP	thiamine pyrophosphate
TS	transition state
TSPP	tetrakis-sulfonatophenyl porphyrin
TTB	tetra-tert-butyl
TTF	tetrathiafulvalene
TX	triple-crossover
UHV	ultrahigh vacuum
UHV-STM	ultrahigh-vacuum scanning tunnelling microscopy

UV	ultraviolet
UV-Vis	ultraviolet and visible
VT-NMR	variable-temperature nuclear magnetic resonance
WLF	Williams–Landel–Ferry
XOR	eXclusive OR
XPS	X-ray photoelectron spectroscopy
XR	X-ray reflectivity
YFP	yellow fluorescent protein

Part I
Molecular Switching

1
Multifunctional Diarylethenes
C. Chad Warford, Vincent Lemieux, and Neil R. Branda

1.1
Introduction

Ever since their development in the late 1980s, molecular switches based on the photoresponsive dithienylethene (DTE) architecture have attracted widespread attention as control elements in molecular devices and chemical systems [1]. This special interest over other classes of photoswitches is well deserved, and is due in part to the high fatigue resistance of the ring-closing and ring-opening photoreactions (Scheme 1.1), which reversibly generate two isomers. Also, the two isomers ('ring-open' and 'ring-closed') tend not to interconvert in the absence of light and, most importantly, possess markedly different optical and electronic properties. The most obvious change is in the colour of solutions, crystals and films containing DTE compounds [2]. However, numerous other useful differences in optical characteristics (emission [3] and optical rotation [4] of light), magnetism [5] and molecular and bulk conductivity [6] have been exploited in a remarkable number of derivatives to exert control over practical molecular systems. A few representative examples are listed in Table 1.1.

Given the large number of reviews already in the literature that extensively cover examples of DTEs having the properties listed above [20–22], we decided to focus this chapter on two under-represented areas where the versatile photoresponsive DTE compounds can be used: *electrochromism* and *controlling chemical/biochemical reactivity*. In this chapter, we first highlight several examples of DTE derivatives that undergo ring-closing and/or ring-opening reactions when oxidized or reduced or irradiated with light (dual-mode photo-/electrochromic systems). The bulk of the chapter then concentrates on illustrating how the DTE backbone can be used to control chemical reactions (or often interactions) between molecules and how, in some cases, the opposite is also true: a chemical reaction or interaction can regulate the photochemistry of the DTE backbone. Several of the changes in molecular structure shown in Scheme 1.1 (such as flexibility, or proximity of pendant functional groups) are responsible for the success of DTEs in these last two areas.

Molecular Switches, Second Edition. Edited by Ben L. Feringa and Wesley R. Browne.
© 2011 Wiley-VCH Verlag GmbH & Co. KGaA. Published 2011 by Wiley-VCH Verlag GmbH & Co. KGaA.

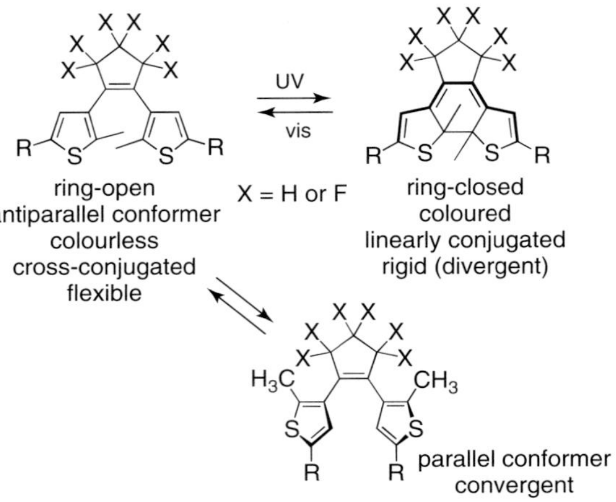

Scheme 1.1 The reversible, photochemical 6π electrocyclization reactions of the 1,3,5-hexatriene and 1,4-hexadiene isomers of the dithienylcyclopentene architecture.

1.2
Electrochemical Ring-Closing and Ring-Opening of DTEs

1.2.1
Electrochemical Behaviour of DTEs

Because the two thiophene heterocycles define many of the properties of the DTE backbone, the ring-open isomers tend to undergo irreversible oxidation at relatively high potentials (greater than 1 V) with accompanying electropolymerisation, as is typical for thiophene derivatives. Due to the creation of the linearly conjugated π-system upon photochemical cyclization, the ring-closed isomers typically undergo reversible oxidations at lower potentials (the absolute value depends on the derivative but are often 400–700 mV less positive), as expected for systems that have higher-energy highest-occupied molecular orbitals. Similarly, the reduction potentials for the ring-closed isomers are less negative than their ring-open counterparts (assuming the π-system created upon photoinduced cyclization is decorated with electron-accepting groups). These differences in redox potentials between the two DTE photoisomers can provide a possible mechanism for the observed selective quenching of fluorescence of pendant emissive dyes, however, only a few reports specifically ascribe electron transfer as the mechanism [23].

What is more pertinent to this chapter is the fact that some DTE derivatives undergo spontaneous ring-closing and ring-opening reactions when an appropriate voltage is applied and they are oxidized or reduced. The first example is compound **1b** shown in Scheme 1.2 [24]. This compound undergoes the typical cyclization reaction when irradiated with UV light (**1a** → **1b**). The reverse reaction is triggered either by exposing a solution of **1b** to visible light or by applying a positive potential,

1.2 Electrochemical Ring-Closing and Ring-Opening of DTEs

Table 1.1 Optical, electronic and bulk properties regulated by DTE derivatives. All DTEs exhibit an intrinsic modulation of their absorption characteristics, therefore, specific examples have not been included.

Structure	Comments	References
Systems that modulate fluorescence		
	Intrinsic fluorescence of the dye is reversibly modulated in a binary response	[7]
	Intrinsic fluorescence of the dye is reversibly modulated in a binary response with excellent fatigue resistance (>10^5 cycles)	[8]
	Fluorescence of a pendant dye is reversibly modulated in a binary response	[9]
	Control of single molecule emission in films is demonstrated	

(continued overleaf)

1 Multifunctional Diarylethenes

Table 1.1 (continued)

Structure	Comments	References
	Fluorescence of the polymeric DTE backbone is reversibly modulated in the solid state	[10]
	Fluorescence of polythiophene is reversibly modulated by pendant DTEs Amplified fluorescence quenching is demonstrated	[11]

Systems that modulate electron and hole transport

Electron transport in the solid state is reversibly modulated with excellent performance [12]

Conductivity in a bulk nanoparticle network is reversibly modulated

The ring-closing or ring-opening reactions of certain DTEs can be suppressed by electronic interactions with metal surfaces. [13]

Hole mobility and photocurrent in a bilayer device is reversibly modulated [14]

(continued overleaf)

Table 1.1 (continued)

Structure	Comments	References
Systems that modulate magnetism		
	Interactions between spin carriers on each end of the DTE backbone are reversibly modulated	[15]
Systems that modulate chirality		
	Diastereoselective ring-closing is demonstrated	[16]

Helical chirality is reversibly created through diastereoselective ring-closing [17]

Glass transition temperature of films is reversibly modulated

Selective metal deposition on crystalline areas is demonstrated [18]

Gelation of organic solvents by chiral helical fibres is reversibly modulated [19]

The effect is amplified via aggregation

Systems that modulate bulk properties

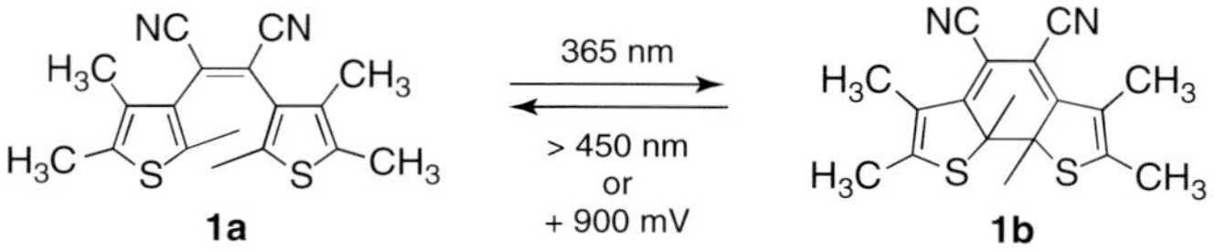

Scheme 1.2 Ring-closing of DTE derivative **1a** with UV light and ring-opening with visible light or electricity.

which oxidizes the ring-closed isomer to its radical cation. Clearly, this species is unstable as it spontaneously ring opens to generate the radical cation of **1a**, which undergoes electron transfer in a catalytic process that will be discussed in more detail.

Other examples of DTE derivatives that show dual-mode photo- and electrochromism are listed in Table 1.2. In each case, how the DTE backbone is decorated with functional groups defines its dual-mode behaviour. For example, when the 'outer positions' (groups 'A' in Table 1.2) are aromatic rings (such as additional thiophenes) and the 'inner positions' (groups 'B' in Table 1.2) are alkyl groups, the ring-open isomers tend to cyclize when they are oxidized. Derivative **3** is an illustrative example and colourless solutions of it become coloured (blue) when a positive potential is applied to them (after the initially produced ring-closed radical cation is reduced) [25]. When the inner alkyl groups are replaced with aromatic rings, the opposite behaviour is observed and the ring-closed isomers undergo ring-opening, as is observed for derivative **2** and related derivatives [26]. In these cases, the initially produced species do not require further reduction or oxidation, which will be explained later in this section. Electron-accepting groups such as cationic pyridinium rings result in reductive ring-closing. For example, a colourless solution of compound **4** changes to blue when a negative potential is applied (after the initially produced ring-closed radical cation is oxidized) [27].

Other examples include derivatives **6** [27], **7** [30] and **8** [31, 32]; the latter two have been used to postulate a mechanism for the electrochemically induced ring-closing and ring-opening reactions [33–35]. In short, when DTEs are oxidized in either the open or closed forms, the electrocyclic reactions of the resulting radical cations will result in the formation of the thermodynamically stable isomer. Though the nature of the cyclopentene (be it electron-withdrawing perfluoro or electron-donating perhydro) plays a role, by and large the behaviour is determined by the electronic nature of the thiophene substituents. In DTEs where the 'external' substituents are donors, the ring-closed isomer is usually favoured, yielding a radical cation that is doubly stabilized by a π-system that brings both donors into conjugation. The opposite is usually observed for DTE derivatives where the 'external' substituents are acceptors – here the less-destabilized ring-open isomer is preferred. In most cases, a reduction must follow to return the neutral DTE product – but nevertheless the isomerization can be redox triggered.

Table 1.2 Dual-mode photo- and electrochromic behaviour of selected DTE derivatives.

Reaction	Group A	Group B	Compound	Property	References
	S (thienyl)	S (thienyl)	2	Ring closes with 365-nm light Ring opens with >450-nm light Ring opens at +900 mV	[28]
	S (thienyl)	–CH$_3$	3	Ring closes with 365-nm light Ring closes at +1.1 V Ring opens with >450-nm light	[29]
	N-CH$_3$ (pyridinium)	–CH$_3$	4	Ring closes with 365-nm light Ring closes at –1.1 V Ring opens with >450-nm light	[26]
	N-CH$_3$ (pyridinium)	S (thienyl)	5	Ring closes with 365-nm light Ring closes at –1.1 V Ring opens with >450-nm light Ring opens at +900 mV	[25]

6
ring-opens when oxidized

7
ring-closes when oxidized

8
ring-closes when oxidized

A closer examination of the electrochemical behaviour of those DTE derivatives that undergo oxidative ring-opening (compounds **1b** and **2**, for example) reveals that less than a stoichiometric number of electrons are required to be removed from the system to effect complete ring-opening and decolour solutions of the compounds. This electrocatalytic process can be rationalized by comparing the cyclic voltammogram of both photoisomers of **2** and compound **9** as illustrated in Figure 1.1. A pale yellow solution of the ring-open isomer (**2**) is irreversibly oxidized at a potential characteristic for terthiophene derivatives (1.41 V). On first inspection, the photostationary state generated when this solution is irradiated with UV light appears to have a cyclic voltammogram identical to that for **2** despite the fact that this deep blue solution contains as much as 85% of the ring-closed isomer (**2c**), as determined by NMR spectroscopy. However, a closer examination reveals that there is a small oxidation peak at a less-positive potential (850 mV), as would be expected for isomer **2c**. The fact that this peak appears at the same potential as the oxidation of bis(dithiophene) derivative, which can be considered as a model for the linearly conjugated π-system (the 'inner' positions), suggests that the ring-closed isomer is too unstable in its oxidized form to be characterized in the lifetime of the experiment. The fact that the peak is very small even at high sweep rates supports the following series of steps:

1) The ring-closed isomer is oxidized to generate its radical cation (**2c**$^{\bullet+}$).
2) This unstable radical cation spontaneously ring opens to generate the corresponding radical cation of **2**.
3) Because species **2**$^{\bullet+}$ accepts an electron at a much more positive potential than is being applied in the experiment (1.41 vs. 0.85 V), electron transfer occurs between it and another species in solution (likely the ring-closed isomer). The result is the neutral ring-open isomer and the regeneration of **2c**$^{\bullet+}$.

The electrocatalytic phenomenon was demonstrated for compound **2** by chemically oxidizing a solution of it with a catalytic amount of a chemical oxidant.

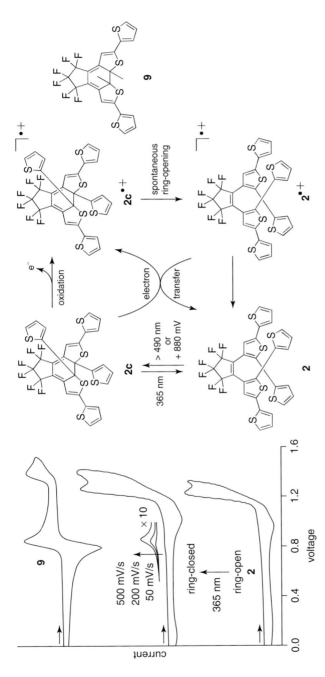

Figure 1.1 Electrocatalytic ring-opening of the ring-closed isomer of DTE derivative **2**. The cyclic voltammogram of a solution of **2** irradiated with UV light shows only a small peak that corresponds to the ring-closed isomer. This peak appears close to the same potential as that for the photostationary state for derivative **9** and increases in height when the sweep rate is increased. Reproduced from reference [26]. Copyright Wiley (2004).

As little as 1% of this oxidant was required to completely decolour a solution of **2** and **2c** at the photostationary state [28]. This mechanism also explains why, unlike for the electrochemically induced ring-closing reactions, the species initially produced in the ring-open reactions do not demand a subsequent redox reaction to generate the original photoresponsive species.

1.2.2
Fully Functional Photo- and Electrochromic DTEs

The final entry in Table 1.2 is an example of a fully photo- and electrochromic DTE system (**5**), which has been rationally designed to include the external bis(pyridinium) substituent of DTE **4**, which was shown to undergo catalytic reductive ring-closing and the internal thiophene substituent of DTE **2**, which undergoes catalytic oxidative ring-opening. As a result **5** can be isomerized either using light or electricity in both directions and constitutes the first dual-mode, photochromic–electrochromic DTE [25]. An unfortunate side effect of the bulky thiophene internal groups and the electron-withdrawing pyridinium groups is a significant rate of thermal ring-opening of the closed isomer. Nevertheless, the fundamental guidelines for designing a molecular system that can be regulated by both electricity and light have been laid down.

The development of bidirectional photo-/electrochromic hybrid molecules would enhance existing molecular-switching strategies by offering a new dimension for regulating the useful properties of these versatile compounds and would stimulate further developments in the innovative design of optical filtering and display technologies.

1.3
Using Dithienylethenes to Modulate How Chemicals React or Interact with Others

1.3.1
Control of Chemical Behaviour through Photoswitching

As already mentioned, the large number of examples of systems that use the DTE backbone to regulate optoelectronic properties for materials science applications does not need justifying or lengthy discussion. We always found it surprising, however, that until recently there were only a few examples of DTE derivatives scattered through the literature designed to modulate chemical reactivity. Our surprise was due to the realization that how molecules react or interact with others are arguably the events most sensitive to the electronic and steric make-up of molecular systems or their environment. Using light to regulate such chemical events would offer huge opportunities to industries that rely on synthesis and to biomedical communities that would benefit from new approaches to photodynamic therapy. The remainder of this chapter highlights the recent progress in these exciting areas.

The relationship between chemical reactivity and photoresponsive DTEs can be thought of in two ways:

1) The photochemistry modulates the chemical reactivity. In this approach, due to the differences in the electronic and geometric structure of the two DTE photoisomers, they influence chemical events uniquely when they act as catalysts or reagents.
2) The chemical reactivity regulates the photochemistry. This approach is the opposite of that described above and relies on a chemical event between molecules to allow or prevent the photoisomerization of the DTE architecture. This phenomenon is often referred to as '*reactivity-gated photochromism*'.

In both approaches, chemical reactivity is the desired 'output'. The renowned changes in the optical properties (colour, fluorescence, refractive index) of the photochromic DTEs should be considered simply as readout signals identifying which isomer is present at any given time.

The very first scheme presented in this chapter summarizes the changes in the DTE backbone that affect chemical reactivity. As already discussed, the creation of a linearly conjugated π-system (i.e. the ring-closed isomer) from one that has π-conjugation localized on each thiophene ring (i.e. the ring-open isomer) results in markedly different electronic properties and is the source of the changes in the colour of the systems. As will be shown, these electronic changes can also be used to dictate how the molecular isomers behave as catalysts, reagents and components in supramolecular systems. The steric changes are equally useful. The ring-open isomer is a flexible system with the ability for both the 'inner' ring positions (the methyl groups in Scheme 1.1) and the 'outer' ring-positions (the 'R' groups in Scheme 1.1) to reside in close proximity to each other. This structural convergence is due to the free rotation around the C–C sigma bonds linking the constituent thiophene rings to the central cyclopentene resulting in the ring-open isomer existing as two rapidly equilibrating conformers (*antiparallel* and *parallel*). Several of the examples of 'gated' systems take advantage of the fact that only the antiparallel conformation can undergo the conrotatory 6π-electron photochemical electrocyclization to the ring-closed isomer. The ring-closed isomer, on the other hand, is rigid and forces the groups at the 'inner' and 'outer' ring positions to diverge away from each other.

In the rest of this section, we discuss representative examples of DTE systems designed to reversibly regulate chemical reactivity by taking advantage of the steric and electronic differences between the two photoisomers. The integration of the DTE backbone into 'gated' systems is discussed in the next section. However, considerable common ground exists between photochemically controlled reactivity and reactivity-gated photochromism, especially in supramolecular systems. Therefore, several of the DTE derivatives highlighted in this chapter can be considered examples of both and will be discussed in this regard.

1.3.2
Controlling Molecular Interactions and Reactions Using the Steric Differences in the DTE Photoisomers

One of the earliest examples of a DTE system that exhibits significant differences in how it reacts due to the steric changes that occur when the photoisomers are interconverted is bis(boronic acid) **10** [36]. The two phenylboronic acid 'arms' have enough conformational freedom that they can converge to chelate D-glucose. The 1 : 1 resulting complex exhibited circular dichroism in the UV region of the spectrum (complexation-induced circular dichroism). Exposing derivative **10** to UV light converted it to its photostationary state mixture, containing 60% of the ring-closed isomer (the remainder is ring-open), which resulted in a reduced binding affinity for D-glucose. This change in association was attributed to the inability of the rigid ring-closed isomer to chelate the carbohydrate. The binding affinity was regained when the photostationary state was exposed to visible light to regenerate the ring-open isomer. DTE **10** also showed 'gated' photochromism. The 1 : 1 complex formed when the ring-open isomer chelated D-glucose did not undergo changes in colour when exposed to UV light because this complex does not allow the hexatriene system to adopt the antiparallel conformation necessary for photoactivity.

Compound **11** and larger homologues show similar behaviour [37–40]. They have enhanced abilities to bind rubidium and cesium ions when in their flexible ring-open forms because the two crown ethers can converge to simultaneously bind to the large cations. Ring-closing using UV light causes the systems to contain two independent ionophores, each being able to bind a cation with the same affinity as a simple, single crown ether.

The geometric effects and the critical dependence of size constraints in the DTE architecture on molecular recognition are exemplified by a DTE derivative containing two α-cyclodextrins (compound **12**) [41]. In a manner similar to derivatives **10** and **11**, only the ring-open isomer has enough flexibility for the two binding sites (in this case the hydrophobic pockets provided by the cyclodextrin rings) to reside in close enough proximity to simultaneously bind a guest molecule. Once again, the rigid backbone in the ring-closed isomer prevents chelation and significantly reduces binding. This difference in binding ability allows the photocontrolled release and uptake of a tetrakis-sulfonatophenyl porphyrin in water. A derivative of **12** containing longer chains linking the cyclodextrin binding pockets to the DTE backbone does not show the same changes in binding when the ring-open isomer is irradiated with UV light, since the longer chain allows the two cyclodextrins to effectively converge upon each other and bind the guest species through the hydrophobic effect.

The $C2$-chiral bis(oxazoline) derivative **13a** (Scheme 1.3) is one of the first examples demonstrating that the DTE system can be used to affect reactions by acting as a photoresponsive catalyst [42]. In this example, two chiral oxazoline rings are tethered to the 'inner' ring positions of the DTE backbone. These two metal binding groups can form a chelating site only when the photoresponsive system is in its ring-open form. A chelated metal ion (Cu^+, for example) is forced to reside in a $C2$-chiral environment in the resulting 1 : 1 coordination complex, which can now act as a chiral catalyst for reactions such as the cyclopropanation of styrene

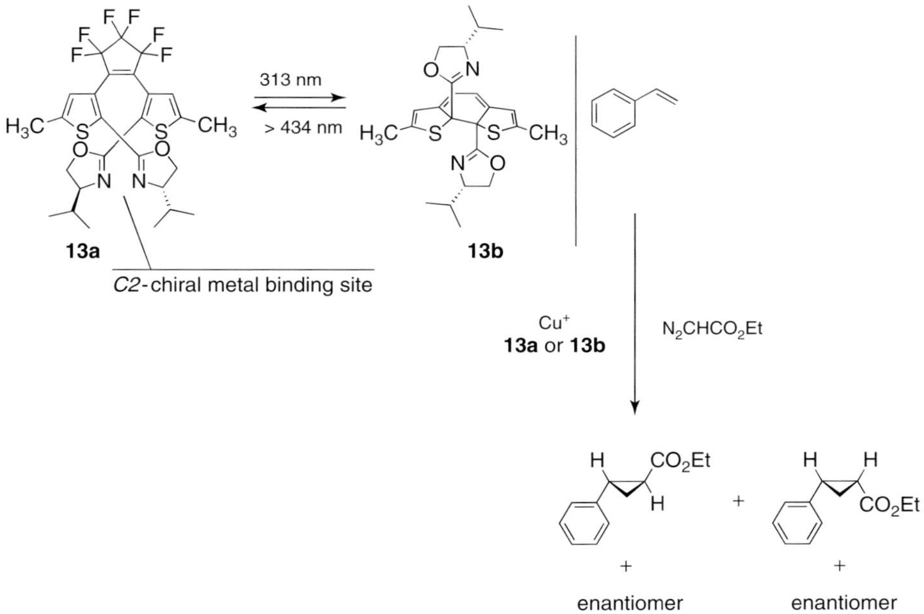

Scheme 1.3 Ring-closing of chiral DTE ligand **13a** with UV light and ring-opening of **13b** with visible light. The stereoselectivity of the cyclopropanation of styrene depends on which photoisomer is present.

with ethyldiazoacetate. The outcome of the stereoselective reaction depends on the DTE isomer present in solution. Both diastereo- and enantioselectivity were observed only when the ligand is in its ring-open form (**13a**). Exposing the system to UV light induces the ring-closing and produces **13b**, in which the two chiral oxazoline rings are inappropriately positioned for chelation, thus reducing the stereoselectivity of the reaction. Visible light resets the system by driving the ring-opening reaction (**13b** → **13a**). Regulation was also demonstrated by irradiating a solution of all components (with the DTE existing as **13a**) with UV light during the course of the reaction.

Once again, there is some 'gated' photochromism observed for **13a**. The 1 : 1 coordination complex formed when Cu^+ is chelated by the ring-open isomer is sufficiently stable to prevent the ring-closing photoreaction in noncoordinating solvents such as dichloromethane. As has already been briefly discussed and will be highlighted in more detail later in this chapter, the photochemistry of several DTE derivatives is prevented due to the hexatriene adopting the nonactive *parallel* conformation. This is not the case for compound **13a**, however, since the geometry of the coordination complex is *antiparallel*. Instead, the 'gated' photochromism is likely due to the driving force of the photochemical reaction (**13a** → **13b**) not being enough to overcome the stability of the coordination complex and allow the two oxazoline rings to move away from each other as they must to form isomer **13b**. The addition of a small amount of a more competitive solvent such as acetonitrile 'loosens' the complex and allows the photochemical ring-closing to proceed. This comes at a price, however, because the lowering of the stability of the coordination complex results in a less-effective chiral catalyst and reduced stereoselectivity in the cyclopropanation reaction.

The differences in binding due to the geometric changes the DTE backbone undergoes during the ring-closing/ring-opening reactions can also be extended to affect biochemical systems, where the Fischer 'lock-and-key' concept is probably the most influential. One example is illustrated in Scheme 1.4 [43]. The enzyme human carbonic anhydrase I (hCAI) has two positions in its active site that can associate to two different inhibiting groups. If either inhibitor (a sulfanilamide or a histidine-binding copper(II) iminodiacetate) is added to the enzyme, micromolar inhibition is observed. The best inhibitors are bivalent and contain both inhibiting groups separated by the appropriate distance (10 Å).

Inhibition can be modulated using photoresponsive bivalent DTE **14** by taking advantage of the changes in size and shape of the system as it undergoes ring-closing and ring-opening [43]. The flexible ring-open isomer (**14a**) can adopt a conformation where the sulfanilamide and copper iminodiacetate are separated by 10 Å, resulting in simultaneous binding in the active site and nanomolar inhibition (IC_{50} = 8 nM). Ring-closing to isomer **14b** with UV light generated rigid **14b**, which places the two inhibiting groups >11 Å apart. Since only one group can bind at any given time in **14b**, the inhibition is nearly identical to that for the other monovalent inhibitors (IC_{50} = 0.46 µM). The enhanced inhibition was regained by ring-opening the system (**14b** → **1a**) with visible light. The additional appeal of this system for use in biological applications is that the activation of the inhibitor is the ring-opening

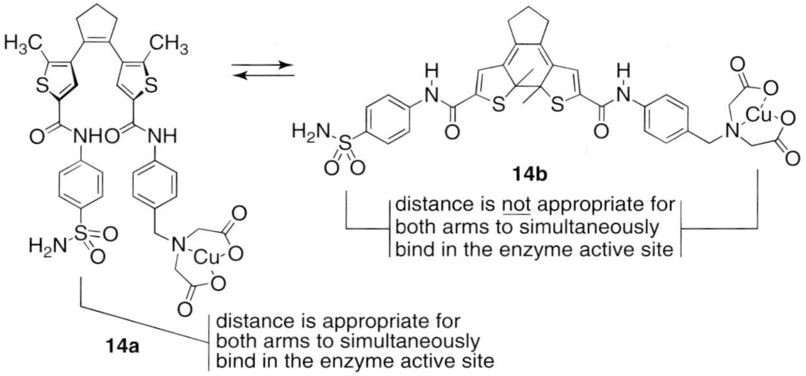

Scheme 1.4 Ring-closing of divalent human carbonic anhydrase inhibitor **14a** with UV light and ring-opening of **14b** with visible light. The enzyme inhibition depends on which photoisomer is present.

reaction, which uses visible light. Visible light is less damaging and has better penetration into tissue than UV light.

1.3.3
Controlling Molecular Reactions and Interactions Using the Electronic Differences in the DTE Photoisomers

The changes that the electron distribution within the DTE backbone undergoes when the ring-open isomer is converted into its ring-closed counterpart have also been used to regulate how molecules react and interact with each other. The localization of the π-electrons on the two thiophene rings in the ring-open isomer implies that groups located at the ends of two 'arms' of the DTE (the 'A' groups in Scheme 1.5) will not be significantly affected by one another's electron-donating or -accepting character and they can be expected to act independently. As emphasized

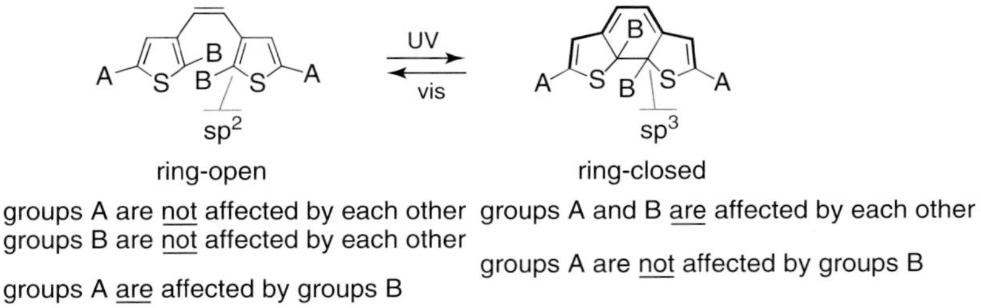

Scheme 1.5 Ring-closing and ring-opening affect how groups A and B communicate by creating a linearly conjugated π-electron backbone and changing the hybridization of the 'internal' carbon atoms on the thiophene rings from sp^2 to sp^3.

several times in this chapter, UV-induced ring-closing creates a linearly conjugated π-system running along the DTE backbone and joining the two 'A' groups. They can then communicate their electronic character to each other through this π-system and affect how each other behave.

On the other hand, the opposite is true for how groups 'A' and 'B' communicate to each other (Scheme 1.5). In the ring-open isomer, groups 'A' and 'B' located on the same thiophene ring are in direct communication and should be influenced by each other. Ring-closing changes the hybridization of the two carbon atoms involved in forming the new C–C bond from sp^2 to sp^3, resulting in a break in communication between 'A' and 'B'. Compounds **15** and **16** are illustrative examples that use each phenomenon.

15 more acidic than in the ring-open isomer

16 break in conjugation — less acidic than in the ring-open isomer

In both cases, the Brönsted acidities of the −OH groups in DTE derivatives **15** and **16** depend on the electronic connection between the phenols and the electron-withdrawing pyridinium groups, which can only effectively stabilize the conjugate phenoxide bases when they are electronically communicating to the phenoxide ions. The two groups in derivative **15** are conjugated along the newly created π-backbone of the ring-closed form; the pKa of the phenol is decreased from 10.5 to 9.3 when the ring-open isomer is exposed to UV light and increases when **15** is irradiated with visible light [44]. The opposite is true for compound **16**, which has a break in the electronic communication between the pyridinium and the phenol in the ring-closed form and increases its acidity when exposed to visible light to generate the ring-open isomer [45]. This isomer now has both groups on the same thiophene ring and in electronic communication. These two compounds illustrate the ability to intelligently engineer the DTE backbone in order to choose which isomer is the active form (and, implicitly, which stimulus is required) provides flexibility for the end user in practical applications.

Basicity (nucleophilicity) has also been regulated by taking advantage of similar changes in the electronic properties of DTE derivatives. Examples include compounds **17** and **18**. Compound **17** is shown as a ring-open isomer, where the azacrown ether can act as an effective ionophore and complex to metal cations such as calcium [46]. When the system is exposed to UV light, it undergoes ring-closing to generate an isomer that now has the ionophore in electronic communication with the electron-withdrawing aldehyde group located at the other end of the conjugated π-system. The result is a decrease in the electron density of the nucleophilic lone-pair electrons on the nitrogen atom in the azacrown ether ring, and a decrease in the binding affinity to Ca^{2+} by 4 orders of magnitude. The nitrogen's lone-pair electrons in compound **18** (shown as a ring-closed isomer) are not as effective as a Lewis base [47] or a nucleophile compared to the ring-open isomer, which is generated using visible light. The former behaviour was demonstrated by showing a reduced binding affinity of the pyridine in **18** to a ruthenium tetraphenylporphyrin to generate **19**. The change in nucleophilicity was shown by examining the rates of alkylation of the pyridines with bromobenzyl bromide, which are lower when the compound is in the ring-closed form [48, 49].

1.3.4
A Specific Approach to Using the Changes in Location of π-Bonds to Control Reactivity

Up to now, we have only discussed examples of DTE derivatives in which the groups most affected by the photoreactions are located on the thiophene rings. Given the creation and destruction of the linearly conjugated π-systems during the photoreactions, this approach is the most straightforward. A more recent approach to harnessing the changes that accompany the ring-closing/ring-opening reactions of DTEs focuses on derivatives that undergo restructuring of π-systems located in the central ring of the photoresponsive backbone. This approach has the distinct advantage that the two thiophene rings are left free to decorate with functional groups that help tune the light absorbed by each isomer and enhance performance of the systems. Two examples are discussed, each relying on the fact that the existence of the linearly conjugated π-system in the ring-closed DTE isomers delocalizes the electrons away from the central ring, where they were originally a key component in determining the compound's properties. This concept can be simplified by considering the central alkene joining the two thiophene rings as acting independently from the heterocycles only in the ring-open isomers.

The first example is DTE **20a** (Scheme 1.6), which features a 1,3,2-dioxaborole ring as a part of the photoresponsive hexatriene system [50]. Because this ring contains six delocalized π-electrons, it can be expected to have significant aromatic character. Because the aromatic system includes the boron atom (and its nominally empty p-orbital), it should not be able to accept a pair of electrons from a Lewis base. Exposing the system to UV light triggers the ring-closing reaction (**20a → 20b**) and creates the linearly conjugated π-system as is typical for DTE derivatives. The π-electrons are now delocalized along the DTE backbone leaving the boron diester to act as a Lewis acid.

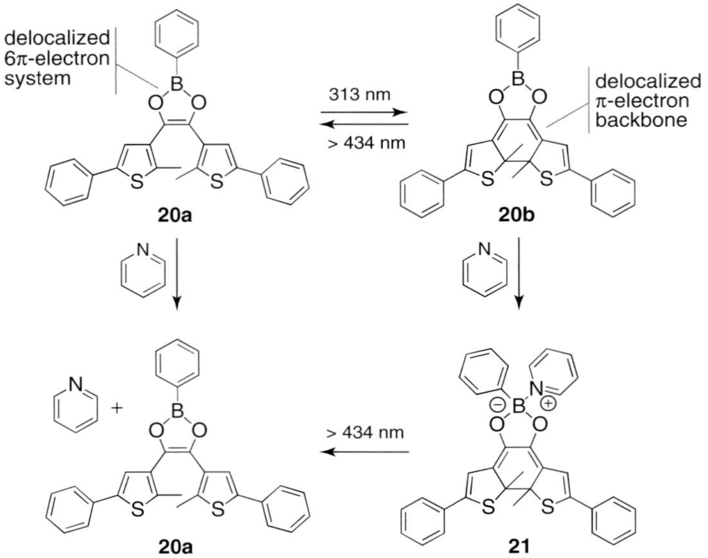

Scheme 1.6 Ring-closing of 1,3,2-dioxaborole DTE **20a** with UV light and ring-opening of **20b** with visible light. The photochemistry affects the delocalization, aromatic character and Lewis acidity of the boron atom. The binding of pyridine depends on which photoisomer is present.

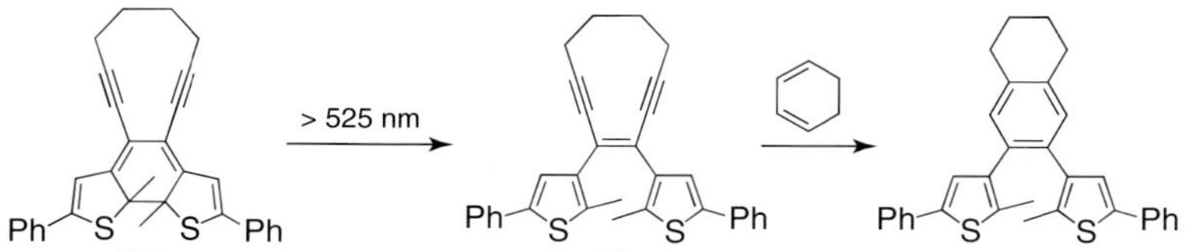

Scheme 1.7 Ring-opening of **22b** with visible light to unmask an enediyne that undergoes spontaneous cyclization and hydrogen abstraction.

The concept was demonstrated by comparing how each isomer (**20a** and **20b**) associated to pyridine. The ring-open isomer (**20a**) did not show any affinity for the Lewis base and even an excess of pyridine did not induce spectral changes typical of its binding to boron diesters. Upon photocyclization using UV light, two of the electrons that were formerly a part of the delocalized dioxaborole ring become delocalized over the backbone of the DTE, the central ring loses its aromatic character, boron can act as a Lewis acid and the binding of pyridine was observed. The spectral changes and association constant were characteristic of forming the complex **21b**. Visible light was used to regenerate the ring-open isomer and release pyridine. This system is currently being examined for its ability to act as a Lewis acid catalyst in chemical synthesis as well as a sequestering agent for pyridine and other nucleophilic catalysts to control chemical processes.

The second example is compound **22b** (Scheme 1.7) [51]. In this case, the DTE is already shown in its ring-closed form. The derivative is decorated with a cyclic system containing two alkynes directly attached to the linearly conjugated π-backbone. This thermally stable compound undergoes ring-opening when exposed to visible light and 'unmasks' a cyclic enediyne (**22a**). Cyclic enediynes are components in natural products that undergo Bergman cyclization reactions [52] to produce highly toxic benzenoid diradicals. In the presence of a hydrogen-transfer agent such as 1,3-cyclohexadiene, the diradical can be trapped as a stable product. In this case, the ultimate product of the reaction is compound **23**. This cyclization did not occur in the case of the ring-closed isomer (**22b**) and no hydrogen abstraction or tetrahydronaphthalene was observed because the active enediyne is masked.

The above is an illustrative example of using light to activate 'masked' chemotherapeutic agents that are broadly toxic, have severe side effects and cannot be administered in their 'un-masked' forms. The well-known Bergman cyclization of enediynes is an important reaction in antitumour activity and one where the presence of a precise arrangement of π-bonds is essential for the reaction to proceed. Unfortunately, the Bergman reaction of **22a** only occurred at elevated temperature (70–80 °C). Future generations of this system will contain activating features to lower the temperature at which the cyclization reaction takes place so the system can operate under physiological conditions. Once again, the use of visible light to trigger **22b** is appealing, especially when the technology is designed to be used in therapeutic applications.

1.4
Gated Photochromism

In Section 1.3 we gave brief mention to systems that were prone to 'gated' photochromism and failed to undergo their characteristic photochemical reactions under conditions where they are conformationally restricted. Although this detrimental phenomenon can be considered an unavoidable side effect when the changes in reactivity are based on strict geometry considerations, the use of these systems in diagnostic applications is not unreasonable and one can envision systems where the changes in the optoelectronic properties is an indication of a change in the chemical environment. Another exciting application is the photorelease of molecules using light, which, as will be shown, would also capitalize on the 'reactivity-gated' condition. Both applications are based on the concept already mentioned in this chapter: instead of the photochemistry dictating how a molecule reacts, a molecule's reactivity can dictate its photochemistry.

In this section of the chapter, we will introduce some DTEs specifically designed so that they do not contain the hexatriene architecture suitable for photocyclization and are not photoresponsive unless they first undergo a spontaneous chemical reaction to generate the appropriate hexatriene system. While very few examples of 'reactivity-gated' photochromism have been reported, there are numerous examples where the photochromism is 'gated' by other means. In these systems, one or both isomers of the photoresponsive structure are reversibly transformed into photoinactive states by a 'gate' signal, which is another external stimulus such as electrochemistry, solvation, temperature and other chemicals (some examples of the latter were already briefly mentioned) [53]. Although our focus is on 'reactivity-gated' photochromism, where irreversible chemical reactions are involved in the 'gating' process, we will include in our discussion a select few examples of systems that rely on proton and electron transfer to affect the photochromic properties of DTEs.

1.4.1
'Gated' Photochromism Based on Steric Effects

We have already mentioned the fact that, due to their flexibility, the ring-open isomers of DTE derivatives rapidly equilibrate between two major conformers (a *parallel* with mirror symmetry and an *antiparallel* with *C2* symmetry as shown in Scheme 1.8). Only the *antiparallel* conformer has its hexatriene in an appropriate geometry to undergo the conrotatory photochemical electrocyclization. Here, we highlight some systems that only exist in their nonphotoactive *parallel* conformation until a reversible chemical reaction or interaction occurs to free geometrical constraints of the system and allows it to adopt its photoactive *antiparallel* form. Both intramolecular and intermolecular examples will be shown. Generally, the former examples rely on the interaction of the groups located on the 'inner' ring positions of the thiophenes (the methyl groups in Scheme 1.8), although compound **10** is an exception. Intermolecular systems tend to utilize the other site on the heterocycles.

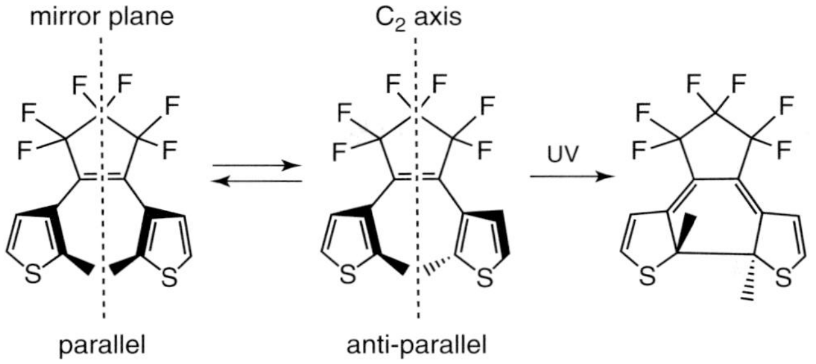

Scheme 1.8 Only the *antiparallel* conformer of the DTE backbone can undergo photochemical ring-closing.

1.4.2
Intramolecular 'Gating'

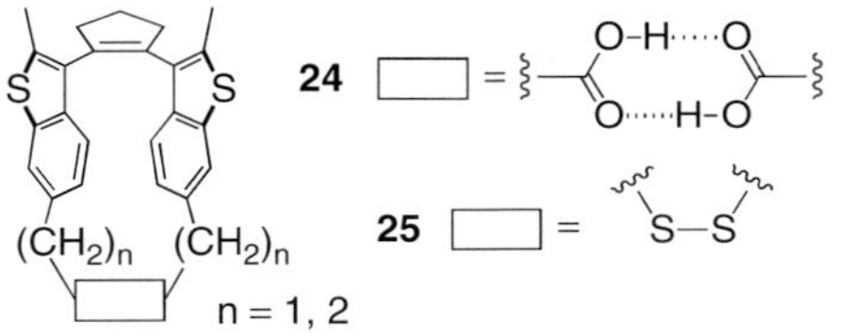

Derivatives **24** and **25** both have interlocking arms, which force them to remain in their *parallel* conformations [54–56]. In the case of diacid **24**, the strong intramolecular hydrogen bonds are responsible for the 'gating' and can only be broken by adding solvents that compete for the hydrogen bonds. Therefore, compound **24** was completely inactive when exposed to UV light when dissolved in apolar solvents such as cyclohexane. The photoactivity was restored by either adding a competitive hydrogen-bonding solvent such as ethanol or propylamine, or by breaking the hydrogen bonds thermally by heating the system above 100 °C in decalin. Compound **25** is locked in its *parallel* conformation by an intramolecular S–S bond and is also nonresponsive to UV light. Its photoactivity is only possible when the intramolecular disulfide bond holding the two arms together is reduced by tris(*n*-butyl)phosphine, which allows freedom of rotation and photochemical ring-closing. Neither example has been used in applications.

1.4.3
Intermolecular 'Gating'

Examples in which the interactions between ligands and metal ions are used to affect the photochromic reactions of DTE derivatives include compounds **26**, **28** and **29**. Although it is technically not a DTE, the similarity between compound **26** and DTEs warrants its inclusion in this discussion [57]. This pyridine derivative contains a hexatriene that undergoes photochemically induced ring-closing, as would be expected. However, its photochromic behaviour can be shut down by introducing a Cu^+ ion into a solution of **26**. The inhibition is due to the chelation of the metal by both bidentate arms of the system in order for it to form the coordination complex **27**, which necessitates a *parallel* arrangement of the hexatriene, rendering the complex photochemically inert.

Compounds **28** and **29** are included in this discussion because their photochemistry, although allowed whether metals ions are present or not, is greatly influenced by their presence. Interestingly, in both cases, complexation improves a photochemical aspect of the photochromic reactions. We have already provided examples that show how the binding of groups located at the 'inner' ring positions of the thiophene heterocycles prevents photocyclization due to locking the hexatriene in the *antiparallel* conformation. The opposite effect is also possible. In compound **28** the pyridine groups located at each end of the DTE backbone bind to Zn^{2+} ions and form a 2 : 2 coordination complex. In order to form this complex, the DTE must adopt an *antiparallel* conformation. This effectively increases the proportion of the photoactive conformation and results in an increased quantum yield for photocyclization [58]. Although the reaction still proceeds in the absence of the metal ion, it is less efficient.

A similar binding mode exists for the coordination compound **29** [59]. Two DTE ligands bind to two metal ions (Cu^+ in this case) to form a 2 : 2 double-stranded helical complex. The chiral oxazoline auxiliaries on the ends of the individual strands are positioned to dictate the handedness of the helicate and bias the formation of only one diastereomeric complex due to the through-space interaction of the benzyl groups across the helical axis. The conformation of both hexatrienes in the 2 : 2 complex is *antiparallel* so the photochemical ring-closing is allowed when solutions of **29** are irradiated with UV light. However, because the self-assembly process produces a double-stranded stereochemically pure helicate, the two thiophene rings are pre-oriented with respect to each other so that photocyclization yields a single diastereomer. There is chiral discrimination during the ring-closing event.

1.4.4
Gating Based on Electronic Effects

Bis(phenol) **30** (shown in its ring-closed form) is an elegant example of how an electrochemical stimulus dictates whether the photochemistry of the DTE backbone occurs [60, 61]. The ring-open isomer undergoes UV-induced cyclization to generate **30**. As is the case for all fully functional DTE derivatives, ring-opening can be triggered by exposing the system to visible light. Electrochemical control of the photochromism is possible because of the significant difference between the oxidation potentials for the two photoisomers and the rearrangement of the π-system as a consequence of the electrochemical event. This is illustrated in Scheme 1.9. While the ring-open isomer is inert to oxidation at 735 mV, its ring-closed counterpart (**30**) is oxidized at this potential due to the linear π-conjugation between the two phenol groups. The product of the oxidation is the extended quinone **31**, which does not have the cyclohexadiene required for photochemical ring-opening. The electrochemistry has 'gated' the photochemistry. The ring-opening photoreaction can be restored only after quinone **31** has been reduced back to the bis(phenol) **30**.

Compound **32a** represents an interesting case where the fate of the excited state of the DTE is dictated by the presence or absence of a phenol –OH group [62]. Typically, the most favourable processes that the excited states (produced when the DTE ring-open isomers are irradiated with UV light) undergo are cyclization to the ring-closed isomers and nonradiative decay. However, in certain systems the rates of other deactivation processes can dominate and prevent electrocyclization. Compound **32a** is one of those cases. Compound **32a** does not undergo photochemical ring-closing to produce **32b** when irradiated with UV light. This phenomenon is attributed to the presence of the acidic –OH, which can serve as a donor for intramolecular proton transfer to one of the carbonyl C=O groups of the maleimide backbone, at a significantly faster rate than photocyclization. The

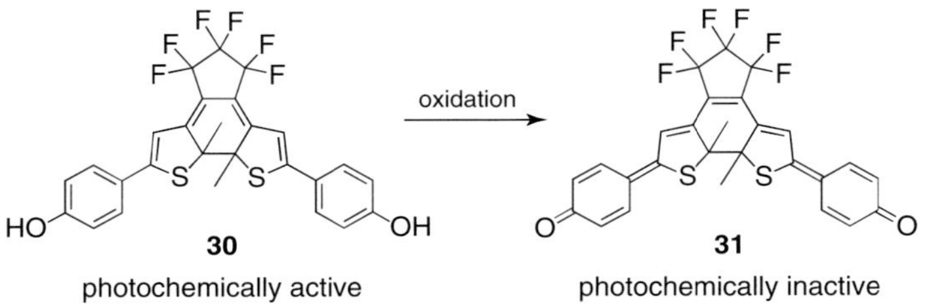

Scheme 1.9 Oxidation of the ring-closed isomer **30** disrupts the conjugation and prevents photochromism of **31**.

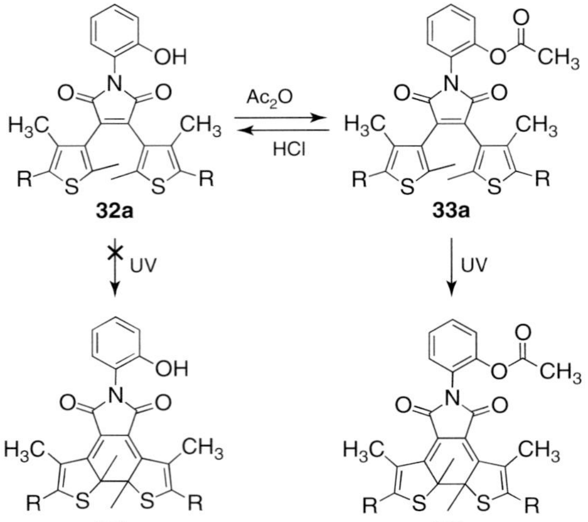

Scheme 1.10 Acylation of the phenol in the photoinactive derivative **32a** produces **33a**, which undergoes photochemical ring-closing.

'reactivity-gated' photochromism is introduced by performing a simple acylation reaction (**32a** → **33a**) as shown in Scheme 1.10. Because the product of the reaction between **32a** and acetic anhydride converts the phenol to a phenoxide ester, proton transfer is not possible and DTE **33a** readily forms its ring-closed isomer (**33b**) when exposed to UV light. The acylating agent turns the system 'on'. The reverse photochemical reaction (**33b** → **33a**) is triggered by visible light. Hydrolysis of the ester regenerates **32a** and turns the system back 'off'.

Other 'reactivity-gated' photochromic systems that rely on chemical protection/deprotection steps as the 'gate' signals are **34**, **35** and **36** [63]. The starting compound is in these cases derivative **34** where the DTE backbone is constructed from a cyclobutene-1,2-dione system, which contributes the central double bond of the hexatriene. This system, however, is not photochromic and the conversion to the corresponding ring-closed isomer is not observed when the system is irradiated with UV light. Protection of only one of the two ketones in **34** as a cyclic acetal had little effect. Once again, the DTE (**35**) was not photoactive and no photochemical ring-closing was observed. Converting both ketones to cyclic acetals did have an effect and derivative **36** proved to be photoresponsive. The ring-closing could be induced with UV light. The postulated reasoning is the involvement of ketenes **37** and **38**, which are favourable structures as they relieve the ring strain of the central ring by breaking the C–C bond between the two carbonyl groups. These ketene structures do not possess the appropriate hexatrienes for the photochemical ring-closing reactions. The ring-closed isomer of bis(acetal) **36** reacts with visible light to regenerate **36**. Moreover, when the ring-closed isomer **36** is singly or doubly deprotected to afford the ring-closed isomers of **35** and **34**, respectively (these cannot be produced from their ring-open counterparts and light), both isomers undergo photochemical ring-opening when exposed to visible light. This system illustrates the close relationship between gating effects and the rearrangement of double bonds, which we will also examine in Section 1.5.

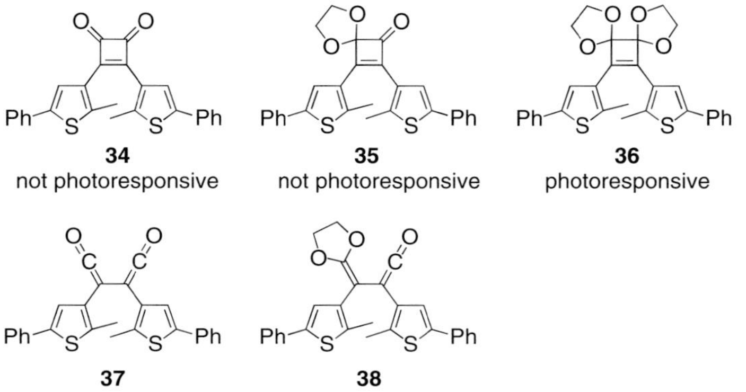

34 not photoresponsive

35 not photoresponsive

36 photoresponsive

37

38

Reversible installation of charge-transfer components in DTE derivatives can also 'gate' their photochromism as illustrated using compounds **39**–**42** (Scheme 1.11). DTE derivatives **39** and **41** were rationally designed to exhibit photoresponsive behaviour only under acidic conditions. As long as they remain in their protonated states, these isomers can be converted to their ring-closed counterparts by

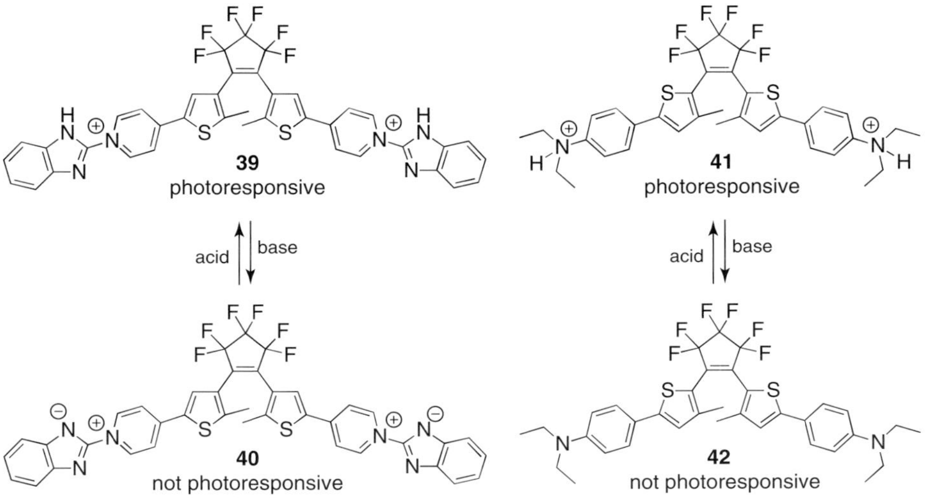

39 photoresponsive

41 photoresponsive

40 not photoresponsive

42 not photoresponsive

Scheme 1.11 Protonation and introduction of charge-transfer species act to 'gate' the photochemical ring-closing reactions.

irradiating them with UV light. They can be regenerated with visible light, as is typical for functional DTE derivatives. The turning 'off' of the photochromism when compounds **39** and **41** are deprotonated (to form pyridinium betaine **40** and the bis(aniline) **42**, respectively) is due to an efficient deactivation of the excited state by an intramolecular charge transfer in **40** and a twisted intramolecular charge transfer in **42**. Since excitation with visible light does not deliver enough excitation energy to produce the charge-transfer state the ring-closed isomer of betaine **40** can be ring opened (but not subsequently ring closed). Protonation of the nitrogen atom in the presence of trifluoroacetic acid allows photocyclization of **41** by chemically transforming it to a non-electron donor [64, 65].

The final example discussed in this section is the related compound **43a** (Scheme 1.12) [48]. Although this dicyanoethylene-thienylethene derivative lacks one of the thiophene rings, it possesses a hexatriene and remains photoresponsive. Irradiating a solution of the ring-open isomer (**43a**) with UV light triggers the ring-closing reaction and produces isomer **43b**. As is the situation for the fully functional DTEs, the reverse reaction (**43b** → **43a**) can be induced with visible light. What is not characteristic is the fact that the coordination compounds formed when either of the isomers is axially coordinated to a ruthenium tetraphenyl porphyrin through the pyridine nitrogen are not photoresponsive and neither ring-closing of **44a** with UV light nor ring-opening of **44b** with visible light is possible. The reason for this inhibition is yet to be determined, however, energy transfer involving triplet states is the likely culprit.

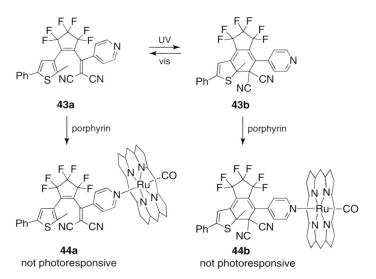

Scheme 1.12 Coordination of the pyridine in isomers **43a** and **43b** turns 'off' the photochemistry of both isomers.

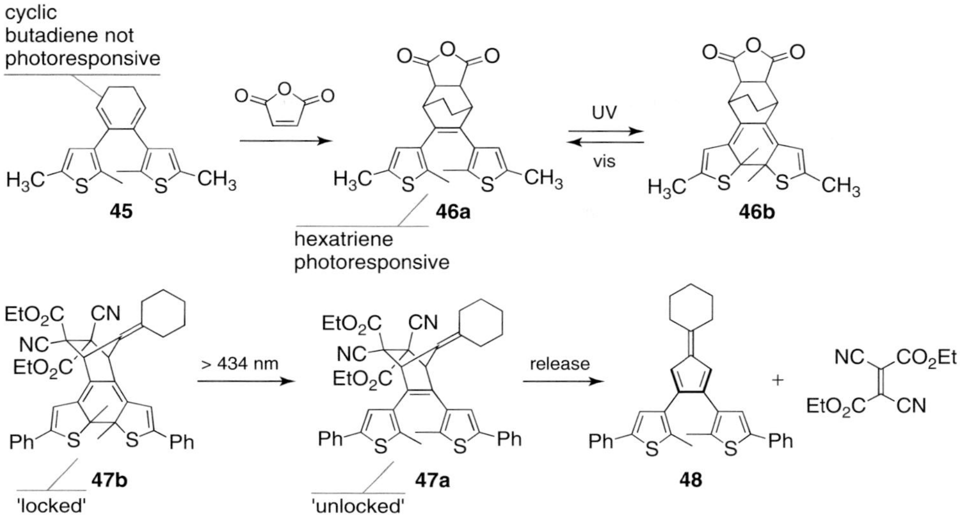

Scheme 1.13 The Diels–Alder reaction of **45** must occur first to enable the photochromism. Visible light can be used to unlock **47b**, which releases a dienophile.

1.5
Reactivity-Gated Photochromism Using the Functional Group Effect

In this chapter, we have presented examples of DTEs that highlight both approaches to integrating chemical reactivity and photochromism: photochemical control of reactivity and chemical control of photochemistry. We will end with a brief discussion of an example that illustrates how both approaches can be combined to create a new photochemical release system. The specific example is the bicyclic system **47b** shown in Scheme 1.13 [66], however, our discussion starts with an examination of the behaviour of a related derivative (**45**) [67].

Compound **45** does not contain the hexatriene required to undergo photochemical ring-closing. In fact, **45** cannot be considered as a DTE at all. It is, therefore, relatively stable and exposing it to UV light does not induce a photochemical reaction. It does, however, contain a cyclic butadiene suitable for reacting with a dienophile such as maleic anhydride in a Diels–Alder reaction. Due to the reorganization of the π-system, the product (**46a**) of this cycloaddition reaction now contains a hexatriene and undergoes ring-closing to produce **46b** when exposed to UV light. As anticipated, the reverse reaction can be triggered with visible light. This is an example of 'reactivity-gated' photochromism and can perhaps be considered for applications in sensing and dosimetry using the changes in colour as the read-out. This particular example is presented in this section of the chapter since it represents a model for a more complex system that combines chemical reactivity and photoreactivity.

It is clear that the rearrangement of the π-system when **45** is converted to **46a** is required to allow the photochromism. What is perhaps less obvious is that the rearrangement of the π-system when **46a** ring-closes to **46b** prevents the reverse thermal Diels–Alder reaction (the cyclohexadiene required for the reverse cycloaddition is removed). Isomer **46b** can be considered as a 'locked' system and would require 'unlocking' with visible light (**46b** → **46a**) before the original dienophile can be released. This is an example of photochemically controlled chemical reactivity. Unfortunately, the reverse Diels–Alder reaction of **46a** to produce **45** and maleic anhydride requires elevated temperatures, which are inappropriate for typical photorelease applications such as drug delivery [68].

The partnership of fulvene **48b** and diethyl dicyanofumarate (Scheme 1.13) is a better choice to demonstrate the combination of 'reactivity-gated' photochromism and photochemically controlled reactivity in a photorelease application. In this Diels–Alder reaction, the equilibrium significantly favours the two starting materials (**48** and the dienophile) at room temperature. By adding an excess of the dienophile, the equilibrium can be displaced to favour bicyclic system **47a**, which can be locked into the ring-closed isomer **47**. This isomer is stable in the dark at room temperature and only undergoes spontaneous reactions when it is 'unlocked' with visible light to regenerate **47a** and release the dienophile (presumably through ring-open isomer **47a**).

1.6
Conclusion

In this chapter, we have highlighted some examples of DTE derivatives designed for use in electrochromic and chemical/biochemical reactivity applications. While the photochromic behaviour of this versatile class of compounds has been well documented and extensively used for numerous applications, extending the stimuli to include electricity has only recently started to be addressed. Multifunctional devices capable of performing several tasks will benefit from materials that respond to more than one stimulus and can be controlled more precisely.

As is hopefully illustrated by the examples presented in this chapter, the use of the DTE architecture – less as a photochromic species, and more as an organic scaffold that can be logically and systematically tuned to influence chemical and biochemical reactivity – is an important effort. Given the impact that these photoresponsive systems will potentially have on advancing industrial processes, sensing and diagnostics and photorelease, significant growth in this area can be expected.

References

1. Irie, M. (2000) *Chem. Rev.*, **100**, 1685–1716.
2. (a) Irie, M., Sakemura, K., Okinaka, M. and Uchida, K. (1995) *J. Org. Chem.*, **60**, 8305–8309; (b) Nakayama, Y., Hayashi, K. and Irie, M. (1991) *Bull. Chem. Soc. Jpn.*, **64**, 789–795.
3. (a) Nakagawa, T., Hasegawa, Y. and Kawai, T. (2008) *J. Phys. Chem. A*, **112**, 5096–5103; (b) Fernandez-Acebes, A.

and Lehn, J.-M. (1999) *Chem. Eur. J.*, **5**, 3285–3292; (c) Murguly, E., Norsten, T. and Branda, N.R. (2001) *Angew. Chem. Int. Ed.*, **40**, 1752–1755; (d) Zhao, H., Al-Atar, U., Pace, T., Bohne, C. and Branda, N. (2008) *J. Photochem. Photobiol.*, **200**, 74–82.

4. (a) Yamaguchi, T., Uchida, K. and Irie, M. (1997) *J. Am. Chem. Soc.*, **119**, 6066–6071; (b) Kodani, T., Matsuda, K., Yamada, T., Kobatake, S. and Irie, M. (2000) *J. Am. Chem. Soc.*, **122**, 9631–9637; (c) Yamamoto, S., Matsuda, K. and Irie, M. (2003) *Org. Lett.*, **5**, 1769–1772; (d) Matsuda, K., Yamamoto, S. and Irie, M. (2001) *Tetrahedron Lett.*, **42**, 7291–7293; (e) Yamagchi, T., Nomiyama, K., Isayama, M. and Irie, M. (2004) *Adv. Mater.*, **16**, 643–645.

5. (a) Matsuda, K. and Irie, M. (2005) *Polyhedron*, **24**, 2477–2483; (b) Tanifuji, N., Irie, M. and Matsuda, K. (2005) *J. Am. Chem. Soc.*, **127**, 13344–13353; (c) Sun, L. and Tian, H. (2006) *Tetrahedron Lett.*, **47**, 9227–9231.

6. (a) Dulic, D., van der Molen, S.J., Kudernac, T., Jonkman, H.T., de Jong, J.J.D., Bowden, T.N., van Esch, J., Feringa, B.L. and van Wees, B.J. (2003) *Phys. Rev. Lett.*, **91**, 207402; (b) Matsuda, K., Yamaguchi, H., Sakano, T., Ikeda, M., Tanifuji, N. and Irie, M. (2008) *J. Phys. Chem. C*, **112**, 17005–17010; (c) Kronemeijer, A.J., Akkerman, H.B., Kudernac, T., van Wees, B.J., Feringa, B.L., Blom, P.W.M. and de Boer, B. (2008) *Adv. Mater.*, **20**, 1467–1473; (d) Whalley, A.C., Steigerwald, M.L., Guo, X. and Nuckolls, C. (2007) *J. Am. Chem. Soc.*, **129**, 12590–12591.

7. Gorodetsky, B. (2008) The design of dual-mode photochromic and electrochromic 1,2-Dithienylcyclopentene dyes. Ph.D. thesis. Department of Chemistry, Simon Fraser University. Burnaby, BC, Canada.

8. Jeong, Y.C., Yang, S.I., Kim, E. and Ahn, K.H. (2006) *Tetrahedron*, **62**, 5855–5861.

9. Irie, M., Fukaminato, T., Sasaki, T., Tamai, N. and Kawai, T. (2002) *Nature*, **420**, 759–760.

10. Jeong, Y.C., Yang, S.I., Kim, E. and Ahn, K.H. (2006) *Macromol. Rapid Commun.*, **27**, 1769–1773.

11. Finden, J., Kunz, T., Branda, N.R. and Wolf, M.O. (2008) *Adv. Mater.*, **20**, 1998–2002.

12. Choi, H., Lee, H., Kang, Y.J., Kim, E., Kang, S.O. and Ko, J. (2005) *J. Org. Chem.*, **70**, 8291–8297.

13. Ikeda, M., Tanifuji, N., Yamaguchi, H., Irie, M. and Matsuda, K. (2007) *Chem. Commun.*, **13**, 1355–1357.

14. Tsuijioka, T., Hamada, Y., Shibata, K., Taniguchi, A. and Fuyuki, T. (2001) *Appl. Phys. Lett.*, **78**, 2282–2284.

15. Matsuda, K. and Irie, M. (2005) *Polyhedron*, **24**, 2477–2483.

16. Yokoyama, Y., Shiozawa, T., Tani, Y. and Ubukata, T. (2009) *Angew. Chem. Int. Ed.*, **48**, 4521–4523.

17. Wigglesworth, T.J., Sud, D., Norsten, T.B., Lekhi, V.S. and Branda, N.R. (2005) *J. Am. Chem. Soc.*, **127**, 7272–7273.

18. Tsujika, T., Sesumi, Y., Tagaki, R., Masui, K., Yokojima, S., Uchida, K. and Nakamura, S. (2008) *J. Am. Chem. Soc.*, **130**, 10740–10747.

19. de Jong, J.J.D., Tiemersma-Wegman, T.D., van Esch, J.H. and Feringa, B.L. (2005) *J. Am. Chem. Soc.*, **127**, 13804–13805.

20. Tian, H. and Wang, S. (2007) *Chem. Commun.*, **8**, 781–792.

21. Cusido, J., Deniz, E. and Raymo, F.M. (2009) *Eur. J. Org. Chem.*, **13**, 2031–2045.

22. Raymo, F.M. and Tomasulo, M. (2005) *Chem. Soc. Rev.*, **34**, 327–336.

23. Odo, Y., Fukaminato, T. and Irie, M. (2007) *Chem. Lett.*, **36**, 240–241.

24. Koshido, T., Kawai, T. and Yoshino, K. (1995) *J. Phys. Chem.*, **99**, 6110–6114.

25. Gorodetsky, B. and Branda, N.R. (2007) *Adv. Funct. Mater.*, **17**, 786–796.

26. Gorodetsky, B., Samachetty, H., Donkers, R.L., Workentin, M.S. and Branda, N.R. (2004) *Angew. Chem. Int. Ed.*, **43**, 2812–2815.

27. Fraysse, S., Coudret, C. and Launey, J.P. (2000) *Eur. J. Inorg. Chem.*, **7**, 1581–1590.
28. Peters, A. and Branda, N.R. (2003) *J. Am. Chem. Soc.*, **125**, 3404–3405.
29. Peters, A. and Branda, N.R. (2003) *Chem. Commun.*, **8**, 954–955.
30. Guirado, G., Coudret, C., Hliwa, M. and Launay, J.-P. (2005) *J. Phys. Chem. B*, **109**, 17445–17459.
31. Zhou, X.-H., Zhang, F.-S., Yuan, P., Sun, F., Pu, S.-Z., Zhao, F.-Q. and Tung, C.-H. (2004) *Chem. Lett.*, **33**, 1006–1007.
32. Areephong, J., Browne, W.R., Katsonis, N. and Feringa, B.L. (2006) *Chem. Commun.*, **37**, 3930–3932.
33. Moriyama, Y., Matsuda, K., Tanifuji, N., Irie, S. and Irie, M. (2005) *Org. Lett.*, **7**, 3315–3318.
34. Browne, W.R., de Jong, J.J.D., Kudernac, T., Walko, M., Lucas, L.N., Uchida, K., van Esch, J.H. and Feringa, B.L. (2005) *Chem. Eur. J.*, **11**, 6414–6429.
35. Browne, W.R., de Jong, J.J.D., Kudernac, T., Walko, M., Lucas, L.N., Uchida, K., van Esch, J.H. and Feringa, B.L. (2005) *Chem. Eur. J.*, **11**, 6430–6441.
36. Takeshita, M., Uchida, K. and Irie, M. (1996) *Chem. Commun.*, **15**, 1807–1808.
37. Takeshita, M. and Irie, M. (1998) *Tetrahedron Lett.*, **39**, 613–616.
38. Takeshita, M. and Irie, M. (1998) *J. Org. Chem.*, **63**, 6643–6649.
39. Kawai, S.H. (1998) *Tetrahedron Lett.*, **39**, 4445–4448.
40. Takeshita, M., Soong, C.F. and Irie, M. (1998) *Tetrahedron Lett.*, **39**, 7717–7720.
41. Mulder, A., Jukovic, A.A., van Leeuwen, F.W.B., Kooijman, H., Spek, A.L., Huskens, J. and Reinhoudt, D.N. (2004) *Chem. Eur. J.*, **10**, 1114–1123.
42. Sud, D., Norsten, T.B. and Branda, N.R. (2005) *Angew. Chem. Int. Ed.*, **44**, 2019–2021.
43. Vomasta, D., Högner, C., Branda, N.R. and König, B. (2008) *Angew. Chem. Int. Ed.*, **47**, 7644–7647.
44. Kawai, S.H., Gilat, S.L. and Lehn, J.-M. (1999) *Eur. J. Org. Chem.*, **9**, 2359–2366.
45. Odo, Y., Matsuda, K. and Irie, M. (2006) *Chem. Eur. J.*, **12**, 4283–4288.
46. Malval, J.-P., Gosse, I., Morand, J.-P. and Lapouyade, R. (2002) *J. Am. Chem. Soc.*, **124**, 904–905.
47. Samachetty, H.D. and Branda, N.R. (2005) *Chem. Commun.*, **22**, 2840–2842.
48. Samachetty, H.D. and Branda, N.R. (2006) *Pure Appl. Chem.*, **78**, 2351–2359.
49. Samachetty, H.D. (2007) Modulation of chemical reactivity using photoresponsive dithienylethene. Ph.D. thesis. Department of Chemistry, Simon Fraser University. Burnaby, BC, Canada.
50. Lemieux, V., Spantulescu, M.D., Baldridge, K.K. and Branda, N.R. (2008) *Angew. Chem. Int. Ed.*, **120**, 5112–5115.
51. Sud, D., Wigglesworth, T.J. and Branda, N.R. (2007) *Angew. Chem. Int. Ed.*, **46**, 8017–8019.
52. Nicolaou, K.C., Zuccarello, G., Riemer, C., Estevez, V.A. and Dai, W.M. (1991) *Angew. Chem. Int. Ed. Engl.*, **30**, 1387–1416.
53. Compiled by McNaught, A. D. and Wilkinson, A. (eds) (1997) IUPAC compendium of chemical terminology, *The "Gold Book"*, 2nd edn, Blackwell Scientific Publications, Oxford, 169.
54. Irie, M., Miyatake, O., Uchida, K. and Eriguchi, T. (1994) *J. Am. Chem. Soc.*, **116**, 9894–9900.
55. Irie, M., Miyatake, O. and Uchida, K. (1992) *J. Am. Chem. Soc.*, **114**, 8715–8716.
56. Irie, M., Miyatake, O., Sumiya, R., Hanazawa, M., Horikawa, Y. and Uchida, K. (1994) *Mol. Cryst. Liq. Cryst.*, **246**, 155–158.
57. Walko, M. and Feringa, B.L. (2005) *Mol. Cryst. Liq. Cryst.*, **431**, 249–253.
58. Qin, B., Yao, R., Zhao, X. and Tian, H. (2003) *Org. Biomol. Chem.*, **1**, 2187–2191.
59. Sud, D., Norsten, T.B. and Branda, N.R. (2004) *Angew. Chem. Int. Ed.*, **44**, 2019–2021.
60. Kawai, S.H., Gilat, S.L. and Lehn, J.-M. (1994) *J. Chem. Soc., Chem. Commun.*, **8**, 1011–1013.
61. Kawai, S.H., Gilat, S.L., Ponsinet, R. and Lehn, J.-M. (1995) *Chem. Eur. J.*, **1**, 285–293.

62. Ohsumi, M., Fukaminato, T. and Irie, M. (2005) *Chem. Commun.*, **31**, 3921–3923.
63. Kühni, J. and Belser, P. (2007) *Org. Lett.*, **9**, 1915–1918.
64. Uchida, K., Matsuoka, T., Kobatake, S., Yamaguchi, T. and Irie, M. (2001) *Tetrahedron*, **57**, 4559–4565.
65. Chen, Z., Zhao, S., Li, Z., Zhang, Z. and Zhang, F. (2007) *Sci. China Ser. B-Chem.*, **50**, 581–586.
66. Lemieux, V., Gauthier, S., Branda, N.R. (2006) *Angew. Chem. Int. Ed.*, **45**, 6820–6824.
67. Lemieux, V. and Branda, N.R. (2005) *Org. Lett.*, **7**, 2969–2972.
68. Raymond Bonnett (2000) *Chemical Aspects of Photodynamic Therapy.*, Gordon and Breach Science Publishers, Amsterdam.

2
Photoswitchable Molecular Systems Based on Spiropyrans and Spirooxazines
Vladimir I. Minkin

2.1
Introduction

During the last few decades a substantial interest has developed in gaining insight into the properties and rearrangements of bistable molecular systems that exist in two thermodynamically stable states and are capable of interconversion under the action of various external stimuli. This interest is highly motivated by the fact that the bistable molecules and molecular systems may serve as the nanoscopic switching elements that have already found diverse applications in the area of molecular electronics and photonics [1]. Equally important are several other areas of application of these compounds related to the role they play in the transport of biochemical information and signal transmission across biological membranes and photochemically switched enzymatic, bio- and chemosensoric systems [2]. An efficient and technologically adaptable way to address bistable molecules and systems from the macroscopic level is the use of light, which represents an excellent means for the supply of energy and circuit operations. Reversible rearrangements of a chemical species between two forms, **A** and **B**, induced in one or both directions by absorption of light and resulted in changes in the absorption spectra (and other physical properties as well) form the basis of the extensively studied phenomenon of photochromism [3]. Although the first examples of photochromic behaviour of organic compounds date back to the nineteenth century [3a], the real interest in the properties and technical application of photochromic compounds that spurred active research on the general phenomenon of photochromism was initiated by the discovery of the photochromism of spiropyrans (SPs) **1** (X = CH, CR) [4a–c] and recognition of significance of their bistability for a 'photochemical erasable memory' [4d,e]. In the background of the intensive studies of photochromic properties of SP, the first synthesis of their closest structural analogues, spirooxazines (SPOs) **1** (X = N), in 1961 (see Ref. [5] for related patents) went almost unnoticed until the discovery of their extraordinarily high photostability that led to their successful use in various applications, especially in eyewear technology.

Molecular Switches, Second Edition. Edited by Ben L. Feringa and Wesley R. Browne.
© 2011 Wiley-VCH Verlag GmbH & Co. KGaA. Published 2011 by Wiley-VCH Verlag GmbH & Co. KGaA.

X = CH, CR' – Spiropyrans; X = N – Spirooxazines; Z – Heteroatom

When compared to other classes of photochromic compounds, SP and SPO have certain advantages and disadvantages. The former include synthetic accessibility and wide possibilities for structural variation aimed at tuning their properties to the requirements of a specific application. SP and SPO photochromic systems are characterized by high values of quantum efficiencies of direct and back photoinitiated rearrangements and possess very high cross sections of two-photon light absorption of ring-opened and ring-closed isomeric forms, the property that is of primary importance for the design of three-dimensional molecular memory media. The principal disadvantages of SP and SPO are associated with the relatively low energy barriers against the thermal back reaction that converts the photoisomers **1B** to their noncoloured spirocyclic forms **1A** resulting in the loss of the light-written information. Although a number of efficient approaches to the improvement of the thermal stability of the ring-opened forms **1B** have been elaborated, by this kinetic parameter SP and SPO are noncompetitive to diarylethenes and fulgides, the ring-opened and ring-closed isomers of which undergo thermally irreversible photochromic rearrangements. This drawback in the properties of the SP and SPO photochromes imposes certain limitations on their use in optical information storage systems. The data on principal spectral and photochemical parameters of SP, SPO and, for comparison, of two other important photochromic systems are collected in Table 2.1.

Table 2.1 Generalized spectral and photochemical parameters of photochromic systems: SP, SPO [5–7], fulgides [8], and diarylethenes [9].

Photochromes	Spiropyrans	Spirooxazines	Fulgides	Diarylethenes
Ring-closed λ_{max} (nm)	320–380	330–435	270–390	230–300
Ring-opened λ_{max} (nm), ε (l mol^{-1} cm^{-1})	440–660 (40–45) × 10^3	540–670 (39–80) × 10^3	470–825 (4.8–26) × 10^3	530–820 (5–18) × 10^3
Quantum yield of photocolouration	0.1–0.9	0.1–0.6	0.02–0.7	0.2–0.5
Quantum yield of back photoreaction	0.001–0.05	0.001–0.05	0.0005–0.9	0.003–0.13
Lifetime of the photoisomer (rt) (s)	1 × 10^{-3} –6 × 10^5	3 × 10^{-1} –5 × 10^3	7 × 10^3 –25 × 10^4	45–years
Two-photon cross section (cm^4 s photon^{-1})	>10^{-48} [10a]	–	10.3 × 10^{-48} [10b]	~10^{-50} [10c]

The chemistry and photochemistry of SP [6] and SPO [5, 7] were the subjects of a number of previous reviews. This chapter is focused on the latest contributions to the study of the mechanisms of the photochromic rearrangements of compounds **1**, properties and application of SP- and SPO-based photoswitchable systems.

2.2 Mechanism of the Photochromic Reaction

The photoinduced reversible conversion between SP and the merocyanine isomeric species has been the subject of numerous experimental [11] and theoretical [12] studies aimed at the understanding of the photophysical behaviour of this photochromic system, which is necessary for its proper utilisation as functionalized molecular switches. The primary step of the photochromic reaction of SP and SPO is the dissociation of a $C_{spiro}-O$ bond occurring in an electronic excited state and leading to the formation of a thermally equilibrated mixture of the coloured merocyanine conformers, 4a-d (Scheme 2.1). These processes are illustrated by the examples of photochemical behaviour of a SP (Figure 2.1) and a SPO (Figure 2.2).

Of eight different stereoisomers from the various possible conformations of the central CC bonds the most realistic are those corresponding to the four configurations pictured in Scheme 2.1, to which four excited merocyanine isomers with lifetimes of 15, 45, 120 ps and 2 ns detected in the time-resolved fluorescence experiments [13] may be assigned. According to the quantum-mechanical calculations carried out for SPs: (at the HF/6-31G* level of approximation for 1′,3′-dihydro-1′,3′,3′-trimethyl-6-nitrospiro[2H-1-benzopyran-2,2′-[2H]indole] **5** [12] (a common acronym is 6-nitroBIPS) and the DFT B3LYP/6-31G** level for 1′,3′,3′-trimethyl-1,2-tetramethylenespiro-[7H-furo(3,2-f)-(2H-1)-benzopyran-7,2′-indoline] [14]) the most stable conformers belong to the **TTC** and **CTC** (or **TTT**) forms, where **T** and **C** indicate *trans* and *cis* conformations of the CC bonds. The same type conformers were found by DFT calculations [12b,c,f] to be energy-preferred forms for the merocyanines formed upon ring-opening reactions of 1′,3′,3′-trimethylspiro[indoline-2′,3-naphtho[2,1-b][1,4]oxazine] **6** and its 2,2′-[2H]-naphtho[1,2-b] isomer **7**.

5 (R = CH₃, 6-nitroBIPS)　　　　**6**　　　　**7**

Modern methods in NMR spectroscopy were successfully applied for elucidation of molecular structures of the ring-opened isomers of SPs and SPOs produced upon irradiation of their solutions [15]. The unprecedented

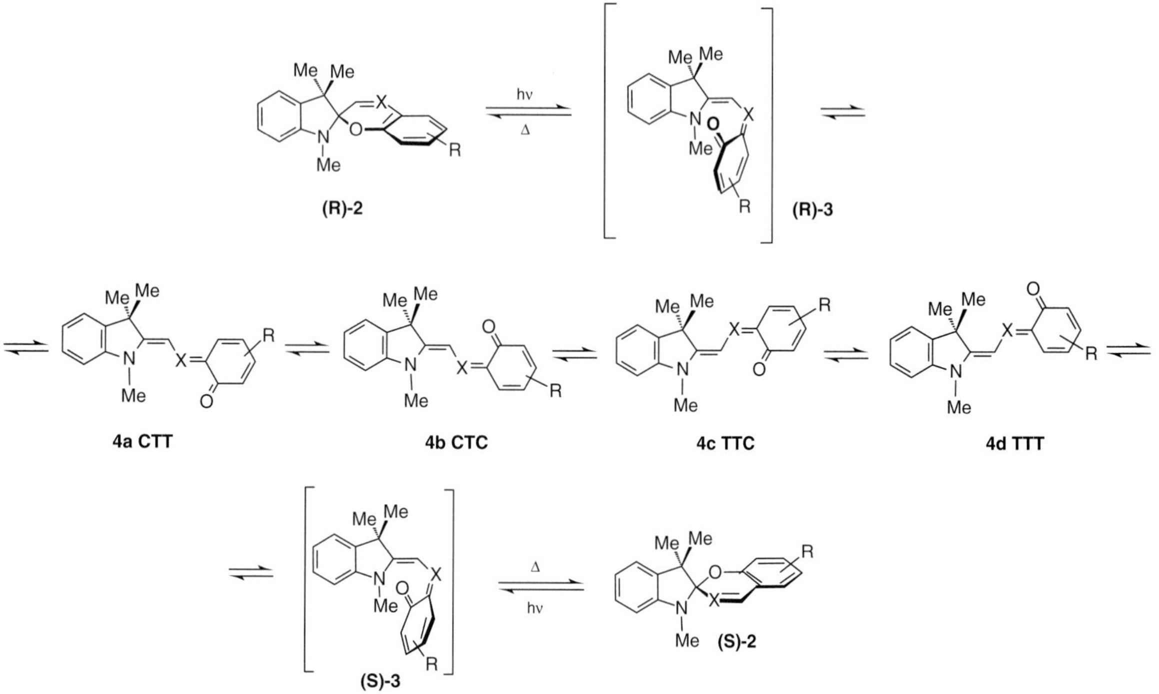

Scheme 2.1 General scheme describing thermal equilibria established in a solution of spiropyrans (X = CH) and spirooxazines (X = N) upon illumination with UV light [6d].

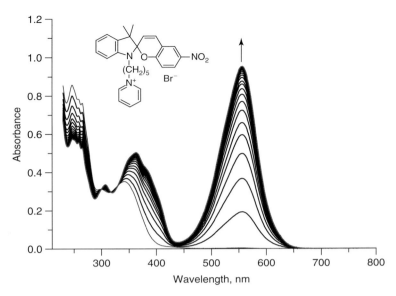

Figure 2.1 UV/visible absorption spectra of 1'-(ω-pyridinioamyl)-3',3'-dimethyl-6-nitrospiro[2H-1-benzopyran-2,2'-indoline] bromide in acetonitrile (20 °C, 3.7×10^{-5} M) before (dashed line) and after irradiation with 365 nm light. The sequence of spectral curves shows growth in the absorbance in the visible region due to the formation of the merocyanine isomer upon irradiation for 10 s (the bottom curve), 20 s, 30 s, and so forth until establishment of the photostationary equilibrium.

coexistence of the ring-closed and all four transoid merocyanine isomers of 5-hydroxy-1',3',3'-trimethylspiro[indoline-2',3-[3H]naphtho[1,2-b][1,4]oxazine] **8** in the equilibrium (Scheme 2.2) established in the dark was reported based on the 2D-ROESY ^{1}H NMR study [15b]. No equilibrium between the ring-closed and the ring-opened forms was observed in the case of a nonsubstituted spironaphthoxazine [16], which fact clearly points to the involvement of a hydroxyl group of **8** in the stabilization of its ring-opened isomers by the intramolecular hydrogen O–H ... O=C bonds.

The polar merocyanine forms of SP and SPO tend to associate into stack-like aggregates [12e, 17]. This tendency is very strong and rather stable associates are formed in very diluted solutions and even in polymeric films. Absorption spectra of J-aggregates, which have the parallel (head-to-head) arrangement of the molecular dipoles, are shifted to longer wavelengths relative to the spectra of the isolated merocyanine molecules. For H-aggregates having head-to-tail arrangement of molecular dipoles the spectra are shifted to shorter wavelengths. An important property of the J-aggregates produced by irradiation of solutions of SPs in nonpolar solvents is that their spectra consist of very narrow absorption bands (absorption peak widths are a few tens of nanometres), which is a necessary condition for the design of wavelength-multiplexed memory systems [18a]. It has been shown that the formation of the SP aggregates causes very large changes in refractive indices

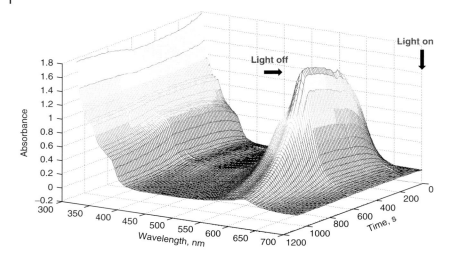

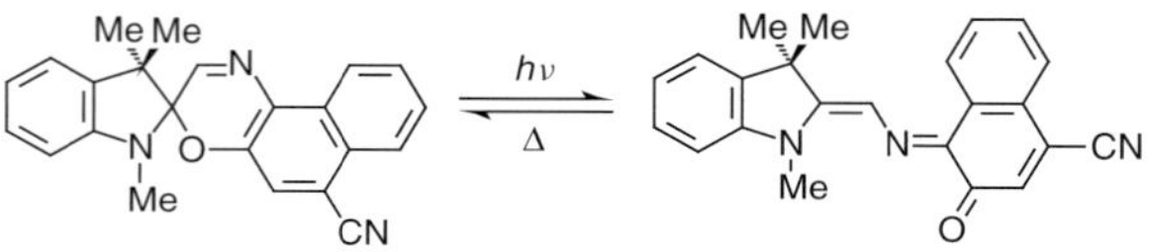

Figure 2.2 Evolution of the absorption spectrum of a toluene solution (1.2×10^{-4} mol l^{-1}, 293 K) of 1′-methyl-8-cyanospiroindoline-2′,3-naphtho[2,1-b][1,4]oxazine under irradiation with 365 nm light and after stopping irradiation [6d]. Reproduced from reference [6d]. Copyright ACS (2004).

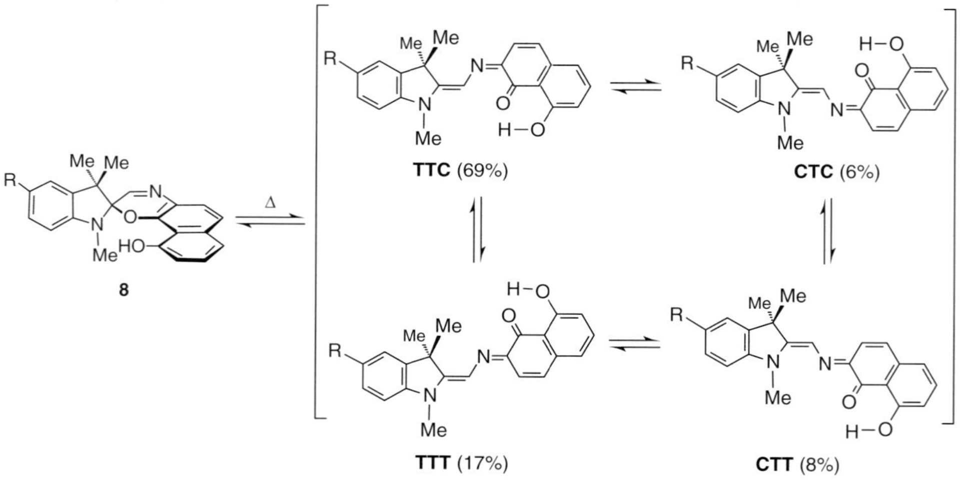

Scheme 2.2 The dark equilibrium between ring-closed and ring-opened forms of SPO **8** (R = OH) in methanol at 243 K.

and, thus, provides a new approach towards the synthesis of tunable photonic bandgap materials [18b]. The J-aggregates of the derivatives of 6-nitroBIPS stable at ambient temperature are characterized by the absorption bands that are sharper and more intense than those of the unassociated ring-opened isomeric forms and show nonlinear photochromism with a threshold, which optical properties make them suitable for use as optically synaptic films and material for memory devices [19].

The comprehension of the photochemical mechanism of a photochromic system is a necessary prerequisite to the enhancement of its properties and, therefore, considerable efforts have been devoted to gaining insight into the nature of the active states and intermediates of the reversible rearrangements of SPs; see Refs. [6c,d, 7c] for recent reviews. The nature of the active excited state of SP and SPO depends on substitution in their benzopyran and naphthoxazine rings. Since the duration of the primary process is very short, the studies of the excited-state reaction dynamics of SPs were mainly performed with the use of time-resolved nano-, pico- and femtosecond time-resolved spectroscopy [11a–d,i,k,m–o,s] and time-resolved resonance Raman spectroscopy [11f, 20]. These studies have shown that for the compounds with a 6-NO_2 group, a triplet state plays a crucial role in the photochemical ring-opening process. The mechanism (Scheme 2.3) explaining the photochemical behaviour of a series of indolinobenzopyrans **2** (X = CH) with a nitro group at position 6 and with or without substituents at position 8 involves the intersystem crossing to the short-lived triplet state of the ring-closed isomer ^{3}**Sp*** directly identified by nanosecond laser photolysis. It serves as the precursor to the triplet **CCC** merocyanine form ^{3}**MC***$_{perp}$ that may be correlated with the nonplanar structure **3** in Scheme 2.1. ^{3}**MC***$_{perp}$ conformation is in an equilibrium with the triplet of the transoid isomer observed as a short-lived transient with absorption maxima at 420–440 and 560–590 nm and lifetime <10 ms. The reaction ends up with quenching the triplet with oxygen and establishment of thermal equilibrium between the most stable merocyanine isomers, presumably **CTC** and **TTC**. This mechanism is evidenced by the extreme sensitivity of the dynamics to the presence of oxygen in solution and identity of the quantum yields of population of the merocyanine triplet state and the overall process of a SP to merocyanine photoconversion [11i,k, 21].

The appearance of a band at 430–450 nm in the spectrum of 6-nitroBIPS **5** within 8 ps after UV excitation was also demonstrated in the flash-photolysis experiments of Krysanov and Alfimov [11a], who also ascribed this band to the intermediate **CCC** form (usually labelled as **X**-form [4c]). The cis-cisoid transient product emerging from the fission of the C_{spiro}–O bond of 6-nitroBIPS after excitation to the S_1 energy level (intermediate **X**) was also invoked as an essential intermediate preceding to

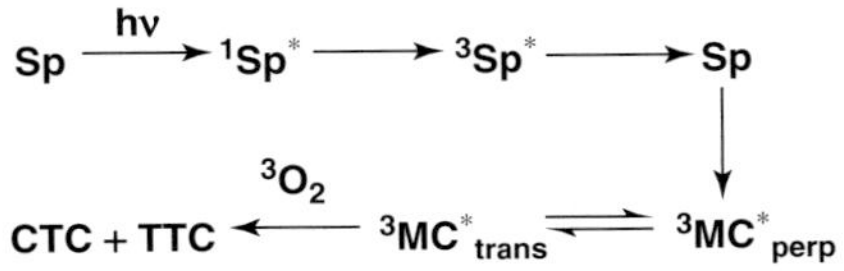

Sp $\xrightarrow{h\nu}$ ^{1}Sp* $\longrightarrow$ ^{3}Sp* $\longrightarrow$ Sp

CTC + TTC $\xleftarrow{^3O_2}$ ^{3}MC*$_{trans}$ $\rightleftharpoons$ ^{3}MC*$_{perp}$

Scheme 2.3 The Görner–Chibisov mechanism for the photocolouration reaction of 6-nitroBIPS.

the stage of conversion to a merocyanine form in the photochromic rearrangement of 6-nitroBIPS induced by two-photon absorption of femtosecond pulses [22].

Another interpretation of the dynamics of the photoexcited ring-opening reaction of 6-nitroBIPS was given on the basis of the femtosecond time-resolved UV/visible and UV/mid-infrared pump-probe spectroscopic study [11p]. The two transient features appearing during the first nanoseconds identified in these experiments were attributed to the ring-closed excited state $S_1 \rightarrow S_n$ absorption and to the second triplet state T_2 of the merocyanine isomer with a lifetime 0.5 ± 0.2 ps. No evidence was found for the intermediate isomeric form **X** or for the involvement of the first singlet excited state ring-opened merocyanine forms in the course of the photochemical ring-opening reaction.

For SPs without nitro groups in the benzopyran rings, the quantum yields of the photocolouration are substantially (on average, by half [23]) lower than those for SPs with nitro substituents. At the same time the participation of the low-energy triplet-state structures in the reaction significantly hastens the photodegradation processes [24a], but they may be substantially dampened in the compounds having no substituents with low-lying π^*- and/or n-orbitals. In contrast to SPs containing nitro groups in the 2H-chromene moiety, the unsubstituted SPs are usually considered to display photochromism only in the excited singlet manifold, although in degassed solution, an existence of a short-lived transient identified as the triplet state has been recently observed for spiropyrans containing 6-chloro and 8-diphenyloxazole substituents [24b]. A generalized mechanism for the photochemical rearrangement is represented by Scheme 2.4 that implies an appearance of the cis-cisoid intermediate **X** at the ground or both ground and first singlet excited states of the products of the photochemical reaction [14, 21b,c].

The existence of the intermediate **X**-form, was first proposed [4c,d] in the study of the photochemistry of SPs in low temperature matrices. On the basis of nano-, pico- and femtosecond spectroscopic data [11a,g,l], the coplanar cis-cisoid structures (corresponding to **CCC** in Scheme 2.1) were assigned to the primary photoproducts that emerge after excitation at the longest wavelength of the ring-closed form of SPs. In the case of 1′,3′,3′-trimethyl-6′-hydroxy-(2H-1)-benzopyran-2,2′-indoline, a part of a metastable species **X**, appearing in less than 100 fs after excitation at 300 nm, reestablishes the broken C_{spiro}–O bond on the timescale of 200 fs, whereas the rest converts to a mixture of merocyanine conformers with a decay time constant of about 100 ps [11g]. The formation of a long-lived cis-cisoid isomer **X** was observed

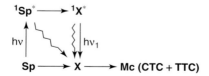

Scheme 2.4 General mechanism for the photochromic rearrangements of spiropyrans and spirooxazines containing no substituents with low-lying π^*-orbitals. The wavy arrows indicate the radiationless intersystem crossing transitions.

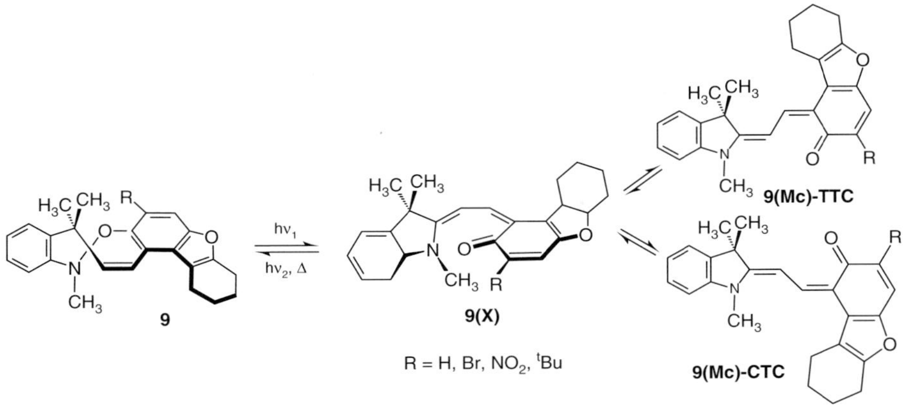

2.2 Mechanism of the Photochromic Reaction | 45

Scheme 2.5 Structural transformations of photochromic spiropyrans **9**.

in the course of the study of the photochromic behaviour of SPs **9** (R = tert-Bu) (Scheme 2.5) [14, 25]. In contrast with the compounds **9** (R = H, Br, NO$_2$) for which UV irradiation of their liquid or solid solutions gives rise to the appearance of long-wavelength absorption bands at 585–625 typical of trans-merocyanines, UV irradiation (365 nm) of rigid isopentane/isopropanol solution of **9** (R = tert-Bu) at 77 K leads to a spectral pattern characterized by a longest-wavelength absorption band at 471 nm. Heating the solution to 178 K or the prolonged irradiation at 77 K transforms the spectrum to that characteristic of other SPs of this series. The lifetime of the detected transient species is long enough to observe its absorption and fluorescence spectra under conditions of continuous irradiation, whereas its attribution to the cis-cisoid intermediate was justified by the good coincidence of its experimental and TD-B3LYP/6-31G** calculated spectra [14]. These data allowed the conclusion to be drawn that in SPs **9** (R = tert-Bu), a bulky tert-butyl group significantly hinders the **9(X)** → **9(Mc)** conformational rearrangement in viscous media. The relaxation time of **9(X)** in ethanol solution reaches 66 s at room temperature.

The primary photochemical processes in the photochemistry of SPOs studied using stationary photolysis [11r, 26] and laser flash photolysis within nano- [11q, 27], pico- [28] and femtosecond [11h,l,m, 29] time domains are generally similar to those of SPs. The formation of the coloured merocyanine isomers of spiroindolinonaphthoxazines and other SPOs occurs entirely from the excited singlet state. The noninvolvement of triplet states explains well the generally significantly higher fatigue resistance of compounds of this class as compared with SPs. An initial photoproduct having the *cis-cisoid* geometry (**X** form) arises within 100–300 fs [11l, 27] and rearranges to a more stable merocyanine on a timescale of several tens of picosecond. A detailed study of spectral properties and kinetics of the intermediate SPO form, to which the geometry characteristic of the *cis-cisoid* **X** isomer was attributed, was undertaken in the work by Glebov *et al.* [11r]. In a rigid frozen methanol matrix, the intermediate isomers **10(X)** have absorption bands in blue (400–500 nm) and far-red (750 nm) regions. The lifetime of the **X**-isomer in the matrix at liquid nitrogen temperature depends strongly on the local environment and varies over a wide range from tens of nanoseconds to milliseconds for **10** (R = CH$_3$) and to hours for **10** (R = C$_{16}$H$_{33}$). The activation energy for the **X** → **10(B)** transition was estimated to fall into the range of 12–32 kJ mol^{-1}. The occurrence of the intermediate **10(X)** form was also detected in the UV-irradiated polystyrene films of 1′,3′,3′-trimethylspiro[indoline-2′,3-[3H]naphth[2,1b][1,4]oxazine **6**, in which case the **10(X)** → **10(Mc)** transition proceeds even at very low temperature, 25 K, with overcoming an energy barrier of only 8 kJ mol^{-1}.

$$10A \underset{h\nu_2, \Delta}{\overset{h\nu_1}{\rightleftharpoons}} [X] \overset{\Delta}{\rightleftharpoons} 10B$$

R = CH$_3$, C$_{16}$H$_{33}$

According to the recent detailed DFT and TD-DFT exploration [12f] of the ground state and low-lying singlet states of the parent spironaphthoxazine **6** the principal course for the photochemical reaction lies through the S$_2$ → S$_1$ deactivation of the spirocyclic form and the subsequent ring opening the oxazine cycle resulting in the formation of the ring-opened *cis-cisoid* intermediate with a low activation barrier. At this point the molecule reaches a conical intersection where fast decay to the ground surface leads to either the initial closed form or the merocyanine isomeric form in one of its *cis*-conformations. The fast kinetics of the ring-opening reaction of SPOs favours the efficiency of their photocolouration, which is one of the most important requirements imposed upon the photochromic compounds with a potential for application to the photonic devices. The photochromic rearrangements of SPO are characterized by sufficiently high quantum yields of the photoreaction (Table 2.1) and higher than SP thermal stability of the photocoloured form, as assessed by energy barriers to the back dark reaction.

Whereas the photochemistry and mechanism of the ring-opening reaction of SP and SPO are now amply studied and generally understood, the reverse photoinitiated ring-closing reaction of the coloured merocyanine isomeric forms of these compounds has been relatively scarcely explored [7c, 11q, 30]. The ultrafast ring-closure kinetics of the merocyanines formed from two 6,8-dinitrosubstituted SPs has been studied in acetonitrile solution under photoexcitation with 390 nm light [30a]. It was concluded that the first singlet excited state of the merocyanines, formed from the higher-energy states for about 500 fs, partitioned between the recovery of the ground state and the formation of the corresponding spiro isomers via intermediates twisted about the central CC bond and, thus, similar to the *cis-cisoid* conformers **X**. The singlet manifold was found to be predominant and no evidence for triplet-state transients presented. This conclusion is in accord with the results of the sophisticated CASSCF/CASPT2 calculations of the mechanism for the photocyclization reaction performed on a model 6-(2-propenylidene)cyclohexadienone system [12g].

2.3
Switching of Physical Properties and Biological Activities via Photochromic Rearrangements of Functionalized Spiropyrans and Spirooxazines

The photoinduced transformation of the ring-closed structures of SP and SPO into their fully π-conjugated isomeric merocyanine forms results not only in the evolution of the absorption spectra (Figures 2.1 and 2.2), but also in the profound alterations of other physical properties of this molecular system, such as dipole moment, polarizability, nonlinear optic properties, emission spectra and macroscopic properties of the corresponding materials, for example refractive index and surface wettability. The rearrangement is also associated with significant changes in the geometry and chemical reactivity of the interconverting isomers that may be exploited in the design of light-responsive biologically active molecules, membranes and surfaces.

2.3.1
Light-Switchable Fluorescence Modulation

Due to the extreme sensitivity of the methods for detection of fluorescence, the fast response time and a possibility to employ fluorescence for nondestructive readout of the optically written information regulation of the emissive behaviour of photochromic compounds is considered as one of the most promising lines of investigation into the photoresponsive compounds and materials based on them [31]. Modulation of a fluorescence switch based on photochromic SP or SPO can be attained in several ways. The rearrangement **1A⇌1B** itself affects the ability of a molecule to emit light. In the case of SP and SPO only one of the isomers, usually the ring-opened one, exhibits fluorescence with a measurable quantum efficiency [32]. Under these conditions, the fluorescence of a photochromic compound can be modulated by switching back and forth between the emissive and nonemissive forms. This method has been employed to demonstrate the feasibility of a bit-oriented three-dimensional optical memory system [6c, 33]. For writing data, application of the method of two-photon absorption spectroscopy first suggested by Barachevsky and coworkers [34] was successfully realized by Rentzepis and coworkers [35] based on a photochromic derivative of 6-nitroBIPS **11**. In two-photon spectroscopy, the energy required for excitation of the ring-closed form in the UV range is delivered by two laser beams intersecting in a certain point of a volumetric polymeric recording medium. The total energy of the two photon energies (532 + 1064 nm) must be equal to or greater than the energy (355 nm) required for the excitation of the photorearrangement leading to the formation of the fluorescent merocyanine isomer. The fluorescence of this form, excited by focusing two 1064 nm laser beams onto the pixel formed, is used for reading data.

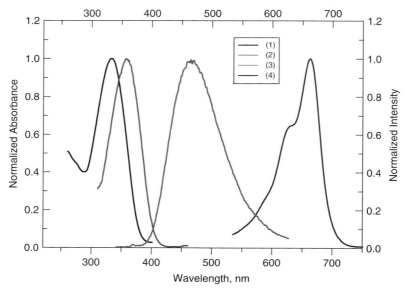

Although this fluorescence readout method is very sensitive, it has various drawbacks including relatively low quantum yields of fluorescence of merocyanines **1B**, their insufficient thermal stability and partial erasure of the written information due to the back photochemical reaction upon exciting for fluorescence. A method for nondestructive readout by detecting the changes in refractive indices of the interconverting forms of SP [33] was proposed, but in general the SP-based three-dimensional optical memory systems are outnumbered by the much more thermally stable diarylethene and fulgide systems [9, 36]. Information on the use of the latter for the development of fluorescent multilayer discs with up to 500 layers and 1 Tb capacity has already appeared in the literature [36b].

Another approach to the fluorescent photochromic switches is based on labelling a photochromic molecule with a strong fluorescent unit, the emission properties of which can be modulated by the photoinduced changes in the polarity of the adjacent moiety, electron- or energy-transfer processes that may occur in one of the interconverting isomers. Spiropyrans **12** labelled by a 3,5-diphenylpyrazoline fluorophore attached to the benzopyran ring exhibit intense green fluorescence

Figure 2.3 Absorption (1), excitation of fluorescence (2) and fluorescence (3) spectra of the ring-closed isomer of SP **12**(d)**A** and (4) absorption spectrum of the ring-opened merocyanine form **12**(d)**B** [37a]. Reproduced from reference [37a]. Copyright Taylor and Francis (1994).

under excitation at 420–430 nm. The photorearrangement induced by the exposing PMMA film of the SP to UV light (365 nm) leads to more polar merocyanine isomers absorbing at 600–700 nm and showing no detectable fluorescence due to either changes in the local polarity around the fluorophore or due to the fluorescence resonance energy transfer (FRET) between the fluorophore and the merocyanine [37]. The system can be used for recording of optical information because the excitation wavelength used for fluorescence spectrum is shifted to longer wavelength with respect to λ_{max} (~360 nm) of **12A** and, therefore, irradiation by the light at the tail of the excitation band (430–440 nm) does not appreciably initiate the A → B rearrangement (Figure 2.3).

Φ_{fl} (A): (a) 0.43 (b) 0.86 (c) 0.47 (d) 0.58 (R=H),
(e) 0.51 (R=CH$_3$)

The FRET mechanism drives the on/off switching of fluorescence of a photochromic dyad **13** composed of two 6-nitroBIPS units and a fluorescein unit [38]. The fluorescence intensity of the dyad is modulated by reversible conversion between the ring-closed spirocyclic and the ring-opened merocyanine forms. In THF solution, the former one representing the ground state of **13** displays strong fluorescence with the maximum at 550 nm typical of fluorescein. The intensity of this fluorescence is significantly decreased on UV irradiation of the solution due to conversion of the SP moieties to the merocyanine forms that have large spectral overlap with the fluorescence spectrum of the fluorescein unit. The energy transfer occurring upon excitation of the fluorescein unit (λ_{exc} = 430 nm) produces the excited merocyanine state **13B** that is responsible for the appearance of a new longer-wavelength (~650 nm) band. In a similar way, photoreversible fluorescence modulation was achieved by supramolecular complexation of a guest rhodamine molecule with 6-nitroBIPS-modified cyclodextrin [39]. Upon UV irradiation the fluorophore (rhodamine) within the cyclodextrin cavity transfers its energy to the merocyanine isomer of SP formed.

Recently, a series of new efficient photoswitches based on 6-nitroBIPS conjugates with the standard fluorescent ATTO dyes has been reported [31b]. The fluorescence emission spectrum of the fluorophore overlaps significantly with the absorption of the merocyanine form of the SP. As long as the SP resides in its uncoloured ring-closed form the fluorophore shows strong fluorescence. Upon switching the SP to its merocyanine form under UV irradiation, fluorescence is strongly quenched by energy transfer. It was demonstrated that modulation of the fluorescence intensity of donor fluorophores occurs reversibly not only at the ensemble, but also at the single-molecule level. This finding gives promise that SPs can be used as effective switchable energy-transfer acceptors for single-molecule-based localization microscopy by stabilization of the coloured merocyanine form through embedment in polymers.

The analogous approach has been recently applied for the preparation of highly stable and fluorescent switching SPOs [40]. The colourless ring-closed forms of SPO **14A** with a fluorescent 1,8-naphthalimide unit incorporated at the naphthoxazine fragment display strong fluorescence characteristic of this fluorophore, whereas the photoisomerization into the polar merocyanine form leads to fading of the fluorescence. The process can be switched 'on' and 'off' by photoinduced rearrangement between the fluorescent ring-closed and nonfluorescent ring-opened forms. A remarkable effect of the hybridization of SPO with a naphthalimide fragment is the significant enhancement of thermal stability of the ring-opened merocyanine isomer. The half-lifetime of **14B** (R = 4,5-benzo) reaches $\sim 10^4$ s in isopropanol at room temperature. The authors [40] explain this effect by the electron-withdrawing effect of the attached naphthalimide fragment lowering the nucleophilicity of the carbonyl oxygen and slowing down the dark cyclization reaction.

14A   R = H, 4,5-Benzo   **14B**

A new trend in the design of the photoswitchable SP-merocyanine systems exhibiting fluorescent response to isomerization is associated with the use of polymer nanoparticles incorporating photochromic compounds into their hydrophobic cavities [41a, d]. In contrast to the generally nonfluorescent character or very low quantum yields of fluorescence emission shown by merocyanine isomers of SP in most environments [32], that of a derivative of 6-nitroBIPS (**5**, R = $CH_2CH_2OC(=O)CH=CH_2$) fluoresces strongly when being encapsulated into 40–400-nm polymeric nanoparticles prepared by the emulsion copolymerization of N-isopropylacrylamide and styrene with a small amount of divinyl benzene. The nanoparticles are colourless in the ring-closed form and turn to dark blue within seconds under 365-nm UV-illumination. The dark blue nanoparticles containing the formed merocyanine isomers emit a strong red fluorescence

($\Phi_{fl} = 0.24$). The fluorescence can be readily switched 'on' and 'off' by alternating UV and visible light. The SP-doped nanoparticles exhibit good reversibility and they are significantly more photostable than the same SPs in solution, which makes them promising materials for potential applications for data storage. An additional application of the photoswitchable nanoparticles was demonstrated by employing them in biological imaging. Using liposomes it was possible to deliver the optically switchable nanoparticles into human embryonic kidney living cells (HEK-293) and to effect *in vivo* optical switching.

By addition of a fluorescent dye (perylene diimide) to the mixture of the polymerizable monomers and SP and applying a slight modification of the emulsion polymerization procedure it becomes possible to prepare optically addressable dual-colour spherical fluorescent nanoparticles with diameters varying from 50 to 110 nm that show high resistance to photobleaching [41b]. When the photoswitchable SP **5** ($R = CH_2CH_2OC(=O)CH=CH_2$) is in its ring-closed form, the perylene chromophore located in the hydrophobic core emits a green fluorescence ($\lambda_{max} = 535$ nm). The UV-initiated ring-opening reaction converts the SP to its merocyanine form whose visible absorption band ($\lambda_{max} = 588$ nm) overlaps the perylene emission band. This leads to quenching the perylene emission by a FRET mechanism and converting the green emission from perylene to red emission from merocyanine when the SP-containing nanoparticles are photochemically switched to merocyanine-containing ones. It is expected that the obtained high-contrast dual-colour hydrophilic nanoparticles incorporating light-driven molecular switches in hydrophobic cavities are particularly suitable for applications in tracking and labelling components of complex biological systems [41b].

The FRET mechanism is also responsible for modulation of a fluorescence switch based on photochromic SPO **8** ($R = OCH_3$) and fluorescent 4-(dicyanomethylene)-2-methyl-6-(*p*-dimethylaminostyryl)-4H-pyran encapsulated into organic nanoparticles. Addition of an emissive-assistant 1,3-bis(pyrene) propane provides for the enhancement of the contrast (20 : 1) of the fluorescence signal between 'on' and 'off' states and also affords a convenient way to tune the excitation wavelength for reading the fluorescence [41c].

2.3.2
Photocontrolled Magnetization

Preparation of optically tunable molecular magnetic compounds and solids that can be used for high-density information storage media and switching devices is currently an active area of research [42]. Photoswitching of magnetization with the use of photochromic compounds may be realized in two principal ways: by intercalation of these compounds into para- or ferromagnetic inorganic materials or by hybridization of photochromes with d^n ($n = 4-7$) transition-metal coordination compounds capable of exhibiting spin crossover between the low-spin and high-spin states. The proper functionalization of magnetic nanoparticles with a photochromic entity may constitute an additional, but not yet realized approach to photocontrolled magnetic materials.

A photochromic molecule-based magnet was prepared by Yu and coauthors [43] by pairing the layered ferromagnetic trioxalate anions $\{M^{II}M^{III}(ox)_3\}_n^{n-}$ with a cationic SP **15**. The compound displays both solution and crystalline-state photochromism resulting in the UV light-induced equilibrium between a yellow ring-closed and a red ring-opened form. Although the rearrangement did not significantly affect the Curie temperature of the powdered samples of the SP, it resulted in pronounced changes in the hysteresis loops of the initial and irradiated samples. The coercive field (H_{coer}) and the remnant magnetization value (M_R) as well were increased by UV irradiation, changing from $H_{coer} = 40$ G and $M_R = 0.3\ \mu_B$ (Bohr magneton) for the nonirradiated sample to $H_{coer} = 290$ G and $M_R = 2\ \mu_B$ for the irradiated sample. It must be noted, however, that the changes in these values are irreversible even though the photoreaction of the cationic part of **15** remains reversible.

The similar approach to the design of photoswitchable molecule-based magnets with the use of the bimetallic oxalate anions as bridging ligands for transferring magnetic interactions was applied by Aldoshin and coauthors [44] who studied the photochromic properties of $[SP^+X^-]$ **17**. UV irradiation of **17** in the polycrystalline state is accompanied by a considerable structural change in crystal structure of the anion X^- due to the separation of the layers by longer distances. At temperatures below 10 K, magnetic susceptibility depends on the magnetic field, thus indicating the transition of the compound to the magnetically ordered ferromagnetic state. The effective magnetic moment μ_{eff} increases from 7.5 μ_B at 300 K to 29 μ_B at 5 K.

15A ⇌ **15B** (UV / Vis, Δ) (SP+)(X−); R=H, R′=i-C$_3$H$_7$

16A ⇌ **16B** (SP+)(Y−); R=CH$_3$; R′=CH$_3$, C$_2$H$_5$, C$_3$H$_7$

17A ⇌ (UV / Vis, Δ) **17B**

Kashima et al. [45] extended the studies of the SP-hybridized photoswitchable two-dimensional magnetic honeycomb structures to the iron mixed-valence complexes with the layered dithiooxalate (dto) networks **16**. They demonstrated that substitution of an ammonium cation in the ferromagnetic complex $(n\text{-}C_3H_7)_4N^+[Fe^{II}Fe^{III}(dto)_3]^-$ by SP cations **16A** gives the formed compounds photochromic properties. UV irradiation of KBr pellets of **16A** results in the **16A → 16B** conversion that significantly changes the ferromagnetic transition temperature and coercive force of the sample before irradiation.

Following the strategy of the intercalation of a photochromic species into the magnetic inorganic structures, N-methylated pyridospiropyran cations **15** (R = H) were inserted into the layered $MnPS_3$ semiconductor compounds [46]. In the $MnPS_3$ intercalation compounds, the guest species densely fill up the interlayer spaces. This feature provides for the formation of closely packed host–guest structures that are subjected to pronounced alterations occurring in the conformation of the guest under the action of an external stimulus. The interconversion of the ring-closed and ring-opened forms of **15** (R = H) within the interlayer $MnPS_3$ galleries leads to a considerable stabilization of the latter. As in the case of the (**15** (R = H))-[MnCr(ox)$_3$] complex [43] intercalation substantially stabilizes the merocyanine form, supposedly because of the formation of J-aggregates within the weakly polar interlayer medium. On the other hand, in contrast to **15** (R = H)-[MnCr(ox)$_3$], the magnetic hysteresis loop that follows UV irradiation of a sample of the intercalate **15** (R = H)-$MnPS_3$ is reversible upon irradiation with visible light. Stable photochromic films of **15** (R = H)-$MnPS_3$ may serve as a prospective material for information storage devices. The preparation of air-stable hybrid organic–inorganic nanoparticles comprising SP **15** (R = H) intercalated in layered $MnPS_3$ has also been reported and photochromic behaviour of the nanoparticles in polyvinylpyrrolidone/chloroform colloid state and solid films observed [47]. No data were communicated, however, on magnetic properties of these nanoparticles. Further investigation into the magnetic behaviour of these and similar types of nanoparticles may be of special interest since each nanoparticle is known to become a single magnetic domain with a fast response to applied magnetic fields making them usable for a wide range of technical applications [48].

The photoinduced aggregation of SPs – derivatives of 6-nitroBIPS containing long alkyl chains in the substituents at positions 1′ and 8 was used to increase magnetization of stable hybrid vesicles formed by these SP on sonicating their and iron oxide (Fe_3O_4) nanoparticles water dispersion [49]. Prolonged UV illumination of the cast composite films of the (SP-Fe_3O_4) vesicles leads to the

J-aggregation due to the formation of the merocyanine isomers of SP and to a significant increase in the magnetization value of magnetic vesicles showing extremely high coercivity and remanence at room temperature. The transformation is reversible, but the back reaction involving separation of the aggregated particles is slow.

A promising way to employ photochromes for switching magnetization may be associated with covalent hybridization of these compounds with d^4–d^7 metal coordination compounds potentially capable of crossover rearrangements. Such a functionalization of a metal complex may induce a spin transition via isomerization of the photoactive ligand that modifies the structural environment of a metal ion and, consequently, ligand field at this ion. The feasibility of this approach, termed ligand-driven light-induced spin change (LD-LISC), was demonstrated by an example of the *trans*-styryl-2,2′-bipyridyl Fe^{II} complex that on UV illumination at 108 K undergoes a sharp spin transition to the low-spin isomer brought about by *trans*–*cis* isomerization of the ligand [42a]. Until now no reports appeared on the LD-LISC rearrangements of the SP-functionalized metal complexes. A study of photochromic behaviour and magnetic properties of a series of such compounds, for example **18** based on the para- and ferromagnetic Schiff base complexes [50] is now in progress in our laboratory. Since magnetization and the appearance of hysteresis loops are cooperative effects the ligand structures must contain molecular fragments providing for sufficiently strong intermolecular interactions (H-bonds, π–π stacking) between the molecules.

18A ⇌ **18B** (hv$_1$ / hv$_2$, Δ)

M = Fe, Co, Mn
X = O, S, NSO$_2$Ar

2.3.3
Photoswitching of Macroscopic Properties of Bulk Materials and Membranes

Modulation of magnetization of solids using inserted or grafted photochromic compounds discussed in the previous section exemplifies the wide area of applications of molecular switches for tuning properties of various macroscopic materials. At the molecular level, the photoinduced rearrangements impose significant structural changes on the interconverting isomeric forms of SP and SPO resulting in profound alterations of their geometry and electronic structure. In bulk materials hosting these photochromes, light-induced modifications of the molecular properties are embodied in certain changes of the macroscopic properties, for example absorption coefficient, refractive index, hydrophobicity, crystal shape and surface morphology. The organized monolayers of bistable photochromic switches can be successfully employed for effecting significant changes and controlling such bulk properties as wettability, conductivity, self-assembly, alignment of liquid crystals and others [51a,b]. Impregnation of properly functionalized SPs and SPOs onto the surfaces of various important materials has opened up a new approach to the design and manufacturing of smart materials with properties controlled by the rearrangements of the photoswitchable compounds. An integrated nanomaterial with photoresponsive behaviour was obtained by coupling SPs to single-walled carbon nanotubes [51c]. The microporous siliceous particles were used as supports for adsorption of a photosensitive 6-nitroBIPS SP, the light-induced transformation of which significantly affects pore hierarchy, morphology and the physical and chemical properties of the particles [51d]. An innovative chip configured as an online photonically controlled self-indicating system for metal-ion accumulation and release was constructed through integrating the characteristics of microfluidic devices and photochromic SPs [51e].

In recent years enormous attention has been directed to reversible photoresponsive surfaces with controllable wettability [52]. Surface wettability is one of the most practically important properties of solid surfaces and many commercial products. For instance, antifogging and self-cleaning titanium-oxide-coated glasses, are based on the light-stimulated regulation of the transitions between the hydrophobicity and hydrophilicity of the surfaces. Polarities of the interconverting forms of SPs are sharply different. The ground-state dipole moments of derivatives of 6-nitroBIPS measured in dioxane are in the range of $(10-15) \times 10^{-30}$ C m, whereas those of the merocyanine forms, $(50-60) \times 10^{-30}$ C m, are four to five times greater [53]. This peculiar property makes the SP photochromes particularly suitable for switching between hydrophobic and hydrophilic characteristics of SP-hybridized materials by modifying the affinity of liquid to the surface. Of all photochrome-based surfaces, only those obtained on the basis of SPs possess enough high-light-triggered surface free energy heterogeneity to provide for the force sufficient for moving water droplets along the surface [54a]. Spiropyrans were successfully used for the conversion of the reversible transformations of initially hydrophobic titanium oxide surfaces to the superhydrophilic ones and reversible amplification of the superhydrophobic properties of the SP-coated silicon nanowire surfaces [54b]. Grafting SP-containing polymers onto synthetic membranes leads to a reversible switching of wettability and protein absorption of the membranes [54c]. Efficient SP-containing photoresponsive surfaces are formed by polymers doped with photochromic SPs [54d]. The mechanism for the light-induced affect on the wettability involves changes in the liquid/air interface of air bubbles in the outermost part of the doped polymer that substantially amplifies the effects rendered by the polarity changes in the isomerized SP molecules.

Light-driven reversible changes of the photochrome-connecting monolayers on surfaces were applied for switching conductivity and controlling supramolecular organization of liquid crystal materials and aggregates of metal nanoparticles [51]. A selective self-assembly of Pt nanoparticles on indium/tin electrode covered with a photoisomerizable monolayer of a derivative of 6-nitroBIPS **5** (R = CH$_3$) was recently used for the photoswitchable electrocatalysed reduction of hydrogen peroxide and the catalytic generation of chemiluminescence at the surfaces modified with this monolayer [55]. The SP monolayer has no affinity for the negatively charged Pt nanoparticles and is catalytically inactive towards the reduction of H_2O_2. The photoisomerization induced by illumination of the monolayer with light $\lambda < 380$ nm converts the ring-closed SP to the merocyanine photoisomer, the subsequent protonation of which leads to the formation of the protonated merocyanine (**5-McH$^+$**) monolayer. The resulting positively charged monolayer attracts the Pt nanoparticles to the surface of thus modified electrode that catalyses the reaction. Back photoisomerization of the **5-McH$^+$**-monolayer to the ring-closed SP-monolayer state occurs under illumination with visible ($\lambda > 475$ nm) light and leads to the detachment of the Pt nanoparticles to restore a catalytically inactive electrode (Figure 2.4). Therefore, the cyclic photoisomerization of the monolayer between the ring-closed and protonated ring-opened SP-monolayer forms allows the reversible switching between inactive 'off' and active 'on' electrocatalytic states, respectively.

A challenging area of research is associated with reversible light-driven switching of photochromic compounds doped within biologic membranes. Due to their very high photosensitivity, fast response to the action of light, proneness to proper structural modification necessary to adjust a photoresponsive unit to the membrane and significant changes in the conformational and polar properties of the rearranging isomeric forms, SP and SPO represent a class of photoactive compounds particularly suitable for design of transmembrane gates. The interior of biological membranes is hydrophobic, and membranes are impermeable to most ions and many hydrophilic molecules, selective transport of which across membranes is of central importance for a large variety of metabolic and signalling processes, for example the transmission of nerve impulses. This process was mimicked using the photoinitiated ring-closed/ring-opened isomerizations of an amphiphilic SP **19** [56b] or SPOs **20**, **21** [56a] inserted into phospholipid bilayers commonly used as cell-membrane models. Addition of small amounts of an

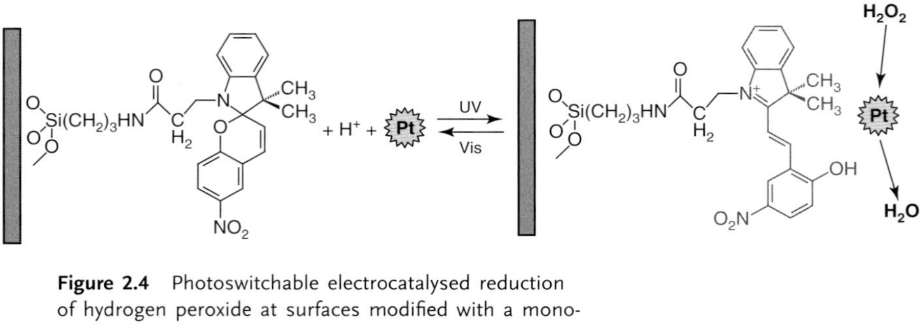

Figure 2.4 Photoswitchable electrocatalysed reduction of hydrogen peroxide at surfaces modified with a monolayer of a derivative of 6-nitroBIPS in the presence of Pt nanoparticles. Adapted from [55].

amphiphilic SP **21** to dihexadecyl phosphate unilamellar vesicles leads to separating the layers within the membrane and substantial increase in the rates of leakage of occluded K^+ and $Co(bpy)_3^{2+}$ ions. UV irradiation generates the polar ring-opened merocyanine form of SP, which subsequently relocates towards the membrane/water interface causing leak rates to revert to the background values. These effects were proved to be fully reversible over several photoisomerization cycles.

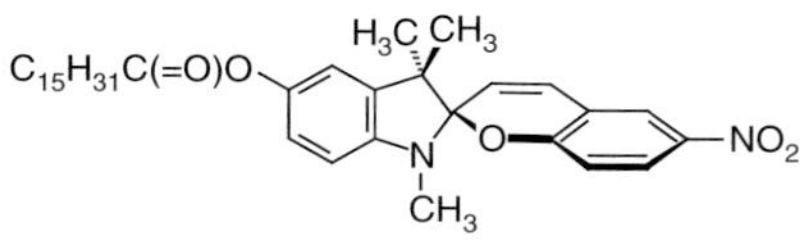

19

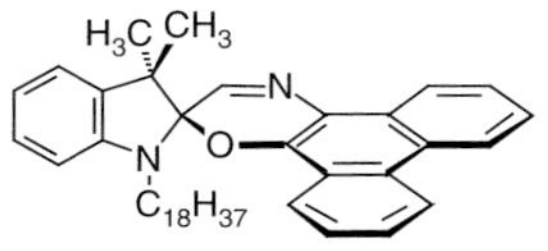

20

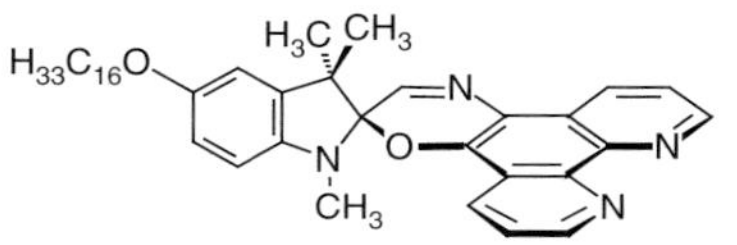

21

The feasibility of appending an addressable light-triggered gate to a naturally occurring membrane channel was recently demonstrated by an elegant construction of a SP-based photoswitchable mechanosensitive channel of large conductance (MscL) from *Esherichia coli* embedded in liposomes [57]. A 6-nitroBIPS core was attached to a cysteine-selective iodoacetate fragment and coupled to MscL to give **22**. The photoinduced rearrangement of **22** provides for the valving of the channel achieved via opening it by UV (366 nm) illumination and closing by irradiation with visible light. The obtained light-actuated nanovalve represents a prototype of the future light-gated nanoscale drug-delivery systems.

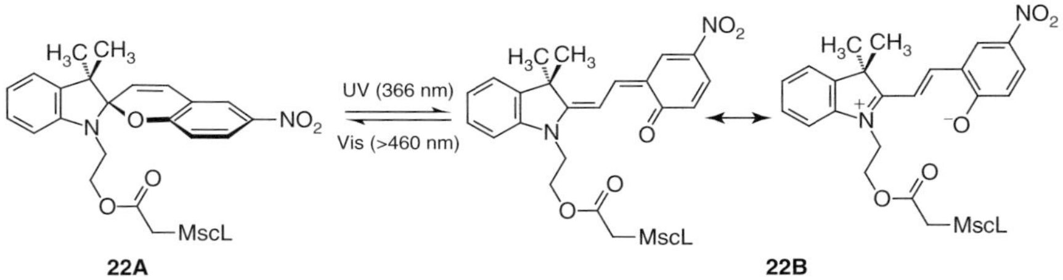

22A   **22B**

Other recent applications of SP/SPO-based photochromic molecular switches chemisorbed, physisorbed or grafted to solid surfaces for controlling various macroscopic properties of bulk materials include photomodulation of the electrode potential of a SP-modified Au electrode [58a], development of molecular shuttles intended for molecular-scale motion on monolayers [58b], light-triggered variation of viscosity and ionic conductivity of polymer solution [58c], artificial photoswitchable nanoporous membranes [58d] and SP-patterned semiconductor fluorescent quantum dots [58e–g]. Figure 2.5 illustrates reversible switching of a hybrid nanosystem obtained by attaching a 6-nitroBIPS derivative via a thiol–metal linkage to a core-shell semiconductive CdSe/ZnS nanocrystal [58e].

2.3.4
Photoswitchable Modulation of Biological Activities

One of the most promising directions of the strategy aimed at achieving spatiotemporal control and enhancement of the selectivity of a certain effect that is caused by a biologically active compound is a light-triggered control of properties of the compound [59]. Along with the so-called 'caged compounds' that undergo irreversible activation under the action of light, much attention has been drawn to the compounds susceptible to controlling their properties when being reversibly switched between an active and an inactive state. The important accomplishments in the rapidly developing area of application of photoswitchable biomaterials, including those related to the SP-containing reversible photochemical systems, have been recently comprehensively reviewed [60], which makes it possible to limit this section to a few illustrative examples.

The reversible random coil/α-helix transitions of poly(L-glutamic acid), the carboxyl groups of which were used to condense the polypeptide with N-(β-hydroxyethyl)benzopyranindoline, is one of the best studied examples of photomodulation of macromolecular conformation by light [60d, 61a]. In hexafluoro-2-propanol solution, the SP units tethered to the polypeptide chain exist in their polar merocyanine forms having strong tendency to dimerization. This effect forces the macromolecules to adopt a disordered structure. The photoisomerization occurring under irradiation with visible light converts the SP units of the conjugate to the less-polar ring-closed form and destroys the merocyanine-formed dimers allowing the SP-containing macromolecules to assume the helical structure. The random coil – α-helix interconversions are fully reversible. This type of photoinduced changes in the size of a 'smart' polymer chain coil is at the origin of the light-triggered activation or inhibition of catalytic activities of some of SP-functionalized enzymes, for example α-amylase [59] and cytochrome P450 3A4 [61].

2.3 Switching of Physical Properties and Biological Activities | 59

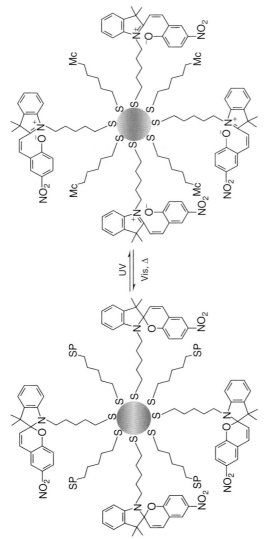

Figure 2.5 Reversible modulation of the intensity of fluorescence emission (at 546 nm) of SP-modified CdSe/ZnS core-shell nanocrystals with alternating cycles of irradiation by UV (350 nm) and visible (588 nm) light. The UV light-initiated formation of the ring-opened merocyanine form (Mc) of SP induces fluorescence resonance energy transfer (FRET) from the nanocrystal to the merocyanine resulting in quenching of its fluorescence. Adapted from Ref. [58e].

In contrast with polyglutamate **23**, polymeric micelles and vesicles composed of amphiphilic block copolymers obtained by functionalization of poly-(2-hydroxypropylmethacrylamide) with 6-bromospiropyranindoline contain the immersed SP-units in their ring-closed form. The rearrangement of these into the ring-opened merocyanine form occurring upon illumination with UV light drastically enhances the polarity of the macromolecular assemblies and leads to their destabilization. Such a destabilization of polymeric micelles or vesicles allows using them as versatile drug carriers with a property to release the drug under the action of an external stimulus [62].

A challenging application of the reversible SP-based system is their use as artificial receptors of biologically important species. By the proper functionalization of SP molecules one of their interconverting isomeric forms can be adapted to the recognition and selective binding of such a species leading to stabilization of this isomer in the form of its complex with the assay and signalling the act of recognition by certain changes in the spectral pattern. The feasibility of this approach to the design of spiropyridopyrans **24** containing molecular recognition sites properly fitted to selective binding of guanine nucleosides and oligonucleotide derivatives was demonstrated by Inoue and coworkers [60b, 63]. Due to the excellent steric conditions providing by the merocyanine form of **24** for the triple hydrogen bond complementarity with guanine derivatives the equilibrium **24A** ⇌ **24B** in the presence of the latter is strongly shifted to the right and the appearance of an intense long-wavelength absorption band ($\lambda_{max} = 550$ nm, CH_2Cl_2) indicates the stabilization of the merocyanine **24B** by its complexation with guanosine. The SP-based receptor **24** well distinguishes guanosine derivatives from both N-methylated guanine bases and other nucleosides that do not contain sites suitable for the triple hydrogen-bond binding.

24A   R = $(t\text{-}C_4H_9)_2(CH_3)_2SiO\text{-}$ **24B**

The ability of the zwitterionic merocyanine forms of SPs to bind polar amino acid molecules via electrostatic interaction makes SPs attractive materials for photocontrolled transfer of amino acids derivatives across bilayers. This property has been recently exploited for the molecular recognition between γ-glutamyl-cysteinyl-glycine (GSH), which is one of the most abundant cellular thiol compounds, and specially designed bis-indolinespiropyrans [60e]. Due to good structure complementary between the opened merocyanine forms of the bis-SP and GSH optimal conditions for the multipoint electrostatic interactions were achieved ensuring for the high affinity and strong fluorescence emission upon binding. An expedient approach to the efficient recognition and quantification of two amino acids, cysteine and homocysteine, in aqueous solution has been recently developed using the interaction of a SP with an amino acid in the presence of metal ions [64]. These interactions have been proven to be highly metal-ion selective with respect to the origin of a metal ion. The most selective and distinctive binding was observed with Cu^{2+} and Hg^{2+} ions. Addition of various amino acids to the ethanol/water solution of SP **25** and the metal salt resulted in the changes of the colour and UV/Vis spectra of the solution, but only cysteine and homocysteine induced a distinct colour change from red-violet to yellow. The sensitivity of the analytical method based on the cooperative interaction between **25**, the amino acid probe and metal ion covers the upper limit of cysteine concentration in normal organisms [64a].

2.4
Spiropyrans and Spirooxazines as Photodynamic Sensors for Metal Ions

One consequence of the thermal or photoinitiated rearrangement of SP and SPO **1A** ⇌ **1B** is the drastic alteration of the electron-donor ability of the ring oxygen centre in their molecules. The phenolate oxygen appeared in the ring-opened merocyanine structure can serve as a powerful ligating centre for coordination to various metal ions. Therefore, switching between the two forms opens interesting possibilities for sensing metal ions under the action of an external stimulus, primarily of light. The first investigation into the area of application of photochromic SPs as chemosensors dates back to the middle of 1960s [65]. Remarkable progress in the SP-based chemosensor technology is associated with the work by Winkler and his group [66] who proposed the use 6-nitro-spiro[indolinepyridobenzopyran] **26** (R = NO_2, R_1 = CH_3, R_2 = H) as a metal-complexation ligand. The merocyanine form obtained upon irradiation of solutions or polymeric films of **26A** offers an additional donor centre, that is engaged in chelating a metal ion and thus provides for much stronger complexing ability of **26B** compared to other merocyanines. Additional advances of spiro[indolinepyridobenzopyrans] **26** as the receptors of metal ions include the dramatic increase in the weak fluorescence exhibited by the noncomplexed merocyanine isomers and significant (30–60 nm) blue shifts in their absorption spectra produced by complexation. The 6-nitro-spiro[indolinepyridobenzopyran]-based metal-ion sensor was shown to be applicable for the selective identification of a number of divalent metal ions (Zn^{2+}, Co^{2+}, Hg^{2+}, Cu^{2+}, Cd^{2+} and Ni^{2+}) with the ability to detect metal ions at the micrograms per litre level [66c]. In a subsequent work [67] the set of spiro[indolinepyridobenzopyrans] has been expanded and a detailed study of the composition and stability constants of their complexes with metal ions performed. It was found that along with the most stable $M(SPP)_2X_2$ complexes solutions of SPP **26** and metal salts MX_2 in polar solvents also contain measurable amounts of 1 : 1 complexes $M(SPP)X_2$. The solutions of spiro[indolinepyridobenzopyrans] containing Mg^{2+}, Zn^{2+} ions exhibit the strongest fluorescence with maximum intensities at 610–625 nm. The ionophoric effects observed on addition of metal ions to solutions of spiro[indolinepyridobenzopyrans]: a significant enhancement of fluorescence of the merocyanine isomer and the hypsochromic shift of the longest-wavelength absorption bands are shown in Figure 2.6 by an example of 6-chloro-spiro[indolinepyridobenzopyran].

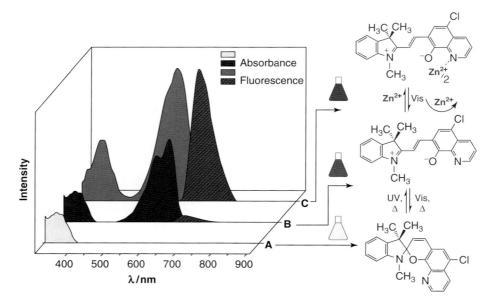

Figure 2.6 The colour changes observed in an acetonitrile solution of SP **26** (R = Cl) before UV irradiation (a) and after UV irradiation in the absence (b) and in the presence (c) of $Zn(ClO_4)_2$. The quantitative data on the sensoric activity of **26** are given in the literature [67, 68a].

2.4 Spiropyrans and Spirooxazines as Photodynamic Sensors for Metal Ions

26

R=H, NO$_2$, Cl; R$_1$-CH$_3$, (CH$_2$)$_2$OH, (CH$_2$)$_2$COOH; R$_2$=H, OC$_{16}$H$_{33}$

The principal idea of the design of selective and sensitive SP-based photodynamic metal ion sensors by furnishing SP molecules with an additional donor centre sterically adapted for coordination to a metal ion was also exploited in the studies of photo- and ionochromism of a series of spiro[indolinenaphthopyrans] **27** [68a,b] and **29** [68c] containing (4,5-diphenyl-1,3-oxazol-2-yl) or, respectively, 2-benzothiazolyl substituent in the *ortho*-position to pyran oxygen atom converting to the phenolate oxygen and complementarily coordinating to metal ions. In contrast with the unsubstituted spiro[indolinenaphthopyrans], compounds **27** exhibit fluorescence of their ring-closed isomers **27A**. Metal ions induce isomerization of **27A** and the formation of intensely coloured complexes with the formed merocyanine isomers the composition and stability of which strongly depends on the nature of a metal ion. Zn^{2+}, Cd^{2+} and Mn^{2+} ions form 1 : 1 complexes **28** (n = 1), whereas with Co^{2+}, Ni^{2+} and Cu^{2+} ions the formation of 1 : 2 complexes was also observed. The complexation causes a hypsochromic shift of the absorption of the merocyanine forms of **27**. The largest shift is observed for solutions containing Cu^{2+} ions. It is noteworthy that an addition of salts of alkali and alkali-earth ions even in 100-fold excess does not exert a significant effect on the spectra. The complexes formed by diamagnetic Cd^{2+} and Zn^{2+} cations have fluorescence with maxima in the region 620–660 nm and exhibit negative photochromism – thermally reversible decolouration under the action of visible light.

27   **28**

R = H, OCH$_3$; M^{2+} = Mg^{2+}, Zn^{2+}, Cd^{2+}, Ni^{2+}, Co^{2+}, Cu^{2+}, Mn^{2+}; n = 1,2

R = CH$_3$, C$_3$H$_7$, CH$_2$CH(CH$_3$)$_2$

29

The ability of properly functionalized SPs to reversibly bind and release metal ions in response to light was suggested to be used for mimicking physiological oscillatory calcium signals observed in excitable and nonexcitable cell types. Based on their finding of selective binding of Ca^{2+} ions in aqueous buffered solutions by SP **30** (R = H) containing an amino acid chelator in the *ortho*-position to the pyran oxygen the authors [69] elaborated a strategy for tuning the photoswitching properties of **30** to the oscillation mechanism through the choice of an appropriate substituent R.

30A = H, CH_3, OCH_3, F, Cl, NO_2 **30B**

Similar to SPs **27**, spironaphthoxazines **31** and **32** containing chelating groups in the *ortho*-position to the oxazine oxygen readily convert to their merocyanine forms upon addition of salts of metal ions (Ni^{2+}, Co^{2+}, Zn^{2+}, Fe^{2+}, Al^{3+}) in acetone or acetone/hexane solutions [70]. The metal chelation reaction occurring even in the dark results in the formation of very stable complexes the visible absorption bands which are hypsochromically shifted (10–40 nm) with respect to those of the free ligand. In contrast with metal complexes formed by SP **27** and **29**, the complexes of SPO – **31** and **32** do not fluoresce.

31 R_1 = H, 5,6-Benzo   **32**

A new class of very efficient photodynamic chemosensors for metal ions was developed using hybridization of photochromic SP and SPO molecules with crown ether fragments. Incorporation of a photoionizable photochromic moiety into a crown ether unit produces significant photoinduced changes in cation-complexing properties of the hybrid receptors that can be rationally employed for the selective recognition of alkali, alkali-earth and transition-metal ions. In recent years this area of research has been extensively explored and amply reviewed [60b, 71]. The design of the photodynamic crowned SP and SPO involves tailoring their molecules in a way providing for the strong interaction between the complexed cations and the phenolate oxygen of the photoinduced coloured merocyanine form that provides for stabilization of the latter. The system **33** represents one of the first preparative realizations of this principle [72a]. The selectivity of SP **33** can be affected by varying the size of the crown ring and the distance between the phenolate donor centre and a metal ion. Thus, solutions of a SP **33** ($n = 1$) with a small ring become

2.4 Spiropyrans and Spirooxazines as Photodynamic Sensors for Metal Ions

coloured on addition of salts of Li^+ ions, whereas an addition of the larger size ions, Na^+ and K^+, which are not captured by that ring, do not shift the equilibrium **33** ⇌ **34** to the complexed merocyanine structure. In the case of SP **33** ($n = 2$) the size of the crown ring fits to the steric demands of Na^+ and K^+ well, which fact ensures the conditions for the formation of stable polydentate complexes **34** resulting in the colouration of solutions of SP **33**. To stabilize the Cs^+-**33** complex it is necessary to have even more space to place the cation between the crown ether and phenolate oxygen donor centres. The required structure was obtained by elongating the N–X–N distance in **33** (X = $(CH_2)_2O(CH_2)_2$).

33   **34**

$M^+ = Li^+, Na^+, K^+, Cs^+$; X=$CH_2CO$, $(CH_2)_2O(CH_2)_2$; n=1,2

In contrast to the behaviour of crowned SPs **33** and their SPO analogues, isomerization of spirobenzothiopyran **35** to its merocyanine form does not proceed by the metal-ion complexation of the crown ether ring under dark conditions. A SP **35** behaves as a true photodynamic sensor on Li^+, Ag^+ and Tl^+ cations. A high affinity of the thiophenolate sulfur centre to soft metal ions, such as Ag^+ and Tl^+, serves as an additional factor facilitating stabilization of the complexed merocyanine forms **36B** [71d].

35   **36A**   **36B**

The crowned SP **33**, **35** with a single SP moiety readily bind monovalent alkali metal by forming 1 : 1 merocyanine–metal-ion complexes stabilized by the intramolecular interaction of a crown-complexed metal ion with the phenolate oxygen or thiolate sulfur donor centres. To provide for the strong and selective binding of multivalent metal ions the crowned SP-based ligand must be modified to offer an additional donor centre to the metal ion. Well suitable to this purpose are crowned bis(SPs), the photoisomerization of which affords the conditions for the formation of stable complexes of the merocyanines with divalent (Mg^{2+}, Ca^{2+}, Zn^{2+}, Ni^{2+}, Sr^{2+}, Ba^{2+}) and tri-valent (Fe^{3+}, Eu^{3+}, Ce^{3+}, La^{3+})

metal ions [71c,d, 73]. The diaza-18-crowned-6 bis-(6-nitroBIPS) demonstrates extremely high affinity to La^{3+} as manifested by the photoinduced ion-selectivity switching **37 – 37-La^{3+} – 37-K$^+$**. UV irradiation of a methanol solution of **37** containing KCl and LaCl$_3$ leads to the selective complexation of La^{3+} ion with the formed merocyanine isomer. The subsequent irradiation with visible light restores the initial ring-closed form of SP **37** and switches back its ion selectivity from the multivalent metal ion to the monovalent one [72b]. A liquid membrane of diaza-12-crown-4-bis(spirobenzopyran) **37** showed excellent Li$^+$ selectivity in the competitive transport of alkali-metal ions and the largest change in Li$^+$ transporting ability and selectivity by photoirradiation [72c].

On the basis of a calix[4]arene derivative bearing two spirobenzopyran moieties in the lower rim a method was recently developed for the recognition of rare-earth metal ions [74]. Alternating irradiation with UV and visible light controls the ligand-to-metal energy transfer between the ligand and the Eu^{3+} ion, thus allowing light-triggered switching of fluorescence of Eu^{3+}. The complex **38** exhibits high thermal stability of the coloured form **38B**.

Photochromic behaviour of a series of photochromic spironaphthoxazines conjugated with aza-15(18)-crown-5(6)-ether moieties **39–42** in the presence of metal ions was extensively studied by a Russian–French group of researchers [75]. It was shown that the addition of Li^+, Mg^{2+}, Ca^{2+}, Sr^{2+} and Ba^{2+} metal cations to solutions of SPO resulted in a hypsochromic shift of the ring-closed forms and a batochromic shift of the long-wavelength absorption band of the ring-opened merocyanine forms.

39 (n = 1,5)   **40**

41   m = 1, 2   **42**

2.5
Modulation of the Intramolecular Energy-Transfer Processes in SP/SPO-Containing Transition-Metal Complexes

Incorporation of a covalently bound photochromic SP or SPO-containing unit into the ligand environment of a transition-metal complex makes it possible to modify the mechanisms and dynamics of the excited-state metal-to-ligand charge transfer (MLCT) processes characteristic of metals with filled d-orbitals and to impart the composite systems with switchable properties [76]. The first observation of the MLCT effect on the behaviour of a (SPO) – (transition-metal complex) dyad caused by strong donation of electron density from the Ru(II) d-orbitals into the SPO π*-orbitals was a dramatic increase in the ring-closing rate for an o-phenanthroline-containing SPO **43** coordinated to a $Ru(bpy)_2^{2+}$ metal centre [56a,b]. Binding transition-metal ions to the bidentate phenanthroline moiety of SPO **43** results in the significantly improved colourability of the parent photochromic compound [56c] and gives rise to the profound stabilization of its photomerocyanine form [56c,d,e]. An intramolecular energy transfer from ^{3}MLCT to the SPO-ligand of the Re(I) tricarbonyl diimine complex **44** was found to be responsible for an efficient photosensization of luminescence of this complex. The luminescence shown by this complex containing a SPO unit in its ring-closed form can be quenched by initiating the photoisomerization of the latter to the ring-opened merocyanine structure [77].

Photoswitched singlet energy transfer was reported to occur in a porphyrin-SP dyad **45** when exciting the SP moiety in the near-UV spectral region and converting it to the merocyanine form absorbing at 600 nm [78]. This leads to singlet–singlet energy transfer to the merocyanine moiety, quenching fluorescence originated from a porphyrin-Zn site with a quantum yield of 0.93 and reducing the lifetime of the porphyrin-Zn excited state 2 orders from its usual value of 1.8 ns. Thus, the photochromic rearrangement in the 6-nitroBIPS moiety of **45** provides light-activated control of the porphyrin excited states. In contrast to the behaviour of the dyad **45**, in **46** the charge-transfer transition from a metal centre (Ru) to bipyridine ligand (^{1}MLCT) is only slightly affected by the photoisomerization of an attached 6-nitroBIPS moiety [79]. At the same time, completing this system with an additional emitting metal centre affords an approach to interesting photonic devices in which two transition-metal complexes are connected by a switchable bridging ligand serving as an interrupter of MLCT transitions. A Ru(II)-SP-Os(II) complex **47** is an example of such a system. By irradiation into the ^{1}MLCT absorption band, an energy-transfer process from the higher-lying Ru(II) centre through the bridging ligand to the Os(II) centre occurs. The emission of light of different wavelengths from the excited states of both metal complexes can be controlled by altering the transmission properties of the switchable SP ligand [79].

47 4 PF6⁻

2.6
Spiropyran-Containing Photoresponsive Polymers

Various technical applications of photochromic materials require embedding photochromic compounds into polymeric matrices to form films, plates, fibres and other practically usable constructions. The incorporation of a photochrome into polymer solids may be achieved in two ways: by doping the polymer with the photochromic compound or by its covalent binding to an appropriate site of the macromolecules. The polymer matrix greatly affects kinetics, the photophysical mechanism of the photochromic reaction and in the case of SPs the tendency of the photoinduced form to aggregate. These kind of matrix effects have been comprehensively analysed and reviewed [6c, 80]. Covalent binding of photochrome moieties to a polymeric chain has considerable advantages in construction of materials with high concentration of switchable units. The tight binding of photochromes hinders its diffusion inside the polymeric matrix and improves the photomodulating influence of the photochrome on the structure and properties of the polymer. In this section, we shall briefly outline the consequences of this kind of matrix effects associated with reversible alterations of the physical properties of polymers that can be controlled by the light-triggered isomerization of the covalently bound SP/SPO-containing photochromic units.

When incorporated into polymer systems, the photochromic rearrangement can control the conformation of the polymer by irradiation. The conformational changes produce a change in viscosity of the SP-based polymer solutions. In the case of the photoresponsive polyelectrolyte of polyacrylic acid with a SP in the side chain **48**, UV irradiation results in the formation of zwitterionic merocyanine that decreases the hydrophobic interaction and leads to a more extended random coiled conformation of the polymer chains with the associated increase in the hydrodynamic volume and thus viscosity [81]. Similar and even more pronounced conformational transformations are characteristic of photoresponsive SP-functionalized polymeric gels [82]. In an acidified medium, the hydrogel **49** prepared by radical copolymerization of *N*-isopropylacrylamide, a vinyl monomer having an SP residue and crosslinker has the grafted SP units in their coloured protonated merocyanine forms. Irradiation with the blue light leads to the immediate shrinking of the gel and the significant deformation occurring in the first 9 s. On the prolonged (for 24 s) irradiation, the hydrogel was decolourized, indicating that most of the SP chromophores were isomerized to their ring-closed deprotonated form. These light-induced processes lead to the dehydration of the hydrogel, the volume of which decreases to about one-fifth of that of the initial state [82a]. The photoinduced generation of the dipolar merocyanine isomeric form in hydrogel matrices of a photoresposive polymer alters the osmotic potential resulting in a positive photomechanical response (i.e. increase in size), increases the porosity, and hence drug diffusion rate, from the polymer. This property, makes such systems adaptable to biomaterials applications [82c].

The polymeric crowned 6-nitroBIPS **50** has a stronger complexing ability with respect to Na^+ and K^+ than monomeric aza-12-crown-4-(6-nitroBIPS). Photoisomerization of **50** in the presence of metal ions brings about remarkable polymer-chain aggregation by forming pairs of the zwitterionic merocyanine. Such a photomodulated behaviour was not observed when the photoisomerization of **50** was carried out in the absence of metal ions [71d]. Some other vinyl polymers bearing crown ether and SP side chains also undergo significant photoinduced rheology changes, that is contraction and extension of their polymer chain. This effect was applied in a material (photochemical valve) for photocontrol systems of solvent permeation [83]. Other applications of photoresponsive polymers carrying covalently linked SP or SPO fragments include control of ionic conductivity and permeability in artificial membranes [84], and surface wettability [59]. Of special interest are the UV light-triggered colour changes of liquid-crystalline phases formed by poly(acrylates) and poly(siloxanes) substituted with SP-containing side chains [80].

2.7
Spiropyran/Spirooxazine-Containing Biphotochromic Systems

The term *'biphotochrome'* was introduced by Dürr [85] to denote molecules containing two covalently bound photochromic units of the same or different types. Depending on the mode of binding of two fragments, biphotochromes can be arbitrarily divided into three structural types **A**, **B** and **C** in which two photochromic units are linked via a nonconjugated chain (**A**), fused through annulation of their aromatic rings (**B**) and linked via a π-conjugated chain (**C**) [86]. In the compounds of type **A**, the photochromic entities included into a composite system behave virtually independently of each other, which makes it possible to develop complex single-molecular systems capable of integration of several switchable functions. A recently studied multiaddressable SP/azobenzene biphotochromic system **51** nicely demonstrates the possibilities for the selective and controlled addressing of individual photochromic units in the combined structure [87]. Depending on the wavelength of irradiation compound **51** exhibits four kinds of isomers, with the ring-closed SP and *trans*-azobenzene fragments, **51a**, being the most stable. The photorearrangements occurring in chloroform solution at room temperature are fully reversible. This four-digit-code multiaddressable biphotochromic system holds promising potential for use in photonic molecular logic gates [88].

By rational selection of the bridge linking two SP or SPO fragments in a biphotochrome it is possible to impart it with the properties unavailable in the individual photochromic compounds. Thus, the polyether chain in SPO **52** and the diazacrown bridge in SP **53** ensure the photomodulated affinity of these compounds for alkali [89] and alkali-earth [73] ions, respectively.

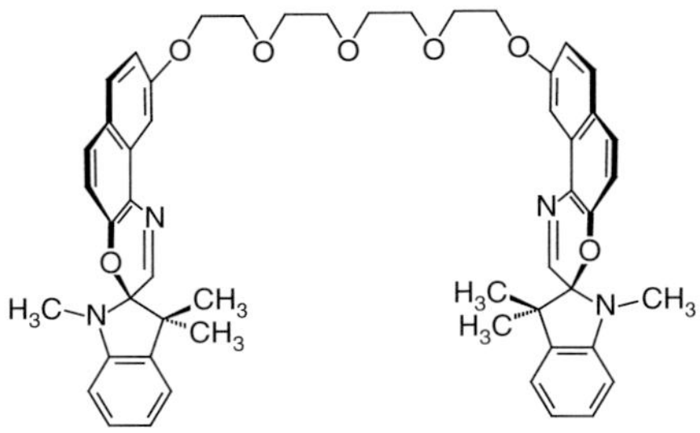

52 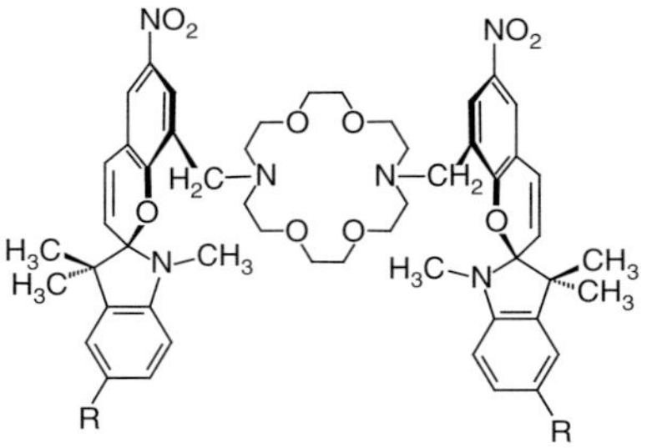 **R=H, CF₃, NO₂** **53**

Special attention has been paid to the synthesis and spectral studies of electronically coupled biphotochromic compounds of types **B** and **C**, the photoisomerization of which, provided it occurred in both photochromic units, should lead to the formation of the extended π-conjugated system with the spectra shifted to the near-IR region. In most cases, however, no formation of photoisomers with both 2H-pyran or 2H-(1,4-oxazine) rings opened has been detected under continuous or pulsed irradiation of solutions or polymeric films of biphotochromic compounds, for example **54** [90] and **55** [5a].

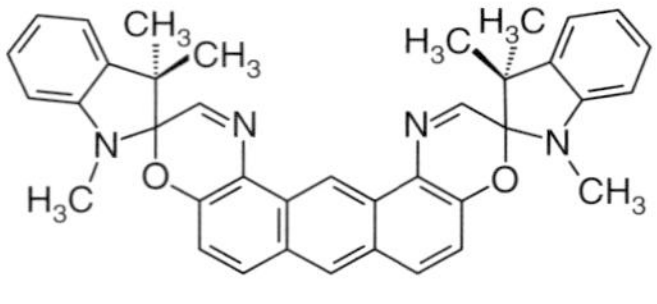

54 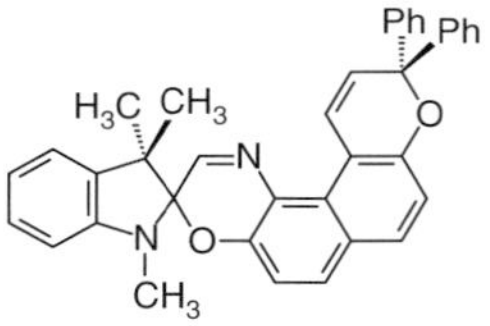 **55**

The presently known cases of the sequential or simultaneous photoinitiated ring-opening in the type **B** biphotochromic SP- or SPO-containing systems include the photochemical behaviour of hybrid SPO-dihydroindolizine **56** [91] and SP-dithienylethene **57** [92] photochromes and also bis-SP **58** [93]. It is interesting that in the latter case the longest-wavelength absorption of the isomeric form with both 2H-pyran rings opened (507 nm) is shifted hypsochromically with respect to the longest-wavelength absorption band (655 nm) of the isomer with only one opened 2H-pyran ring. This finding was corroborated by the results of TD DFT calculations.

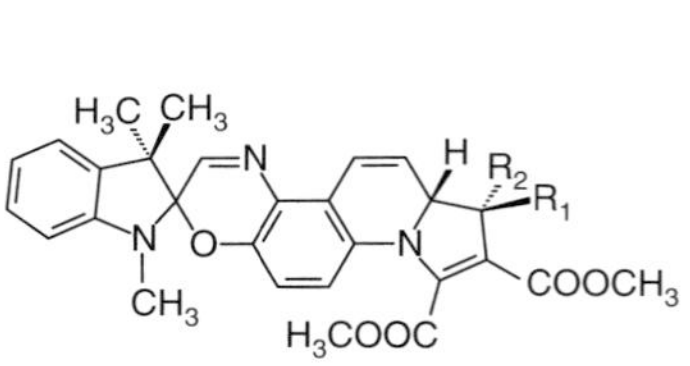

56 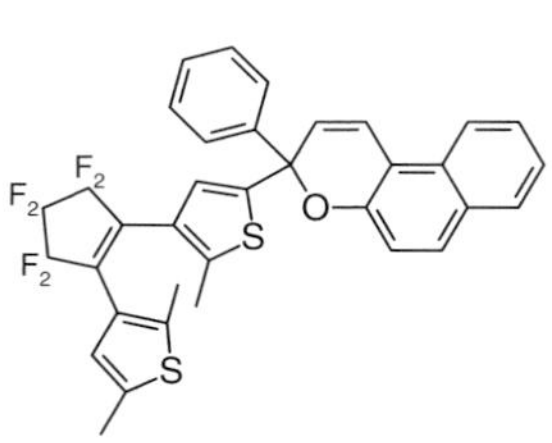 **57** 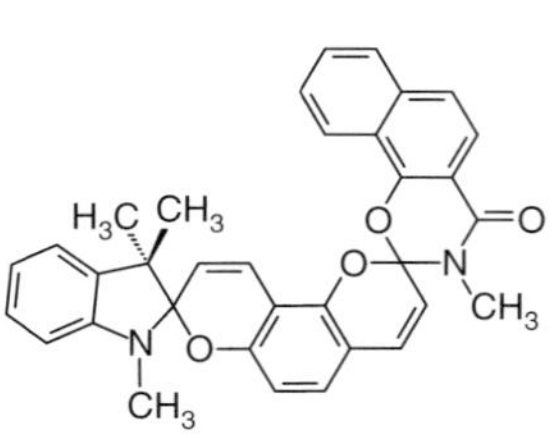 **58**

The successive opening of both oxazine rings accompanied by the Z → E isomerization and conrotatory $2_\pi + 2_\pi + 2_\pi$ electrocyclic reaction was reported to occur on UV irradiation of a solution of type **C** biphotochrome **59** [86].

2.8
Concluding Remarks

The study of photochromism of SPs and somewhat later of SPOs pioneered the intensive exploration of the properties and technical applications of photoswitchable organic compounds. Along with diarylethenes and fulgides SPs and SPOs constitute now the most amply studied classes of the light-triggered molecular switches. Whereas first applications of these photochromes were oriented towards their use as photoresponsive optical filters, in particular, in light-sensitive ophthalmic lenses (which still remains the major commercial application of photochromic dyes) the further continuous investigation into the broadly varied spectral characteristics, photophysical and photochemical mechanisms and kinetics of the light-driven rearrangements of SP and SPO allowed a variety of new areas to be revealed for the expedient use of these compounds and, especially, the molecular assembles including SP or SPO as highly efficient photoresponsive units for triggering diverse molecular and macroscopic properties of the hybrid materials. These areas include the use of SP and SPO as photodynamic chemosensors and their use for photocontrol and triggering of magnetic properties and surface wettability of bulk materials, permeability of membranes, rheologic properties of polymers, activity of enzymes, alignment of liquid crystals, and so forth. All these applications exploit the most valuable characteristics of SP and SPO as the light-induced molecular switches, such as high quantum efficiency of the photoreaction, high contrast in the electronic properties of the isomeric states and the significant changes in the geometry of the interconverting isomers. At the same time, relatively low thermal stability of the ring-opened merocyanine forms of SP and SPO remains to be the principal drawback of these photochromes that imposes restrictions on certain areas of their application, in particular, those requiring long-term storage of light-written information. It must be mentioned, however, that significant progress has been achieved recently in the preparation of SPs [6d,e, 94] and SPOs [40, 56c, 95] with the stable merocyanine species both in solution and in the solid state, through either appropriate structural modification of their molecules, complexation with metal ions and protons or grafting these to polymeric backbones and inclusion into the supramolecular assembles.

The potential of SPs and SPOs for their use in various molecular devices, such as molecular memories, sensors, logic gates, bioswitchable materials are far from being exhausted and there is good reason to believe that new-generation nanometric scale electronic circuits and molecular machines will benefit from the incorporation of the SP- and SPO-based molecular switches and hybrid materials derived on their basis.

Acknowledgement

Support by the grants (N. Sch. 3233.2010.3 and 02.740.11.0456) of the Ministry of Education and Science of Russian Federation is gratefully acknowledged.

References

1. (a) Aviram, A. (1998) *J. Am. Chem. Soc.*, **110**, 5687; (b) Irie, M. Ed Photochromism: Memories and Switches, Special issue of (2000) *Chem. Rev.*, **100**, 1683; (c) Feringa, B.L. (2001) *Molecular Switches*, Wiley-VCH Verlag GmbH, Weinheim; (d) Pease, A.R., Jeppesen, J.O., Stoddart, J.F., Luo, Y., Collier, C.P. and Heath, J.R. (2001) *Acc. Chem. Res.*, **34**, 433; (e) Weibel, N., Grunder, S. and Mayor, M. (2007) *Org. Biomol. Chem.*, **5**, 2343; (f) Heath, J.R. and Ratner, M.A. (2003) *Phys. Today*, **56**, 43; (g) Green, J.E., Choi, J.W., Boukai, A., Bunimovich, Y., Johnston-Halpern, E., Delonno, E., Luo, Y., Sheriff, B.A., Xu, K., Shin, Y.S., Tseng, H.-R., Stoddard, J.F. and Heath, J.R. (2007) *Nature*, **445**, 414; (h) Minkin, V.I. (2008) *Russ. Chem. Bull. Int. Ed.*, **57**, 687.

2. (a) Willner, I. and Willner, B. (2000) *Angew. Chem. Int. Ed.*, **29**, 1180; (b) Willner, I. and Katz, E. (2001) *Trends Biotech.*, **19**, 222; (c) Goeldner, M. and Givens, R. (2005) *Dynamic Studies in Biology*, Wiley-VCH Verlag GmbH, Weinheim.

3. (a) *Organic Photochromism* (IUPAC Technical Report). Prepared by Bouas-Laurent, H. and Dürr, H. (2001) *Pure Appl. Chem.*, **73**, 639; (b) Brown, H.G. (ed.) (1971) *Photochromism, Techniques of Chemistry*, vol. 3, John Wiley & Sons, Inc., New York; (c) Barachevsky, V.A., Lashkov, G.L. and Tsekhomsky, V.A. (1977) *Photochromism and its Application*, Khimiya, Moscow (in Russian); (d) Eltsov, A.V. (1990) *Organic Photochromes*, Consultants Bureau, New York; (e) Dürr, H. and Bouas-Laurent, H (eds) (1990) *Photochromism. Molecules and Systems*, Elsevier, Amsterdam; (f) Crano, J.C. and Guglielmetti, R. (eds) (1999) *Organic Photochromic and Thermochromic Compounds*, vols. 1, 2, Plenum Press, New York.

4. (a) Fischer, E. and Hirshberg, Y. (1952) *J. Chem. Soc.*, 4522; (b) Chaudé, O. and Rumpf, P. (1953) *C. R. Acad. Sci.*, **236**, 697; (c) Heligman-Rim, R. and Hirshberg, Y. (1962) *J. Phys. Chem.*, **66**, 2470; (d) Hirshberg, Y. (1956) *J. Am. Chem. Soc.*, **78**, 2304; (e) Hirshberg, Y. (1960) *New Scientist*, **7**, 1243.

5. (a) Lokshin, V., Samat, A. and Metelitsa, A.V. (2002) *Russ. Chem. Rev. Int. Ed.*, **71**, 893; (b) Lokshin, V., Samat, A., Metelitsa, A.V., Minkin, V.I. (2007) in *Modern Approaches to the Synthesis of O- and N-Heterocycles*, vol. 1, (eds T.S. Kaufman and E.L. Larghi), Research Signpost, Kerala, p. 187.

6. (a) Guglielmetti, R. (1990) in *Photochromism. Molecules and Systems* (eds H. Dürr and H. Bouas-Laurent), Elsevier, Amsterdam, p. 314; (b) Bertelson, R.C. (1999) in *Organic Photochromic and Thermochromic Compounds*, vol. 1, (eds J.C. Crano and R. Guglielmetti), Plenum Press, New York, p. 11; (c) Berkovic, G., Krongauz, V. and Weiss, V. (2000) *Chem. Rev.*, **100**, 1741; (d) Minkin, V.I. (2004) *Chem.*

Rev., **104**, 2751; (e) Lukyanov, B.S. and Lukyanova, M.B. (2005) *Chem. Het. Comp. Int. Ed.*, **41**, 281.

7. (a) Chu, N.Y.C. (1990) in *Photochromism. Molecules and Systems* (eds H. Dürr and H. Bouas-Laurent), Elsevier, Amsterdam, pp. 493–509, 879–882; (b) Maeda, S. (1999) in *Organic Photochromic and Thermochromic Compounds* (eds J.C. Crano and R. Guglielmetti), vol. 1, Plenum Press, New York, pp. 85–110; (c) Feng, C.-G. and Wang, J.-Y. (2006) *Chinese J. Org. Chem.*, **26**, 1012.

8. Fan, M.-G., Yu, L. and Zhao, W. (1999) in *Organic Photochromic and Thermochromic Compounds*, vol. 1, (eds J.C. Crano and R. Guglielmetti), Plenum Press, New York, p. 141.

9. (a) Irie, M. (2000) *Chem. Rev.*, **100**, 1685; (b) Kobatake, S. and Irie, M. (2003) *Ann. Rep. Prog. Chem. Sect. C.*, **99**, 277; (c) Matsuda, K. and Irie, M. (2004) *J. Photochem. Photobiol. C.*, **5**, 169.

10. (a) Akimov, D.A., Fedotov, A.V., Koroteev, N.I., Levich, E.V., Magnitskii, S.A., Naumov, A.N., Sidorov-Biryukov, D.A., Sokolyuk, N.T. and Zheltokov, A.M. (1997) *Opt. Mem. Neural Netw.*, **6**, 31; (b) Belfield, K.D., Liu, Y., Negres, R.A., Fan, M., Pan, G., Hagan, D.J. and Hernandez, F.E. (2002) *Chem. Mater.*, **14**, 3663; (c) Saita, S., Yamaguchi, T., Kawai, T. and Irie, M. (2005) *Phys. Chem. Chem. Phys.*, **6**, 2300.

11. (a) Krysanov, S.A. and Alfimov, M.V. (1982) *Chem. Phys. Lett.*, **91**, 77; (b) Kalisky, Y., Orlowski, T.E. and Williams, D.J. (1983) *J. Phys. Chem.*, **87**, 5333; (c) Lenoble, C. and Becker, R.S. (1986) *J. Phys. Chem.*, **90**, 62; (d) Ernsting, N.P., Dick, B. and Arthen-Engeland, Th. (1990) *Pure Appl. Chem.*, **62**, 1483; (e) Ernsting, N.P. and Arthen-Engeland, T. (1991) *J. Phys. Chem.*, **95**, 5502; (f) Aramaki, S. and Atkinson, G.H. (1992) *J. Am. Chem. Soc.*, **114**, 438; (g) Zhang, J.Z., Schwartz, B.J., King, J.C. and Harris, C.B. (1992) *J. Am. Chem. Soc.*, **114**, 10921; (h) Tamai, N. and Masuhara, H. (1992) *Chem. Phys. Lett.*, **191**, 189; (i) Tamai, N. and Masuhara, H. (2000) *Chem. Rev.*, **100**, 1875; (j) Chibisov, A.K. and Görner, H. (1997) *J. Photochem. Photobiol. A. Chem.*, **105**, 261; (k) Görner, H. (2001) *Phys. Chem. Chem. Phys.*, **3**, 416; (l) Antipin, S.A., Petrukhin, A.N., Gostev, F.E., Marevtsev, V.S., Titov, A.A., Barachevsky, V.A., Strokach, Yu.P. and Sarkisov, O.M. (2000) *Chem. Phys. Lett.*, **331**, 378; (2002) *Khim. Phys.*, **21**, 10; (m) Asahi, T., Suzuki, M. and Masuhara, H. (2002) *J. Phys. Chem. A*, **106**, 2335; (n) Hobley, J., Pfeifer-Fukumura, U., Bletz, M., Asahi, N., Masuhara, H. and Fukumura, H. (2002) *J. Phys. Chem A*, **106**, 2265; (o) Rini, M., Holm, A.-K., Nibbering, E.T.J. and Fidder, H. (2003) *J. Am. Chem. Soc.*, **125**, 3028; (p) Holm, A.-K., Mohammed, O.F., Rini, M., Mukhtar, E., Nibbering, E.T.J. and Fidder, H. (2005) *J. Phys. Chem. A.*, **109**, 8962; (q) Wohl, C.J. and Kuklauskas, D. (2005) *J. Phys. Chem. B.*, **109**, 22186; (r) Glebov, E.M., Vorobyev, D.Yu., Grivin, V.P., Plyusnin, V.F., Metelitsa, A.V., Voloshin, N.A., Minkin, V.I. and Micheau, J.-C. (2006) *Chem. Phys.*, **323**, 490; (s) Poisson, L., Rafael, K.D., Soep, B., Mestdagh, J.-M. and Buntinx, G. (2007) *J. Am. Chem. Soc.*, **128**, 3169.

12. (a) Celani, P., Bernardi, F., Olivucci, M. and Robb, M.A. (1997) *J. Am. Chem. Soc.*, **119**, 10815; (b) Horii, T., Abe, Y. and Nakao, R. (2001) *J. Photochem. Photobiol. A.*, **144**, 119; (c) Maurel, F., Aubard, J., Rajzmann, M., Guglielmetti, R. and Samat, A. (2002) *J. Chem. Soc., Perkin Trans. 2.*, 1307; (d) Futami, Y., Chin, M.L.S., Kudoh, S., Takayanagi, M. and Nakata, M. (2003) *Chem. Phys. Lett.*, **370**, 460; (e) Sheng, Y., Leszcynski, J., Garcia, A.A., Rosario, R., Gust, D. and Springer, J. (2004) *J. Phys. Chem. B.*, **108**, 16233; (f) Maurel, F., Aubard, J., Millie, P., Dognon, J.P., Rajzmann, M., Guglielmetti, R. and Samat, A. (2006) *J. Phys. Chem. A.*, **110**, 4759; (g) Go'mez, I., Reguero, M. and Robb, M.A. (2006) *J. Phys. Chem. A.*, **110**, 3986.

13. Bahr, J.L., Kodis, G., Garza, L., Lin, S. and Moore, A.L. (2001) *J. Am. Chem. Soc.*, **123**, 7124.

14. Minkin, V.I., Metelitsa, A.V., Dorogan, I.V., Lukyanov, B.S., Besugliy, S.O. and

Micheau, J.-C. (2005) *J. Phys. Chem. A.*, **109**, 9605.

15. (a) Delbaere, S. and Vermeersch, G. (2008) *J. Photochem. Photobiol. C.*, **9**, 61; (b) Berthet, J., Delbaere, S., Carvalho, L.M., Vermeersch, G. and Coelho, P. (2006) *Tetrahedron Lett.*, **47**, 4903; (c) Yee, L.H., Hanley, T., Evans, R.A., Davis, T.P. and Ball, G.E. (2010) *J. Org. Chem.*, **75**, 2851.

16. Christie, R.M., Li, L.-J., Spark, R.A., Morgan, K.M., Boyd, A.S.F. and Lycka, A.J. (2005) *J. Photochem. Photobiol. A,* **169**, 37.

17. (a) Krongauz, V. (1979) *Isr. J. Chem.*, **18**, 3; (b) Li, Y., Zhou, J., Wang, Y., Zhang, F. and Song, X. (1998) *J. Photochem. Photobiol. A. Chem.*, **113**, 65; (c) Uznanski, P. (2000) *Synth. Met.*, **109**, 281.

18. (a) Hibino, Y., Moriyama, K., Suzuki, M. and Kishimoto, K. (1992) *Thin Solid Films*, **210/211**, 562; (b) Gu, Z.-Z., Hayami, S., Meng, B.-O., Iyoda, T., Fujishima, A. and Sato, O. (2000) *J. Am. Chem. Soc.*, **122**, 10730.

19. (a) Miyata, A., Matsushima, T., Ohki, H., Unuma, Y. and Higashigaki, Y. (1995) *Adv. Mater. Opt. Electron.*, **5**, 37; (b) Kobayashi, T. (ed.) (1996) *J-Aggregates*, World Scientific, Singapore.

20. Yuzawa, T. and Takahashi, H. (1994) *Mol. Cryst. Liq. Cryst.*, **246**, 279.

21. (a) Görner, H., Atabekyan, L.S. and Chibisov, A.K. (1996) *Chem. Phys. Lett.*, **260**, 59; (b) Görner, H. (1998) *Chem. Phys. Lett.*, **282**, 381; (c) Görner, H. and Chibisov, A.K. (2004) in *Organic Photochemistry and Photobiology*, 2nd edn, (eds W. Horspool, and F. Lenci), CRC Press, Boca Raton, p. 36-1.

22. Konorov, S.O., Sidorov-Biryukov, D.A., Bugar, I., Chorvat, C., Chorvat, D. and Zheltikov, A.M. Jr (2003) *Chem. Phys. Lett.*, **381**, 572.

23. (a) Kellmann, A., Tfibel, F., Pottier, R., Guglielmetti, R., Samat, A. and Rajzmann, M. (1993) *J. Photochem. Photobiol. A. Chem.*, **76**, 77; (b) Ait, O.A., Barachevsky, V.A., Alfimov, M.V. and Baskin, I.I. (1997) *Mol. Cryst. Liq. Cryst.*, **297**, 271.

24. (a) Malkin, Ya.N., Krasieva, T.B. and Kuzmin, V.A. (1989) *J. Photochem. Photobiol. A. Chem.*, **49**, 75; (b) Zakharova, M.I., Coudret, C., Pimienta, V., Micheau, J.C., Sliwa, M., Poizat, O., Buntinx, G., Delbaere, S., Vermeersch, G., Metelitsa, A.V., Voloshin, N., Minkin, V.I. (2011) *Dyes and Pigments.*, **89**, 324.

25. Metelitsa, A.V., Dorogan, I.V., Lukyanov, B.S., Minkin, V.I., Besugliy, S.O. and Micheau, J.-C. (2005) *Mol. Cryst. Liq. Cryst.*, **430**, 45.

26. (a) Ortica, F. and Favaro, G. (2000) *J. Phys. Chem.*, **104**, 12179; (b) Metelitsa, A.V., Micheau, J.-C., Voloshin, N.A., Voloshina, E.N. and Minkin, V.I. (2001) *J. Phys. Chem. A.*, **105**, 8417; (c) Metelitsa, A.V., Lokshin, V., Micheau, J.-C., Samat, A., Guglielmetti, R. and Minkin, V.I. (2002) *Phys. Chem. Chem. Phys.*, **4**, 4340; (d) Berthet, J., Delbaere, S., Lokshin, V., Bochu, C., Samat, A., Guglielmetti, R. and Vermeersch, G. (2002) *Photochem. Photobiol. Sci.*, **1**, 333; (e) Voloshin, N.A., Metelitsa, A.V., Micheau, J.-C., Voloshina, E.N., Besugliy, S.O., Shelepin, N.E., Minkin, V.I., Tkachev, V.V., Safoklov, B.B. and Aldoshin, S.M. (2003) *Russ. Chem. Bull. Int. Ed.*, **52**, 2038.

27. (a) Kellmann, A., Tfibel, F. and Guglielmetti, R. (1995) *J. Photochem. Photobiol. A.*, **91**, 131; (b) Chibisov, A.K. and Görner, H. (1999) *J. Photochem. Photobiol. A.*, **103**, 5211; (c) Chibisov, A.K., Marevtsev, V.S. and Görner, H. (2003) *J. Photochem. Photobiol. A.*, **159**, 233.

28. (a) Schneider, S. (1987) *Z. Phys. Chem. Neue Folge.*, **154**, 1222; (b) Aramaki, S. and Atkinson, G.H. (1990) *Chem. Phys. Lett.*, **170**, 181; (c) Wilkinson, F., Worrall, D.R., Hobley, J., Jansen, L., Williams, S.L., Langley, A.J. and Matousek, P. (1996) *J. Chem. Soc. Faraday Trans.*, 1331.

29. (a) Suzuki, M., Asahi, T. and Masuhara, H. (2002) *Phys. Chem. Chem. Phys.*, **4**, 185; (b) Buntinx, G., Foley, S., Lefumeux, C., Lokshin, V., Poizat, O. and Samat, A. (2004) *Chem. Phys. Lett.*, **391**, 33.

30. (a) Hobley, J., Pfeifer-Fukumura, U., Bletz, M., Asahi, T., Masuhara, H. and Fukumura, H. (2002) *J. Phys. Chem. A.*, **106**, 2265; (b) Gaeva, E.B., Pimenta, V., Metelitsa, A.V., Voloshin, N.A., Minkin, V.I. and Micheau, J.-C. (2004) *J. Phys. Org. Chem.*, **18**, 315.
31. (a) Raymo, F.M. and Tomasulo, M. (2005) *J. Phys. Chem. A.*, **109**, 7343; (b) Seefeldt, B., Kasper, R., Beining, M., Mattay, J., Arden-Jacob, J., Kemnitzer, N., Drexhage, R.H., Heilemann, M. and Sauer, M. (2010) *Photochem Photobiol. Sci.*, **9**, 213.
32. (a) Kuzmin, M.G. and Kozmenko, M.V. (1990) in *Organic Photochromes*, Chapter 6 (ed. A. Eltsov), Consultants Bureau, New York; (b) Minami, T., Tamai, N., Yamazaki, T. and Yamazaki, L. (1991) *J. Phys. Chem.*, **95**, 3988; (c) Winkler, J.D., Bowen, C.M. and Michelet, V. (1998) *J. Am. Chem. Soc.*, **120**, 3237.
33. Kawata, S. and Kawata, Y. (2000) *Chem. Rev.*, **100**, 1777.
34. (a) Manjikov, V.F., Darmanyan, A.P., Barachevsky, V.A. and Gerulaitis, Yu.N. (1972) *Opt. Spectrosc. (Russ.)*, **32**, 412; (b) Manjikov, V.F., Murin, V.A. and Barachevsky, V.A. (1973) *Quantum Electron. (Russ.)*, **2**, 66.
35. (a) Parthenopoulos, D.A. and Rentzepis, P.M. (1989) *Science*, **24**, 842; (b) Parthenopoulos, D.A. and Rentzepis, P.M. (1990) *J. Appl. Phys.*, **68**, 814; (c) Dvornikov, A.S., Cokgor, I., Wang, M., McCormick, F.B., Esener, S.C. and Rentzepis, P.M. Jr (1997) *IEEE Trans. Comp. Pack. Manufact. Technol. A*, **20**, 203.
36. (a) Liang, Y.C., Dvornikov, A.S. and Rentzepis, P.M. (2003) *Proc. Natl. Acad. Sci.*, **100**, 8109; (b) Dvornikov, A.S., Liang, Y.C., Cruse, C.S. and Rentzepis, P.M. (2004) *J. Phys. Chem. B.*, **108**, 8652.
37. (a) Minkin, V.I. (1994) *Mol. Cryst. Liq. Cryst.*, **246**, 9; (b) Voloshin, N.A., Volbushko, N.V., Trofimova, N.S., Shelepin, N.E. and Minkin, V.I. (1994) *Mol. Cryst. Liq. Cryst.*, **246**, 41.
38. Guo, X., Zhang, D., Zhou, Y. and Zhu, D. (2003) *J. Org. Chem.*, **68**, 5681.
39. Wu, S., Luo, Y., Zeng, F., Chen, J., Chen, Y. and Tong, Z. (2007) *Angew. Chem. Int. Ed.*, **46**, 7015.
40. Meng, X., Zhu, W., Guo, Z., Wang, J. and Tian, H. (2006) *Tetrahedron*, **62**, 9840.
41. (a) Zhu, M.-Q., Zhu, L., Han, J.J., Wu, W., Hurst, J.K. and Li, A.D.Q. (2006) *J. Am. Chem. Soc.*, **128**, 4303; (b) Zhu, L., Wu, W., Zhu, M.-Q., Han, J.J., Hurst, J.K. and Li, A.D.Q. (2007) *J. Am. Chem. Soc.*, **129**, 3524; (c) Sheng, X., Peng, A., Fu, H., Liu, Y., Zhao, Y., Ma, Y. and Yao, J. (2007) *Nanotechnology*, **18**, 145707; (d) Rlajin, R., Stoddart, F.J. and Grzybowski, B.A. (2010) *Chem. Soc. Rev.*, **39**, 2203.
42. (a) Gütlich, P., Garcia, Y. and Woike, Th. (2001) *Coord. Chem. Rev.*, **219-221**, 839; (b) Sato, O. (2003) *Acc. Chem. Res.*, **36**, 692; (c) Einaga, Y. (2006) *J. Photochem. Photobiol. A.*, **7**, 69; (d) Sato, O., Tao, J. and Zhang, Y.-Z. (2007) *Angew. Chem. Int. Ed.*, **46**, 2152; (e) Halcrow, M.A. (2008) *Chem. Soc. Rev*, **37**, 278.
43. Bénard, S., Rivière, E., Yu, P., Nakatani, K. and Delois, J.F. (2001) *Chem. Mater.*, **13**, 159.
44. (a) Aldoshin, S.M., Sanina, N.A., Minkin, V.I., Voloshin, N.A., Ikorskii, V.N., Ovcharenko, V.I., Smirnov, V.A. and Nagaeva, N.K. (2007) *J. Mol. Struct.*, **826**, 69; (b) Aldoshin, S.M., Sanina, N.A., Nadtochenko, V.A., Yurieva, E.A., Minkin, V.I., Voloshin, N.A. and Ikorskii, V.N. (2007) *Russ. Chem. Bull. Int. Ed.*, **56**, 1095; (c) Sanina, N.A., Aldoshin, S.M., Shilov, G.V., Kurganova, E.V., Yurieva, E.A., Voloshin, N.A., Minkin, V.I., Nadtochenko, V.A. and Morgunov, R.B. (2008) *Russ. Chem. Bull. Int. Ed.*, **57**, 1451; (d) Aldoshin, S.M. (2008) *Russ. Chem. Bull. Int. Ed.*, **57**, 718.
45. Kashima, I., Okubo, M., Ono, Y., Ito, M., Kida, N., Hikita, M., Enomoto, M. and Kojima, N. (2005) *Synth. Met.*, **153**, 473.
46. Bénard, S., Léaustic, A., Rivière, E., Yu, P. and Clément, R. (2001) *Chem. Mater.*, **13**, 3709.
47. Zhang, H., Yi, T., Li, F., Delahaye, F., Yu, P. and Clément, R. (2007) *J. Photochem. Photobiol. A.*, **186**, 173.
48. Lu, A.H., Salabas, E.L. and Schüth, F. (2007) *Angew. Chem. Int. Ed.*, **46**, 1222.

49. Einaga, Y. (2006) *Bull. Chem. Soc. Japan.*, **79**, 361.
50. Garnovskii, A.D., Ikorskii, V.N., Uraev, A.I., Vasilchenko, I.S., Burlov, A.S., Garnovskii, D.A., Lyssenko, K.A., Vlasenko, V.G., Shestakova, T.E., Koshchienko, Y.V., Kuz'menko, T.A., Divaeva, L.N., Bubnov, M.P., Rybalkin, V.P., Korshunov, O.Y., Pirog, I.V., Borodkin, G.S., Bren, V.A., Uflyand, I.E., Antipin, M.Y. and Minkin, V.I. (2007) *J. Coord. Chem.*, **60**, 1493.
51. (a) Katsonis, N., Lubomska, M., Pollard, M.M., Feringa, B.L. and Rudolf, P. (2007) *Prog. Surf. Sci.*, **82**, 407; (b) Tsujioka, T. and Irie, M. (2010) *J. Photochem. Photobiol. C: Photochem. Rev.*, **11**, 1; (c) Del Canto, E., Flavin, K., Natali, M., Perova, T. and Giordani, S. (2010) *Carbon*, **48**, 2815; (d) Dominguez, J.M., Rosas, R., Aburto, J., Terrés, E., López, A. and Martinez-Palou, R. (2009) *Microporous Mesoporous Mater.*, **118**, 121; (e) Benito-Lopez, F., Scarmagnani, S., Walsh, Z., Paull, B., Macka, M. and Diamond, D. (2009) *Sens. Actuators B: Chem.*, **140**, 295.
52. (a) Sung, T., Feng, L., Gao, X. and Jiang, L. (2005) *Acc. Chem. Res.*, **38**, 644; (b) Wang, S., Song, Y. and Jiang, L. (2007) *J. Photochem. Photobiol. C.*, **8**, 18.
53. Bletz, M., Pfeifer-Fukumura, U., Kolb, U. and Baumann, W. (2002) *J. Phys. Chem. A.*, **106**, 2232.
54. (a) Bunker, B., Kim, B., Houston, J., Rosario, R., Garcia, A., Hyes, M., Gust, D. and Picraux, S. (2003) *Nano Lett.*, **3**, 1723; (b) Rosario, R., Gust, D., Garcia, A.A., Hayes, M., Taraci, J.L., Dailey, J.W. and Picraux, S.T. (2004) *J. Phys. Chem. B*, **108**, 12640; (c) Nayak, A., Liu, H. and Belfort, G. (2006) *Angew. Chem. Int. Ed.*, **45**, 4094; (d) Athanassiou, A., Klyva, M., Lakiotaki, K., Georgiou, S. and Fotakis, C. (2005) *Adv. Mater.*, **17**, 988.
55. Niazov, T., Shlyahovsky, B. and Willner, I. (2007) *J. Am. Chem. Soc.*, **129**, 6374.
56. (a) Khairutdinov, R.F., Giertz, K., Hurst, J.K., Voloshina, T.N., Voloshin, N.A. and Minkin, V.I. (1998) *J. Am. Chem. Soc.*, **120**, 12707; (b) Khairutdinov, R.F. and Hurst, J.K. (2001) *Langmuir*, **17**, 6881; (c) Kopelman, R.A., Snyder, S.M. and Frank, N.L. (2003) *J. Am. Chem. Soc.*, **125**, 13684; (d) Zhang, C., Zhang, Z., Fan, M. and Yan, W. (2008) *Dyes Pigments*, **76**, 832; (e) Zhang, Z., Zhang, C., Fan, M. and Yan, W. (2008) *Dyes Pigments*, **77**, 469.
57. Koçer, A., Walko, M., Meijberg, W. and Feringa, B.L. (2005) *Science*, **309**, 755.
58. (a) Wen, G., Yan, J., Zhou, Y., Zhang, D., Mao, L. and Zhu, D. (2006) *Chem. Commun.*, 3016; (b) Zhou, W., Chen, D., Li, J., Xu, J., Lv, J., Liu, H. and Li, Y. (2007) *Org. Lett.*, **9**, 3929; (c) Kim, S.-H., Park, S.-Y., Shin, C.-J. and Yoon, N.-S. (2007) *Dye Pigments.*, **72**, 299; (d) Vlassiouk, I., Park, C.D., Vail, S.A., Gust, D. and Smirnov, S. (2006) *Nano Lett.*, **6**, 1013; (e) Zhu, L., Zhu, M.-Q., Hurst, J.K. and Li, A.D.Q. (2005) *J. Am. Chem. Soc.*, **127**, 8968; (f) Tomasulo, M., Yildiz, I. and Raymo, F.M. (2007) *Inorg. Chim. Acta*, **360**, 938; (g) Medintz, I.L., Trammell, S.A., Mattousi, H. and Mauro, J.M. (2004) *J. Am. Chem. Soc.*, **126**, 30.
59. Mayer, G. and Heckel, A. (2006) *Angew. Chem. Int. Ed.*, **45**, 4900.
60. (a) Willner, I. (1997) *Acc. Chem. Res.*, **30**, 347; (b) Inoue, M. (1999) in *Organic Photochromic and Thermochromic Compounds*, vol. 2, Chapter 9 (eds J.C. Crano and R. Guglielmetti), Plenum Press, New York, p. 391; (c) Willner, I. and Willner, B. (2001) in *Molecular Switches*, Chapter 6 (ed. B.L. Feringa), Wiley-VCH Verlag GmbH, Weinheim, p. 165; (d) Pieroni, O., Fissi, A., Angelini, N. and Lenci, F. (2001) *Acc. Chem. Res.*, **34**, 9; (e) Shao, N., Jin, J., Wang, H., Zheng, J., Yang, R., Chan, W. and Abliz, Z. (2010) *J. Am. Chem. Soc.*, **132**, 725.
61. (a) Ciardelli, F., Fabbri, D., Pieroni, O. and Fissi, A. (1989) *J. Am. Chem. Soc.*, **111**, 3470; (b) Gartner, C.A., Wen, B., Wan, J., Becker, R.S., Jones, G., Gygi, S.P. and Nelson, S.D. (2005) *Biochem.*, **44**, 1846.
62. Rijcken, C.J.F., Soga, O., Hennink, W.E. and van Nostrum, C.F. (2007) *J. Control. Release*, **120**, 131.
63. Inoue, M., Kim, K. and Kitao, T. (1992) *J. Am. Chem. Soc.*, **114**, 778.
64. (a) Shao, N., Jin, J.Y., Cheung, S.M., Yang, R.H., Chan, W.H. and Mo, T.

(2006) *Angew. Chem. Int. Ed.*, **45**, 4944; (b) Shao, N., Gao, X., Wang, H., Yang, R. and Chan, W. (2009) *Anal. Chim. Acta*, **655**, 1.

65. (a) Phillips, J.P., Mueller, A. and Przystal, F. (1965) *J. Am. Chem. Soc.*, **87**, 420; (b) Taylor, L.D., Nickolson, J. and Davis, R.B. (1967) *Tetrahedron Lett.*, **8**, 1585; (c) Atabekyan, L.S., Roitman, G.P. and Chibisov, A.K. (1982) *J. Anal. Chem. (Russ.).*, **37**, 293.
66. (a) Winkler, J.D., Bowen, C.M. and Michelet, V. (1998) *J. Am. Chem. Soc.*, **120**, 778; (b) Collins, G.E., Choi, L.S., Ewing, K.J., Bowen, C.M. and Winkler, J.D. (1999) *Chem. Commun.*, 321; (c) Evans, L., Collins, G.E., Shaffer, R.E., Michelet, V. and Winkler, J.D. III (1999) *Anal. Chem.*, **71**, 5322.
67. (a) Chernyshev, A.V., Metelitsa, A.V., Gaeva, E.B., Voloshin, N.A., Borodkin, G.S. and Minkin, V.I. (2007) *J. Phys. Org. Chem.*, **20**, 908; (b) Zacharova, M.I., Pimienta, V., Metelitsa, A.V., Minkin, V.I. and Micheau, J.C. (2009) *Russ. Chem. Bull. Int. Ed.*, **58**, 1329.
68. (a) Voloshin, N.A., Chernyshev, A.V., Raskita, I.M., Metelitsa, A.V. and Minkin, V.I. (2005) *Russ. Chem. Bull. Int. Ed.*, **54**, 705; (b) Chernyshev, A.V., Voloshin, N.A., Raskita, I.M., Metelitsa, A.V. and Minkin, V.I. (2006) *J. Photochem. Photobiol. A.*, **184**, 289; (c) Zacharova, M.I., Coudret, C., Pimienta, V., Micheau, J.C., Delbaere, S., Vermeersch, G., Metelitsa, A.V., Voloshin, N. and Minkin, V.I. (2010) *Photochem. Photobiol. Sci*, **9**, 199.
69. Lu, N., Nguyen, V.N., Kumar, S. and McCurdy, A. (2005) *J. Org. Chem.*, **70**, 9067.
70. (a) Minkovska, S., Kolev, K., Jeliazkova, B. and Deligeorgiev, T. (1998) *Dyes Pigm.*, **39**, 25; (b) Minkovska, S., Fedieva, M., Jeliazkova, B. and Deligeorgiev, T. (2004) *Polyhedron*, **23**, 3147; (c) Jeliazkova, B., Minkovska, S. and Deligeorgiev, T. (2005) *J. Photochem. Photobiol. A.*, **171**, 153; (d) Alhashimy, N., Byrne, R., Minkovska, S. and Diamond, D. (2009) *Tetrahedron Lett.*, **50**, 2573.
71. (a) Valeur, A. and Leray, I. (2000) *Coord. Chem. Rev.*, **205**, 3; (b) Bren, V.A. (2001) *Russ. Chem. Rev. Int. Ed.*, **70**, 1017; (c) Alfimov, M.A., Fedorova, O.A. and Gromov, S.P. (2003) *J. Photochem. Photobiol. A.*, **158**, 183; (d) Kimura, K., Sakamoto, H. and Nakamura, M. (2003) *Bull. Chem. Soc. Jpn.*, **76**, 225; (e) Gokel, G.W., Leevy, W.M. and Weber, M.E. (2004) *Chem. Rev.*, **104**, 2723; (f) Minkin, V.I., Dubonosov, A.D. and Bren, V.A. (2008) *Arkivoc*, (4), 90.
72. (a) Kimura, K., Yamashita, T., Kaneshige, M. and Yokoyama, M. (1992) *Chem. Commun.*, 969; (b) Kimura, K., Mizutani, R., Yokoyama, M. and Arakawa, R. (1997) *J. Am. Chem. Soc.*, **119**, 2062; (c) Sakamoto, H., Takagaki, H., Nakamura, M. and Kimura, K. (2005) *Anal. Chem.*, **77**, 1999.
73. Roxburgh, C.J. and Sammes, P.G. (2006) *Eur. J. Org. Chem.*, (4), 1050.
74. (a) Liu, Z., Jiang, L., Liang, Z. and Gao, Y. (2005) *Tetrahedron Lett.*, **46**, 885; (b) Liu, Z., Jiang, L., Liang, Z. and Gao, Y. (2005) *J. Mol. Struct.*, **737**, 267.
75. (a) Fedorova, O.A., Gromov, S.P., Strokach, Yu.P., Pershina, Yu.V., Sergeev, S.A., Barachevsky, V.A., Pepe, G., Samat, A. and Guglielmetti, R. (1999) *Russ. Chem. Bull.*, **48**, 1950; (b) Fedorova, O.A., Gromov, S.P., Strokach, Yu.P., Koshkin, A.V., Valoya, T.M., Alfimov, M.V., Feofanov, A.V., Alaverdian, I.S., Lokshin, V.A., Samat, A., Guglielmetti, R., Girling, R.B., Moore, J. and Hester, R.E. (2002) *New J. Chem.*, **26**, 1137; (c) Fedorova, O.A., Gromov, S.P., Koshkin, A.V., Strokach, Yu.P., Barachevsky, V.A., Lokshin, V.A., Samat, A., Guglielmetti, R. and Alfimov, M.V. (2002) *Russ. Chem. Bull.*, **51**, 58; (d) Korolev, V.V., Vorobyev, D.Yu., Glebov, E.M., Grivin, V.P., Plyusnin, V.F., Koshkin, A.V., Fedorova, O.A., Gromov, S.P., Alfimov, M.V., Shklyaev, Yu.V., Vshivkova, T.S., Rozhkova, Yu.S., Tolstikov, A.G., Lokshin, V.A. and Samat, A. (2007) *J. Photochem. Photobiol. A.*, **192**, 75; (e) Paramonov, S., Delbaere, S., Fedorova, O., Fedorov, Yu., Lokshin, V., Samat, A. and Vermeersch, G. (2010) *J. Photochem. Photobiol. A: Chem.*, **209**, 111.
76. (a) Juris, A., Balzani, V., Barigelletti, F., Campagna, S. and Belser, P. (1988)

Coord. Chem. Rev., **84**, 85; (b) Meyer, T.J. (1989) Acc. Chem. Rev., **22**, 163; (c) Otsuki, J., Akasaka, T. and Araki, K. (2008) Coord. Chem. Rev., **252**, 32; (d) Kume, S. and Nishihara, H. (2008) Dalton Trans., 3260.

77. Yam, V.W.-W., Ko, C.-C., Wu, L.-X., Wong, K.M.-C. and Cheung, K.-K. (2000) Organometallics, **19**, 1820.

78. Bahr, J.L., Kodis, G., de la Garza, L., Lin, S., Moore, A.L., Moore, T.A. and Gust, D. (2001) J. Am. Chem. Soc., **123**, 7124.

79. (a) Querol, M., Bozic, B., Salluce, N. and Belser, P. (2003) Polyhedron, **22**, 655; (b) Belser, P., de Cola, L., Hartl, F., Adamo, V., Bosic, B., Chriqui, Y., Iyer, V.M., Jukes, R.T.F., Kühni, J., Querol, M., Roma, S. and Salluce, N. (2006) Adv. Funct. Mater., **16**, 195.

80. Ichimura, K. (1999) in Organic Photochromic and Thermochromic Compounds, vol. 2, Chapter 1 (eds J.C. Crano and R. Guglielmetti), Plenum Press, New York, pp. 9–62.

81. Moniruzzaman, M., Sabey, C.J. and Fernando, G.F. (2007) Polymer, **48**, 255.

82. (a) Sumaru, K., Ohi, K., Takagi, T., Kanamori, T. and Shinbo, T. (2006) Langmuir, **22**, 4353; (b) Nishikori, H., Tanaka, N., Takagi, K. and Fujii, T. (2007) J. Photochem. Photobiol. A., **189**, 46; (c) Kumar, A., Srivastava, A., Galaev, I.Yu. and Matiasson, B. (2007) Prog. Polymer Sci., **32**, 1205; (d) McCoy, C.P., Donnelly, L., Jones, D.S. and Gorman, S.O. (2007) Tetrahedron Lett., **48**, 657.

83. Ubukata, T., Hara, M., Ichimura, K. and Seki, T. (2004) Adv. Mater., **16**, 220.

84. Kim, S.-H., Park, S.-Y., Dhin, C.-J. and Yoon, N.-S. (2007) Dyes Pigments, **72**, 299.

85. Dürr, H. (1990) in Photochromism. Molecules and Systems (eds H. Dürr and H. Bouas-Laurent), Elsevier, Amsterdam, p. 223.

86. Samat, A., Lokshin, V., Chamontin, K., Levy, D., Pepe, G. and Guglielmetti, R. (2001) Tetrahedron, **57**, 7349.

87. Kinashi, K., Furuta, H., Harada, Y. and Ueda, Y. (2006) Chem. Lett., **35**, 298.

88. (a) de Silva, A.P., McGlenaghan, N.D. and McCoy, C.P. (2001) in Molecular Switches, Chapter 11 (ed. B.L. Feringa), Wiley, Wienheim, pp. 339–362; (b) de Silva, A.P. and McGlenaghan, N.D. (2004) Chem. Eur. J., **10**, 574; (c) Callan, J.F., de Silva, A.P. and Magri, D.C. (2005) Tetrahedron, **61**, 8551; (d) Pischel, U. (2007) Angew. Chem. Int. Ed., **46**, 4026.

89. Kang, T.J., Chang, S.H. and Kim, D.J. (1996) Mol. Cryst. Liq. Cryst., **278**, 181.

90. (a) Vlasenko, T.Ya., Marevtsev, V.S., Zaichenko, N.L. and Cherkashin, M.I. (1990) Izv. Akad. Nauk SSSR (Ser. Khim.)., 2179; (b) Luchina, V.G., Sychev, I.Yu., Marevtsev, V.S., Vlasenko, T.Ya., Khamchukov, Yu.D. and Cherkashin, M.I. (1992) Izv. Akad. Nauk SSSR (Ser. Khim.)., 2718.

91. Dürr, H. (1999) in Organic Photochromic and Thermochromic Compounds, vol. 1 (eds J.C. Crano and R. Guglielmetti), Plenum Press, New York, pp. 231, 255.

92. (a) Frigoli, M. and Mehl, G.H. (2005) Angew. Chem. Int. Ed., **44**, 5048; (b) Delbaere, S., Vermeersch, G., Frigoli, M. and Mehl, G.H. (2006) Org. Lett., **8**, 4931.

93. (a) Mukhanov, E.L., Alexeenko, Yu.S., Lukyanov, B.S., Ryabuchin, Yu.I., Ryashchin, O.N. and Lukyanova, M.B. (2007) Khim. Get. Soed., (1), 129; (b) Mukhanov, E.L., Alexeenko, Yu.S., Dorogan, I.V., Tkachev, V.V., Lukyanov, B.S., Aldoshin, S.M., Besugliuy, S.O., Minkin, V.I., Utenyshev, A.N. and Ryaschzin, O.I. (2010) Khim. Get. Soed., 357.

94. (a) Guo, X., Zhou, Y., Zhang, D., Yin, B., Liu, Z., Liu, C., Lu, Z., Huang, Y. and Zhu, D. (2004) J. Org. Chem., **69**, 8924; (b) Liu, Z., Jiang, L., Liang, Z. and Gao, Y. (2005) J. Mol. Struct., **737**, 267; (c) Cho, M.J., Kim, G.W., Jun, W.G., Lee, S.K., Jin, J.-I. and Choi, D.H. (2006) Thin Solid Films, **500**, 52; (d) Gaeva, E.B., Pimienta, V., Delbaere, S., Metelitsa, A.V., Voloshin, N.A., Minkin, V.I., Vermeersch, G. and Micheau, J.C. (2007) J. Photochem. Photobiol. A: Chem., **191**, 114.

95. (a) Nakao, R., Horii, T., Kushino, Y., Shimaoka, K. and Abe, Y. (2002) Dyes Pigments., **52**, 95; (b) Li, X., Li, J., Wang, Y., Matsuura, T. and Meng, J. (2004) J. Photochem. Photobiol. A., **161**, 201.

3
Fulgides and Related Compounds

Yasushi Yokoyama, Tsuyoshi Gushiken, and Takashi Ubukata

3.1
Introduction – Fulgides

Fulgides were first synthesized by Stobbe at the beginning of twentieth century [1]. Therefore fulgides have a research history of more than 100 years. The structural definition of a fulgide molecule is (i) having a basic skeleton of bismethylenesuccinic anhydride (or succinimide, known as '*fulgimide*') and (ii) having at least one aromatic ring directly connected to the methylene carbon atom.

Photochromism of fulgides occurs between the colourless (or yellow, if coloured) open isomer and the coloured (from yellow to green) closed isomer. The colourless isomer may be divided further into two geometrical isomers with regard to the double bond connecting the acid anhydride and the aromatic ring. As the priority of the aromatic substituent that takes part in photocyclization is usually (but not always) higher than the other substituent on the double bond, the isomer with E geometry cyclizes upon photoirradiation, while the other with Z geometry cannot cyclize. Of course, the geometrical isomerization between E and Z isomers usually occurs, though it is regarded as an energy-wasting process [2].

The prototype structure of today's 'fulgide' (**1**) was developed by Heller and coworkers in 1981 [3]. It was proven that (i) **1C** was thermally stable and (ii) almost all molecules of **1** can be transformed to the coloured form **1C** at a

Molecular Switches, Second Edition. Edited by Ben L. Feringa and Wesley R. Browne.
© 2011 Wiley-VCH Verlag GmbH & Co. KGaA. Published 2011 by Wiley-VCH Verlag GmbH & Co. KGaA.

photostationary state with UV irradiation, though some **1Z** is produced from **1E** also [4].

1Z ⇌ (UV/UV) **1E** ⇌ (UV/Vis, UV) **1C**

As comprehensive reviews have appeared to date [2] and a chapter dealing with the fulgides as a molecular switch has been written in the first edition of this book in 2001, the development of the science related to molecular switches with fulgides in recent years will be described here. In addition, as photochromic systems closely related to fulgides have appeared in this period, they will be introduced briefly.

3.2
Reviews Dealing with Fulgides

After the publication of a comprehensive review [2a] and a book [2b] for each photochromic family, several minireviews that contain a section of fulgides were published: for supramolecular photochromic compounds [5] and for liquid crystals [6]. Fugide derivatives work as photochemical switches to control supramolecular action or cholesteric liquid-crystalline properties.

3.3
Introduction of New Fulgides towards Molecular Switches

Although the first fulgides synthesized by Stobbe, such as bisbenzylidenesuccinic anhydride, were C_2-symmetric, molecular evolution to give thermal irreversibility and to remove undesired side reactions required an alkylidene group to be introduced instead of a benzylidene group. Returning to the original system, Yokoyama, Kiji and coworkers [7] synthesized fulgide **2** possessing C_2 symmetry by way of a novel Pd-catalysed synthetic method, which is useful for the synthesis of sterically congested fulgides. Although **2** was supposed to exhibit complicated photochromism, the main C-form observed at the photostationary state was **2C$_{EE}$**, which was generated from **2EE**. The photocyclization occurred with 90% diastereomeric excess because of the large **2C$_{EZ}$** to **2EZ** quantum yield of UV irradiation.

3.3 Introduction of New Fulgides towards Molecular Switches

[Structures: 2EE ⇌ 2EZ ⇌ 2ZZ; 2EE ⇌ 2C_EE; 2EZ ⇌ 2C_EZ]

Matsushima and coworkers [8, 9] made a very important discovery. In their report on an oxazolylfulgide **3**, they prepared a glass cell by sticking two glass plates face to face, on both of which a fulgide–polymer mixture was spin coated, in order to avoid exposure of the fulgide to oxygen. The fatigue resistance of fulgides became unprecedentedly high. After an initial 10 cycles of colouration–decolouration, 15% of the compound seemed to be lost. However, the absorbance of the coloured form remained unchanged over the following 5000 cycles (Figure 3.1). Similarly, they also reported a method to avoid decomposition of a spin-coated fulgide–polymer

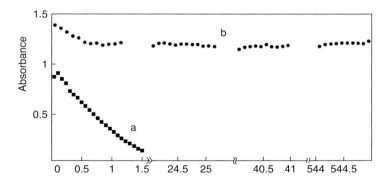

Figure 3.1 Photochemical fatigue resistivity of poly(styrene) film containing an oxazolylfulgide **3**. (a) In a naked single film and (b) sealed film pair, on alternate irradiation with UV (3 min) and visible (3 min) light beams. (Reprinted with permission from Ref. [8]. Copyright 2000 Taylor & Francis.)

film on a glass plate simply by coating the surface with poly(vinyl alcohol) (PVA) film to avoid contact with oxygen [10].

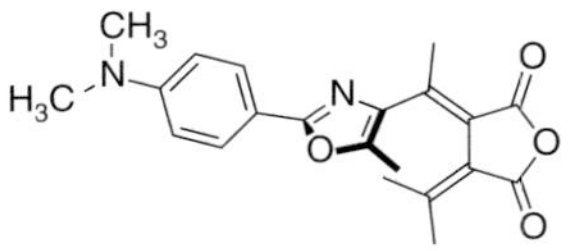

3

Although fulgides are expected to have a succinic anhydride moiety, several fulgides have been reported with other functional groups such as binaphthol, cyanomethylene, diester and lactone [2a]. In an effort to obtain better fulgide-related photochromic switch molecules, Kose and coworkers [11] synthesized the novel compounds **4** and **5** from fulgides. The synthesis of novel fulgides and the investigation of there properties have also been reported recently [12].

Yokoyama and Takahashi [13] reported a fatigue resistive trifluoromethyl-substituted fulgide **6** in 1996. Based on this report, Wolak and coworkers [14] developed a series of fluorine-substituted fulgides.

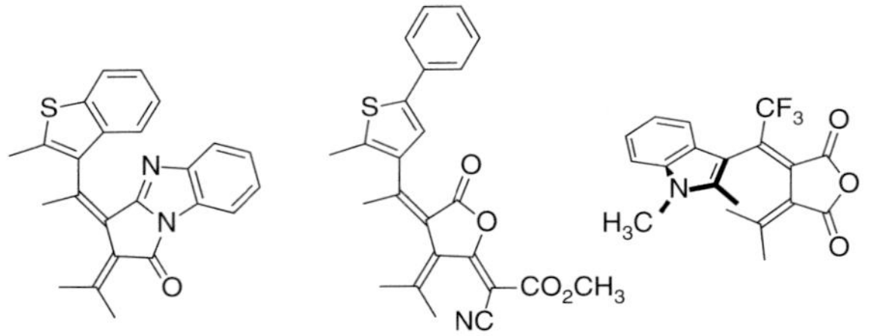

4   **5**   **6**

3.4
Photophysics of Molecular Switches

3.4.1
Investigations into Reaction Pathways

Inspection of photophysical processes of photochromic reactions often affords a key to novel application, technology or materials. From this viewpoint, examination of the nature of the excited state is important in the basic research towards application.

Port and coworkers [15] investigated the transient absorption of the excited state of several fulgides including **1** in poly(methyl methacrylate) (PMMA) film and in toluene on the subpicosecond timescale. They found that, whereas the reaction in solution proceeds via two reaction pathways and one of them has an intermediate, reaction in PMMA film proceeds with one pathway that has no intermediate. Therefore, the photocyclization reaction in PMMA proceeds faster than that in solution.

3.4.2
Two-Photon-Absorption Excitation

Two-photon-absorption excitation of photochromic molecules is an important method to achieve a photochromic reaction only at the point where the photon density is high. It can be applied to three-dimensional optical recording and three-dimensional imaging. It can be achieved either by one beam (focused) or by two beams (meet at one point) of laser light. Belfield and coworkers [16] used an indolylfulgide **7**, and Liao and coworkers [17] used a pyrroylfulgide **8**, both for ring closure by two-photon absorption.

7   8

As reported by Miyasaka and coworkers [18, 19], it is noteworthy that the quantum yield of the ring-opening reaction of fulgide **9** from the higher excited state brought about by two-photon absorption using picosecond laser pulse excitation (532 nm), but not by a femtosecond laser pulse (480 nm), was much larger (0.45) than with one-photon excitation (0.066) (Figure 3.2). Although this is different from the usual 'two-photon-absorption excitation' that implies an excitation by the simultaneous absorption of two lower energy photons that cannot induce photoexcitation alone, this finding is important to open a way to efficient switching of ring-opening quantum yields of fulgides between small and large values.

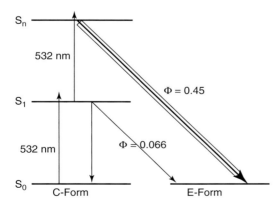

Figure 3.2 Schematic diagram of one-photon and two-photon excitation and ring opening of fulgide **9**.

3.4.3
Energy Transfer

Since the report of switching of the energy trap ability of the fulgimide core that can allow or block energy transfer from an antenna group to a fluorescent moiety by Port and coworkers [20], research on application of photochromic molecules to the control of energy flow became active, mainly in relation to the artificial photosynthesis. Gust and coworkers [21] connected a fulgimide to prepare **10**. Excitation of the coloured form of the fulgimide moiety by 450 nm light caused efficient energy transfer (not less than 95% efficiency) to the porphyrin moiety. Because the open form does not have absorption at 450 nm, energy transfer cannot occur. They also synthesized a porphyrin connecting a bisthienylethene molecule, and found that the bisthienylethene worked differently to the fulgimide, as the energy acceptor [22]. Then they linked both fulgimide and bisthienylethene to a porphyrin (**11**). When both photochromic moieties are coloured, energy transfer from the fulgimide to the diarylethene occurred. However, when the diarylethene takes the colourless form, the excited energy stays on the porphyrin moiety, which fluoresces. When the fulgimide is colourless, no uptake of photons by fulgimide takes place so that no energy transfer can occur. This system was applied as a logic gate with NOR or XOR functions [23].

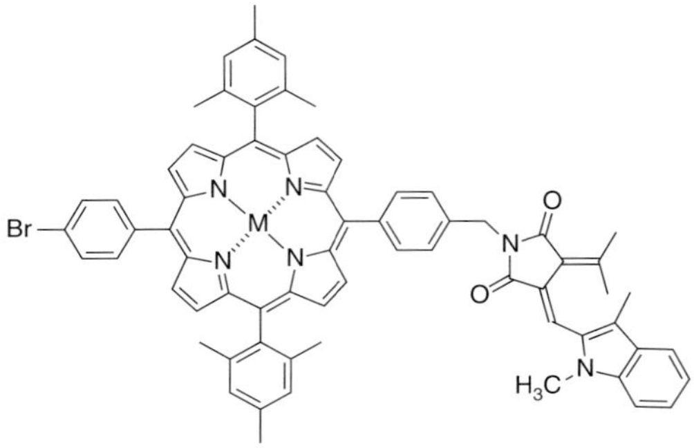

10

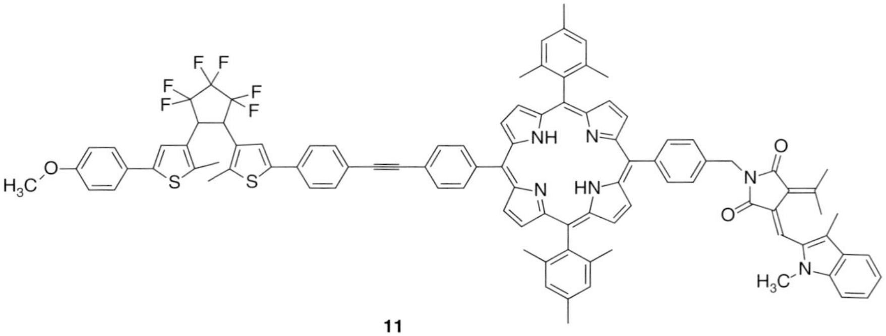

11

The indolylfulgide and fulgimide used in this research have interesting properties towards non-destructive readout (see Section 3.5.3).

3.5 Towards Optical Recording

Because a photochromic reaction generates two distinguishable states by photoirradiation, they can be regarded as 'written' and 'erased' states. If the photochromic compound used is thermally irreversible, then the record is retained until illuminated with light of a different wavelength. The problem is how to readout the recorded information quickly. As the readout process should not induce photochromic reactions, light that causes a photochromic reaction cannot be used, even though readout with light is quick and reliable. Although fluorescence readout is a candidate because of the high sensitivity, it is a slow-destructive readout because fluorescent light comes from the excited state that also affords the photochromic reaction product. Additional ideas are necessary.

In this section, several papers describing new ideas for recording, regardless of the nondestructive readout, are discussed.

3.5.1 Wavelength-Multiplied Recording

Chen, Fan and coworkers [24] prepared a recording media composed of PMMA and two fulgides **3** and **12** possessing different absorption in the visible region. After colouration with UV light, two different patterns were drawn on the same position of the film by two different laser beams, and the two patterns were readout independently by scanning with the laser used for writing, with much weaker power. The crosstalk was negligible, and the readout was repeated for more than 200 times.

12

3.5.2 Incident-Angle-Multiplied Recording

Multiplexed recording on a recording unit is an important technique to increase the recording density. Kurita, Yokoyama and coworkers [25] proved that polystyrene powder containing C-form of a binaphthol-condensed fulgide derivative **13** has a unique property. The C-form fluoresces, whereas the E-form does not. When the powder was irradiated with 488-nm light, which the C-form absorbs, the C-form

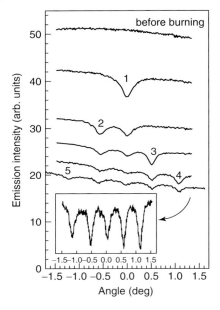

Figure 3.3 Profile of successively burned five holes. Emission intensity was measured as a function of the incident angle. After the measurement of the initial profile before the burning, holes were burned at different incident angles indicated by the numerals (wavelength 488 nm, intensity 0.5 mW/mm^2, burning time 20 s for each hole) and then the profiles were measured. The final profile with five holes is enlarged in the inset. (Reprinted with permission from Ref. [25]. Copyright 2000 Taylor & Francis.)

molecules in the grains of polystyrene in the light path were bleached so that the fluorescence became weaker than at other places. When the angle of the incident light to the same spot on the powder was altered slightly, the incident light takes a different path, the C-form in this way was not yet bleached. Therefore, bleaching C-form at a certain angle made a hole that depended on the angle of the incident light. This can be readout by the intensity of the fluorescence. Hole burning of fivefold multiplex recording at the same spot of the powder was achieved by changing the incident angle of the light by 0.33° steps (Figure 3.3).

13

3.5.3
Nondestructive Readout with Fluorescence

If the state of a photochromic compound switches the ability of the neighbouring fluorescent molecule, then the fluorescence can be used for nondestructive readout. Rentzepis and coworkers [26] realized this in a novel fulgimide. Fulgimides with the same skeleton, such as **14**, were also employed by Gust *et al.* [21, 23] in their energy-transfer experiments, which were introduced in Section 3.4.3.

The novel fulgimide **15** can be recognized as the 'Fulgide of the Decade'. When the fulgimide moiety of **15** takes the C-form, the oxazine moiety scarcely emits fluorescent light. When it is changed to the E-form by visible-light irradiation, the oxazine starts to emit strong fluorescence. Because the absorption band of oxazine is located in the longer-wavelength region than the absorption band of the C-form, excitation of oxazine does not induce photochromism. Therefore, the photochromic reaction is induced only in the shorter-wavelength region, while reading out by the change in intensity of fluorescence can be achieved in the longer-wavelength region [27]. As the oxazine is fluorescent under nonpolar conditions, and the polarity of the fulgimide is stronger when it takes the C-form, the attached oxazine is fluorescent only when it is in the E-form (Figure 3.4).

A similarly behaving fulgimide **16** with the same skeleton, but that is this time fluorescent when it is placed in a polar environment, was used in fluorescence switching in a living cell [28].

3.5.4
Recording with Optical Anisotropy

If an optical anisotropy is induced by a photochromic reaction, then it can be a hidden record. Specific information, which can be identified only by polarized light, may be embedded in the isotropic bulk of the valueless records. Yao and coworkers [29] found that during the bleaching of the C-form of a pyrrolylfulgide **17** in a PMMA film by a polarized laser light, an anisotropic alignment of the molecules in the film

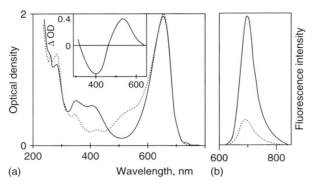

Figure 3.4 Spectra of **15** dissolved in 1-propanol: (a) absorption and (b) fluorescence spectra of the read form (E-form: solid line) and write form (C-form: dashed line). Inset in (a) differential spectra of the write and read forms. (Reprinted with permission from Ref. [27b]. Copyright 2004 American Chemical Society.)

was induced. The axis of the induced anisotropy was the same as the polarization of the irradiating light. Writing with a laser at a different polarization axis generated a different pattern that can be read only when the polarization of the reading laser light is the same as that of the writing light used. Non-destructive all optical writing and readout has been achieved by Malkmus et al. recently using IR light [30].

17

3.5.5
Formation of Nanostructures of Fulgides

Nanostructures formed on a film by photochromic reactions can be imaged by atomic force microscopy (AFM) or scanning near-field optical microscopy (SNOM) techniques. Jansson and coworkers [31] first observed the formation of dots on the surface of spin-coated films composed of only the E-form fulgides (**18**, **19**) prepared on a silicon wafer after several days of storage in the dark. They monitored the formation of dots for several days by AFM, and presumed that they were composed of the oxidation products of fulgide molecules by oxygen. They did not consider, however, the possibility of aggregation (or crystallization) of fulgide molecules, as shown in the report of Port et al. described below.

Port and coworkers [32, 33] reported a similar observation. When they prepared a film of fulgide **20** by high-vacuum deposition, a spontaneous formation of dot-like pattern of fulgide molecules was observed. The dot exhibited photochromism upon irradiation of light, and the SNOM images were changed by iterative photoreactions. Formation of similar patterns was observed when the deposition was carried out at 10 K on a quartz substrate.

18   **19**   **20**

Formation and photoreaction of these dots may be important in high-density optical information storage in the future.

3.6
Understanding of Molecular Structures from Calculations

Calculations can, in principle, predict the results of 'real' experiments in the future when the accuracy of such calculations reaches maturity. Until then, comparison of calculation results with real experimental results is important.

Yoshioka and coworkers [34] have been involved in the calculation of photochromic molecules. They reported on the stable conformations of E-, Z- and C-forms of several simple fulgides by *ab initio* calculations. It was shown that while C-forms of furyl- and thienylfulgides are more stable than their E-forms, the C-form of pyrroylfulgide is less stable than its E-form.

They also estimated the feasibility of thermal reactions between E- and C-forms and E- and Z-forms by *ab initio* calculations, and succeeded in explaining the experimental results that heating of the E-form of a furoylfulgide **1** at an elevated temperature yielded its Z-form [35].

Sakakibara, Yokoyama and coworkers [36] showed that the CD spectra of the optically resolved E-form of indolylfulgide **21** and that of E- and C-forms of chiral binaphthol-condensed indolylfulgide **22** were reproduced by semiempirical molecular orbital calculations, by taking the population of stable conformers into account. Recently methods for the calculation of conical intersections have been developed for fulgides [37, 38].

21   **22**

3.7
Development of Photochromic Switches Closely Related to Fulgides

Two major representatives of classes of thermally irreversible photochromic compounds are fulgides [2a] and diarylethenes [39]. Because their photochromism are based on 6π-electrocyclization, other structurally related compounds that undergo 6π-electrocyclization may be thermally irreversible. In this context, several reports on such photochromic compounds that do not belong either to fulgides or diarylethenes have appeared recently.

Yokoyama and the coworkers [40] and Branda and coworkers [41] reported independently on the preparation of arylbutadienes such as **23** [40] and **24** [41]. They have one aromatic group and one olefinic functional group attached to a hexafluorocyclopentene. These compounds showed thermally irreversible photochromism. Takami and Irie [42] have synthesized a thiazole-substituted compound also.

As a sort of spin-off of this research, thermally *reversible* photochromic compounds such as **25** and **26** have been obtained [43]. They can be classified as members of the heliochromic compounds such as **27** [44, 45].

3.8
Perspectives of Research with Fulgides

Over the past 20 years, the most frequently investigated thermally irreversible photochromic family of switches was the diarylethenes, because of high fatigue resistance. In contrast, fulgides were generally perceived to be rather unstable towards repeated photochromic reactions. However, due to the efforts of Matsushima and coworkers, fatigue resistivity as well as thermal stability was improved dramatically by a simple method – just prevent contact with oxygen [8, 10].

Fulgides have several excellent properties. (i) Attachment of functional groups in the vicinity of the fulgide core can be done easily without the formation of C–C bonds – for instance attached to the acid anhydride moiety by changing it to an imide group [27]. (ii) For most of the compounds, the absorption band of the coloured form is symmetric in nature [10]. (iii) The steric effects of the substituents and the electronic effects of the aromatic ring are predictable so that control of quantum yield, absorption maximum wavelength and molar absorption coefficient is feasible [2a]. Therefore, efforts to improve their properties and to produce novel fulgide-related compounds are quite important to increase the possibility of application in the future.

References

1. (a) Stobbe, H. (1905) *Berichte*, **38**, 3673; (b) Stobbe, H. (1905) *Berichte*, **40**, 3372; (c) Stobbe, H. (1911) *Annalen*, **380**, 1.
2. Reviews: (a) Yokoyama, Y. (2000) *Chem. Rev.*, **100**, 1717; (b) Fan, M.G., Yu, L. and Zhao, W. (1999) in *Organic Photochromic and Thermochromic Compounds*, Main Photochromic Families, Vol. 1 (eds J.C. Crano and R.J. Guglielmetti), Plenum Press, New York, p. 141.
3. Darcy, P.J., Heller, H.G., Strydom, P.J. and Whittall, J. (1981) *J. Chem. Soc., Perkin Trans. 1*, 202.
4. Yokoyama, Y., Goto, T., Inoue, T., Yokoyama, M. and Kurita, Y. (1988) *Chem. Lett.*, 1049.
5. (a) Alfimov, M.V., Fedorova, O.A. and Gromov, S.P. (2003) *J. Photochem. Photobiol. A: Chem.*, **158**, 183; (b) Yokoyama, Y. and Kose, M. (2004) *J. Photochem. Photobiol. A: Chem.*, **166**, 9.
6. Eelkema, R. and Feringa, B.L. (2006) *Org. Biomol. Chem.*, **4**, 3729.
7. (a) Yokoyama, Y., Sagisaka, T., Yamaguchi, Y., Yokoyama, Y., Kiji, J., Okano, T., Takemoto, A. and Mio, S. (2000) *Chem. Lett.*, 220; (b) Kiji, J., Okano, T., Takemoto, A., Mio, S., Konishi, T., Kondou, Y., Sagisaka, T. and Yokoyama, Y. (2000) *Mol. Cryst. Liq. Cryst.*, **344**, 235.
8. Matsushima, R., Hayashi, T. and Nishiyama, M. (2000) *Mol. Cryst. Liq. Cryst.*, **344**, 241.
9. Kohno, Y., Tamura, Y. and Matsushima, R. (2009) *J. Photochem. Photobiol. A: Chem.*, **201**, 98–101.
10. Matsushima, R., Ito, Y., Yamashita, K., Yamanoto, K. and Kouno, Y. (2005) *Chem. Lett.*, **34**, 574.
11. (a) Kose, M. and Orhan, E. (2006) *J. Photochem. Photobiol. A: Chem.*, **177**, 170–176; (b) Kose, M., Orhan, E. and Buyukgungor, O. (2007) *J. Photochem. Photobiol. A: Chem.*, **188**, 358.
12. (a) Luyksaar, S.I., Migulin, V.A., Nabatov, B.V. and Krayushkin, M.M. (2010) *Russ. Chem. Bull.*, **59**, 446–451; (b) Balenko, S.K., Makarova, N.I., Rybalkin, V.P., Shepelenko, E.N., Popova, L.L., Tkachev, V.V., Aldoshin,

S.M., Metelitsa, A.V., Bren, V.A. and Minkin, V.I. (2010) *Russ. Chem. Bull.*, **59**, 954–959; (c) Baghaffar, G.A. and Asiri, A.M. (2008) *Arabian J. Chem.*, **1**, 25–35; (d) Balenko, S.K., Makarova, N.I., Karamov, O.G., Rybalkin, V.P., Dorogan, I.V., Popova, L.L., Shepelenko, E.N., Metelitsa, A.V., Tkachev, V.V. and Aldoshin, S.M. (2007) *Russ. Chem. Bull.*, **56**, 2400–2406; (e) Baghaffar, G.A. and Asiri, A.M. (2008) *Pigm. Resin Tech*, **37**, 140–144; (f) Baghaffar, G.A. and Asiri, A.M. (2008) *Pigm. Resin Tech*, **37**, 145–150; (g) Menke, N., Yao, B., Wang, Y., Dong, W., Lei, M., Chen, Y., Fan, M. and Li, T. (2008) *J. Modern Optics*, **55**, 1003–1011; (h) Rupp, R.A., Zheng, Y., Wang, Y., Dong, W., Menke, N., Yao, B., Zaho, W. Chen, Y., Fan, M. and Xu, J. (2007) *Phys. Stat. Solidi B*, **244**, 1363–1375.

13. Yokoyama, Y. and Takahashi, K. (1996) *Chem. Lett.*, 1037.
14. (a) Wolak, M.A., Gillespie, N.B., Thomas, C.J., Birge, R.R. and Lees, W.J. (2001) *J. Photochem. Photobiol., A: Chem.*, **144**, 83; (b) Wolak, M.A., Gillespie, N.B., Thomas, C.J., Birge, R.R. and Lees, W.J. (2002) *J. Photochem. Photobiol., A: Chem.*, **147**, 39; (c) Wolak, M.A., Finn, R.C., Rarig, R.S., Thomas, C.J., Hammond, R.P., Birge, R.R., Zubieta, J. and Watson, J. Jr (2002) *Acta Crystallogr., Sect. C: Cryst. Struct. Commun.*, **C58**, o389; (d) Wolak, M.A., Thomas, C.J., Gillespie, N.B., Birge, R.R. and Watson, J.L. (2003) *J. Org. Chem.*, **68**, 319.
15. Port, H., Gartner, P., Hennrich, M., Ramsteiner, I. and Schoeck, T. (2005) *Mol. Cryst. Liq. Cryst.*, **430**, 15.
16. (a) Belfield, K.D., Schafer, K.J., Liu, Y., Liu, J., Ren, X. and Van Stryland, E.W. (2000) *J. Phys. Org. Chem.*, **13**, 837; (b) Belfield, K.D., Liu, Y., Negres, R.A., Fan, M., Pan, G., Hagan, D.J. and Hernandez, F.E. (2002) *Chem. Mater.*, **14**, 3663.
17. Liao, N., Gong, M., Xu, D., Qi, G. and Zhang, K. (2001) *Chin. Sci. Bull.*, **46**, 1856.
18. Ishibashi, Y., Murakami, M., Miyasaka, H., Kobatake, S., Irie, M. and Yokoyama, Y. (2007) *J. Phys. Chem. C*, **111**, 2730.
19. Ishibashi, Y., Katayama, T., Ota, C., Kobatake, S., Irie, M., Yokoyama, Y. and Miyasaka, H. (2009) *New J. Chem.*, **33**, 1409–1419.
20. Walz, J., Ulrich, K., Port, H., Wolf, H.C., Wonner, J. and Effenberger, F. (1993) *Chem. Phys. Lett.*, **213**, 321.
21. Straight, S.D., Terazono, Y., Kodis, G., Moore, T.A., Moore, A.L. and Gust, D. (2006) *Aust. J. Chem.*, **59**, 170.
22. Liddell, P.A., Kodis, G., Moore, A.L., Moore, T.A. and Gust, D. (2002) *J. Am. Chem. Soc.*, **124**, 7668.
23. Straight, S.D., Liddell, P.A., Terazono, Y., Moore, T.A., Moore, A.L. and Gust, D. (2007) *Adv. Funct. Mater.*, **17**, 777.
24. Chen, Y., Xiao, J.P., Yao, B. and Fan, M.G. (2006) *Opt. Mater.*, **28**, 1068.
25. Kurita, A., Kanematsu, Y., Kushida, Y., Sagisaka, T. and Yokoyama, Y. (2000) *Mol. Cryst. Liq. Cryst.*, **344**, 205.
26. (a) Liang, Y., Dvornikov, A.S. and Rentzepis, P.M. (1999) *J. Photochem. Photobiol. A: Chem.*, **125**, 79; (b) Liang, Y., Dvornikov, A.S. and Rentzepis, P.M. (2002) *Macromolecules*, **35**, 9377.
27. (a) Liang, Y.C., Dvornikov, A.S. and Rentzepis, P.M. (2003) *Proc. Natl. Acad. Sci.*, **100**, 8109; (b) Dvornikov, A.S., Liang, Y., Cruse, C.S. and Rentzepis, P.M. (2004) *J. Phys. Chem. B*, **108**, 8652.
28. Berns, M.W., Krasieva, T., Sun, C.-H., Dvornikov, A. and Rentzepis, P.M. (2004) *J. Photochem. Photobiol. B: Biol.*, **75**, 51.
29. (a) Wang, Y.L., Yao, B.-L., Menke, N., Chen, Y., Li, T.-K., Zheng, Y., Lei, M., Dong, W.-B., Fan, M.-G. and Chen, G.-F. (2004) *Chin. Phys. Lett.*, **214**, 679; (b) Yao, B., Wang, Y., Lei, M., Menke, N., Chen, G., Chen, Y., Li, T. and Fan, M. (2005) *Opt. Exp.*, **13**, 20; (c) Menke, N., Yao, B., Wang, Y., Zheng, Y., Lei, M., Ren, L., Chen, G., Chen, Y., Fan, M. and Li, T. (2006) *J. Opt. Soc. Am., A*, **23**, 267.
30. Malkmus, S., Koller, F.O., Draxler, S., Schrader, T.E., Schreier, W.J., Brust, T., DiGirolamo, J.A., Lees, W.J., Zinth, W. and Braun, M. (2007) *Adv. Func. Mater.*, **17**, 3657–3662.

31. Jansson, R., Zangooie, S., Kugler, T. and Arwin, H. (2001) *J. Phys. Chem. Solids*, **62**, 1219.
32. (a) Rath, S., Mager, O., Heilig, M., Strauss, M., Mack, O. and Port, H. (2001) *J. Lumin.*, **94**, 157; (b) Rath, S. and Port, H. (2006) *Chem. Phys. Lett.*, **421**, 152.
33. Rath, S., Heilig, M., Port, H. and Wrachtrup, J. (2007) *Nano Lett.*, **7**, 3845–3848.
34. Yoshioka, Y., Usami, M. and Yamaguchi, K. (2000) *Mol. Cryst. Liq. Cryst.*, **345**, 405.
35. Yoshioka, Y., Usami, M., Watanabe, M. and Yamaguchi, K. (2003) *J. Mol. Struct.*, **623**, 167.
36. Ankai, E., Sakakibara, K., Uchida, S., Uchida, Y., Yokoyama, Y. and Yokoyama, Y. (2001) *Bull. Chem. Soc. Jpn.*, **74**, 1101.
37. Tomasello, G., Bearpark, M.J., Robb, M.A., Orlandi, G. and Garavelli, M. (2010) *Angew. Chem., Int. Ed.*, **49**, 2913–2916.
38. Voll, J., Kerscher, T., Geppert, D. and De Vivie-Riedle, R. (2007) *J. Photochem. Photobio., A: Chem.*, **190**, 352–358.
39. Irie, M. (2000) *Chem. Rev.*, **100**, 1685.
40. (a) Shrestha, S.M., Nagashima, H., Yokoyama, Y. and Yokoyama, Y. (2003) *Bull. Chem. Soc. Jpn.*, **76**, 363; (b) Yokoyama, Y., Shrestha, S.M. and Yokoyama, Y. (2005) *Mol. Cryst. Liq. Cryst.*, **431**, 433.
41. (a) Peters, A., Vitols, C., McDonald, R. and Branda, N.R. (2003) *Org. Lett.*, **5**, 1183; (b) Wüstenberg, B. and Branda, N.R. (2005) *Adv. Mater.*, **17**, 2134.
42. Takami, S. and Irie, M. (2004) *Tetrahedron*, **60**, 6155.
43. (a) Yokoyama, Y., Nagashima, H., Shrestha, S.M., Yokoyama, Y. and Takada, K. (2003) *Bull. Chem. Soc. Jpn.*, **76**, 355; (b) Yokoyama, Y., Nagashima, H., Takada, K., Moriguchi, T., Shrestha, S.M. and Yokoyama, Y. (2005) *Mol. Cryst. Liq. Cryst.*, **430**, 53; (c) Eguchi, N., Ubukata, T. and Yokoyama, Y. (2007) *J. Phys. Org. Chem.*, **20**, 851.
44. Yokoyama, Y., Nakata, H., Sugama, K. and Yokoyama, Y. (2000) *Mol. Cryst. Liq. Cryst.*, **344**, 253.
45. Whittall, J. (1990) in *Photochromism: Molecules and Systems*, Chapter 1 (eds H. Dürr and H. Bouas-Laurent), Elsevier, Amsterdam, p. 467.

4
Transition Metal-Complexed Catenanes and Rotaxanes as Molecular Machine Prototypes

Christian Tock, Julien Frey, and Jean Pierre Sauvage

4.1
Introduction

In the course of the last decade, many dynamic molecular systems, for which the movements are controlled from the outside, have been elaborated. These compounds are generally referred to as *'molecular machines'* [1–22]. Transition-metal-containing catenanes and rotaxanes [23–61] are ideally suited to build such systems. In the present chapter, we will discuss a few examples of molecular machines mostly, but not exclusively, elaborated and studied in Strasbourg.

The first section is devoted to the archetype of copper-based molecular machines, a [2]catenane, set in motion electrochemically.

In the second section we will discuss the kinetic inertness of the first generation of copper-based molecular machines, its possible origin and how to speed up the ligand-exchange reactions.

A linear molecular muscle will be presented in the third section, the driving force in this case is a metal-exchange reaction, stretching or contracting the system, thus mimicking the activity of a natural muscle.

In the fourth section we will discuss a heterodinuclear bismacrocycle transition-metal complex in which intramolecular motion is controlled by an electrochemical stimulus.

In the last section we will discuss light-driven machines, consisting of ruthenium(II)-complexed rotaxanes or catenanes. For these latter systems, the synthetic approach is based on the template effect of an octahedral ruthenium(II) centre. Two 1,10-phenanthroline (phen) ligands are incorporated in a ring, affording the precursor to the catenane. Ru(diimine)$_3$$^{2+}$ complexes display the universally used metal-to-ligand charge transfer (^{3}MLCT) excited state and another interesting excited state, the ligand field (^{3}LF) state, which is strongly dissociative. By taking advantage of this latter state, it has been possible to propose a new family of molecular machines, which are set in motion by populating the dissociative ^{3}LF state, thus leading to ligand exchange in the coordination sphere of the ruthenium(II) centre. These systems are of particular interest, because their external stimulus is readily available, easy to switch on and off and is free of waste side-products.

Molecular Switches, Second Edition. Edited by Ben L. Feringa and Wesley R. Browne.
© 2011 Wiley-VCH Verlag GmbH & Co. KGaA. Published 2011 by Wiley-VCH Verlag GmbH & Co. KGaA.

4.2
Copper-Complexed [2]Catenanes in Motion: the Archetypes

4.2.1
A Copper-Complexed [2]Catenane in Motion with Two Distinct Geometries

Bistability is an essential property for imaging and information storage. The first molecular motor elaborated and studied in our group was based on a catenane containing two different interlocking rings and two different coordination possibilities [62]. The interconversion between both forms of the complex is electrochemically triggered and corresponds to the sliding motion of one ring within the other. It leads to a profound rearrangement of the compound and can thus be regarded as a complete metamorphosis of the molecule. The principle of the process is explained in Figure 4.1. Essential is the difference of preferred coordination number (CN) for the two different redox states of the metal: CN = 4 for copper(I) and CN = 5 (or 6) for copper(II).

The organic backbone of the asymmetric catenane consists of a 2,9-diphenyl-1, 10-phenanthroline (dpp) bidentate chelate included in one cycle and, interlocked to it, a ring containing two different subunits: a dpp moiety and a terdentate ligand, 2,2′,6′,2″-terpyridine (terpy). Depending on the mutual arrangement of both interlocked rings, the central metal atom (copper, for instance) can be tetrahedrally complexed (two dpps) or five-coordinated (dpp + terpy). Interconversion between these two complexing modes results from a complete pirouetting of the two-site ring. It can, of course, be electrochemically induced by taking advantage of the different geometrical requirements of the two redox states of the copper(II)/copper(I) couple. From the stable tetrahedral monovalent complex, oxidation leads to a four-coordinate Cu(II) state that rearranges to the more stable five-coordinate compound. The process can be reversed by reducing the divalent

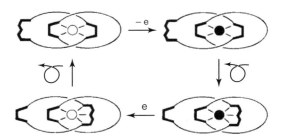

Figure 4.1 Principle of the electrochemically triggered rearrangement of an asymmetric [2]-catenane. The stable four-coordinate monovalent complex (top left, the white circle represents **Cu(I)**) is oxidized to an intermediate tetrahedral divalent species (top right, the black circle represents **Cu(II)**). This compound undergoes a complete reorganization process to afford the stable four-coordinate **Cu(II)** complex (bottom right). Upon reduction, the five-coordinate monovalent state is formed as a transient (bottom left). Finally, the latter undergoes the conformational change, which regenerates the starting complex.

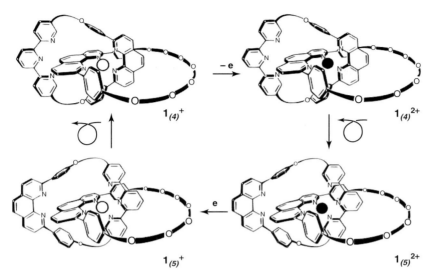

Figure 4.2 Electrochemically induced molecular rearrangements undergone by the copper catenane $1_{(4)}{}^{+}$.

state to the five-coordinate Cu(I) complex obtained as a transient species before a changeover process takes place to afford back the starting tetrahedral monovalent state (Figure 4.2).

4.2.2
A Copper-Complexed [2]Catenane in Motion with Three Distinct Geometries

Multistage systems seem to be uncommon, although they are particularly challenging and promising in relation to nanodevices aimed at important electronic functions and, in particular, information storage [63]. Among the few examples that have been reported in recent years, three-stage catenanes are particularly significant since they lead to unidirectional rotary motors [11]. In the mid-1990s, our group described a particular Cu-complexed [2]catenane, which represents an example of such a multistage compound [64]. The molecule displays three distinct geometries, each stage corresponding to a different CN of the central complex (CN = 4, 5 or 6). The principle of the three-stage electrocontrollable catenane is represented in Figure 4.3.

Similar to the very first and simpler catenane discussed in the previous paragraph, the gliding of the rings in the present system relies on the important differences of stereochemical requirements for coordination of Cu(I) and Cu(II). For the monovalent state the stability sequence is CN = 4 > CN = 5 > CN = 6. By contrast, divalent Cu is known to form stable hexacoordinate complexes, with pentacoordinate systems being less stable and tetrahedral Cu(II) species being even more strongly disfavoured.

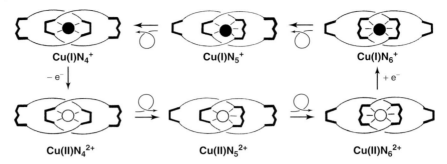

Figure 4.3 A three-geometry Cu(I) catenane whose general molecular shape can be dramatically modified by oxidizing the central metal (Cu(I) to Cu(II)) or reducing it back to the monovalent state. Each ring of the [2]-catenane incorporates two different coordinating units: the bidentate dpp unit (dpp) 2,9-diphenyl-1,10-phenanthroline) is symbolized by a U, whereas the terpy fragment (2,2′:6′,2″-terpyridine) is indicated by a stylized W. Starting from the tetracoordinate monovalent Cu complex (Cu(I)N$_4^+$; top left) and oxidizing it to the divalent state (Cu(II)N$_4^{2+}$), a thermodynamically unstable species is obtained that should first rearrange to the pentacoordinate complex Cu(II)N$_5^{2+}$ by gliding of one ring (down; middle) within the other and, finally, to the hexacoordinate stage Cu(II)N$_6^{2+}$ by rotation of the second cycle (down; right) within the first one. Cu(II)N$_6^{2+}$ is expected to be the thermodynamically stable divalent complex. The double ring-gliding motion following oxidation of Cu(I)N$_4^+$ can be inverted by reducing Cu(II)N$_6^{2+}$ to the monovalent state (Cu(I)N$_6^+$; top right), as represented on the top line of the figure.

The synthesis of the key catenane [Cu(I)N$_4$](PF$_6^-$) = **2**$_{(4)}^+$ (Figure 4.4a) (one should notice that, as usual, the subscripts 4, 5 and 6 indicate the CN of the copper centre) derives from the usual three-dimensional template strategy [65].

The visible spectrum of this deep red complex shows a MLCT absorption band (λ_{max} = 439 nm, ε = 2570 mol^{-1} L cm^{-1}, MeCN). Cyclic voltammetry (CV) of a MeCN solution shows a reversible redox process at +0.63 V (vs. Saturated calomel electrode (SCE)). Both the CV data and the UV-Vis spectrum are similar to those of other related species [65, 66]. The reaction of **2**$_{(4)}^+$ with KCN afforded the free catenane (not represented), which was subsequently reacted with Cu(BF$_4$)$_2$ to give **2**$_{(6)}^{2+}$ as a very pale green complex (Figure 4.4c). The hexacoordinate structure of this species was evidenced by UV-Vis spectroscopy and electrochemistry. The cyclic voltammogram shows an irreversible reduction at −0.43 V (vs SCE, MeCN). These data are similar to the ones obtained for the complex Cu(diMe-tpy)$_2$(BF$_4$)$_2$ (diMe-tpy = 5,5″-dimethyl-2,2′:6′,2″-terpyridine).

When a dark red MeCN solution of **2**$_{(4)}^+$ was oxidized by an excess of NO$^+$ BF$_4^-$, a green solution of **2**$_{(4)}^{2+}$ was obtained. The CV is the same as for the starting complex, and the visible absorption spectrum shows a band at λ_{max} = 670 nm, ε = 810 l mol^{-1} cm^{-1}, in MeCN, typical of these tetrahedral Cu(II) complexes [66]. A decrease of the intensity of this band was observed when monitoring it as a function of time. This fact is due to the gliding motion of the rings to give the penta- (Figure 4.4b) and (Figure 4.4c) hexacoordinate Cu(II) complexes, whose molar absorptivity are lower as compared to that of **2**$_{(4)}^{2+}$ (circa 125 and 100,

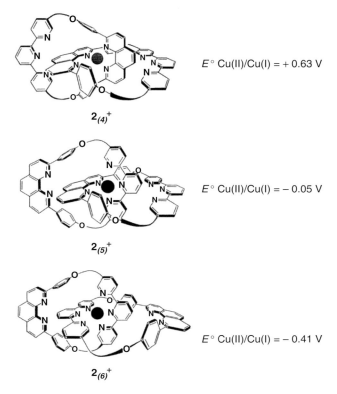

Figure 4.4 (a–c) The four-, five- and six-coordinate copper complexes involved. The corresponding Cu(II)/Cu(I) redox potentials are also indicated. They clearly show the sequence of preferred stabilities for copper (II) versus copper (I), the hexacoordinate complex producing the most stable divalent complex.

respectively). The final product is $2_{(6)}{}^{2+}$, as indicated by the final spectro- and electrochemical data. A similar behaviour was observed when a solution of $2_{(4)}{}^+$ was electrochemically oxidized.

When either the $2_{(6)}{}^{2+}$ solution resulting from this process or a solution prepared from a sample of isolated solid $2_{(6)} \cdot (BF_4)_2$ were electrochemically reduced at -1 V, the tetracoordinate catenane was quantitatively obtained. The cycle depicted in Figure 4.3 was thus completed. The changeover process for the monovalent species is faster than the rearrangement of the Cu(II) complexes, as observed for the previously reported simpler catenane [65]. In fact, the rate is comparable to the CV timescale and three Cu species are detected when a CV of a MeCN solution of $2_{(6)} \cdot (BF_4)_2$ is performed. The waves at $+0.63$ V and -0.41 V correspond, respectively, to the tetra- and hexacoordinate complexes mentioned above. By analogy with the value found for the previously reported copper-complexed catenane [65], the wave at -0.05 V is assigned to the pentacoordinate couple (Figure 4.4b).

4.3
Fighting the Kinetic Inertness of the First Copper-Based Machines; Fast-Moving Pirouetting Rotaxanes

The rate of the motion is an important factor, which has been determined in a limited number of examples. Depending on the nature of the movement, it can range from microseconds, as in the case of organic rotaxanes acting as light-driven molecular shuttles [67], to seconds or even minutes in other systems involving threading–dethreading reactions [68, 69] or transition-metal-centred redox processes [66]. Metal hopping between two distinct sites can be relatively fast (hundreds of milliseconds) when the motion is triggered by a pH change but it seems to be much slower when it involves ligand exchange following a redox process. The main weak point of our molecular machines based on the Cu(II)/Cu(I) couple is certainly the long response time of the system. A first rotaxane whose ring can pirouette between two positions around an axle was published 1999 by Raehm et al. [70] (Figure 4.5; $3_{(n)}^+$).

The rotaxane $3_{(n)}^+$ has two stable states: the four-coordinate copper(I) complex $3_{(4)}^+$ and the five-coordinate copper(II) species $3_{(5)}^{2+}$ – again, the subscripts refer to the CN of the metal centre. The interconversion between these two states is performed electrochemically. Each interconversion process involves two steps: (i) an electron-transfer step (oxidation of Cu(I) or reduction of Cu(II) either chemically or, better, electrochemically) and (ii) a rearrangement reaction corresponding to the pirouetting of the ring around the axle from one position to the other. The rate of the motion depends strongly on the oxidation state of the copper centre:

$$3_{(4)}^+ \xrightarrow{-e^-} 3_{(4)}^{2+} \xrightarrow{k_{4/5}=8\times10^{-3}\,\text{s}^{-1}} 3_{(5)}^{2+} \tag{4.1}$$

$$3_{(5)}^{2+} \xrightarrow{-e^-} 3_{(5)}^+ \xrightarrow{k_{4/5}=2\times10^{2}\,\text{s}^{-1}} 3_{(4)}^+ \tag{4.2}$$

The five-coordinate copper (I) complex rearranges relatively fast (~50 ms) but it takes about 2 min for the four-coordinate divalent copper complex to reach its stable form. These rates are obviously far too low if one wants to elaborate practical systems (switches or mechanical devices) based on rotaxanes containing the same fragments as 3. In order to improve the rate of the motions, we reasoned that lowering steric hindrance and thus making the metal centre as accessible as possible should certainly be very favourable since ligand exchange within the coordination sphere of the copper centre must be facilitated as much as possible. It is very likely that the rate-limiting step of each motion ($3_{(4)}^{2+} \rightarrow 3_{(5)}^{2+}$ and $3_{(5)}^+ \rightarrow 3_{(4)}^+$) involves decoordination of the metal centre. To verify this hypothesis, a new bistable rotaxane $4_{(n)}^+$ was prepared [71]. Its two forms, $4_{(4)}^+$ and $4_{(5)}^{2+}$, are depicted in Figure 4.5. The molecular axis contains a 2,2′-bipyridine motif, which is at the same time thinner and less rigid than a phen fragment and thus is expected to spin more readily within the cavity of the ring. In addition, and probably more importantly, the bipy chelate does not bear substituents in the α-position to the nitrogen atoms in contrast with the corresponding phen fragment of $3_{(n)}^+$.

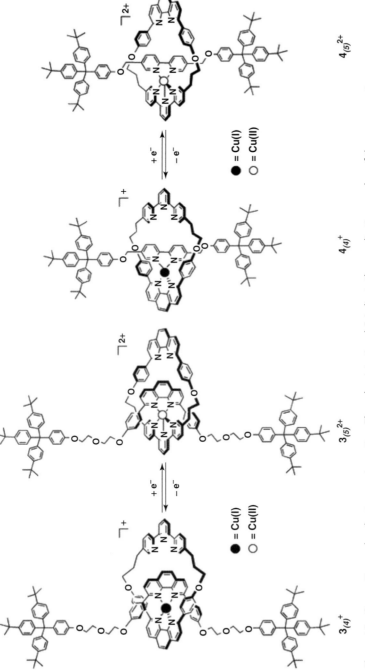

Figure 4.5 Copper(I)-complexed rotaxanes in motion. The subscripts 4 and 5 indicate the coordination number of the copper centre.

As shown in Figure 4.5, it is expected that $4_{(4)}^+$ rearranges to the five-coordinate species $4_{(5)}^{2+}$ after oxidation and *vice versa*. The electrochemically driven motions were studied by CV, which, as usual, turned out to be the technique of choice to set the molecule in motion, to monitor the movements and to measure their rates.

The cyclic voltammograms of $4_{(4)}^+$ were recorded at two different scan rates, for example at 100 mV s^{-1} and at 3000 mV s^{-1}. By starting the CV at a potential value of −0.4 V (*vs.* a silver quasireference electrode), no current is observed since $4_{(4)}^+$ is electrochemically inactive below the oxidation potential at which Cu(I) starts to be oxidized. On increasing the potential towards positive values, an oxidation peak at +0.45 V (at 100 mV s^{-1}) is observed, as expected for the $4_{(4)}^+ \rightarrow 4_{(4)}^{2+}$ redox process. By comparison with the potential values found for related Cu(II)/Cu(I) couples with dpp-based ligands [72], a significant cathodic shift for the $4_{(4)}^+ \rightarrow 4_{(4)}^{2+}$ process is observed.

After the peak potential, Cu(II) is obtained and the current intensity decreases. By inverting the scan potential, starting from 0.9 V to the cathodic region, it is expected that the Cu(II) species be reduced to Cu(I). If the Cu(II) complex is still four-coordinate, the return wave should be observed around 0.4 V, corresponding to the $4_{(4)}^{2+} \rightarrow 4_{(4)}^+$ process. By contrast, if the pirouetting motion is faster than the potential sweep, the return wave corresponding to the reduction of the four-coordinate Cu(II) complex $4_{(4)}^{2+}$ is no longer observed. Instead, it is replaced by another wave corresponding to the reduction of the rearranged complex $4_{(5)}^{2+}$ at a slightly negative potential ($E_p = -0.04$ V). A second scan between −0.4 and +0.9 V will allow estimation of the rate of the five-coordinate copper(I) complex rearrangement: the reoxidation wave expected after reduction of $4_{(5)}^{2+}$ should be observed around 0 V if the pirouetting process is slow but at a substantially higher potential, corresponding to $4_{(4)}^+ \rightarrow 4_{(4)}^{2+}$, if this process is fast. In this case, as well as in other studies at higher scan rates, $4_{(5)}^+$ is never observed. This is a clear demonstration that the five-coordinate copper(I) complex rearranges rapidly.

A lower limit for the rate constant of the process can be estimated using the procedure reported by Nicholson and Shain [73]:

$$4_{(5)}^+ \xrightarrow{k > 500 \text{ s}^{-1}} 4_{(4)}^+$$

Using this k value it can be calculated that $t < 2$ ms ($\tau = k^{-1}$).

By applying the same treatment on the wave observed around 0.5 V, an estimate of the rearrangement rate for the slower four-coordinate Cu(II) complex is obtained:

$$4_{(4)}^{2+} \xrightarrow{k = 5 \text{ s}^{-1}} 4_{(5)}^{2+}$$

The measured k value of 5 s^{-1} (corresponding to t = 200 ms for $4_{(4)}^{2+}$) for the $4_{(4)}^{2+} \rightarrow 4_{(5)}^{2+}$ process shows that $4_{(n)}^+$ is nearly 3 orders of magnitude faster to rearrange than its sterically hindered parent compound $3_{(n)}^+$. These results also confirm that Cu(I) complexes are substitutionally much more labile than Cu(II) species. Clearly, the use of a nonsterically hindering chelate in the rotaxane axis allows fast motion. Subtle structural factors can have a very significant influence

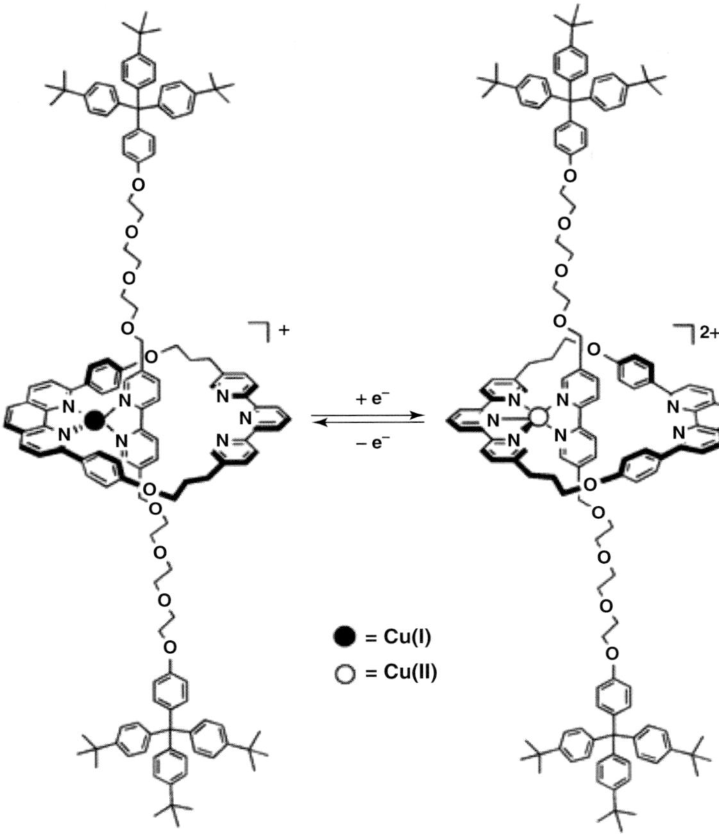

Figure 4.6 The two stable states of the pirouetting rotaxanes $5_{(n)}{}^+$.

on the general behaviour (rate and reversibility, in particular) of artificial molecular machines.

In order to make the copper centre as easy as possible to reach for entering ligands, the bulky stoppers should be located far away from the central complex. Therefore, the bistable rotaxane $5_{(n)}{}^+$ has been prepared and studied [74]; its general structure is similar to that of the previously discussed compound, but its stoppers are now very remote from the copper centre (see Figure 4.6).

Experimentally, an oxidation peak can be observed at a potential of +460 mV, corresponding to the expected oxidation of Cu(I) to Cu(II) in agreement with the closely related copper-complexed rotaxanes $4_{(n)}{}^+$, whereas on the reverse scan, no corresponding reduction peak can be observed. However, another reduction peak corresponding to the reduction of the pentacoordinated species $5_{(5)}{}^{2+} \rightarrow 5_{(5)}{}^+$ is observed at −140 mV. These results clearly show that, upon oxidation or reduction of the metal centre, the macrocycle is set in motion.

$$5_{(4)}^+ \underset{k > 1.2\times10^3\,\text{s}^{-1}}{\overset{k=12\,\text{s}^{-1}}{\rightleftarrows}} 5_{(5)}^{2+}$$

The reaction rate measured for the rearrangement $5_{(4)}^{2+} \to 5_{(5)}^{2+}$ gave a constant $k = 12\,\text{s}^{-1}$ at $-40\,^\circ\text{C}$ in acetonitrile, and the corresponding half-lifetime was calculated to be $t_{1/2} = 60$ ms. In agreement with previous studies on related compounds, this step is about 2–3 orders of magnitude slower than the rearrangement $5_{(5)}^+ \to 5_{(4)}^+$, and the half-lifetime of $5_{(5)}^+$ was estimated to be in the range of 100 μs.

Compared to the previously reported rotaxane $4_{(n)}^+$, the motion of the macrocycle around the axle in $5_{(n)}^+$ is 2 orders of magnitude faster.

4.4
Molecular Motions Driven by Chemical Reactions – Use of a Chemical Reaction to Induce the Contraction/Stretching Process of a Muscle-Like Rotaxane Dimer

In order to make a roughly rod-shaped compound whose overall length can be controlled and modified at will, a system whose topology is that of a rotaxane dimer (see Figure 4.7) was envisaged [12].

The synthesis of this rotaxane dimer is a challenge in itself. The chosen strategy to obtain the target molecule is the synthesis of a ring-and-string conjugate, expected to undergo the desired double-threading reaction under certain circumstances, followed by the attachment of additional chemical groups, including stoppers, at the two ends of the threaded dimer.

After substantial synthetic work, the conjugate 7 of Figure 4.8 was obtained and tested in the gathering and threading process. This ring-and-string conjugate incorporates a bidentate chelate (phen) in the macrocyclic unit and another analogous coordinating unit in the small filament attached to the ring. In view of the potential complexity and variety of complexation reactions that could be envisaged by mixing copper(I) and ligand 7 of Figure 4.7, it was not certain that the doubly threaded topology of Figure 4.8 would be obtained. However, as

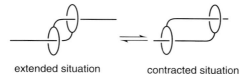

extended situation   contracted situation

Figure 4.7 The stretching/contraction motion of the rotaxane dimer is induced by gliding filaments along one another instead of using mechanical strain as in springs. This functioning principle is reminiscent of biological muscles, for which thick filaments (mostly myosin) glide along thin filaments (actin).

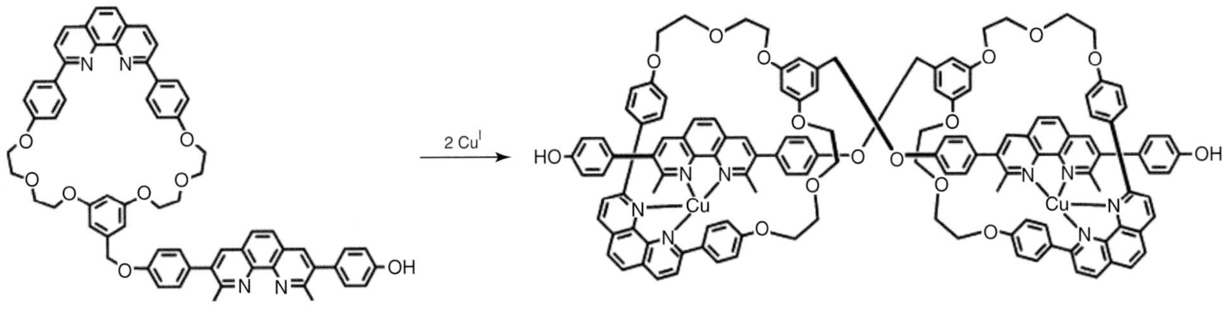

7

8^{2+}

Figure 4.8 Copper(I)-directed formation of the rotaxane dimer **8**$^{2+}$, precursor to the muscle.

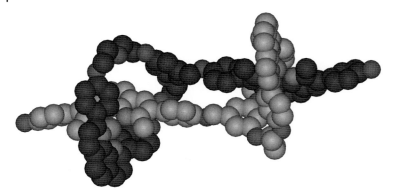

Figure 4.9 Crystallographic view of the rotaxane dimer 8^{2+}.

represented in Figure 4.8 the desired hermaphrodite-like complex was formed quantitatively [75].

This complexation reaction represents an interesting assembly process in itself. Immediately after mixing the two components **7** and Cu(I) in the stoichiometric proportion, a complex mixture of products is obtained that probably consists of threaded and nonthreaded complexes with various nuclearities. After a few days at room temperature, the system finds its way to the thermodynamically most stable situation by a series of decoordination/recoordination reactions, so as to afford compound 8^{2+}. The quantitative formation of 8^{2+} is certainly driven by translational entropy. Since the formation of a tail-biting structure consisting of a mononuclear copper(I) complex of **7** is impossible, the dimer 8^{2+} is the smallest imaginable assembly and it is thus highly favoured.

An X-ray structure of the complex was obtained [75] and is shown in Figure 4.9. Interestingly, the length of this incomplete 'muscle' is already respectable, being ∼3.6 nm from one end to the other. The Cu–Cu intramolecular distance is also large (1.8 nm), precluding any electronic interaction between the two metal centres. In order to make a muscle-like compound and thus to be able to modify the length of the molecule in a controlled fashion, additional functions have to be added. As easily visualized on the structure of Figure 4.9, if the distance between the two copper centres is increased, the effect on the overall length will be opposite: the molecule will be shortened by moving the two metals away from one another.

An attractive way to induce lengthening of the metal–metal distance is to attach other coordinating units at both ends of compound 8^{2+} and, subsequently, send an external signal to the compound, which will trigger ligand exchange so as to force the newly attached ligands to replace the phen units in the metal-coordination sphere. terpy was selected because it is a *tri*dentate ligand, expected to form five-coordinate complexes when used in conjunction with the *bi*dentate phen ligand inscribed in the ring. The principle of the motion is explained in the cartoon of Figure 4.10. Two triggering signals can be envisioned to set the molecule in motion:

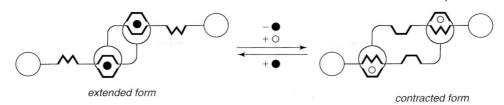

extended form   *contracted form*

Figure 4.10 Contraction and stretching of the muscle-like doubly threaded species obtained by anchoring a terpy unit and a stopper at both ends of **8^{2+}**.

- An electrochemical signal, converting copper(I) to copper(II) and thus favouring five-coordinate situations [Cu(II)] over tetrahedral situations [Cu(I)]. The reversible nature of metal-localized redox processes makes electrochemical signals particularly appealing.
- A chemical reaction, leading to reversible metal exchange allowing to convert the four-coordinate situation to the five-coordinate binding mode and *vice versa*.

The first approach did not prove successful since the four-coordinate copper(II) complex, although formed very readily by electrochemical oxidation of the monovalent copper(I) complex, was kinetically too stable and did not lead to the thermodynamically more stable five-coordinate species. Fortunately, metal exchange takes place easily and quantitatively at room temperature, allowing to interconvert both forms, the four- and the five-coordinate species. Copper(I) is expelled from its coordination sites by CN^- and addition of Zn^{2+} leads instantaneously to the five-coordinate complex **9-$2Zn^{4+}$**. In order to regenerate the four-coordinate species **9-$2Cu^{2+}$**, excess copper(I) is added to the bis-zinc complex, the metal-exchange reaction being again very fast. The two forms of the complete rotaxane dimer are depicted in Figure 4.11 [76].

Clearly, the four-coordinate situation corresponds to the 'stretched' geometry, with a Cu–Cu distance of 1.8 nm (from the X-ray structure of 8^{2+}; see Figure 4.9) whereas the pentacoordinated species is significantly contracted compared to the bis-copper(I) complex. Paradoxically, the Zn–Zn distance is larger than the Cu–Cu distance of the stretched form. It can be estimated on Corey Pauling Koltun (CPK) models as ∼4 nm.

However, a chemical signal does not seem to be the best means to set molecular systems in motion, although most of the biological motors are chemically driven (ATP hydrolysis). Electrochemical or, better, photochemical signals are certainly more promising in terms of potential applications. It is reasonable to assume that, in the future, systems derived from that of Figure 4.11 will afford light-driven muscle-like machines or electrochemically addressable molecules. Interestingly, reactive functional groups will replace the chemically inert stoppers, allowing attachment of the 'muscle' to a large variety of substrates including molecular species (chromophores, biological systems, tags, etc.), organic beads, inorganic and metal surfaces (electrodes). It will also be of particular interest to incorporate stretchable/contractible molecules in polymers and thus to fabricate real muscle-like fibres.

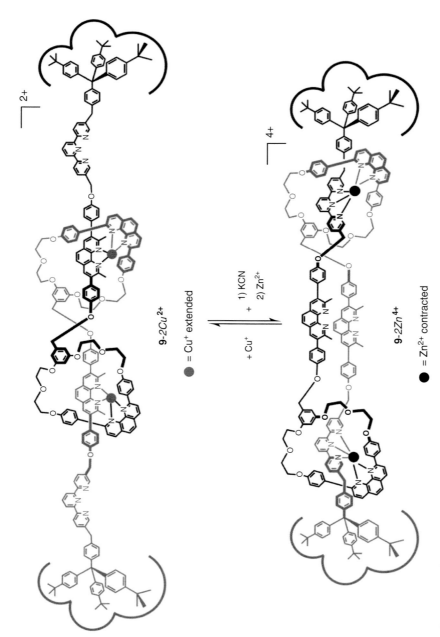

Figure 4.11 The two states of the muscle. The bis-zinc species is approximately 2 nm shorter than the bis-copper(I) complex, these lengths being estimated on CPK models.

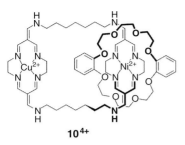

Figure 4.12 Heterodinuclear [2]catenane **10⁴⁺**.

4.5
Electrochemically Controlled Intramolecular Motion within a Heterodinuclear Bismacrocycle Transition-Metal Complex

Wozniak and coworkers [77] recently described the first catenane incorporating a heterodinuclear bismacrocyclic transition-metal complex **10⁴⁺** (Figure 4.12) that exhibits potential-driven intramolecular motion of the interlocked crown-ether unit. Although the system contains transition metals, the main interaction between the various subunits, which also allowed the construction of the catenane **10⁴⁺**, is an acceptor–donor interaction of the charge-transfer type.

The reported heterodinuclear catenane should allow a controlled translocation of the crown-ether unit back and forth between two different metal centres in response to an external stimulus, specifically a potential applied to the electrode (Figure 4.13).

The present system can be set in motion using two consecutive redox signals. The main feature of the machine-like catenane is that the preferred conformation will be such that the most electrodeficient transition metal macrocyclic complex will lie in between the two aromatic donor fragments of the crown ether.

The bis-macrocyclic ring is positively charged because of the presence of Ni(II) and Cu(II). The crown ether and the bis-azamacrocyclic ring form a sandwich-like structure in such a way that one of the crown-ether aromatic rings is located between the two metal-coordinated macrocyclic rings. The second

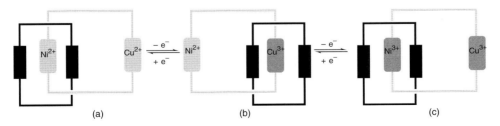

Figure 4.13 (a–c) Schematic representation of electrochemically controlled molecular motion.

aromatic ring is located almost parallel to the previous one outside the two linked macrocycles.

As nickel(II) is a better acceptor than copper(II), the situation at the beginning of the process is the one depicted in Figure 4.13a. Then, upon oxidation of the molecule, the copper(II) centre turns into copper(III) since oxidation of the nickel(II) centre is more difficult. But Cu(III) being of course a better acceptor than Ni(II), the crown ether relocates from the nickel(II) centre to the copper(III) centre (Figure 4.13b). By increasing the potential, the nickel(II) centre is finally oxidized, and the new nickel(III) centre is, as expected, a stronger acceptor than the copper(III) centre. Hence, the crown-ether ring moves for the second time, yielding the third situation (Figure 4.13c).

4.6
Ru(II)-Complexes as Light-Driven Molecular Machine Prototypes

Among the light-driven molecular machine prototypes that have been described in the course of the last few years, a very distinct family of dynamic molecular systems takes advantage of the dissociative character of ligand-field states in Ru (diimine)$_3^{2+}$ complex [78–84]. In these compounds, one part of the system is set in motion by photochemically expelling a given chelate, the reverse motion being performed simply by heating the product of the photochemical reaction so as to regenerate the original state. In these systems, the light-driven motions are based on the formation of dissociative excited states. Complexes of the [Ru(diimine)$_3$]$^{2+}$ family are particularly well adapted to this approach. If distortion of the coordination octahedron is sufficient to significantly decrease the LF, which can be realized by using one or several sterically hindering ligands, the strongly dissociative LF state (^{3}d–d state) can be efficiently populated from the ^{3}MLCT state to result in expulsion of a given ligand. The principle of the whole process is represented in Figure 4.14.

It is thus essential that the ruthenium(II) complexes, which are to be used as building blocks of future machines, contain sterically hindering chelates so as to force the coordination sphere of the metal to be distorted from the perfect octahedral geometry. As a consequence, the dissociative ^{3}d–d state will be easily accessible from the ^{3}MCLT state. We will discuss the synthesis of a catenane of this family and describe the photochemical reactivity of these molecules [17, 85]. The complexes made and studied incorporate encumbering ligands, which will indeed facilitate the light-induced motions.

Complex **12**$^{2+}$ was formed by reacting **11** and Ru(DMSO)$_4$Cl$_2$, (DMSO: dimethylsulfoxide) followed by refluxing the dichloro-intermediate complex in CH$_3$CN and H$_2$O. **12**.(PF$_6$)$_2$ was obtained as an orange *solid* in 46% yield after anion exchange. **12**$^{2+}$ is a rare example of a bis-phen, or, more generally, a bis-bidentate octahedral complex with a cis-arrangement inscribed in a ring. The next step was carried out using **13** and the macrocyclic complex **12**$^{2+}$. Threading of the 'filament' **10** does take place under relatively harsh conditions (ethylene glycol, 140 °C)

4.6 Ru(II)-Complexes as Light-Driven Molecular Machine Prototypes | 113

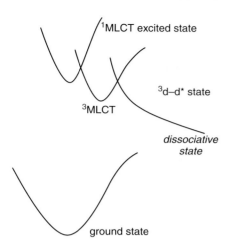

Figure 4.14 The ligand-field state d–d* can be populated from the MLCT state, provided the energy difference between these two states is not too large: formation of this dissociative state leads to dissociation of a ligand.

and the catenane precursor **14²⁺** was obtained in good yield (56%). The final compound, catenane **15²⁺**, was prepared from **14²⁺**, in a 68% yield by ring-closing metathesis (RCM). **15·(PF₆)₂** is a red-orange solid that has been fully characterized by various spectroscopic techniques. The electrospray mass spectroscopy (ES-MS) and ^{1}H NMR measurements provide clear evidence for the structure of **15²⁺**.

The [2]catenane **15²⁺** was synthesized as described in the previous paragraph. The other [2]catenane of Figure 4.15, **16²⁺** was prepared using a slightly different procedure [85]. Compound **15²⁺** consists of a 50-membered ring that incorporates two phen units and a 42-membered ring that contains the bipy chelate. Compound **16²⁺** contains the same bipy-incorporating ring as **15²⁺**, but the other ring is a 63-membered ring. Clearly, from CPK model considerations, **16²⁺** is more adapted than **15²⁺** to molecular motions in which both constitutive rings would move with respect to one another, since the situation is relatively tight for the catenane.

The light-induced motion and the thermal back reaction carried out with **15²⁺** or **16²⁺** are represented in Figure 4.16. They are both quantitative, as shown by UV/Vis measurements and by ^{1}H NMR spectroscopy. The photoproducts, [2]catenanes **15'** and **16'**, contain two disconnected rings since the photochemical reaction leads to decomplexation of the bipy chelate from the ruthenium(II) centre. In a typical reaction, a degassed CH_2Cl_2 solution of **16²⁺** and $NEt_4^+Cl^-$ was irradiated with visible light, at room temperature. The colour of the solution rapidly changed from red (**16²⁺** : λ_{max} = 458 nm) to purple (**16'** : λ_{max} = 561 nm) and after a few minutes the reaction was complete. The recoordination reaction **16'** → **16²⁺** was carried out by heating a solution of **16'**. The quantum yield for the photochemical reaction **16²⁺** → **16'** at 25 °C and λ = 470 nm (±50 nm) can be very roughly estimated as 0.014 ± 0.005. One of the weak points of the present system is certainly the limited control over the shape of the photoproduct,

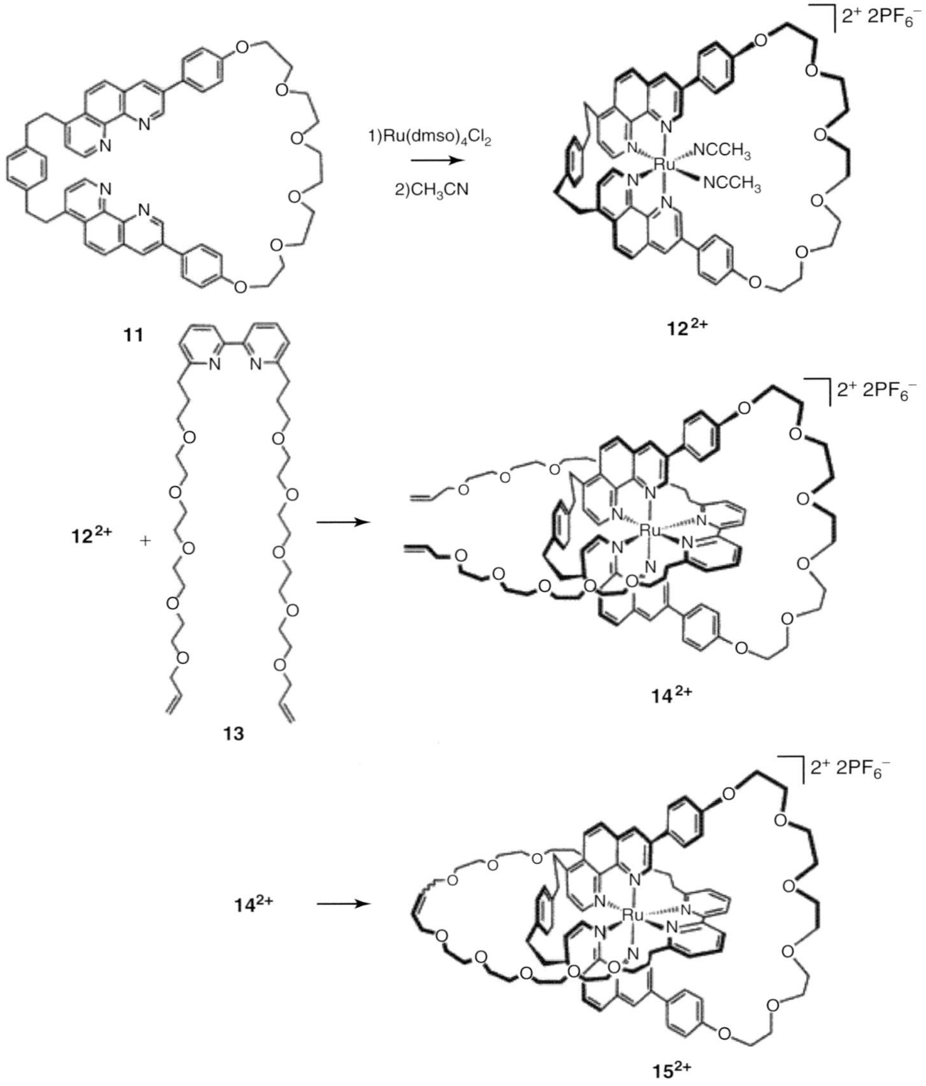

Figure 4.15 Sequence of reactions affording the ruthenium(II)-complexed [2]catenane **15²⁺**.

since the decoordinated ring can occupy several positions. It is hoped that in the future, an additional tunable interaction between the two rings of the present catenanes, **15'** or **16'**, will allow better control over the geometry of the whole system. In parallel, two-colour machines could be elaborated, for which both motions will be driven by photonic signals operating at different wavelengths.

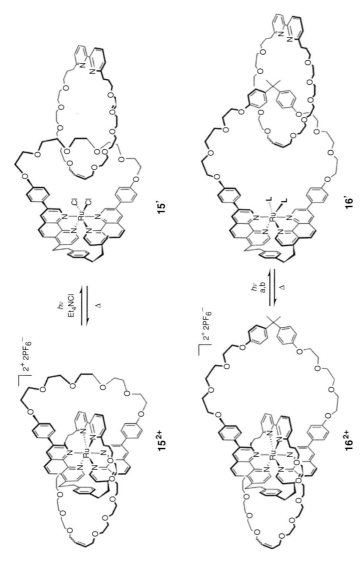

Figure 4.16 Catenanes **15²⁺** or **16²⁺** undergo a complete rearrangement by visible light irradiation: the bipy-containing ring is efficiently decoordinated in the presence of Cl⁻. By heating the photoproducts **15'** or **16'**, the starting complexes **15²⁺** or **16²⁺** are quantitatively regenerated.

4.7
Conclusion and Prospective

In this chapter, a few examples of transition-metal-containing molecular machines of the catenane and rotaxane family have been discussed. The input is often electrochemical, involving the transition metal rather than the organic ligands. The systems can be oxidized or reduced: Cu(II)/Cu(I) or Ni(III)/Ni(II) in the examples discussed here. Alternatively, a photonic stimulus can be used to set the system in motion. Here again, the role of the metal centre is essential since the excited state responsible for the absorption of light is an MLCT excited state. This state is rapidly converted to a dissociative LF state, which is responsible for the first step of the movement. The third type of stimulus is chemical. A metal-exchange reaction allows to profoundly modify the shape of the compound (molecular 'muscle').

From a purely scientific viewpoint, the field of molecular machines is particularly challenging and motivating: the fabrication of dynamic molecular systems, with precisely designed dynamic properties, is still in its infancy and will certainly experience a rapid development during the next decades. Such a research area is highly multidisciplinary and requires a high level of competence in synthetic chemistry as well as in physical and materials sciences, which makes it an excellent school for young and ambitious scientists. It is indeed very challenging to reproduce some of the simplest functions of the natural biological motors (motor proteins, DNA polymerase, bacterial flagella, etc.) using synthetic molecular systems. It must nevertheless be kept in mind that the presently accessible molecular machines and motors are extremely primitive compared to the beautiful and exceedingly complex molecular machines of nature.

It is still not certain whether the field will lead to applications in a short-term prospective, although spectacular results have been obtained in the course of the last few years in relation to information storage and processing at the molecular level [86] or to the elaboration of macroscopic devices based on molecular nanomachines [16]. Other ambitious and futuristic practical outcomes could be considered, such as the fabrication of 'microrobots' or even 'nanorobots' able to perform various functions: transport molecules or ions through a membrane, sort different molecules, store energy, just to cite a few.

References

1. Kay, E.R., Leigh, D.A. and Zerbetto, F. (2007) *Angew. Chem. Int. Ed.*, **46**, 72.
2. Balzani, V., Venturi, M. and Credi, A. (2003) *Molecular Devices and Machines – A Journey into the Nanoworld*, Wiley-VCH Verlag GmbH, Weinheim.
3. Raehm, L. and Sauvage, J.-P. (2001) *Struct. Bond. (Berlin, Germany)*, **99**, 55.
4. Balzani, V., Credi, A., Raymo, F. and Stoddart, J.F. (2000) *Angew. Chem. Int. Ed.*, **39**, 3348.
5. Balzani, V., Credi, A. and Venturi, M. (2003) *Pure. Appl. Chem.*, **75**, 541.
6. Ballardini, R., Balzani, V., Credi, A., Gandolfi, M.T. and Venturi, M. (2001) *Acc. Chem. Res*, **34**, 445.
7. Feringa, B.L. (2001) *Molecular Switches*, Wiley-VCH Verlag GmbH, Weinheim, Germany.

8. Fabbrizzi, L., Licchelli, M. and Pallavicini, P. (1999) *Acc. Chem. Res.*, **32**, 846.
9. Kelly, T.R., De Silva, H. and Silva, R.A. (1999) *Nature*, **401**, 150.
10. Koumura, N., Zijistra, R.W.J., Van Delden, R.A., Harada, N. and Feringa, B.L. (1999) *Nature*, **401**, 152.
11. Leigh, D.A., Wong, J.K.Y., Dehez, F. and Zerbetto, F. (2003) *Nature*, **424**, 174.
12. Jimenez-Molero, M.C., Dietrich-Buchecker, C. and Sauvage, J.-P. (2003) *Chem. Commun.*, 1613.
13. Katz, E., Lioubashevsky, O. and Willner, I. (2004) *J. Am. Chem. Soc.*, **126**, 15520.
14. Shipway, A.N. and Willner, I. (2001) *Acc. Chem. Res.*, **34**, 421.
15. Harada, A. (2001) *Acc. Chem. Res.*, **34**, 456.
16. Eelkema, R., Pollard, M.M., Vicario, J., Katsonis, N., Ramon, B.S., Bastiaansen, C.W.M., Broer, D.J. and Feringa, B.L. (2006) *Nature*, **440**, 163.
17. Bonnet, S., Collin, J.-P., Koizumi, M., Mobian, P. and Sauvage, J.P. (2006) *Adv. Mater.*, **18**, 1239.
18. van den Heuvel, M.G.L. and Dekker, C. (2007) *Science*, **317**, 333.
19. Kelly, T.R., Cai, X., Damkaci, F., Panicker, S.B., Tu, B., Bushell, S.M., Cornella, I., Piggott, M.J., Salives, R., Cavero, M., Zhao, Y. and Jasmin, S. (2007) *J. Am. Chem. Soc.*, **129**, 376.
20. Saha, S. and Stoddart, J.F. (2007) *Chem. Soc. Rev.*, **36**, 77.
21. Liu, Y., Flood, A.H., Bonvallet, P.A., Vignon, S.A., Northrop, B.H., Tseng, H.-R., Jeppesen, J.O., Huang, T.J., Brough, B., Baller, M., Magonov, S., Solares, S.D., Goddard, W.A., Ho, C.-M. and Stoddart, J.F. (2005) *J. Am. Chem. Soc.*, **127**, 9745.
22. Berna, J., Leigh, D.A., Lubowska, M., Mendoza, S.M., Pérez, E.M., Rudolf, P., Teobaldi, G. and Zerbetto, F. (2005) *Nature Mater.*, **4**, 704.
23. Schill, G. (1971) *Catenanes, Rotaxanes and Knots*, Academic Press, New York.
24. Dietrich-Buchecker, C. and Sauvage, J.-P. (1999) *Catenanes, Rotaxanes and Knots. A Journey Through the World of Molecular Topology*, Wiley-VCH Verlag GmbH, Weinheim.
25. Fujita, M. (1999) *Acc. Chem. Res.*, **32**, 53.
26. Voegtle, F., Duennwald, T. and Schmidt, T. (1996) *Acc. Chem. Res.*, **29**, 451.
27. Hoshino, T., Miyauchi, M., Kawaguchi, Y., Yamaguchi, H. and Harada, A. (2000) *J. Am. Chem. Soc.*, **122**, 9876.
28. Kawaguchi, Y. and Harada, A. (2000) *Org. Lett.*, **2**, 1353.
29. Bogdan, A., Vysotsky, M.O., Ikai, T., Okamoto, Y. and Boehmer, V. (2004) *Chem. Eur. J.*, **10**, 3324.
30. Stanier, C.A., O'Connell, M.J., Anderson, H.L. and Clegg, W. (2001) *Chem. Commun.*, 493.
31. Stanier, C.A., Alderman, S.J., Claridge, T.D.W. and Anderson, H.L. (2002) *Angew. Chem. Int. Ed.*, **40**, 1769.
32. Wisner, J.A., Beer, P.D., Drew, M.G.B. and Sambrook, M.R. (2002) *J. Am. Chem. Soc.*, **124**, 12469.
33. Chichak, K.S., Walsh, C. and Branda, N.R. (2000) *Chem. Commun.*, 847.
34. Breault, G.A., Hunter, C.A. and Mayers, P.C. (1999) *Tetrahedron*, **55**, 5265.
35. Durr, H. and Bossmann, S. (2001) *Acc. Chem. Res.*, **34**, 905.
36. Iwamoto, H., Itoh, K., Nagamiya, H. and Fukazawa, Y. (2003) *Tetrahedron Lett.*, **44**, 5773.
37. Fitzmaurice, D., Rao, S.N., Preece, J.A., Stoddart, J.F., Wenger, S. and Zaccheroni, N. (1999) *Angew. Chem. Int. Ed.*, **38**, 1147.
38. Gibson, H.W., Yamaguchi, N., Hamilton, L. and Jones, J.W. (2002) *J. Am. Chem. Soc.*, **124**, 4653.
39. Hodge, P., Monvisade, P., Owen, G.J., Heatley, F. and Pang, Y. (2000) *New J. Chem.*, **24**, 703.
40. Chang, S.-Y., Choi, J.S. and Jeong, K.-S. (2001) *Chem. Eur. J.*, **7**, 2687.
41. Fujimoto, T., Nakamura, A., Inoue, Y., Sakata, Y. and Kaneda, T. (2001) *Tetrahedron Lett.*, **42**, 7987.
42. Kim, K. (2002) *Chem. Soc. Rev.*, **31**, 96.
43. Park, K.-M., Kim, S.-Y., Heo, J., Whang, D., Sakamoto, S., Yamaguchi, K. and Kim, K. (2002) *J. Am. Chem. Soc.*, **124**, 2140.
44. Korybut-Daszkiewicz, B., Wieckowska, A., Bilewicz, R., Domaga-la, S. and Wozniak, K. (2001) *J. Am. Chem. Soc.*, **123**, 9356.

45. Chen, L., Zhao, X., Chen, Y., Zhao, C.-X., Jiang, X.-K. and Li, Z.-T. (2003) *J. Org. Chem.*, **68**, 2704.
46. Davidson, G.J.E. and Loeb, S.J. (2003) *Angew. Chem. Int. Ed.*, **42**, 74.
47. Coumans, R.G.E., Elemans, J.A.A.W., Thordarson, P., Nolte, R.J.M. and Rowan, A.E. (2003) *Angew. Chem. Int. Ed.*, **42**, 650.
48. Arduini, A., Ferdani, R., Pochini, A., Secchi, A. and Ugozzoli, F. (2000) *Angew. Chem. Int. Ed.*, **39**, 3453.
49. McArdle, C.P., Vittal, J.J. and Puddephatt, R.J. (2000) *Angew. Chem. Int. Ed.*, **39**, 3819.
50. Udachin, K.A., Wilson, L.D. and Ripmeester, A. (2000) *J. Am. Chem. Soc.*, **122**, 12375.
51. Gunter, M.J., Bampos, N., Johnstone, K.D. and Sanders, J.K.M. (2001) *New J. Chem.*, **25**, 166.
52. Ghosh, P., Mermagen, O. and Schalley, C.A. (2002) *Chem. Commun.*, 2628.
53. Andrievsky, A., Ahuis, F., Sessler, J.L., Voegtle, F., Gudat, D. and Moini, M. (1998) *J. Am. Chem. Soc.*, **120**, 9712.
54. Belaissaoui, A., Shimada, S., Ohishi, A. and Tamaoki, N. (2003) *Tetrahedron Lett.*, **44**, 2307.
55. Shukla, R., Deetz, M.J. and Smith, B.D. (2000) *Chem. Commun.*, 2397.
56. Smukste, I. and Smithrud, D.B. (2003) *J. Org. Chem.*, **68**, 2547.
57. Simone, D.L. and Swager, T.M. (2000) *J. Am. Chem. Soc.*, **122**, 9300.
58. MacLachlan, M.J., Rose, A. and Swager, T.M. (2001) *J. Am. Chem. Soc.*, **123**, 9180.
59. Watanabe, N., Yagi, T., Kihara, N. and Takata, T. (2002) *Chem. Commun.*, 2720.
60. Watanabe, N., Kihara, N., Furusho, Y., Takata, T., Araki, Y. and Ito, O. (2003) *Angew. Chem. Int. Ed.*, **42**, 681.
61. Willner, I., Pardo-Yissar, V., Katz, E. and Ranjit, K.T. (2001) *J. Electroanal. Chem.*, **497**, 172.
62. Dietrich-Buchecker, C. and Sauvage, J.-P. (1987) *Chem. Rev.*, **87**, 795.
63. Irie, M., Miyatake, O. and Uchida, K. (1992) *J. Am. Chem. Soc.*, **114**, 8715.
64. Cardenas, D.J., Livoreil, A. and Sauvage, J.-P. (1996) *J. Am. Chem. Soc.*, **118**, 11980.
65. Dietrich-Buchecker, C., Sauvage, J.-P. and Kintzinger, J.-P. (1983) *Tetrahedron Lett.*, **24**, 5095.
66. Livoreil, A., Dietrich-Buchecker, C.O. and Sauvage, J.-P. (1994) *J. Am. Chem. Soc.*, **116**, 9399.
67. Brouwer, A.M., Frochot, C., Gatti, F.G., Leigh, D.A., Mottier, L., Paolucci, F., Roffia, S. and Wurpel, G.W.H. (2001) *Science*, **291**, 2124.
68. Ashton, P.R., Ballardini, R., Balzani, V., Baxter, I., Credi, A., Fyfe, M.C.T., Gandolfi, M.T., Gomez-Lopez, M., Martinez-Diaz, M.V., Piersanti, A., Spencer, N., Stoddart, J.F., Venturi, M., White, A.J.P. and Williams, D.J. (1998) *J. Am. Chem. Soc.*, **120**, 11932.
69. Balzani, V., Credi, A., Mattersteig, G., Matthews, O.A., Raymo, F.M., Stoddart, J.F., Venturi, M., White, A.J.P. and Williams, D.J. (2000) *J. Org. Chem.*, **65**, 1924.
70. Raehm, L., Kern, J.-M. and Sauvage, J.-P. (1999) *Chem. Eur. J.*, **5**, 3310.
71. Poleschak, I., Kern, J.-M. and Sauvage, J.-P. (2004) *Chem. Commun.*, 474.
72. Dietrich-Buchecker, C., Sauvage, J.-P. and Kern, J.-M. (1989) *J. Am. Chem. Soc.*, **111**, 7791.
73. Nicholson, R.S. and Shain, I. (1964) *Anal. Chem.*, **36**, 706.
74. Letinois-Halbes, U., Hanss, D., Beierle, J.M., Collin, J.-P. and Sauvage, J.-P. (2005) *Org. Lett.*, **7**, 5753.
75. Jiménez, M.C., Dietrich-Buchecker, C., Sauvage, J.-P. and De Cian, A. (2000) *Angew. Chem. Int. Ed.*, **39**, 1295.
76. Jimenez, M.C., Dietrich-Buchecker, C. and Sauvage, J.-P. (2000) *Angew. Chem. Int. Ed.*, **39**, 3284.
77. Korybut-Daszkiewicz, B., Wieckowska, A., Bilewicz, R., Domagala, S. and Wozniak, K. (2004) *Angew. Chem. Int. Ed.*, **43**, 1668.
78. Adelt, M., Devenney, M., Meyer, T.J., Thompson, D.W. and Treadway, J.A. (1998) *Inorg. Chem.*, **37**, 2616.
79. Van Houten, J. and Watts, R.J. (1978) *Inorg. Chem.*, **17**, 3381.
80. Suen, H.F., Wilson, S.W., Pomerantz, M. and Walsh, J.L. (1989) *Inorg. Chem.*, **28**, 786.

81. Pinnick, D.V. and Durham, B. (1984) *Inorg. Chem.*, **23**, 1440.
82. Gleria, M., Minto, F., Beggiato, G. and Bortolus, P. (1978) *J. Chem. Soc., Chem. Commun.*, 285.
83. Durham, B., Caspar, J.V., Nagle, J.K. and Meyer, T.J. (1982) *J. Am. Chem. Soc.*, **104**, 4803.
84. Tachiyashiki, S. and Mizumachi, K. (1994) *Coord. Chem. Rev.*, **132**, 113.
85. Mobian, P., Kern, J.-M. and Sauvage, J.-P. (2004) *Angew. Chem. Int. Ed.*, **43**, 2392.
86. Green Jonathan, E., Choi Jang, W., Boukai, A., Bunimovich, Y., Johnston-Halperin, E., DeIonno, E., Luo, Y., Sheriff Bonnie, A., Xu, K., Shin Young, S., Tseng, H.-R., Stoddart, J.F. and Heath James, R. (2007) *Nature*, **445**, 414.

5
Chiroptical Molecular Switches
Wesley R. Browne and Ben L. Feringa

5.1
Introduction

Chirality is a distinct feature of the essential building blocks that make up the materials and molecular systems of life [1]. Chiral phenomena occur at various hierarchical levels ranging from the homochirality of amino acids and sugars to the supramolecular chirality of DNA and the macromolecular chirality of peptides [2–6]. Although the absolute configuration is normally fixed at the level of the individual amino acid or sugar molecules, nature has fully exploited 'dynamic' chiral conformational space, as is most elegantly seen in the chirality of catalytic centres in enzymes and the supramolecular chirality in the DNA double helix and protein α-helices. In fact, numerous key functions in cells are associated with the phenomenon of dynamic chirality including molecular recognition and signal transduction, assembly and organization, catalysis, replication, transport and motion [7]. As precise expression of molecular chirality is intrinsic to all these phenomena, it is evident that the control of chirality through an external trigger signal holds great promise as a powerful tool in the design of functional systems and smart materials [8]. Switching of chirality has been explored in novel approaches towards a broad range of applications including information storage systems [9, 10], responsive materials and liquid-crystalline (LC) devices [11] and molecular motors and machines [12] operating at the nanoscale.

In this chapter, recent advances in the design, functioning and application of chiral optical (*chiroptical*) molecular switches will be discussed [13, 14]. The focus will be on chiroptical switches based on overcrowded alkenes and more advanced light-driven molecular motors derived from these switches, but pertinent examples of other chiroptical systems will be discussed to cover the most important aspects of the field. For earlier examples the reader is referred to our chapter in the previous volume [15]. It should also be noted that the examples discussed here are not exhaustive but serve to illustrate the main principles.

First, some basic principles and concepts of bistable molecular systems and different types of chiroptical switching are discussed. Subsequently, control of molecular chirality and function through various chiral switching elements is

Molecular Switches, Second Edition. Edited by Ben L. Feringa and Wesley R. Browne.
© 2011 Wiley-VCH Verlag GmbH & Co. KGaA. Published 2011 by Wiley-VCH Verlag GmbH & Co. KGaA.

presented. Next, we will describe the evolution of chiroptical molecular switches into unidirectional light-driven molecular motors and elaborate on the various design features and parameters that govern motor behaviour. This is followed by applications of switches and motors in the dynamic control of properties of polymers, liquid crystals and gels. Finally, conclusions and a brief perspective will be given.

5.2
Molecular Switching

Switching between – chiral – molecular states is based on bistability, that is a molecule exists in two distinct forms A and B and each state can be converted to the other state reversibly upon application of an external stimulus (Figure 5.1a). A crucial feature is that switching between the two states must be controlled and not proceed spontaneously (within the time frame of the experiment) [15]. In designing bistable systems it should be realized that any molecular material that has two stable identifiable states can, in principle be used as a switching element to control functions, including chirality.

Photochromic compounds, where the reversible switching process is based on photochemical interconversion, usually with two different wavelengths of light, are the most prominent in this field [17]. Arguably the most elegant and delicate optical molecular switch is the retinal system in the process of vision (Figure 5.1b) [2]. Nature has achieved a remarkable balance in switching time, bistability, nondestructive readout, cooperative responsiveness and reversibility in the use of retinal. Employing this chiroptical switch, comprising a protein-bound retinal molecule, the basic process of visions involves a fast *cis–trans* photoisomerization of the retinal unit that triggers a conformational change in the protein that is readout via an ion cascade initiated as a result of the photoisomerization. Reversibility is achieved by a slower enzymatic *trans* to *cis* isomerization event.

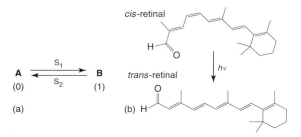

Figure 5.1 (a) The principle of bistability in molecular switching. A and B are two distinct states and S_1 and S_2 are applied stimuli. (b) Photoinduced *cis–trans* isomerization of retinal is the first step in the process of vision [16]. The isomerization results in a change in shape from a bent to a linear structure that affects the rhodopsin protein. This triggers a cascade of events eventually leading to a signal transmitted to the brain.

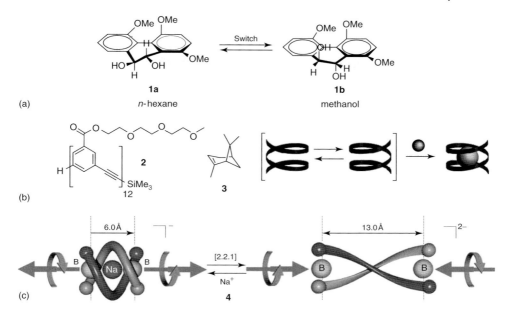

Figure 5.2 Conformational changes in a chiroptical switch induced by changes in (a) solvent, reproduced from Ref. [18], copyright ACS 2007, (b) a chiral guest reproduced from Ref. [19], copyright ACS 2001 and (c) charged guest. Reproduced from Ref. [20]. Copyright Nature 2010.

While light is arguably an attractive noninvasive way to address chiral bistable states, it should be remembered that several other ways to change chirality in responsive molecular systems are known. The responsive effect might be triggered by changes in solvent, pH, molecules or ion binding, stirring, temperature, and so on [9, 15]. An elegant example of a chiral molecular bistable system has recently been reported by Reichert and Breit [18], in which the axial chirality of bisphenol and the dynamic conformational changes in a six-membered ring were employed (Figure 5.2a). The chirality of the system is determined by the relative stabilities of the *pseudo-bis*-equatorial **1a** and *bis*-axial **1b** atropisomers. The perturbation, a solvent change, brings about a reversible change in molecular conformation and the shift in equilibrium is expressed as a change in chirality.

Oligomer and polymer systems, in which achiral monomeric units are connected through covalent bonds, can adopt a helical secondary structure that can be responsive to various modulators through noncovalent interactions. An illustrative early example is seen in the so-called foldamers introduced by Moore, for example amphiphilic oligo(*m*-phenylene ethynylene) **2**, which folds into helical conformations in polar solvents due to solvophobic effects [19]. In the absence of a chiral input, the left- and right-handed helical conformation are in equilibrium (Figure 5.2b). In polar solvents the complexation of chiral apolar monoterpene **3** within the hydrophobic cavity of the foldamer results in strongly preferred handedness of the helical structure as observed by circular dichroism (CD) spectroscopy. Often strong cooperative effects are seen in the expression and modulation of chirality and these principles involve shifting of equilibriums between (a-) chiral states have been widely explored in other polymeric and supramolecular systems [6, 21].

The winding and unwinding of a double-stranded helicate **4** by reversible ion binding as reported by Yashima and coworkers [20] recently, is a beautiful example of the reversible change in chirality induced in a supramolecular complex (Figure 5.2c). The binding and removal of the central sodium ion triggers the contraction and extension of the helix without changing the handedness. This spring-like motion illustrates the link between chiroptical switching and molecular mechanical machines, as will be discussed below.

In these examples the change in equilibrium position and the direction of switching of chirality is controlled by changing the free energy of ground states. In this chapter the focus will be on switchable systems in which external stimuli (in particular light, but also change in redox potential and chemical conversion) are employed not to change the relative ground-state energy of the system but instead to overcome barriers to interconversion that are sufficiently large to preclude the reversible processes occurring thermally under ambient conditions.

This relative rarity of bistability has proven a major hurdle in several of the earlier developments exploiting photochromic materials [17], in particular for applications in molecular-based data-storage systems [10, 15]. Another crucial issue is nondestructive readout, implying that the individual states should be sufficiently distinct so that each state can be 'read' without interfering with the other state of the system. For various photochromic materials used so far, UV/Vis spectroscopy is the most common detection technique [17, 22]. However, this involves sampling at the absorption bands that frequently leads to undesired side effects like partial reversal of the photochromic switching used to store information [23]. Efforts to avoid these problems [24] resulted in light-switchable molecules in which photochromism is accompanied by changes in other properties, such as complexation of ions [25], change in refractive index [26], redox behaviour [27] or conformational changes in polymers [15, 28]. Other ways that are explored in order to avoid destructive readout are based on the modulation of the organization of an ensemble of molecules, for example gels [29], liquid crystals [30] and Langmuir–Blodgett films [31].

Chiroptical molecular switches, where the switching process is based on photochemical interconversion, offer unique opportunities not only to trigger changes in function, chirality and organization but also as potential memory elements in data-storage devices based on binary logic [9, 32]. In particular, the optical and structural features of chiral bistable molecules are well suited for nondestructive readout using a variety of chiroptical techniques, including optical rotary dispersion (ORD), circular dichroism (CD) and circular polarised luminescence (CPL).

To achieve such goals several basic requirements for a chiral optical molecular switch have to be fulfilled including:

- photochemical switching between two chiral forms of the molecule (or two forms of the supramolecular ensemble or macromolecule) should be possible;

- no thermal isomerization over a large temperature range (typically −20 to 100 °C) to avoid loss of information stored by thermal induced reversal of the system (thermochromism);
- the photochemical interconversion should show fast response times;
- high selectivity; especially at the molecular level high stereoselectivity is essential;
- the quantum yield of photochemical interconversion should be high allowing for efficient switching;
- high fatigue resistance allowing numerous switching cycles;
- detection should be sensitive, discriminative and nondestructive;
- all these properties should be retained when the switching unit is incorporated in a larger ensemble, macromolecular structure or nanodevice.

For data storage and optical devices there are of course various additional technical requirements to be fulfilled [33]. One can distinguish several approaches in the design of light-triggered chiral bistable molecules that can operate as switching elements in molecular and nanodevices and photonic materials. In a chiral photochromic system P and M represent two different chiral forms of a bistable molecule (Figure 5.3). Changes in chirality, that is from P to M or from M to P, can be induced by the absorption of light. The nature of the overall change depends on the structure and chirality of the molecule or the response of the entire system embedding the chiroptical switch.

A possible classification of chiroptical molecular switches is given in Figure 5.3 and the following systems can be distinguished:

1) Switching between enantiomers. Two enantiomers have identical absorption spectra and irradiation of a chiroptical switch as a single enantiomer (P or M) using normal, nonpolarized light will result in full racemization (P, M) irrespective of the wavelength used. Interconversion between enantiomers can, however, be achieved by irradiation at distinct wavelengths using left-or

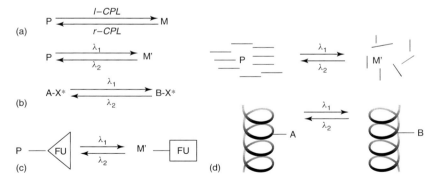

Figure 5.3 Chiral switching between: (a) enantiomers, (b) diastereoisomers (X* = chiral auxiliary), (c) functional chiral switches (FU = functional unit) and (d) macromolecular switch and switching of the organization of the matrix. P and M denote right- and left-handed helical structures, A and B denote two bistable forms of a switchable molecule. l- and r-cpl indicate left and right circularly polarized light, respectively.

right-circularly polarized light (*l*- or *r*-CPL). This allows switching in either direction in an enantioselective manner.

2) Switching of diastereoisomers. If the molecule has a fixed stereocentre (R or S) as well as a photochromic unit, the stereocentre itself is not affected by the photochromic reaction but the overall chirality might change. Typically this can be accomplished by linking a chiral auxiliary X* to a photochromic unit A, which itself can be achiral or chiral. In these A–X*, B–X* systems the change in chirality during the switching process is controlled by the subtle interplay of the auxiliary X* and the photochromic unit. Alternatively, the compound can exist in one of two intrinsically chiral but diastereomeric forms, for instance a P (right-handed) and M' (left-handed) helical structure, which are thermally stable but can be interconverted at two different wavelengths λ_1 and λ_2.

3) Functional chiral switches. All these photochromic systems incorporate at least one other function or frequently they are multifunctional in nature. The induced change in chirality triggers a concomitant change in a particular function resulting in the modulation of, for example fluorescence, molecular recognition or motion.

4) Switching of macromolecular or supramolecular organization. In these systems a photobistable unit is part of a larger supramolecular ensemble or incorporated into a polymer and the chirality change is amplified over different hierarchical levels. Typically, the photoisomerization of the switch unit induces a change in the organization at the macro- or supramolecular level. For instance, this can be achieved by controlling the helical twist sense of the backbone of a polymer or the organization of the surrounding matrix, as seen in the chiral phase of an LC material or a gel. In these systems the chiral response due to the light switching is detected indirectly as a change in chiral structure or organization of the material.

In the following sections the various approaches to chiroptical switching will be discussed and pertinent examples given to illustrate how these dynamic chiral systems have been applied in the control of function.

5.2.1
Chiroptical Switches Based on Overcrowded Alkenes

Inherently chiral overcrowded alkenes with a stable helical structure have been at the basis of many chiroptical molecular switches and light-driven molecular rotary motors [34]. In these molecules, switching between the two chiral states is achieved through light-induced *cis–trans* isomerization reminiscent of the retinal optical switch, leading to a change in molecular helicity (Figures 5.4 and 5.5) [13, 14]. Either enantiomers can be used, which allows for switching between two enantiomeric states P and M (Figure 5.4), or pseudoenantiomers that allows for switching between two diastereomeric states P and M' (Figure 5.5). The former have seen very limited use so far but the diastereomeric chiroptical switches, due to their excellent switching properties, have been applied in the control of a variety of functions.

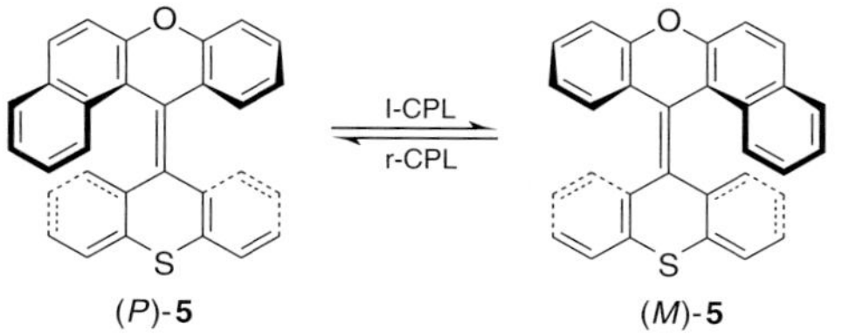

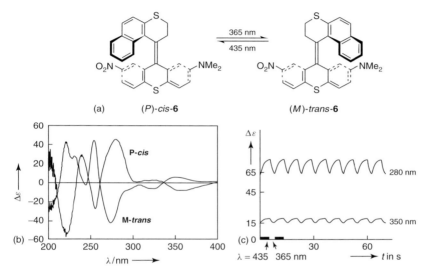

Figure 5.4 Overcrowded alkene-based chiroptical switch.

Figure 5.5 (a) Chiroptical molecular switch based on pseudoenantiomers (P and M') as a binary storage element. Irradiation of **6** at different wavelengths (λ_1 and λ_2) results in interconversion between P and M helicity. (b) CD spectra of the *pseudo*enantiomers (P)-*cis*-**6** and (M)-*trans*-**6**. (c) Repetitive switching of **6** as detected by CD spectroscopy. Reproduced from Ref. [35a]. Copyright Wiley 1993.

5.2.1.1 Enantiomeric Photochromic Switches

Overcrowded alkene **5** satisfied the requirements that proved necessary to demonstrate that with left- or right-CPL irradiation switching in either direction is possible and that an excess of a particular enantiomer over the other could be achieved (Figure 5.3a, Figure 5.4) [36]. It should be noted that upon irradiation with non-circularly polarized light, photochemical isomerization between the enantiomers occurs, but this always results in a racemic mixture.

Demonstrating that the use of circularly polarized light could shift the photoequilibrium out of the 50 : 50 regime to achieve an excess of one enantiomer proved a fundamental challenge to molecular design. The enantiomers of **5** are stable at ambient temperature ($\Delta G_{rac} = 25.9$ kcal mol^{-1}), fatigue resistant and undergo rapid photoracemization upon irradiation at 300 nm with nonpolarized light with high quantum yield ($\Phi_{rac} = 0.40$, n-hexane). Switching between two photostationary states (PSSs), with enantiomeric excess (ee) values of 0.07% for P-**5** or M-**5**, was achieved upon irradiation with right and left CPL. The experimental Kuhn anisotropy factor g (-6.4×10^{-3} at 314 nm) indicates that under optimal conditions an ee value of 0.3% is possible [37]. This chiroptical molecular system has three distinct states; racemic (P,M)-**5**, P-enriched **5** and M-enriched **5**, and all three states can be addressed by a *single wavelength* simply by switching the chirality of the light from nonpolarized to *l*-CPL or *r*-CPL.

5.2.1.2 Diastereomeric Photochromic Switches

When diastereomeric (or so-called pseudoenantiomeric) bistable overcrowded alkenes P and M' are used (Figure 5.3b), much higher stereoselectivities can be reached compared to enantiomeric systems [13, 14, 32]. The reason is that two different wavelengths of light can be selected in such a way that the PSSs of the forward and reverse reaction are maximized. Large changes in chirality between the two states offer distinct advantages in practical applications of such chiroptical switches, in particular, facilitating readout and amplification of the preferred chirality. A typical example of a highly selective diastereomeric chiroptical switch based on an overcrowded alkene is shown in Figure 5.5.

The near mirror image CD spectrum of P-*cis*-**6** and M-*trans*-**6** reflect the pseudoenantiomeric nature of the two isomers in which the inherent helical chromophore dictates the overall chiroptical response. The optimal wavelength for the forward and reverse photoisomerization and the PSSs (and the ratio of M and P helices) was tuned through the introduction of appropriate nitro-acceptor and amine-donor moieties [35, 38]. The switching of helical architecture and chiral properties is simply achieved by changing the wavelength of the light used and repetitive switching cycles can be monitored readily by CD spectroscopy (Figure 5.5b and c). The ratios of P-*cis*-**6** and M-*trans*-**6** of 30:70 (at 365 nm) and 90:10 (at 435 nm) revealed the large difference in isomer composition at the two PSS but also show that clear reversal from a preferred M to P helicity in this system occurs.

In the systems discussed so far distinct upper and lower halves were present in the overcrowded alkene. In a modified structural design two identical helical moieties as well as two additional stereogenic centres with fixed (S,S)- chirality, are present (Figure 5.6) [39]. This molecule is a modified version of a first-generation molecular motor (*vide infra*) and can adopt an M,M or P,P helical structure. This system functions as a perfect chiroptical molecular switch showing >99% selectivity for MM-**7** or PP-**7** isomers upon irradiation at 376 or 303 nm, respectively.

5.2.2
Azobenzene-Based Chiroptical Switching

Azobenzenes are among the most widely used photoresponsive compounds and this popularity can be attributed to the ability to switch between two geometrical isomers, *cis* and *trans*, with large amplitude changes in size. Typically, the distance between the peripheral groups is modulated by up to 0.7 nm (Figure 5.7)

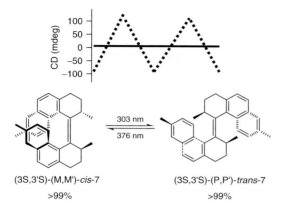

Figure 5.6 Pseudoenantiomeric forms of an overcrowded alkene-based molecular switch **7**. Reproduced from Ref. [39]. Copyright RSC 2004.

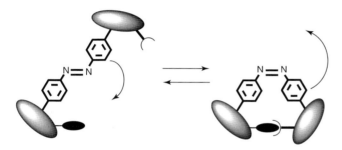

Figure 5.7 Large amplitude change in molecular size driven by the *cis–trans* isomerization of azobenzene.

[40, 41]. The photochemical switching has been exploited in the control of a variety of functions for instance binding events in host–guest systems [41], polymer structure [42] and LC organization [43].

Azobenzenes are not inherently chiral and chiroptical switches based on these systems typically comprise a chiral unit (or chiral auxiliary) and an achiral bistable azobenzene (Figure 5.3b). A large change in chirality is not always observed since the *cis–trans* isomerization itself is not associated with an intrinsic change in chirality. Frequently azobenzenes suffer from low thermal stability of the energetically less-stable *cis* isomer leading to thermal isomerization back to the *trans*-state. The large geometrical changes that accompany photoisomerization, however, can drastically affect the stereochemical properties of appending chiral units.

An illustrative recent example comes from Wang's group in which a pentahelicene (**8**) with an α-phenylethylimide chiral auxiliary moiety is functionalized with an azobenzene photochromic unit (Figure 5.8) [44]. This chiroptical system can be addressed both electrochemically and by light, while readout is achieved by UV-Vis

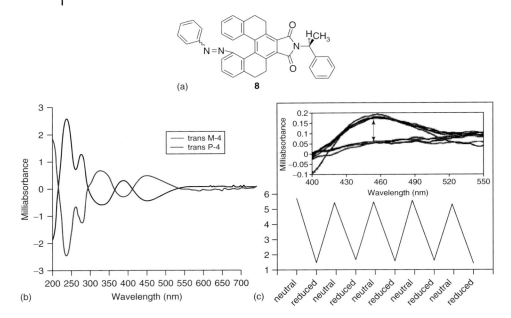

Figure 5.8 (a) The structure switchable pentahelicene **8**. (b) CD spectra of M- and P-form in acetonitrile (10^{-5} M). (b) Electrochemically induced variation in the ellipticity of M-form monitored at 454 nm. (Inset) CD spectra obtained during the electrochemical modulation study. Reproduced from Ref. [44]. Copyright ACS 2005.

and CD spectroscopy. Photochemical control is achieved through cis–trans isomerization of the azobenzene unit while electrochemical control of the chiroptical properties is possible through reduction and oxidation of the same unit.

In an alternative approach, enantiomerically pure azobenzenes with an ortho-, meta- or para-tolylsulfoxide chiral moiety (**9**) were prepared [45]. Different chiral responses were observed due to conformational effects controlled by the stereogenic sulfoxide moiety. Furthermore, transfer of chirality from the sulfoxide to the azobenzene was observed by CD measurements both for the cis and trans isomers (Figure 5.9). Axially chiral bis(azo)benzene photochemical switches **10** and **11** have been designed for high responsivity and as multiple switches [46]. This is based on the premises that chiral binaphthyl units have strong exciton CD effects and large LC helical twisting powers. However, the switching efficiency and selectivity for **11** is relatively low, as little as 50% of the trans azo-units in **11** switch at the optimal wavelength of 402 nm. Furthermore, a PSS was found with 29% trans-trans, 49% trans-cis and 22% cis-cis-**11** accompanied by a distinct change in CD and a reversal of LC helicity when used as a chiral switchable dopant in a nematic liquid crystal. Similarly, an axial chiral bis(azo)-derivative designed by the Gottarelli group, allows for modulation of the pitch of a cholesteric phase at low dopant concentration with good thermal stability although

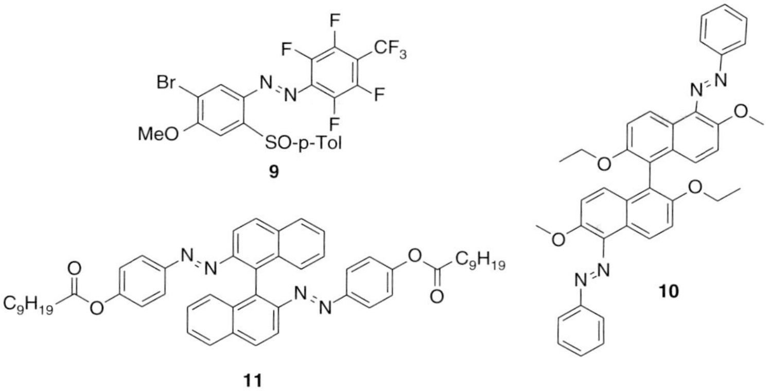

Figure 5.9 Structures of azobenzene-based chiroptical switches.

response times still need substantial improvement [47]. Related approaches are based on binaphthyl structures with single azobenzene switch units, leading in some cases to excellent responsive behaviour, in particular, in the control of cholesteric LC phases [48].

Tian and coworkers [49] have reported an azobenzene-based rotaxane system (**12**) comprised of a cyclodextrin, which functions as the torus of the rotaxane, and a covalently attached azobenzene with a naphthalimide stopper unit (Figure 5.10). The chiral cyclodextrin induces a CD effect in the chromophores and the states of this switching system can be monitored by chiroptical techniques. Multiple switching cycles by photoisomerization of the azobenzene unit could be achieved.

Yashima's group [50] reported the first switchable self-assembled optically active double helix based on the quanidine–carboxylate binding motif. By incorporating photoresponsive azobenzene moieties in each unit and chiral quanidines (**14**) in one of the units, light-induced changes in supramolecular organization and chirality were achieved (Figure 5.11). The overall helical chirality is controlled by stereocentres relatively remote from the photoactive azobenzene group. The helicity of the hydrogen-bonded supramolecular structure can be controlled by photoisomerization of the azobenzenes. In the *cis* state multiple hydrogen bonding is sufficient to maintain a helical structure, while in the *trans* state the helix inverts partly due to a change in hydrogen bonding as a result of geometrical constraints.

The systems discussed here are illustrative of the control of chiral phenomena through azobenzene switches. In addition, azobenzenes allow reversible photochemical modification of monolayers and thin films [51], aggregation and solubility [52], nonlinear optical behaviour [53], conformational control of (cyclic) peptides [54] but have also been used to construct photoresponsive amphiphiles [55] and membranes [56]. Numerous other applications of azobenzenes have been reported, although only in a limited number of cases, such as those mentioned here, have the special features associated with control of chirality.

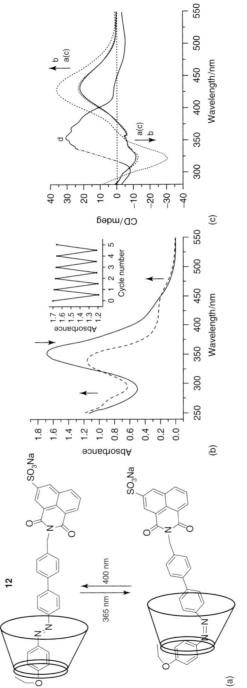

Figure 5.10 (a) Reversible configuration change of the **[1]**rotaxane **12** under different stimuli. (b) The absorption spectra of rotaxane **12** in aqueous solution before (—) and after (---) irradiation at 365 nm for 15 min at room temperature. Insert: changes in the absorption spectra of rotaxane **12** (absorption value at around 350 nm) for several cycles. In one cycle, (c) CD spectra before: (a) and after: (b) irradiation at 365 nm and (c) after irradiation at 430 nm, (d) CD spectrum of cis-**12**. Reproduced from Ref. [49a]. Copyright RSC 2007.

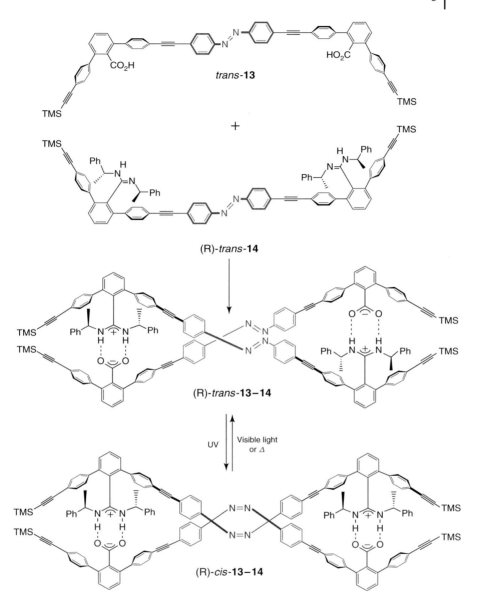

Figure 5.11 Double-helix formation and subsequent *trans–cis* photoisomerization of supramolecular switch pair **13–14**. Reproduced from Ref. [50]. Copyright RSC 2007.

5.2.3
Diarylethene-Based Chiroptical Switches

A particularly versatile class of photochromic compounds that can be used to build chiroptical switches are the diarylethenes [57]. The most widely used are dithienylethenes (see Chapter 1) that upon exposure to UV light undergo a reversible photochemical ring closure of a hexatriene to a cyclohex-1,3-diene unit in the core of the molecule. Irradiation with visible light reverses the process resulting in ring opening. As the conjugation length in the molecule changes dramatically, a large bathochromic shift in the UV-Vis absorption spectra is seen upon ring closure. Furthermore, the open form shows large conformational flexibility while the closed form is rigid. The open antiparallel form exists as two rapidly interconverting helical structures (Figure 5.12a).

The introduction of hetero-arene moieties eliminates the low thermal stability of the dihydro-form, which is a recurrent problem and a severe limitation in

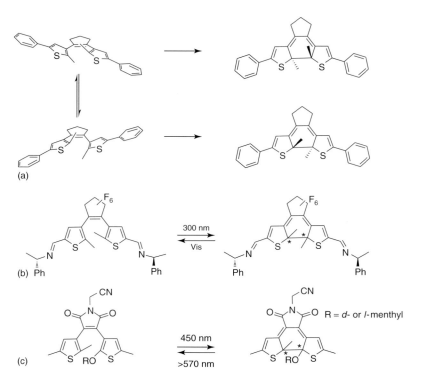

Figure 5.12 (a) The dynamic equilibrium of the open state of a dithienylethene between a P and M helix is locked upon photochemical ring closure. When a chiral substituent is placed on the periphery (b) or at the reactive carbon centres and (c) then the ring closing reaction leads to a diastereomeric excess.

the application of the reversible photocyclization of stilbene derivatives. A variety of hetero-arenes have been used. Bridging the central alkene bond, to prevent unwanted *cis–trans* isomerization, is another important structural improvement [58]. The groups of Irie and Lehn [59] have developed a series of diarylethenes that cover the whole visible spectrum by tuning the conjugation length via the introduction of donor and acceptor substituents.

The open form of a diarylethene compound consists of a dynamic system of helical conformers in rapid equilibrium. Photochemical cyclization is a concerted process and hence a C2-symmetric racemic molecule is obtained with a *trans* orientation of the methyl substituents at the two newly formed stereogenic centres [60]. This class of photochemical switches can be readily modified with chiral groups to bias the photochemical ring closure in favour of one of the diastereoisomers (Figures 5.12b,c). The chiral auxiliary group can be either attached to one or both of the prochiral carbon atoms of the thienyl rings [61] or to the peripheral carbon-5 of the thienyl rings not involved in the ring-closing reaction [62]. Both approaches have been successful in detection of the switching event with chiroptical techniques, although the first approach, in which the directing chiral group is close to the reactive carbon (the prochiral C2), provides for much higher diastereomeric excess.

Recently, we were able to isolate the individual atropisomers of photochromic diarylethenes **15** (Figure 5.13) [63]. Instead of a (perfluoro) cyclopentene bridging moiety, a phenanthrene unit was introduced that resulted in a high barrier of rotation ($\Delta G = 109.6$–111.5 kJ mol^{-1}) about the single bonds connecting thiophene and phenanthrene units leading to robust chiroptical switching behaviour. The switching process of this intrinsically chiral dithienylethene between open and closed form could be followed by CD spectroscopy readily. Alternating irradiation with UV ($= 313$ nm) and visible light ($= 460$ nm) allowed for several switching cycles without any racemization.

Helicenes are inherently dissymmetric and due to the helical nature of the chromophore usually show large chiroptical effects. Branda and coworkers [64] have designed a helicene **16o** with a built-in photochromic diarylethylene unit that allows the reversible formation of the chiral helicene (Figure 5.14a). In the absence of a chiral auxiliary group a racemic mixture of **16c** is obtained. When the helicene-based dithienylethene photochromic switch was annulated with homochiral pinene units as seen in structure **17**, chiroptical switching can readily be achieved (Figure 5.14b) [65]. Upon ring closure, one of the diastereomeric helicenes is formed preferentially and a large change in the CD spectrum compared with the open system is observed (Figure 5.14c and d).

Related approaches to chiral helicene type diarylethenes with large chiroptical responses have been followed by Yokoyama and coworkers [66]. In diarylethene **18** a benzothiophene unit is present with a chiral methoxyethyl group at the prochiral C-3 position (Figure 5.15) [67]. The diastereoselective photochemical ring closure to the corresponding 7-heterohelicene structure was found to proceed with 47% diastereomeric excess. Notably, there is a large change in optical rotation (1300° at 589 nm); a wavelength where both closed and open form do not absorb. This allows

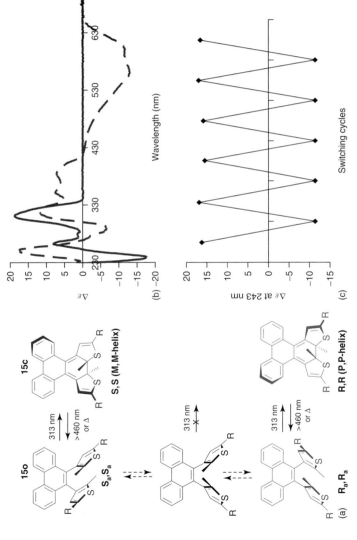

Figure 5.13 (a) Stereoisomers and thermal and photochemical processes of atropisomeric dithienylethenes (**12o**). (b) CD spectra of the *M,M*-helical isomer in the open (thick line) and photostationary state (dashed line) form in heptane. (c) Switching cycles observed upon alternated irradiation with UV light (λ = 313 nm) and visible light (λ = 460 nm) as detected by CD spectroscopy at 243 nm. Reproduced from Ref. [63]. Copyright RSC 2007.

5.2 Molecular Switching | 137

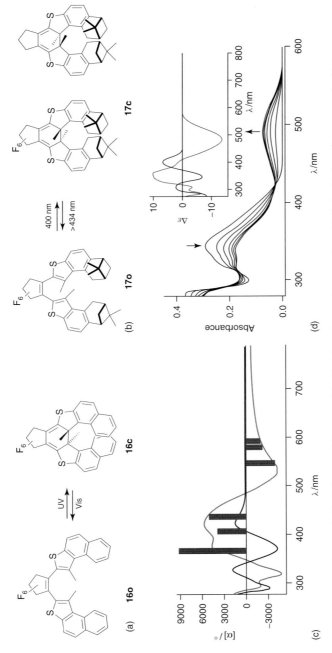

Figure 5.14 (a) The switchable helicenes (**16**) reported by Branda and coworkers together with their chiral auxiliary approach. (b) A dithienylethene-based photochromic switch (**17o**) is modified with two chiral units. (c) The experimental and calculated ORD spectra of **17o** and **17c**. (d) Changes to the UV/Vis absorption spectrum upon irradiation with UV light and (inset) CD spectra in the open and closed states. Upon ring closure one of the helical forms is preferred. Reproduced from Ref. [65]. Copyright ACS 2005.

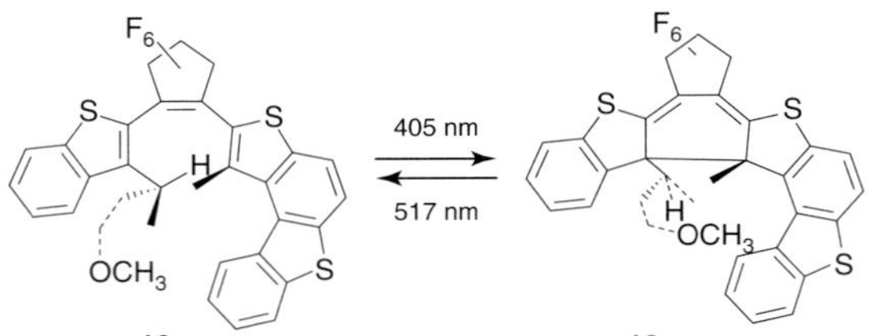

Figure 5.15 Chiroptical switch **18** reported by Yokoyama and coworkers.

for facile nondestructive readout of the chiroptical response without interference with the switching processes.

These photochromic helicene systems, as with the overcrowded alkenes discussed above, satisfy the requirements for successful chiroptical switches, that is they are thermally stable in both open and closed states, show high stereoselectivity in the photocyclization reaction and/or large changes in CD or ORD spectra allowing for facile detection.

5.3
Chiral Fulgides

Following the discovery of the photochromic behaviour of fulgides by Stobbe in 1904 [68] there has been considerable interest in these molecules as potential candidates for erasable and rewritable optical memory systems and as photochromic trigger elements [69]. The bistability, as in the previous systems, is based on the reversible photochemical conrotatory electrocyclization of a 1,3,5-hexatriene moiety. An example of a chiral indolylfulgide **19** is shown in Figure 5.16. The photochromic reaction involves the open-colourless and conformational mobile *trans*-form **19a**, which is in equilibrium with the *cis*-form **19b**, and the closed and rigid C-form **19c**. All three isomers are chiral due to a stereogenic centre in the closed C-form and a helical conformation in the open (*cis* and *trans*) structures. The presence of an isopropyl group reduces the propensity for the undesired *trans-cis* isomerization dramatically and allows the enantiomers of the *trans* isomer to be resolved by chiral high performance liquid chromatography (HPLC) [70]. Irradiation at 405 nm in toluene resulted in a PSS with a high excess of the coloured closed form (ratio

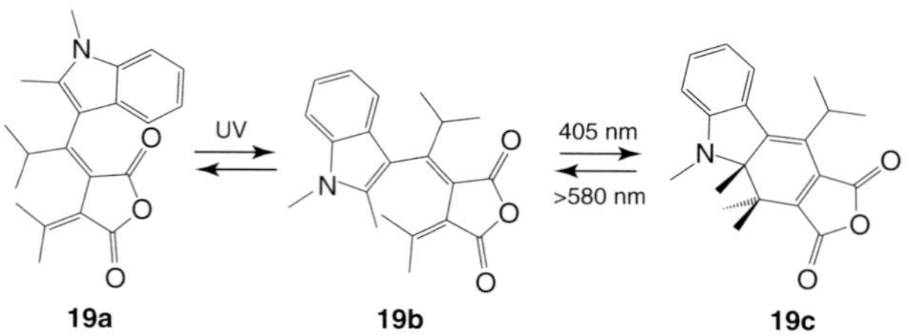

Figure 5.16 Molecular structural changes that accompany switching of fulgide **19**.

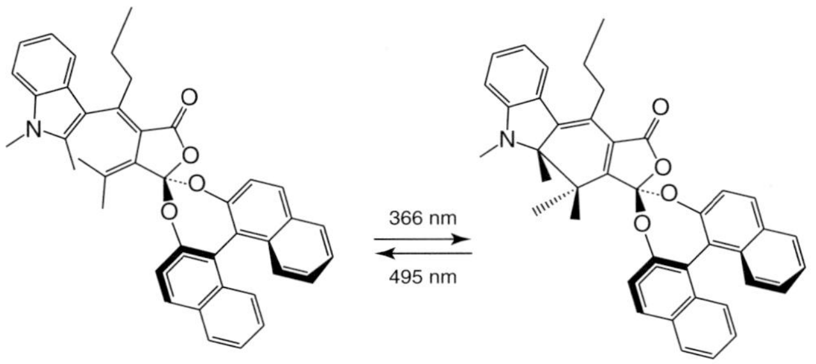

Figure 5.17 Molecular structural changes that accompany switching of fulgide **20**.

open–closed 19–81) without formation of the unwanted *cis* isomer. Irradiation with visible light (>580 nm) provided exclusively the ring-opened *trans* form. The chiroptical switching process can readily be followed by CD although gradual photoracemization was observed.

In an alternative approach to chiral fulgides, binaphthol was used as a chiral auxiliary group (Figure 5.17) [71]. Diastereoselective photoswitching is observed between the open (p)-*trans*-**20** form and the closed (9aS)-**20** form. In this case, the dynamics of the chiral photochromic system are complicated by the presence, under ambient conditions, of two major conformers (a and b) in a 57 : 43 ratio in rapid equilibrium. Only the a-conformer adopts the proper geometry for rapid and diastereoselective photocyclization. Irradiation at 360 nm generates a PSS comprised of 86% of the closed form with a 95 : 5 diastereomeric ratio. The reverse switching process, to regenerate the open form, is unfortunately quite inefficient with this system. Fulgides have also been explored as chiroptical switches in polymeric liquid crystals [72].

5.3.1
Redox-Based Chiroptical Molecular Switching

Several groups have explored the effect of a change in the redox state of a chiral molecule on the geometry and electronic structure and hence the chiroptical properties, in particular, the CD spectra [73]. Various approaches are discussed throughout the chapters compiled in this volume. We will restrict ourselves to selected recent examples to illustrate the concepts explored. The redox-active group can either be a metal complex or a redox-active organic moiety. Canary and coworkers [73–75] have designed a series of redox-triggered chiroptical switches based on the change in coordination at a metal centre upon oxidation or reduction. The system shown in Figure 5.18 illustrates this concept. The chiral ligand **21** is derived from the amino acid methionine, and forms a tetradentate complex with Cu(II) via coordination of three nitrogen and one carboxylate donor moieties. The resulting chiral complex with a propeller-type twist can be reduced to the Cu(I) complex, which is accompanied by ligand reorganization. Instead of binding the

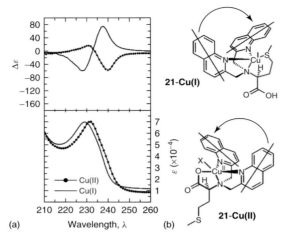

Figure 5.18 (a) UV/Vis and CD spectra and (b) structure of complex in the Cu(I) and Cu(II) oxidation states. Reproduced from Ref. [74a]. Copyright ACS 2003.

carboxylate, the Cu(I) centre preferentially binds the sulfide group, resulting in an inversion in the helical orientation of the two quinoline moieties. As a consequence, inversion of the CD absorption is observed and this chiroptical switching can be modulated by alternating addition of ascorbate (reducing agent) and ammonium persulfate (oxidant). The change in coordination mode can be employed to drive large-amplitude structural changes in much the same way as has been achieved in the azobenzene-based systems discussed earlier.

An elegant example of an organic redox-driven chiroptical switch is shown in Figure 5.19. A relatively simple approach is followed coupling a chiral unit with two electroactive units. Reduction of both pyrene moieties in the chiral trans-cyclohexanediol bispyrene esters (1R,2R)-**22a** and (1S,2S)-**22b** (Figure 5.19) to their radical anions resulted in a strong change in the UV-Vis and CD spectra [76]. The neutral compounds absorb only below 450 nm, whilst the radical anions show a strong absorption band at 510 nm with a split CD due to strong intramolecular exciton coupling between the chromophores. Interestingly, the corresponding bisamide compounds trans-cyclohexane bispyrene amides (1R,2R)-**23a** and (1S,2S)-**23b** failed to show any CD signals showing exciton coupling. A related approach was followed with a chiral Binol-based boron dipyrromethene that functions as a highly reversible on-off redox-active chiroptical switch [77]. Switching between states with a strong CD signal and a complete disappearance of CD activity in a certain wavelength regime can be considered a distinct advantage to most chiroptical switches and is particularly appealing for detection.

A geometrical change due to the interaction between identical units when in a partially oxidized state has also been observed in a bis (catecholketal) system in which planar chirality is present [78]. Upon one-electron oxidation, the molecule

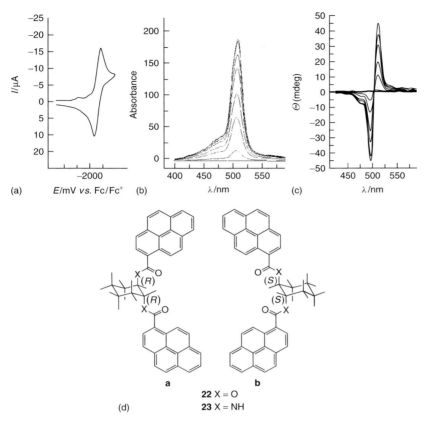

Figure 5.19 (a) Cyclic voltammetry of **18**. (b) Changes in UV/Vis absorption. (c) CD spectra upon reduction of **22**. (d) Structure of **22a/b** and **23a/b**. Reproduced from Ref. [76]. Copyright RSC 1999.

adopts a folded instead of the extended conformation and a new visible absorption band appears that enables this system to be an on/off switch of chirality. In these multicomponent chiroptical systems the chiral unit does not necessarily need to be the same as the photoresponsive unit [79, 80]. This approach offers the advantage that a broad range of functional/responsive units can be employed. A pertinent example is also seen in the redox-active chiral system comprising two anthracene units, that can dimerize reversibly, which are tethered to a chiral binaphthyl unit [79]. Zhou et al. [81] have reported a simple chiroptical system **24** based on a chiral binaphthyl unit connected covalently to two tetrathiafulvalene (TTF) redox-active units **24** (Figure 5.20). Changes in the strength and nature of the interaction between the TTF units in the neutral, partially and fully oxidized states are the driving force behind the electrochemical control of the chiroptical output signal of the binaphthyl moiety, which functions as the chiral reporter unit.

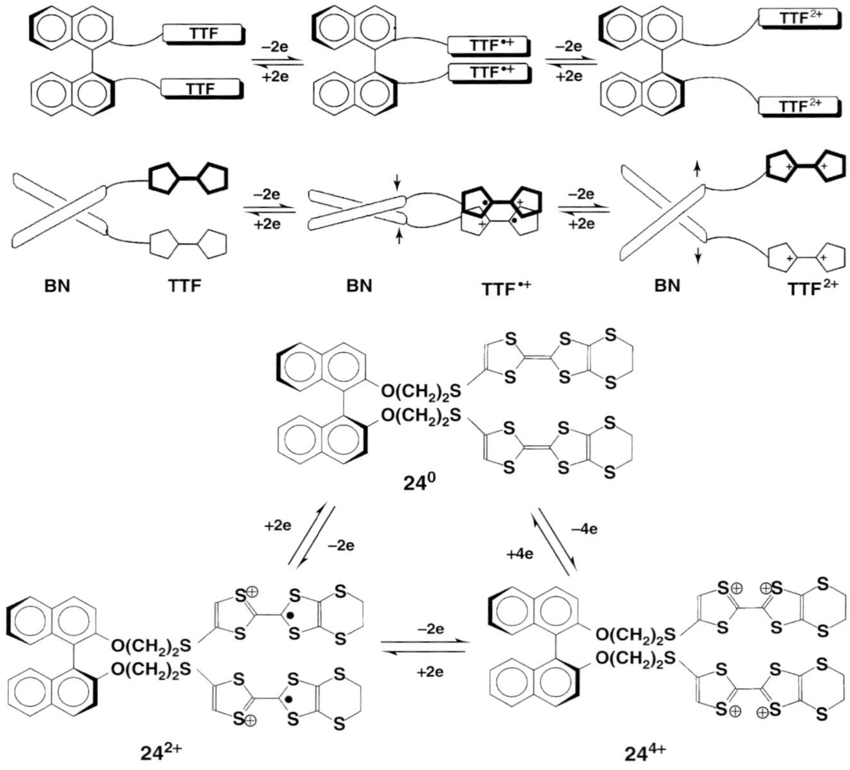

Figure 5.20 Electrochiroptical switch **24** based on interacting tetrathiofulvalene redox units. BN = binaphthyl, TTF = tetrathiofulvalene (neutral state), TTF$^{\bullet+}$ and TTF^{2+} = tetrathiofulvalene (partial or fully oxidized states).

In the partially oxidized state the two TTF units are attracted to one another, reducing the dihedral angle between the naphthalene units, whereas in the fully oxidized state the TTF^{2+} units show strong electrostatic repulsion, increasing the dihedral angle with respect to the neutral state. A related approach was followed by Deng et al. [83] using the reductive switching of the dialkyl-4,4'-bipyridinium unit in place of oxidative switching of the TTF unit as in the previous case. The changes in the CD spectrum in both these systems when addressed electrochemically are modest. Nevertheless, these systems represent clear examples of a bicomponent functional system where the intrinsic properties of the chiroptical unit are controlled by external perturbation.

Other redox-driven chiroptical switches recently developed include electrochromic systems based on axially chiral biphenyl-2,2'-diyl-type dicationic dyes [82, 83] and multi-input-output responsive systems based on the dynamic redox behaviour of tetraaryldihydrohelicenes (resulting in electrochromism, fluorescence and chirality modulation) [84].

Near-infrared chiroptical switching has recently attracted attention in view of the future applications foreseen for near-infrared (NIR) chiroptical materials [85], such a NIR bio-imaging and sensing. NIR detection of chiroptical signals has been reported for a number of cobalt, nickel and lanthanide complexes [86]. The redox-active single enantiomer of the dinuclear Ru-complex represent a recently developed redox-induced NIR chiroptical switch [87]. Electrochemical switching between Ru (II)/Ru (II) and Ru(III)/Ru(III) states shows a distinct change in the metal-to-metal charge transfer (MMCT) transition reflected in the modulation of the CD band at 890 nm.

5.3.2
Miscellaneous Chiroptical Switches

Following earlier approaches [9, 15] a variety of other chiroptical molecular switches have been designed in recent years. Figure 5.21 summarizes several of the systems in which theoretical or experimental studies towards (photo-)chemical modulation of chirality have been reported. These include binaphthopyrans [88], *cis*-fluoroethenyl-2-fluorobenzene [89], dendritic systems [90], cyanine dye/poly(L-glutamatic acid) [91], spiropyrans [92] and helicene-based bis-azobenzene dual-mode switches [93]. Although not all systems fulfil the requirements mentioned above for a practical chiroptical switch, for instance due to low chiroptical response or lack of thermal stability, the structural diversity demonstrates the enormous potential for the dynamic control of chirality.

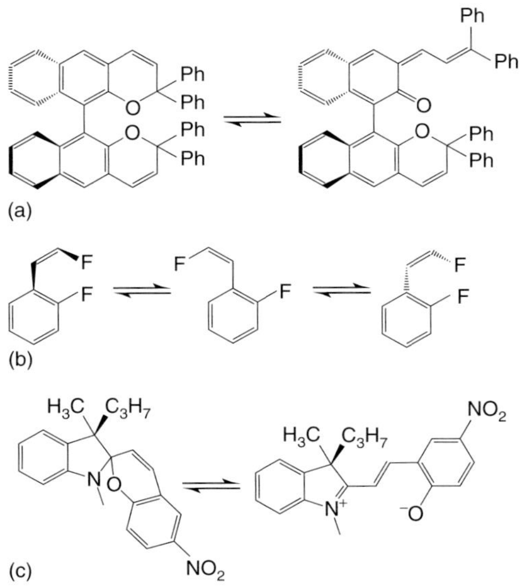

Figure 5.21 Molecular structures of (a) binaphthopyrans, (b) *cis*-fluoroethenyl-2-fluorobenzene and (c) spiropyrans.

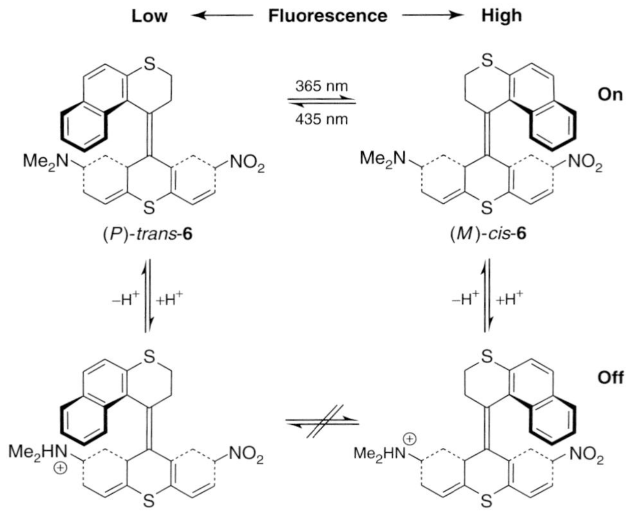

Figure 5.22 (pH)-Gated dual-mode photoswitching of donor–acceptor molecular switch **6**.

5.3.3
Chiroptical Switching of Luminescence

Modulation of fluorescence is a particularly attractive property in the context of molecular logic functions and sensing (see Chapter 18) [94]. The typical high sensitivity of detection of fluorescence output has been exploited in numerous applications [95]. Several chiroptical photochromic systems show, besides their propensity for light-induced molecular switching, reversible changes in fluorescence. In particular, chiral overcrowded alkenes are promising candidates for multistage luminescent switches. A typical example is shown in Figure 5.22 [96]. P-*trans*-**6**, which shows low fluorescence, can be converted to M-*cis*-**6** by irradiation at 365 nm with a concomitant enhancement in the fluorescence significantly. Upon irradiation at 435 nm the process is reversed, going from a high to a low fluorescent state. In addition, the dimethylamine moiety in **6** can be protonated, which leads to nonfluorescent molecules, while the switching is also blocked.

This system operates as a four-stage switching system in which both switching and luminescence can be modulated through irradiation and protonation. The presence of the dimethylamine-donor substituent and nitro-acceptor group are the cause of a large difference in electronic interaction of the upper naphthalene chromophore with the lower part of these molecules allowing highly selective switching. Amine protonation generates a nonphotoactive acceptor (ammonium)-acceptor (nitro) substituted system. Circularly polarized luminescence studies on this system revealed that a single enantiomer of these overcrowded alkenes, that is (P)-*cis*-**6**, can emit left- or right-circularly polarized

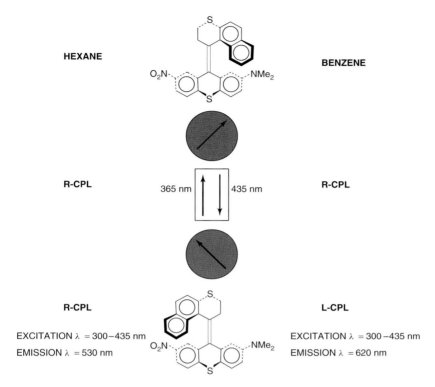

Figure 5.23 Switching of chirality and CPL emission and solvent dependence of CPL emission. In hexane, both *P* and *M* helical switch molecules emit *r*-CPL, whereas in benzene switching of molecular helicity results in switching of the sense of the CPL emission between *r*-CPL and *l*-CPL.

light (Figure 5.23) [97]. Surprisingly, the handedness of the emitted light was found to depend on the solvent used. In addition, the sense of chirality of the emitted CPL light could be modulated through the photochemical switching process.

5.4
Light-Driven Molecular Rotary Motors

In nearly every important biological process including transport, cell division and muscle movement, molecular motors play a crucial role [12]. Fascinating examples of nature's rotary motors are the ATPase [98] and bacterial flagellar motors [99] inducing proton transport and bacterial movement, respectively, using ATP as a fuel. From the outset of the development of nanotechnology the design of artificial molecular motors has always been one of the grand challenges taking inspiration from nature's numerous elegant examples but also stimulated by the prospects for the construction of nanoscale machines and robots powered by such motors [100, 101].

Before taking the step from switches to motors and starting our discussion on the design and functioning of molecular motors let us define the minimum requirements for a molecular rotary motor [102, 103]. The following features can be identified in order to classify a molecular system as a rotary motor: (i) rotary motion, (ii) input of energy/consumption of fuel, (iii) unidirectional rotation and (iv) repetitive movement [104, 105].

In addition, one should realize that at the nanoscale, one has to operate against or take advantage of the random Brownian motion [106]. In this context it is not surprising that nature has taken advantage of the possibilities to restrict motion and Brownian motion is an integral part of biological (Brownian ratchet-type) molecular motors. Furthermore, one should realize that in contrast to motors operating at the macroscopic level, nanoscale motors are not affected by gravity and mass but function in a high-viscosity environment and the interactions between components is governed by noncovalent interactions. So which steps have to be taken from a chiral molecular switch to a unidirectional rotary motor and how do they operate? The basic principles and the design features in order to control their proper functioning are outlined in the next sections.

5.4.1
First- and Second-Generation Motors

The prototypical first- and second-generation light-driven unidirectional molecular motors are shown in Figure 5.24 [107, 108]. The molecules comprise a chiral

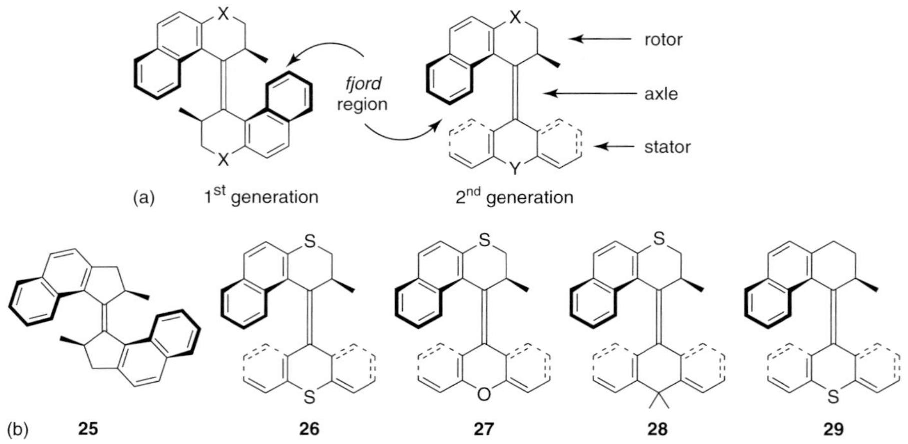

Figure 5.24 (a) Terminology used and (b) Molecular structures of first- and second-generation molecular rotary motors discussed in the text.

overcrowded alkene structure in which two halves are connected by a central olefinic bond. One half of the molecule functions as a rotor unit and the other functions as a stator part, while the connecting olefin operates as the axle of the rotational system. As with the overcrowded alkene-based chiroptical switches these molecules undergo a stilbene-like E–Z isomerization upon irradiation.

However, in contrast to the molecular switches the first- and second-generation molecular rotary motors have both an intrinsic helical chiral structure and stereogenic centres in allylic positions with respect to the axis of rotation. The subtle interplay of these chiral features is crucial for the operation of the motor systems. Typically, a full unidirectional rotary cycle comprises four steps; two photochemical – energy supplying – isomerization steps, that generate high-energy isomers, each followed by a thermally activated helix inversion to relax the system to a more stable isomer. Upon continuous irradiation and at the appropriate temperature to allow for thermal helix inversion, the rotor part of these molecules displays unidirectional rotary motion relative to the stator part.

In the first-generation molecular motor based on overcrowded alkenes, two stereogenic centres, bearing methyl substituents, are present that dictate the direction of rotation [107]. These methyl groups adopt a pseudoaxial orientation to minimize steric repulsion with the other half of the molecule. The photochemical and thermal isomerization processes of the full 360° unidirectional rotary cycle of 1R,1R'-(P,P)-*trans*-30 are shown in Figure 5.25a. The formation of each of the isomers during the cycle can be followed by ^{1}H-NMR and UV/Vis spectroscopy, while CD spectroscopy (Figure 5.25b) is especially powerful in determining the helix inversion in each step (Figure 5.26).

Irradiation of stable (P,P)-*trans*-30 with UV light ($\lambda > 280$ nm) initiates a *trans–cis* isomerization around the central double bond with a concomitant helix inversion (from P,P to M,M) generating a less-stable isomer (step 1). In the PSS an unstable (M,M)-*cis* to stable (P,P)-*trans* isomer ratio of 95 : 5 is observed. In the less-stable isomer (M,M)-*cis*-30, the methyl substituents are forced to adopt a strained pseudoequatorial orientation. A subsequent thermal helix inversion steps occurs spontaneously under ambient conditions releasing the strain in the molecule providing stable (P,P)-*cis*-30 ($\tau_{1/2} = 32$ min at 20 °C in n-hexane, $\Delta G^{\#} = 91$ kJ mol^{-1}) (step 2) [109]. This step is unidirectional and the driving force for the formation of the stable *cis* isomer is the energy gain due to the fact that less sterically hindered pseudoaxial orientation of the methyl groups is re-established. Molecular modelling showed that the energy difference between unstable (M,M)-*cis*-30 and stable (P,P)-*cis*-30 is 46 kJ mol^{-1} and this gain in Gibbs energy going from unstable to stable isomers is responsible for the unidirectional nature of the first half of the rotary process [101, 110]. Two similar photochemical (step 3) and thermal (step 4) isomerization steps convert stable (P,P)-*cis*-30 into unstable (M,M)-*trans*-30 and subsequently stable (P,P)-*trans*-30 completing a full 360° rotary cycle. The barrier for thermal helix inversion in step 4 ($\Delta G^{\#} = 107$ kJ mol^{-1}, $\tau_{1/2} = 439$ h at 20 °C in n-hexane) is much higher than in step 2 and this step is rate limiting in the entire rotary cycle. This large barrier to helix inversion requires that the system has to be heated to 60 °C, to allow repetitive rotation upon continuous irradiation.

148 | 5 Chiroptical Molecular Switches

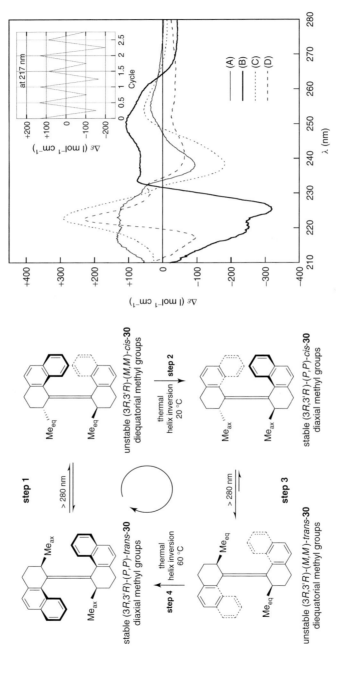

Figure 5.25 (a) Photochemical and thermal isomerization processes of molecular motor **30**. (b) Circular dichroism (CD) spectra in each of four stages of switching (see text). Trace A, (*P,P*)-*trans*-**30**; trace B (*M,M*)-*cis*-**30**; trace C, (*P,P*)-*cis*-**30**; trace D, (*M,M*)-*trans*-**30**. Inset, change in CD signal during full rotation cycle of (*P,P*)-*trans*-**30** monitored at 217 nm. Adapted with permission of Nature from Ref. [107]. Copyright Nature 1999.

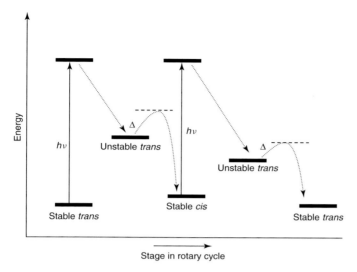

Figure 5.26 A Gibbs energy profile of the rotary cycle involved in the functioning of molecular rotary motors based on overcrowded alkenes.

The input of energy is achieved through photoexcitation of stable *trans* and stable *cis* isomers, providing the corresponding higher-energy unstable *cis* and *trans* isomers, respectively. These unstable isomers undergo an energetically downhill helix-inversion step to provide the stable *cis* and stable *trans* isomers. These steps are rendered irreversible by the large energy difference between unstable and stable isomers. It should be noted that in the first-generation rotary motor the photochemical equilibrium is highly favourable towards the unstable form, which is not always the case with these light-driven molecular motors. Although this is an attractive feature regarding the efficiency of the system, it is not essential for the operation of the motor as the subsequent thermal helix inversion (following the photoisomerization) removes the unstable form quantitatively and irreversible. The structures and sequence of helix inversion have been studied by molecular modelling and molecular dynamics studies [111, 112].

Insight into the mechanism of the helix inversions has been obtained via structural modification in combination with quantitative isomerization studies [113]. The introduction of the sterically more demanding isopropyl-group instead of a methyl at the stereogenic centre (Figure 5.27) allowed the observation of unstable (P,M)-*trans*-**31** during the thermal isomerization of unstable (M,M)-*trans*-**31** to (P,P)-*trans*-**31**. This finding clearly points to a stepwise thermal helix-inversion process from unstable to stable forms.

In the design of the second-generation rotary motors (Figures 5.24 and 5.28) the rotor and stator part are distinctly different from each other and only a single stereogenic centre is present in the rotor moiety [108]. Major advantages of the second-generation motors are that the barrier for helix inversion in both thermal

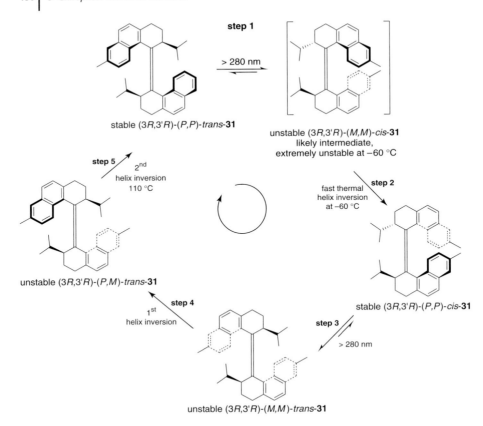

Figure 5.27 Rotary cycle of **31**. The structural modification (introduction of isopropyl group at the stereogenic centre) resulted in acceleration of the helix inversion in step 2 and retardation of helix inversion in step 5. This shows that the thermal isomerization is a stepwise process proceeding through an (M,P) intermediate. Reproduced from Ref. [113a]. Copyright ACS 2005.

steps is nearly identical and that the stator and rotor part can be functionalized individually. For instance, the lower stator part can be anchored to a surface (*vide infra*) without compromising the rotary motion of the upper rotor part. It should be noted that the directionality of the system is governed now by the absolute configuration at a single stereogenic centre. The four-step unidirectional rotary cycle of a typical second-generation motor stable (S)-(M)-*trans*-**32** is shown in Figure 5.28.

As with the first-generation molecular motors the methyl group at the stereogenic centre adopts a pseudoaxial orientation in the stable isomers, while a crowded pseudoequatorial conformation is found in the unstable forms [114, 115]. Again, the photochemical isomerization of stable to unstable forms and the subsequent release of strain from the unstable forms drive the overall unidirectional rotary motion.

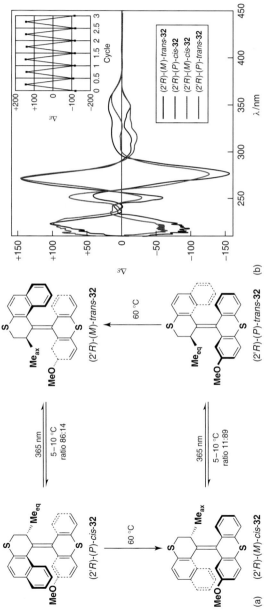

Figure 5.28 (a) Photochemical and thermal isomerization processes of motor **32**. (b) CD spectra of each of the four stages of rotation. Black line, (2'R)-(M)-trans-**32**; blue line, (2R)-(P)-trans-**32**; red line, (2R)-(P)-cis-**32**; green line, (2R)-(M)-cis-**32**; (2R)-(P)-trans-**32**. Inset, changes in $\Delta\varepsilon$ value over a full rotational cycle, monitoring at 272 nm. Reproduced from Ref. [108]. Copyright ACS 2002.

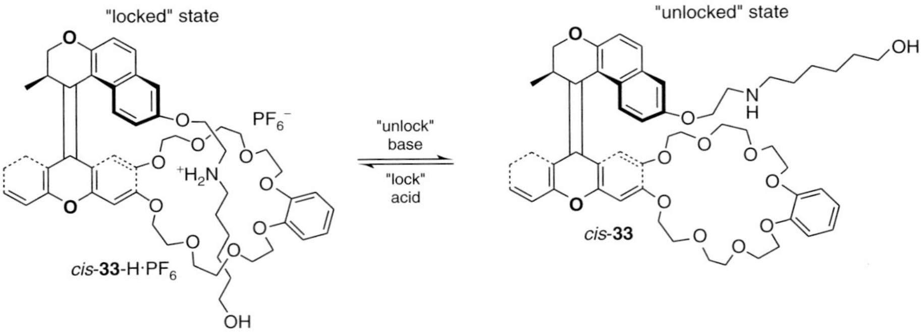

Figure 5.29 Molecular brake **33** based on a switchable rotaxane/rotary motor bifunctional molecule [116].

It is obvious that the rotary motion of these molecular motors stops when the system is not irradiated as the energy input is removed. Recently, a system **33** was designed based on a second-generation motor, a so-called 'molecular brake', which features a self-complexing lock to control the on-off state of rotary motion (Figure 5.29) [116].

To achieve such multifunctional behaviour a pseudorotaxane structure was integrated with a second-generation molecular motor. The stator part has an annelated crown ether and the rotor part features an 'arm' with an dialkylammonium crown ether binding unit. In the *locked state* the protonated arm binds to the crown-ether moiety providing a pseudorotaxane structure that prevents any photochemical isomerization of the motor part to take place. In the *unlocked state*, obtained by simple deprotonation of the amine unit, the pseudorotaxane dissociates, liberating the arm and allowing the rotor to function normally when the molecule is irradiated. In this way by changing the pH threading and unthreading of the pseudorotaxane is controlled and as a consequence the rotor functioning. This motor with self-complexing lock provides a potential stepping stone to more elaborate functional motors like systems in which a translational and rotary motion are coupled ('molecular crankshaft').

Several biological motors have the ability to reverse the direction of rotation depending on the specific dynamic function, that is performed [12, 98, 99]. For instance, the flagellar motor can reverse the direction of rotation when the bacteria has to change direction from forward to backward translational movement [117]. In a light-driven molecular rotary motor, as discussed above, the directionality is controlled by the absolute configuration of the system. So a trivial way to change directionality of rotary motion is to take the other enantiomer of a stable motor as a starting point. Due to the homochirality of the amino acids in protein motors this is not an option in biological systems. To some extent reversing the direction of rotary motion has been achieved using multistep reaction sequences in an interlocked system as well as with a biaryl propeller system [102, 104]. Very recently, a system was designed in which the direction of rotation could be switched from continuous forward to backward rotation in a light-driven motor (Figure 5.30) [118].

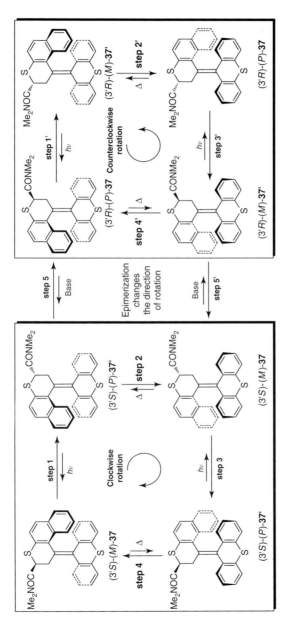

Figure 5.30 Reversal of the direction of the rotary motor **37** using base-catalysed epimerization of the stereogenic centre. Reproduced from Ref. [118]. Copyright Nature 2011.

The basic stereochemical principle involves a photochemical helix inversion from a stable to unstable form, which is followed by a base-mediated epimerization at the stereogenic centre. The inversion of the stereocentre relieves the strain in the system and this energetically downhill isomerization simultaneously results in overall helix inversion of the entire motor molecule. Further irradiation now leads to a rotation in the opposite direction to that of the original motor. Therefore, starting in a clockwise rotary cycle one has at each stage where an unstable form is generated the choice to continue the forward clockwise rotary cycle via thermal isomerization or switch to the reverse counterclockwise rotary cycle by base-induced epimerization/helix inversion. Two key features need to be mentioned; (i) the stereogenic centre with the electron-withdrawing amide moiety present in **37**, essential to allowed base-mediated epimerization, had to be placed in the homoallylic position to prevent an undesired alkene shift and (ii) as a consequence of this change in design a competing thermal reverse isomerization is observed. The adaptive mechanical behaviour in this new motor, where the direction of rotation can be reversed, offers attractive possibilities in rotary motors exhibiting multiple functions.

The ability to control the speed of rotation of these molecular motors is a crucial aspect, particularly in view of a diverse range of potential future applications [119]. The original rotary motors were exceedingly slow, taking hours to complete a full rotary cycle but the low speed allowed one to observe and characterize all intermediate stages of the rotary cycle [107, 108].

Before starting our discussion on the structural modifications, which were introduced to enhance the speed of rotation, we should realize that the thermal isomerization steps are rate determining and control the speed of the overall rotary process. The photochemical E/Z isomerization of these overcrowded alkenes is in the picosecond regime as was established by time-resolved spectroscopy [120].

So how can one enhance the speed of these rotary motors?

The most important parameter is the reduction of the steric hindrance in the so-called 'fjord region' of the molecule that lowers the energy barrier for thermal helix inversion. Four approaches have been investigated up to now, which are based on modulating steric and electronic parameters.

1) In the first-generation molecular motors the rotary process was accelerated by contracting the six-membered rings at the central olefin to the corresponding five-membered rings (Figure 5.31), reducing the steric crowding in the fjord region [121]. Furthermore, the 1,3-diaxial strain between methyl groups at the stereogenic centres and the benzylic protons in **30** is relieved to some extent in the five-membered analogue **25**. This results in additional lowering of the relative energy of the stable forms compared to the unstable forms in these overcrowded alkenes and therefore a more favourable thermal isomerization process and higher speed is observed. It turned out that the structural modification from **30** to **25** did not significantly alter the barrier for thermal helix inversion in step 2 going from unstable (M,M)-*cis*-**25** to stable (P,P)-*cis*-**25** ($\Delta G^{\#} = 93 \text{ kJ mol}^{-1}$, $\tau_{1/2}(20\,^\circ\text{C}) = 74 \text{ min}$). In contrast, the barrier to thermal helix inversion in step 4 of unstable (M,M)-*trans*-**25** to stable (P,P)-*trans*-**25**

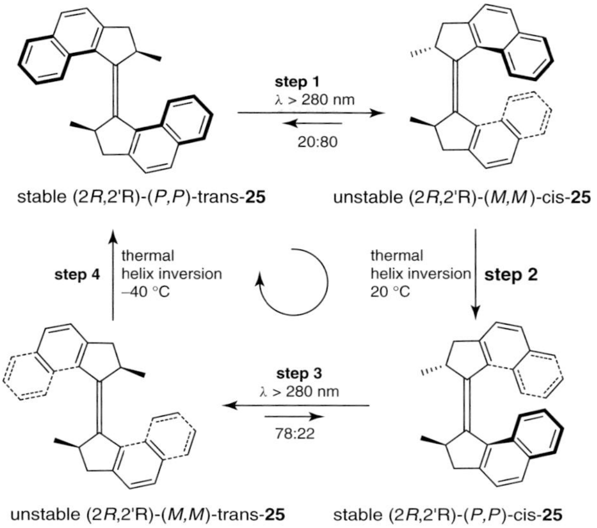

Figure 5.31 Unidirectional rotary cycle for first-generation molecular motor **25**.

was 30 kJ mol^{-1} lower ($\tau_{1/2}$ of only 18 s at 20 °C) than the barrier found for the corresponding step with motor **30** (110 kJ mol^{-1}, $\tau_{1/2}$ is 439 h at 20 °C). From these experiments it is clear that structural modification such as the use of five-membered rings fused to the axis of rotation instead of six-membered rings lead to a dramatic, circa 10^5, acceleration of the helix inversion step in the *trans* isomer but only a minor change for the *cis* isomerization step. It also highlights that the structural features that govern the Gibbs free energy barriers for *cis* and *trans* helix-inversion steps can be quite different.

2) For the second-generation motors a series of structures with different X and Y moieties were examined anticipating that shortening of d1 or d2 would facilitate the helix inversion reducing the steric hindrance at the fjord region (Table 5.1) [108]. This study revealed that even the change of a single atom can have a dramatic effect on the barrier for thermal helix inversion. For instance, exchanging the bridging atom X in the upper half in **26** from a larger sulfur atom (C–S bond length = 1.77 Å) to a methylene (C–C bond length 1.54 Å) in **28** results in a 300-fold acceleration of the thermal isomerization step at room temperature; from 215 h to only 40 min. From the kinetic studies it became apparent that it is not only the distances d_1 and d_2 that determine the barrier to thermal helix inversion but other factors, like the nature of X and Y, are also of influence. For the bridging group X in the rotor part the size of X appears to correlate roughly with the barrier of thermal helix inversion but for the Y group in the stator part there is no such correlation between the barrier and size of Y. Fluorene stators, lacking the Y bridging group, display the highest barriers, presumably due to their rigid planar structure. On the other hand, it was found that substituting the six-membered ring in the rotor (X=CH$_2$) for

Table 5.1 Kinetic parameters for the thermal helix inversion of molecular motors of the second-generation type with differing bridging atoms (X and Y).

Motor	X	Y	R	k (293 K) (s)	$\Delta^{\ddagger}G°$ (kJ mol^{-1})	$\tau_{1/2}$ (293 K) (h)
trans-32	S	S	OMe	1.04×10^{-6}	105	184
26	S	S	H	8.95×10^{-7}	106	215
27	S	O	H	7.32×10^{-6}	101	26.3
28	S	C(CH$_3$)$_2$	H	8.26×10^{-7}	106	233
29	CH$_2$	S	H	2.89×10^{-4}	91.8	0.67
31	CH$_2$	C(CH$_3$)$_2$	H	9.59×10^{-5}	94.4	2.01
32	CH$_2$	CH=CH	H	3.21×10^{-6}	103	60.1

a five-membered ring (X = –) while maintaining the fluorene moiety in the stator part (Y = –) results in a dramatic decrease of the thermal barrier by 10^8 (from $\tau_{1/2}$ = 1300 years to 3.2 min at room temperature) [122].

3) Another approach to control the speed of the motor is via modulation of the size of the substituent at the stereogenic centre [123]. The influence of the size of this group was evaluated in a series of second-generation motors featuring a fluorene stator unit (Table 5.2). It was found that increasing the size of the substituent R decreases the barrier to thermal helix inversion. The origin of this effect was evaluated using density functional theory (DFT) calculations of stable and unstable isomers as well as transition-state structures for thermal helix inversion pathways. It appears that an increase in the size of the R group results in a more twisted geometry adopted by the unstable form raising its energy relative to the stable form. This is apparent from torsion angles in the unstable forms and increased bond length of the central double bond due to less-efficient π-overlap with enhanced torsion. By contrast, the energy of the transition state and stable forms is less affected and as a consequence the thermal helix inversion is more favourable. When R = t-butyl the lowest barrier was found resulting in a half-lifetime lifetime of only 5.7 ms at room temperature (ΔG = 60 kJ mol^{-1}).

4) An alternative to control isomerization barriers is via the change of electronic effects, in particular, those that influence directly the central double bond (rotary axle) [124]. The design **37** shown in Figure 5.32a features an electronic push-pull π system involving the central double bond. The lone pairs of the amine donor moiety in the rotor can delocalize by direct conjugation along

Table 5.2 The barrier to thermal helix inversion is dependent on the size of the substituent at the stereogenic centre. The larger the substituent the lower the barrier to thermal helix inversion.

Motor	R	k (293 K) (s)	$\Delta^{\ddagger}G°$ (kJ mol^{-1})	$\Delta G_{St \to TS}$ (kJ mol^{-1})[a]	$\Delta G_{Unst \to TS}$ (kJ mol^{-1})[a]	Length of central double bond of unstable form (Å)[a]	$\tau_{1/2}$ (293 K) (h)
33	Phenyl	1.18 × 10^{-3}	88	108.2	94.6	1.3515	587
34	Methyl	3.64 × 10^{-3}	85	102.8	89.4	1.3525	190
35	i-Propyl	7.32 × 10^{-3}	84	98.3	88.2	1.3535	95
36	t-Butyl	1.21 × 10^{2}	60	88.6	67.3	1.3569	5.74 × 10^{-3}

[a] The values of ΔG and the bond lengths were calculated by DFT (B3LYP hybrid density functional, 6-31G(d) basis set. St = stable, Unst = unstable, TS = transition state.

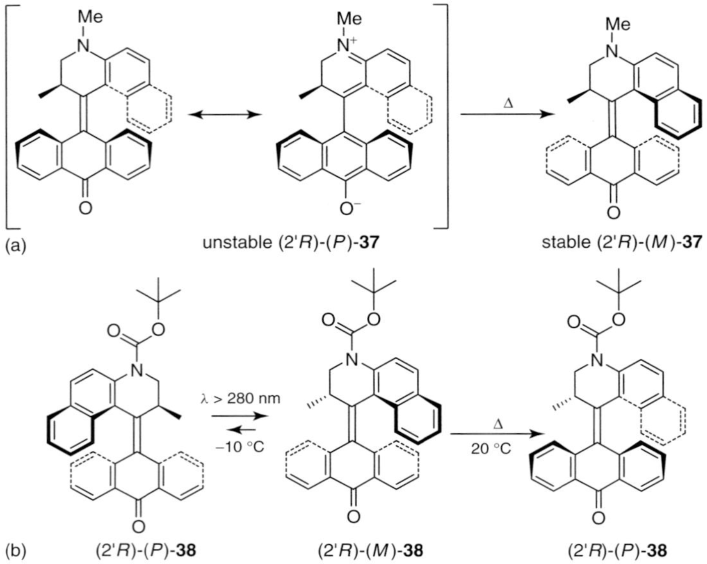

Figure 5.32 (a) Proposed resonance structure and mechanism for enhancement of rate of thermal helix inversion in **37**. (b) Thermal and photochemical isomerization of **38**.

the central double bond into the ketone acceptor group in the stator part. This electronic push-pull effect was anticipated to create a large effect as a resonance structure with a single bond axle connecting rotor and stator can be drawn. The bond lengthening and therefore reduced steric hindrance will lower the barrier to thermal helix inversion. Indeed, the N-Boc protected motor **38** (Figure 5.32b) operates similarly to the previously discussed second-generation molecular motors but shows a large increase in rotary speed [124]. The barrier for thermal helix inversion is lowered ($\Delta G = 8.7$ kJ mol^{-1}) and the half-lifetime at room temperature was reduced to 40 s.

Figure 5.33 summarizes some of the approaches taken to accelerate the second-generation molecular motors. Systematic modification of the rotor and stator parts, taking the approaches discussed above and in addition by decreasing the size of the arene moiety in the rotor part (Figure 5.33), has resulted in the fastest system allowing in principle rotation in a unidirectional sense at 3.3 MHz at room temperature [125]. It should be noted that in order to be able to study the dynamic behaviour of these ultrafast rotors nanosecond transient spectroscopy was used to access the highly unstable isomers and follow the thermal isomerization. Based on these extensive series of studies on the control of the speed of rotary motion, molecular rotary motors are now available that cover the entire range from hours to microseconds.

Of course, several parameters dictate how efficient the rotary motion in the light-driven motors is including the quantum yield, the effective input of light

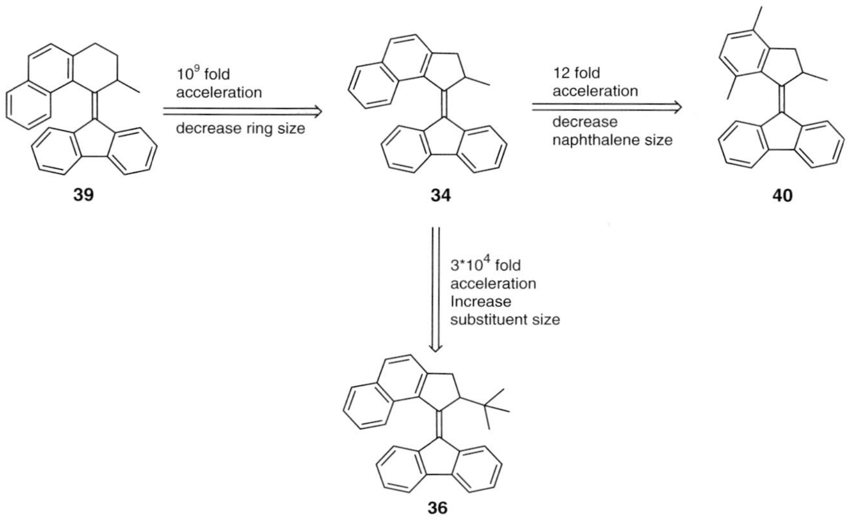

Figure 5.33 Acceleration of the rate of thermal helix inversion in fluorene-based rotary molecular motors.

energy, interference with surfaces or medium effects like the nature or viscosity of the solvent. Furthermore, in particular, when forward thermal isomerization is slow, competing photoisomerization pathways will interfere with the motor functioning. The dynamic behaviour under these conditions has been analysed with the help of a Markov model and a unidirectionality parameter for the rotary motion defined [126]. The most important outcome is that for each motor there is an optimal regime of energy input and kinetic behaviour and temperature range to have full unidirectional rotary motion.

5.4.2
Light-Driven Motors on Surfaces

An important issue both for molecular switches and molecular motors is to interface the photoactive molecular systems with the micro-and macroworld by assembly on surfaces. Considerable efforts have been devoted to assemble molecular switches on conducting surfaces in attempts towards molecular electronic devices [127] (see also Chapter 20 for an extensive discussion). The assembly and positional ordering of light-driven molecular on various surfaces is of course a key step towards integrating motors with other functions and constructing nanoscale machinery and devices, but it is also of prime importance to get out of the Brownian regime.

The first approach towards surface-immobilized systems comprised covalent binding of a second-generation motor in an azimuthal orientation to gold nanoparticles (Figure 5.34) [128]. The initial choice for nanoparticles for the surface assembly of motors was made on the premise that – as in solution studies – all spectroscopic

160 | 5 Chiroptical Molecular Switches

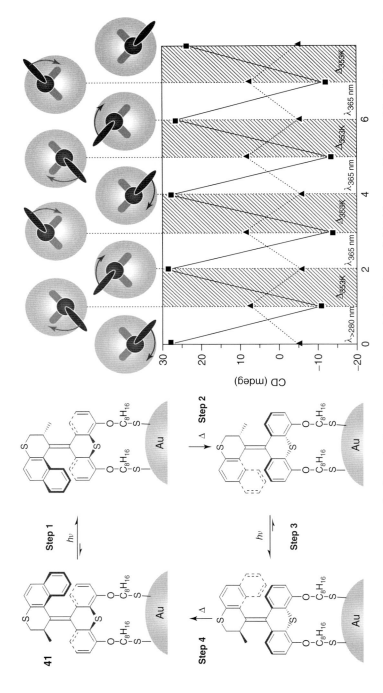

Figure 5.34 (a) Four step rotary cycle of a molecular rotary motor **41** immobilized with two alkylthiol legs to gold nanoparticles. (b) Top: Unidirectional rotary motion of the propeller part of the motor viewed along the rotation axis and two four-stage 360° rotary cycles. Bottom: The change in CD absorption at 290 nm (solid) and 320 nm during the sequential photochemical and thermal steps. Reproduced from Ref. [128]. Copyright Nature 2005.

techniques can still be used to prove the molecular motors unidirectional rotary behaviour.

The stator part of a second-generation motor was functionalized with two thiol-terminated 'legs' allowing self-assembly of a monolayer of motors on gold surfaces. The use of two attachment points in **41** prevents uncontrolled free rotation of the entire system with respect to the surface. The length and nature of the tethers (legs) is crucial to the proper functioning for two reasons: (i) first, sufficient conformational freedom should be present as the four-step rotary cycle is accompanied by major conformational changes and (ii) secondly, the motor unit should not be too close to the surface as otherwise quenching of the photochemically generated excited states occurs prior to isomerization. This latter phenomenon has been observed frequently with photoactive molecular switches. Using the Au-nanoparticle bound motors and ^{1}H-NMR spectroscopy it was confirmed that two light-induced *cis–trans* isomerizations each followed by a thermal helix inversion occur, resulting in a full and unidirectional 360° rotation of the rotor with respect to the surface-mounted stator. A slightly higher barrier for the thermal helix inversion was found for these surface-bound motors compared to the same system operating in solution. This effect was attributed to a reduction in the degrees of freedom when grafted on a surface. Related close-packing effects of surface assembled switches have been observed.

Having proven that the rotary motor works properly when connected to a surface these studies were recently extended to azimuthal motors **42** assembled on thin films of Au [129] as well as grafted to quartz (Figure 5.35) [130]. This is an important step compared to the study of rotation with respect to nanoparticles as

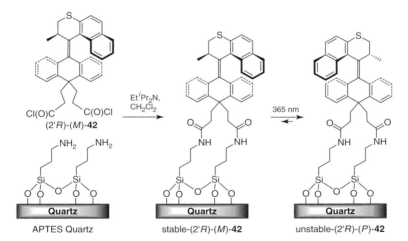

Figure 5.35 A unidirectional molecular rotary motor ((2′R)-(M)-**42**) was immobilized on to APTES modified quartz, allowing for photochemical isomerization to be studied on a surface. Adapted from Ref. [130]. Copyright Wiley 2007.

these can still undergo free rotational and translational motion in solution due to Brownian motion. Bound to Au or quartz surfaces the absolute rotation of the rotor upper half of the motor relative to the flat surface can be examined. The study of such monolayers proved to be much harder compared to solution studies due to low signal-to-noise ratios but X-ray photoelectron spectroscopy (XPS), to verify surface attachment, and CD, to assess the helix inversions, was particularly useful. The photochemical and thermal isomerization behaviour of the motor in solution and on both surfaces could be correlated through UV/Vis and CD studies and the dynamic behaviour was consistent with the unidirectional rotary cycle of the motor.

Recently, surface-bound motors with altitudinal orientation were also studied and the rotary cycle confirmed (Figure 5.36) [131]. In this case, a new click approach was followed to attach a monolayer of motors with the appropriate orientation to quartz. First, the surface was modified with azides via short spacers while the motor's legs were functionalized with alkyne groups. Copper-catalysed 1,3 dipolar cycloaddition provided a monolayer of altitudinal rotary motors and XPS and CD spectroscopy were especially useful in determining to what extent the azide-based click chemical connection has worked. The assembly of these rotary motors both in azimuthal and altitudinal orientations on various surfaces sets the stage for future studies in which the collective operation of a large ensemble of molecular motors can be exploited.

Finally, an approach is briefly discussed in which switches and motors are assembled at the solid/liquid interface through self-assembly in monolayers on surfaces. An illustrative example is seen in Figure 5.37 [132]. In this case,

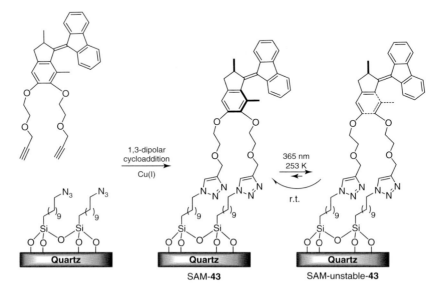

Figure 5.36 Grafting **43** to azide-modified quartz surface in altitudinal fashion.

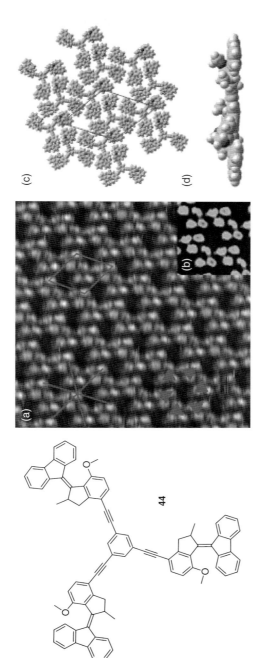

Figure 5.37 Molecular structure and STM image of trimer motor **44**. Adapted from Ref. [132]. Copyright Wiley 2009.

a trimer **44** of an ultrafast second-generation motor is deposited on a highly ordered pyrolytic graphite (HOPG) surface forming a large array monolayer. Due to the close packing in the hexagonal array on the surface and the presence of the underlying HOPG layer photoisomerization has not been observed yet. Surface quenching effects are most probably also the reason that the prototype of a molecular nanocar designed by the Tour group, featuring carborane wheels connected to a molecular frame integrated with a second-generation rotary motor, did not show autonomous light-driven motion on a surface to date [133].

With the basic principles of chiroptical molecular switches and motors discussed, the application in the dynamic control of the organization and properties of polymers, liquid crystals and gels are illustrated in the next sections. In particular, the focus is on the dynamic transmission of chiral information from the molecular level to the macromolecular, and supramolecular level.

Light-induced modulation of chirality is an attractive noninvasive way to change geometry and properties at the molecular level and it is intriguing to establish if sufficient changes in interaction with other molecules can be achieved to allow dynamic control of supra/macromolecular organization and assembly. It should be remembered that in numerous biological systems one finds dynamic self-assembly with simultaneous control of organization at different hierarchical levels. Mimicking such behaviour in artificial systems is one of the major challenges in the field of molecular switches and motors.

5.4.3
Transmission of Molecular Chiroptical Switching from Bicomponent Molecules to Polymers

Transmitting the mechanical action of a switching unit to other components of a large molecular or polymer system is challenging and must consider the distance over which the information can be communicated, that is the persistence length in polymers. An elegant example of transmission of molecular switching in chiroptical systems has been reported by Aida and coworkers [134]. In their system **45** the geometric change, which accompanies the *cis–trans* isomerization, is harnessed to produce larger changes in other parts of large molecular structures through mechanical action, that is a molecular 'scissors' action (Figure 5.38a). These systems use the change in structure upon switching of an azobenzene unit coupled to a ferrocene rotor unit. The use of the ferrocene unit, furthermore, allows for redox chemistry to be exploited also in controlling the PSS of the azobenzene unit, that is reached (i.e. in the reduced state the *cis* isomer is favoured ($\lambda_{exc} = 350$ nm), while in the oxidized state the *trans* isomer is favoured. This system **46** can be taken a step further by coupling the pivoting action induced by azobenzene isomerization to the movement of several molecular components in a so-called light-powered 'molecular pedal' 135, in which *cis–trans* photoisomerization is transmitted via a ferrocene to change the relative orientation of two porphyrin units. This motion induces a clockwise or counterclockwise rotary motion in a bound rotor guest (Figure 5.38b).

5.4 Light-Driven Molecular Rotary Motors | 165

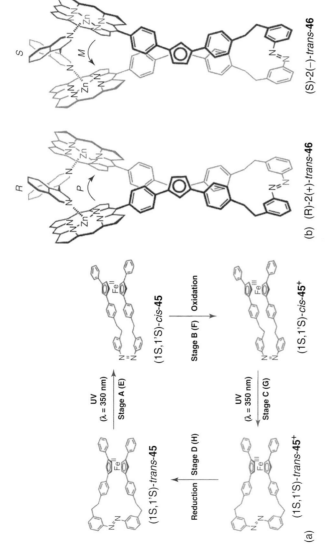

Figure 5.38 (a) Sequential operation of a ferrocene–azobenzene scissor **45**. Sequential irradiation with UV light and redox changes of the ferrocene move the systems in a scissors-like action. (b) Application of the molecular scissors **46**: the ferrocene pivot facilitates actuation of the movement of the Zn-porphyrins and hence the chiral guest, resulting in an inversion of the stereochemistry of the guest molecule.

The incorporation of photoactive components in macromolecules is an important step towards smart materials for diverse applications including light-responsive systems, coatings and thin films and materials for information storage. Numerous photochromic polymer materials have been investigated and several pertinent examples are presented in this volume [15, 21b, 136]. The use of chiroptical molecular switches, either covalently attached to a polymer backbone or embedded in polymer matrices, have been earlier discussed in the context of molecular approaches to digital data storage [15, 137]. Here, we will focus on a different aspect, namely, the use of a unidirectional rotary molecular motor and chiroptical switch to control the helicity (twist sense) of a polymer in a fully dynamic way.

The system (Figure 5.39) comprises a polyisocyanate with a pending molecular motor **47** [138]. The amplification of chiral information takes place from a single motor molecule to the helical polyisocyanate. This polymer adopts a helical conformation and the P and M helices are in rapid equilibrium. As shown by Green and coworkers [139] a strong preference for a particular handedness of the polymer can be obtained by a subtle chiral influence due to a large cooperative effect between achiral subunits and the resulting infrequent occurrence of helix reversals along the chain. For instance, a small chiral bias generated by the substitution of a hydrogen for a deuterium to generate a single stereocentre in each unit of the polymer is sufficient to achieve amplification of chirality. The strong amplification of chirality involves the 'sergeant-soldiers' and 'majority' rules effects [139, 140].

In the molecular design shown in Figure 5.39 a single enantiomer of a molecular motor with an amide group at the stator part was used as an initiator for

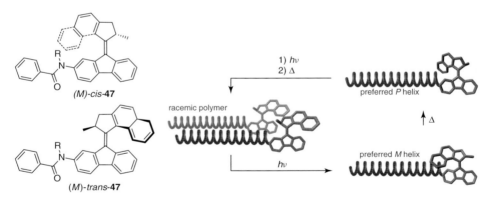

Figure 5.39 The second-generation rotary motor **47** is used to modified a polyisocyanate polymer, illustrated schematically on the right. Reversible induction and inversion of the helicity of the polymer backbone is induced by a single light-driven molecular motor positioned at the terminus. Irradiation of the motor unit results in the preference for a helical sense in the polymer. Thermal isomerization of the motor unit inverts this preferred helicity. A photochemical and thermal isomerization step returns the system to its original form with random helicity of the polymer backbone. Adapted from Ref. [138]. Copyright Wiley 2007.

the polymerization of hexylisocyanate. Starting from stable (2′S)-(M)-*trans*-**47**, a polymer was obtained with a random distribution of helices, as reflected in the absence of a CD signal due to the polyisocyanate. The photoisomerization of the motor unit to the unstable (2′S)-(P)-*cis* form results in a polymer with excess of M helicity. Subsequent thermal isomerization of the motor unit to (2′S)-(M)-*cis*-**47** results in helix reversal of the polymer now showing excess P helicity. Finally, photochemical and thermal isomerization steps re-establish the original (2′S)-(M)-*trans*-**47** motor and simultaneously a random mixture of left- and right-handed polymers is observed. In this system a single terminal motor molecular governs the preferred helicity of a long helical polymer rod in a fully dynamic sense. Amplification of chirality occurs therefore from the molecular to the macromolecular level. Importantly, this system is fully dynamic and as long as light energy is supplied to the system rotational motion takes place and as a consequence helix inversion of the polymer occurs. Although a steady state is observed the system is not in thermodynamic equilibrium. This design nicely illustrates how light-powered motors can be used to control out-of-equilibrium systems.

In a recent design the helical polymer chain was attached to the rotor part and this change in structure resulted in a stronger chiral amplification effect from motor to polymer [141]. In this case, motor function was not observed but the system operates as a two stage P = M chiroptical switch.

An intriguing phenomenon was observed when thin films of these polymers were deposited on a variety of surfaces including quartz, gold and mica [142]. Spontaneous large-range microscopic patterning was found most probably by a dewetting mechanism favouring the formation of toroidal structures. It turned out that the presence of the chiral switch unit at the end of the polymer chain is essential for the toroid formation to be observed. These chiral polymers have recently also been shown to form twisted nematic LC films that are responsive to light.

5.5
Liquid Crystals

Soft materials based on LC compounds have found widespread application in LC devices, sensors and displays [143]. For decades LC materials have played a prominent role in attempts to demonstrate how molecular switches can be used to control supramolecular organization and properties. A variety of chiral dopants was introduced as guests in nematic liquid crystals to induce cholesteric phases and inherent dissymmetric binaphtyl compounds are among the most potent small molecule chiral dopants currently available [144].

Overcrowded alkenes are particularly attractive as they combine high helical twisting power with the propensity to change chirality upon photoisomerization. Doping of LC films with chiral photoswitchable guest molecules enables control of organization and hence optical properties of these LC films through dynamic

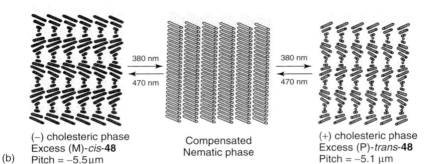

Figure 5.40 (a) Photoisomerization of N-hexyl substituted cis- and trans isomers of chiroptical switch **48**, and structure of calamitic mesogen. (b) Schematic representation of the switching of the handedness of a doped cholesteric liquid crystal film.

changes in molecular orientation of the strongly anisotropic host molecules [145]. A prominent example is shown in Figure 5.40.

Enantiomerically pure (P)-*cis*-**48** is used as a dopant (1 wt%) in nematic 4′-(pentyloxy)-4-biphenylcarbonitrile resulting in the formation of a twisted nematic (cholesteric) phase. The helicity of the chiroptical switch **48** can be readily modulated at the molecular level by irradiation with 470 nm (excess M helix) or 380 nm (excess P helix). As a consequence of this change in chirality of the dopant, the helical screw sense and the pitch of the supramolecular cholesteric phase are modulated upon irradiation at these two different wavelengths. Two properties of this system are particularly noteworthy: (i) a complete reversal of the handedness of the cholesteric phase is achieved and (ii) depending on irradiation time and wavelength used a mixture of (P)-*cis*-**48** and (M)-*trans*-**48** with a nearly 1 : 1 ratio (pseudoracemate) can be generated, resulting in a compensated nematic (achiral) phase. The chirality of the mesophase can be switched on again by shifting the photoequilibrium upon prolonged irradiation using either 380- or 470-nm light.

Overcrowded alkene **5** was also used as a chiroptical switch in nematic liquid crystals [36]. In this case a small chiral preference (0.07%) for either the P or M enantiomer was obtained by irradiation with CPL (*vide supra*). This tiny chiral bias was subsequently amplified in a nematic liquid crystal providing a cholesteric phase the helical screw sense of which is determined by the chirality of the CPL used for molecular switching (Figure 5.41). The chiral amplification mechanism comprises

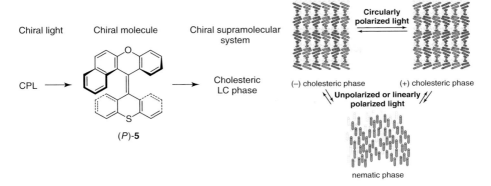

Figure 5.41 (Left) Amplification of chirality by hosting **5** in an LC phase. (Right) Schematic representation of switching between different liquid-crystalline phases.

therefore two stages; first, the chirality of the circularly polarized light controls the small excess of one enantiomer of the molecular switch and secondly when this compound is used as a guest in a LC host material the chirality is amplified in the helical organization of the induced cholesteric phase.

Light-driven molecular motors based on overcrowded alkenes can also serve as excellent chiral dopants for LC films to induce cholesteric phases [146]. For instance, molecular motor **33** with a phenyl-substituent at the stereogenic centre, was designed as a switchable chiral dopant for a range of mesogenic materials (Figure 5.42). The change in molecular helicity upon photoisomerization is accompanied by a large change in helical twisting power. When used in E7, indeed a cholesteric phase is observed and upon irradiation the colour of the LC film can be readily tuned over the entire visible spectrum due to the change in helical organization of the LC phase. The system is fully reversible and this opens new avenues in the dynamic generation of LC colour pixels and sensing.

This system was further explored to demonstrate that a molecular motor can perform work, in particular, by moving microscopic objects (Figure 5.43) [147]. Indeed, it could be shown that changes in helicity of this rotary molecular motor are sufficient to change the surface texture of the LC films and induce microscopic rotary motion. Motor **33** is a highly effective chiral dopant generating a twisted nematic organization with planar anchoring at the LC/air interface. Atomic force microscopy (AFM) and optical profilometry revealed a surface relief of 20 nm and the orientation and nature of this surface relief responded to the changes in topology (helicity) of the motor molecule **33** used as chiral dopant. As with the chiral switches discussed above the photoisomerization changes the helical twisting power and chirality of the host motor molecule, which results in a reorganization of the polygonal LC texture [148]. The concomitant change in surface relief (observed as a rotation and widening of the lines) generates sufficient torque to rotate a

(a)   6.8 wt% motor in E7, irradiation at 365 nm

(b)   Same sample, thermal step back at $T = 22\ °C$.

Figure 5.42 Molecular motor **33** undergoes photochemical and thermal isomerization to produce colour changes in the LC forming E7. Adapted from Ref. [146]. Copyright Wiley 2006.

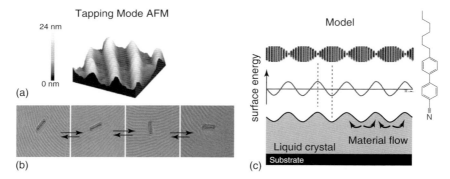

Figure 5.43 (a) AFM image (tapping mode) of the surface of an LC film that has been doped with molecular motor **33**. (b) Micrometer-size glass rod objects resting on the LC surface can be physically moved by the reorganization induced by light. (c) Model showing the orientation of the LC component in the twisted nematic phase, surface energy profile and material flow accompanying the change in helical twisting power (HTP) of dopant **33** upon irradiation. Adapted from Refs. [147, 148]. Copyright Nature 2006 and American Chemical Society 2006.

microscopic object, for example a glass rod placed on top of the LC film with full control of directionality of the rotation.

A distinct feature with these motors is that the reverse rotation can be induced upon thermal helix inversion of the motor molecules. In this system, the change in chirality of the motor is transmitted through the change in the organization of the LC phase and the surface texture to the moving microscopic object. For instance, using (M)-**33** as a chiral dopant irradiation of the LC matrix results in clockwise rotation of the glass rod. When the PSS is reached the rotation stops. The next step, the thermal isomerization of the motor molecule results in a reverse counterclockwise rotation of the rod. Therefore, the direction of rotation of the rod reflects the change in chirality of the motor due to the rotor movement. Although there is no direct transmission of motion it is remarkable that, using light as an energy source, a collection of molecular motors is capable of rotating microscopic objects 10 000 times their own size.

5.6
Gels

Gels are fascinating soft materials with a remarkable range of applications in the food, chemical, pharmaceutical and personal-care industries. Low molecular weight (LMW) organogelators and hydrogelators typically assemble into supramolecular aggregates and fibre networks in a range of solvents [149]. These physical gels are highly dynamic and the incorporation of molecular switches makes these gels responsive. A variety of trigger functions have been introduced in LMW gels to control the self-assembly including pH-, redox- and light-responsive units. Chiroptical molecular switches are particularly attractive as many gels are highly sensitive to changes in molecular chirality and light offers, of course, a noninvasive way to achieve reversible control over the process of self-assembly. This can be a major advantage in, for instance, gel-based controlled-release systems.

Towards this goal diarylethene photochromic switches were functionalized with amide groups based on (R)-1-phenylethylamide [150]. The amide groups have a strong propensity to form multiple hydrogen bonds and not unexpectedly these molecules aggregate to form fibre-like structures. In photochromic LMW gelator **49** the open form is flexible but after switching with UV light a highly rigid core structure is obtained. Visible-light irradiation leads to ring opening going from a more rigid to a more flexible core structure. Furthermore, the open form features two rapidly interconverting helices in dynamic equilibrium, while in the closed form the chirality of the core part is locked. The ability to form multiple hydrogen bonds and the control by light of the dynamics and chirality of the switch unit is the basis for the multiple addressable gelator system developed with **49** (Figure 5.44).

When a dilute solution of the photochromic switch **49** is irradiated with UV light, the closed form is obtained as a 50 : 50 mixture of diastereoisomers. However, in the gel state a highly stereoselective switching process is observed resulting in a 98 : 2 ratio of stereoisomers of the closed form. Once the dynamic open form of these compounds aggregates, helical fibres are formed (determined by electron microscope (EM) and CD measurements) and the equilibrium shifts, resulting in unique molecular chirality in the fibres. Upon photochemical ring closure the molecular chirality is locked. The gel with **49** in its closed form is metastable and when a heating–cooling cycle is applied the system undergoes

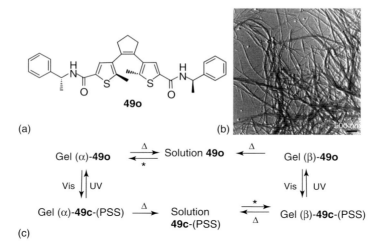

Figure 5.44 (a) A bis-amide substituted dithienylethene low molecular weight gelator molecule. (b) Transmission electron microscope (TEM) image of the gel fibres formed upon gelation of an organic solvent. (c) The four gel state cycle. **49o** and **49c** are the ring open and ring-closed states, respectively. Gel-α and gel-β indicate distinct gelation states, Δ indicates heating, * indicates cooling.

rearrangement to give a new gel with inverted helicity at the supramolecular level. Subsequent photochemical ring opening leads to another metastable gel and finally a heating–cooling cycle provides the original stable gel state. The four-stage switching cycle is shown in Figure 5.44.

It is remarkable that near-complete stereocontrol is only reached in the gel state where the supramolecular chirality (fibre helicity) dictates the molecular chirality of the switch unit. In its turn, after photochemical switching, the chiral switching unit dictates the supramolecular chirality. Due to this mutual stereochemical control metastable chiral aggregates can be obtained in a fully reversible manner. It should be emphasized that the presence of the chiroptical switches allows for control of self-assembly, supramolecular chirality and viscoelastic properties of these gels. In addition, the control of mass transport with light was explored in holographic patterning [151]. These photoresponsive chiral gels nicely illustrate how dynamic control at different hierarchical levels in a self-assembled system can be achieved with a subtle interplay of chirality and molecular interactions taking advantage of the unique properties of molecular switches.

5.7
Conclusions and Perspectives

Chirality is a symmetry property intrinsic to life on the earth and it is not surprising that the ability to control chirality using molecular switches offers superb opportunities in the design of smart responsive materials. A small change in molecular chirality can be amplified through different hierarchical levels in a system and this usually has a profound influence on several parameters varying from the propensity for self-assembly to the structure and properties of a surface. The presence of optical molecular switches offers the advantage that chirality can be controlled in a noninvasive but dynamic and reversible sense. This allows the triggering and control of a myriad of functions, as illustrated in this chapter.

Much of the early work on chiroptical molecular switches [15] focused on data storage, taking advantage of chirality for nondestructive readout. In recent years, the emphasis has shifted to control of supramolecular (gels, liquid crystals) and macromolecular systems as well as surface properties through molecular switches. For these systems response time is often not as critical as in materials for information storage. Furthermore, an important step has been taken in going from molecular switches to rotary molecular motors. The ability to control unidirectional rotary motion not only offers a marvellous illustration of how subtle stereochemical effects can be exploited, but these developments also pave the way to more complex switching behaviours.

A key lesson learned from the studies discussed in this chapter is that the translation of molecular switching and motor systems from homogeneous solution to surfaces might affect the dynamic behaviour, in particular, regarding competing photochemical pathways. On the other hand, matrix effects can play a prominent role regarding quantum yield, kinetics, switching times, stereoselectivity, and so on. The same holds for more complex multicomponent systems with integrated switch and motor functions where mutual interference often occurs.

The switchable, responsive or adaptive behaviour that can be introduced through chiroptical molecular switches offers, of course, tremendous opportunities with the ability to control chirality as a great bonus. Finally, the control of metastable systems using chiroptical switches or the design of out-of-equilibrium systems powered by molecular motors are only two illustrations of the many fascinating challenges ahead.

References

1. Gardner, M. (1974) *The Ambidextrous Universe*, Penquin, London.
2. Berg, J.M., Tymoczko, J.L. and Stryer, L. (2002) *Biochemistry*, 5th edn, W.H. Freeman, New York.
3. Crick, F. (1981) *Life Itself*, McDonald & Co, London.
4. Brunner, H. (1999) *Rechts Oder Links*, Wiley-VCH Verlag GmbH, Weinheim.
5. McManus, C. (2002) *Right Hand, Left Hand, the Origins of Asymmetry in Brains, Bodies, Atoms and Cultures*, Weidenfeld & Nicolson, London.
6. Amabilino, D.B. (ed.) (2009) *Chirality at the Nanoscale*, Wiley-VCH Verlag GmbH, Weinheim.
7. Alberts, B., Johnson, A., Lewis, J., Raff, M., Roberts, K. and Walter, P. (2002) *Molecular Biology of the Cell*, Garland Science, New York.
8. Goodsell, D.S. (1996) *Our Molecular Nature, the Body's Motors, Machines and Messages*, Springer, Heidelberg.
9. Feringa, B.L., Jager, W.F. and de Lange, B. (1993) *Tetrahedron*, **49**, 8267.

10. Irie, M. (ed.) (1994) *Photoreactive Materials for Ultrahigh Density Optical Memory*, Elsevier, Amsterdam.
11. Bahadur, B. (ed.) (1991) *Liquid Crystals: Applications and Uses*, vol. I-III, World Scientific, Singapore.
12. Schliwa, M. (ed.) (2003) *Molecular Motors*, Wiley-VCH Verlag GmbH, Weinheim.
13. Feringa, B.L. (2001) *Acc. Chem Res.*, **34**, 504.
14. Feringa, B.L., van Delden, R.A., Koumura, N. and Geertsema, E.M. (2000) *Chem. Rev.*, **100**, 1789.
15. Feringa, B.L. (ed.) (2001) *Molecular Switches*, Wiley-VCH Verlag GmbH, Weinheim.
16. Mathies, R.A. and Lugtenburg, J. (2000) in *Molecular Mechanisms of Visual Transduction* (eds D.G. Stavenga, W.J. DeGrip and E.N. Pugh Jr), Elsevier, Amsterdam, the Netherlands, p. 55.
17. (a) Durr, H. and Bouas-Laurent, H. (2003) *Photochromism : Molecules and Systems*, Elsevier; (b) Irie, M. (ed.) (2000) *Chem. Rev.*, **100** (5), 6–1890 thematic issue on photochromism.
18. Reichert, S. and Breit, B. (2007) *Org. Lett.*, **9**, 899–902.
19. Hill, D.J., Mio, M.J., Prince, R.B., Hughes, T.S. and Moore, J.S. (2001) *Chem. Rev.*, **101**, 3893–4011.
20. Miwa, K., Furusho, Y. and Yashima, E. (2010) *Nature Chem.*, **2**, 444.
21. (a) Crego-Calama, M. and Reinhoudt, D.N. (eds) (2006) *Top. Curr. Chem.*, 265; (b) Pijper, D. and Feringa, B.L. (2008) *Soft Matter*, **4**, 1349; (c) Qiu, Y., Chen, P., Guo, P., Li, Y. and Liu, M. (2008) *Adv. Mater.*, **20**, 2908–2913.
22. Wasielewski, M.R. (1996) *Science*, **274**, 584.
23. Durr, H. (1989) *Angew. Chem. Int. Ed. Engl.*, **28**, 413.
24. Irie, M. and Mohri, M. (1988) *J. Org. Chem.*, **53**, 803.
25. Shinkai, S., Ogawa, T., Jusano, Y., Manabe, O., Kikukawa, K., Goto, T. and Matsuda, T. (1982) *J. Am. Chem. Soc.*, **104**, 1960.
26. Wendorff, J.H., Eich, M., Reck, B. and Ringsdorf, H. (1987) *Macromol. Chem. Rapid Commun.*, **8**, 59.
27. (a) Daub, J., Fischer, C., Knochel, T., Kunkely, H., Rapp, K.M. and Salbeck, J. (1989) *Angew. Chem. Int. Ed. Engl.*, **28**, 1494; (b) Gobbi, L., Seiler, P. and Diederich, F. (1999) *Angew. Chem. Int. Ed. Engl.*, **38**, 674.
28. Schmidt, H.W. (1989) *Adv. Mater.*, **1**, 940.
29. Shinkai, S., Matsuo, K., Harada, A. and Manabe, O. (1982) *J. Chem. Soc. Perkin Tr. 2*, 1261.
30. (a) Kreysig, D. and Stumpe, J. (1990) in *Selected Topics in Liquid Crystalline Research* (ed. H.D. Koswig), Wiley-VCH Verlag GmbH, Weinheim; (b) Ikeda, T. and Tsutsumi, O. (1995) *Science*, **268**, 1873.
31. (a) Tachibana, H., Goto, A., Nakamura, T., Matsumoto, M., Manda, E., Niino, H., Yabe, A. and Kawabata, Y. (1989) *Thin Solid Films*, **179**, 207; (b) Sawodny, M., Schmidt, A., Urban, C., Ringsdorf, H. and Knoll, W. (1992) *Prog. Colloid. Polymer Sci.*, **89**, 165; (c) Guo, P., Zhang, L. and Liu, M. (2006) *Adv. Mater.*, **18**, 177.
32. Feringa, B.L., Jager, W.F., de Lange, B. and Meijer, E.W. (1991) *J. Am. Chem. Soc.*, **113**, 5468.
33. Tomlinson, W. (1984) *J. Appl. Opt.*, **23**, 4609.
34. Feringa, B.L. and Wynberg, H. (1977) *J. Am. Chem. Soc.*, **99**, 602.
35. (a) Jager, W.F., de Jong, J.C., de Lange, B., Huck, N.P.M., Meetsma, A. and Feringa, B.L. (1995) *Angew. Chem. Int. Edn. Engl.*, **34**, 348; (b) Feringa, B.L., Jager, W.F. and de Lange, B. (1993) *Chem. Commun.*, 288.
36. Huck, N.P.M., Jager, W.F., de Lange, B. and Feringa, B.L. (1996) *Science*, **273**, 1686.
37. Inoue, Y. (1992) *Chem. Rev.*, **92**, 741.
38. Feringa, B.L., Schoevaars, A.M., Jager, W.F., de Lange, B. and Huck, N.P.M. (1996) *Enantiomer*, **1**, 325.
39. van Delden, R.A., ter Wiel, M.K.J. and Feringa, B.L. (2004) *Chem. Commun.*, 200.
40. (a) Shinkai, S. (2001) in *Molecular Switches*, Chapter 9 (ed. B.L. Feringa), Wiley-VCH Verlag GmbH, Weinheim; (b) Shinkai, S. and Manabe, O. (1984) *Top. Curr. Chem.*, **121**, 67.

41. (a) Shinkai, S., Nakaji, T., Ogawa, T., Shigematsu, K. and Manabe, O. (1981) *J. Am. Chem. Soc.*, **103**, 111; (b) Shinkai, S., Ogawa, T., Kusano, Y., Manabe, O., Kikukawa, K., Goto, T. and Matsuda, T. (1982) *J. Am. Chem. Soc.*, **104**, 1960.
42. Ikeda, T. and Tsutsumi, O. (1995) *Science*, **268**, 1873.
43. (a) Goodman, M. and Kossey, A. (1966) *J. Am. Chem. Soc.*, **88**, 5010; (b) Pieroni, O., Houben, J.L., Fissi, A., Costantino, P. and Ciardelli, F. (1980) *J. Am. Chem. Soc.*, **102**, 5913; (c) Ueno, A., Anzai, J., Osa, T. and Takahashi, K. (1981) *J. Am. Chem. Soc.*, **103**, 6410.
44. Wang, Z.Y., Todd, E.K., Meng, X.S. and Gao, J.P. (2005) *J. Am. Chem. Soc.*, **127**, 11552.
45. Carreno, M.C., Garcia, I., Nunez, I., Merino, E., Ribagorda, M., Pieraccini, S. and Spada, G.P. (2007) *J. Am. Chem. Soc.*, **129**, 7089.
46. van Delden, R.A., Mecca, T., Rosini, C. and Feringa, B.L. (2004) *Chem. Eur. J.*, **10**, 61.
47. Pieraccini, S., Masiero, S., Spada, G.P. and Gottarelli, G. (2003) *Chem. Commun.*, 598.
48. (a) Kawamoto, M., Aoki, T., Shiga, N. and Wada, T. (2009) *Chem. Mater.*, **21**, 564; (b) Takaishi, K., Kawamoto, M., Tsubaki, K. and Wada, T. (2009) *J. Org. Chem.*, **74**, 5723.
49. (a) Ma, X., Qu, D., Ji, F., Wang, Q., Zhu, L., Xu, Y. and Tian, H. (2007) *Chem. Commun.*, 1409; see also (b) Harada, A. (2006) *J. Polym. Sci. Part A-Polym. Chem.*, **17**, 5113.
50. Furusho, Y., Tanaka, Y., Maeda, T., Ikeda, M. and Yashima, E. (2007) *Chem. Commun.*, 3174.
51. (a) Malcolm, B.R. and Pieroni, O. (1990) *Biopolymers*, **29**, 1121; (b) Menzel, H., Weichart, A., Schnidt, S., Paul, S., Knoll, W., Stumpe, J. and Fischer, T. (1994) *Langmuir*, **10**, 1926.
52. Pieroni, O., Fissi, A., Houben, J.L. and Ciardelli, F. (1985) *J. Am. Chem. Soc.*, **107**, 2990.
53. Ciardelli, F., Aglietho, M., Carlini, C., Chiellini, E. and Solaro, R. (1982) *Pure Appl. Chem.*, **54**, 521.
54. Ulysse, L., Cubillos, J. and Chmielewski, J. (1995) *J. Am. Chem. Soc.*, **117**, 8466.
55. Higuchi, M., Nimoura, N. and Kinoshita, T. (1994) *Chem. Lett.*, 227.
56. Kinoshita, T. (1995) *Prog. Polym. Sci.*, **20**, 527.
57. (a) Irie, M. (2000) *Chem. Rev.*, **100**, 1685; (b) Uchida, K. and Irie, M. (1995) *Chem. Lett.*, 969; (c) Gilat, S.L., Kawai, S.H. and Lehn, J.-M. (1993) *J. Chem. Soc. Chem. Commun.*, 1439; (d) Irie, M. (2001) in *Molecular Switches* (ed. B.L. Feringa), Wiley-VCH Verlag GmbH, Weinheim, pp 37–62.
58. Irie, M. and Nakamura, S. (1988) *J. Org. Chem.*, **53**, 6136.
59. (a) Gilat, S.L., Kawai, S.H. and Lehn, J.-M. (1995) *Chem. Eur. J.*, **1**, 275; (b) Irie, M., Sakemura, K., Okinaka, M. and Uchida, K. (1995) *J. Org. Chem.*, **60**, 8305; (c) Tsivgoulis, G.M. and Lehn, J.-M. (1996) *Chem. Eur. J.*, **2**, 1399; (d) Tsivgoulis, G.M. and Lehn, J.-M. (1995) *Angew. Chem., Int. Ed. Engl.*, **34**, 1119.
60. Woodward, R.B. and Hoffmann, R. (1970) *The Conservation of Orbital Symmetry*, Wiley-VCH Verlag GmbH, Weinheim.
61. Yamaguchi, T., Uchida, K. and Irie, M. (1997) *J. Am. Chem. Soc.*, **119**, 6066.
62. Denekamp, C. and Feringa, B.L. (1998) *Adv. Mater.*, **10**, 1080.
63. Walko, M. and Feringa, B.L. (2007) *Chem. Commun.*, 1745.
64. Norsten, T.B., Peters, A., McDonald, R., Wang, M. and Branda, N.R. (2001) *J. Am. Chem. Soc.*, **123**, 7447.
65. Wigglesworth, T.J., Sud, D., Norsten, T.B., Lehki, V.S. and Branda, N.R. (2005) *J. Am. Chem. Soc.*, **127**, 7272.
66. (a) Yokoyama, Y., Shiraishi, H., Tani, Y., Yokoyama, Y. and Yamaguchi, Y. (2003) *J. Am. Chem. Soc.*, **125**, 7194; (b) Kose, M., Shinoura, M., Yyokoyama, Y. and Yomoyama, Y. (2004) *J. Org. Chem.*, **69**, 8403; (c) Yokoyama, Y. (2004) *Chem. Eur. J.* **10**, 4389.
67. Okuyama, T., Tani, Y., Miyake, K. and Yokoyama, Y. (2007) *J. Org. Chem.*, **72**, 1634.

68. (a) Stobbe, H. (1904) *Chem. Ber.* **37**, 2232; (b) Stobbe, H. (1905) *Chem. Ber.*, **38**, 3673.
69. (a) Heller, H.G. (1978) *Chem. Ind.*, (6), 193; (b) Heller, H.G. and Oliver, S. (1981) *J. Chem. Soc. Perkin Tr. 1*, 197; (c) Yokoyama, Y., Shimizu, Y., Uchida, S. and Yokoyama, Y. (1995) *Chem. Commun.* 785.
70. (a) Yokoyama, Y., Shimizu, Y., Uchida, S. and Yokoyama, Y. (1995) *Chem. Commun.*, 785; (b) Yokoyama, Y., Iwai, T., Yokoyama, Y. and Kurita, Y. (1994) *Chem. Lett.*, 225.
71. Yokoyama, Y., Uchida, S., Yokoyama, Y., Sugawara, Y. and Kurita, Y. (1996) *J. Am. Chem. Soc.*, **118**, 3100.
72. Ringsdorf, H., Cabrera, I. and Dittrich, A. (1991) *Angew. Chem. Int. Ed. Engl.*, **30**, 76.
73. Canary, J.W. (2009) *Chem. Soc. Rev.*, **38**, 747–756.
74. (a) Zhang, S. and Canary, J.W. (2006) *Org. Lett.*, **8**, 3907; (b) Zahn, S. and Canary, J.W. (2000) *Science*, **288**, 1404.
75. (a) Zahn, S. and Canary, J.W. (1998) *Angew. Chem. Int. Ed.*, **37**, 305; (b) Zahn, S. and Canary, J.W. (2002) *J. Am. Chem. Soc.*, **124**, 9204; (c) Das, D., Dai, Z.H., Holmes, A. and Canary, J.W. (2008) *Chirality*, **20**, 585; (d) Barcena, H.S., Holmes, A.E., Zahn, S. and Canary, J.W. (2003) *Org. Lett.*, **5**, 709; (e) Holmes, A.E., Das, D. and Canary, J.W. (2007) *J. Am. Chem. Soc.*, **129**, 1506.
76. Westermeier, C., Gallmeier, H.-C., Komma, M. and Daub, J. (1999) *Chem. Commun.*, 2427.
77. Beer, G., Niederalt, C., Grimme, S. and Daub, J. (2000) *Angew. Chem. Int. Ed.*, **39**, 3252.
78. Fukui, M., Mori, T., Inoue, Y. and Rathore, R. (2007) *Org. Lett.*, **9**, 3977.
79. Nishida, J., Suzuki, T., Ohkita, M. and Tsuji, T. (2001) *Angew. Chem. Int. Ed.*, **40**, 3251.
80. Gomar-Nadal, E., Veciana, J., Rovira, C. and Amabilino, D.B. (2005) *Adv. Mater.*, **17**, 2095.
81. Zhou, Y., Zhang, D., Zhu, L., Shuai, Z. and Zhu, D. (2006) *J. Org. Chem.*, **71**, 2123.
82. Suzuki, T., Iwai, T., Ohta, E., Kawai, H. and Fujiwara, K. (2007) *Tetrahedron Lett.*, **48**, 3599.
83. Deng, J., Song, N., Zhou, Q. and Su, Z. (2007) *Org. Lett.*, **9**, 5393.
84. Suzuki, T., Ishigaki, Y., Iwai, T., Kawai, H., Fujiwara, K., Ikeda, H., Kano, Y. and Mizuno, K. (2009) *Chem. Eur. J.*, **15**, 9434.
85. Zheng, J., Qiao, W., Wan, J.P., Gao, J.P. and Wang, Z.Y. (2008) *Chem. Mater.*, **20**, 6163.
86. (a) Bari, L.D., Lelli, M., Pintacuda, G. and Salvadori, P. (2002) *Chirality*, **14**, 265; (b) Dickens, R.S., Aime, S., Batsanov, A.S., Beeby, A., Botta, M., Bruce, J.I., Howard, J.A.K., Love, C.S., Parker, D., Peacock, R.D. and Puschmann, H., (2002) *J. Am. Chem. Soc.*, **124**, 12697; (c) Lelli, M., Bari, L.D. and Salvadori, P. (2007) *Tetrahedron Asymmetry*, **18**, 2876.
87. Li, D., Wang, Z.Y. and Ma, D. (2009) *Chem. Commun.*, 1529.
88. Kickova, A., Donovalova, J., Kasak, P. and Putala, M. (2010) *New. J. Chem.*, **34**, 1109.
89. (a) Kroner, D., Klaumunzer, B. and Klamroth, T. (2008) *J. Phys. Chem. A*, **112**, 9924; (b) Kroner, D. and Klaumunzer, B. (2007) *Phys. Chem. Chem. Phys.*, **9**, 5009.
90. Jiang, X., Lim, Y.-K., Zhang, B.J., Opsitnick, E.A., Baik, M.-H. and Lee, D. (2008) *J. Am. Chem. Soc.*, **130**, 16812.
91. Zhang, G. and Liu, M. (2009) *J. Mater. Chem.*, **19**, 1471.
92. Eggers, L. and Bush, V. (1997) *Angew. Chem. Int. Ed. Engl.*, **36**, 881.
93. Chen, W.-C., Lee, Y.-W. and Chen, C.-T. (2010) *Org. Lett.*, **12**, 1472.
94. de Silva, A.P., McClenaghan, N.D. and McCoy, C.P. (2001) in *Molecular Switches*, Chapter 11 (ed. B.L. Feringa), Wiley-VCH Verlag GmbH, Weinheim.
95. Magri, D.C., Brown, G.J., Mcclean, G.D. and de Silva, A.P. (2006) *J. Am. Chem. Soc.*, **128**, 4950.
96. Huck, N.P.M. and Feringa, B.L. (1995) *J. Am. Chem. Soc.*, **107**, 1095.
97. van Delden, R.A., Huck, N.P.M., Piet, J.J., Warman, J.M., Meskers, S.C.J.,

Dekkers, H.P.J.M. and Feringa, B.L. (2003) *J. Am. Chem. Soc.*, **125**, 15659.
98. Walker, J.E. (1998) *Angew. Chem. Int. Ed.*, **37**, 2308.
99. Berg, H.C. and Anderson, R.A. (1973) *Nature*, **245**, 380.
100. (a) *Sci. Am.*, special issue: Nanotech: the science of small gets down to business, September 2001; (b) Feynman, R.P. (1971) in *Miniturization* (ed. Gilbert, H.D.), Reinhold, New York; (c) Drexler, K.E. (1992) *Nanosystems: Molecular Machines, Manufacturing and Computation*, John Wiley & Sons, Inc., New York; (d) Astumian, R.D. (2001) *Sci. Am.*, **285**, 56.
101. Browne, W.R. and Feringa, B.L. (2006) *Nature Nanotech.*, **1**, 125.
102. Feringa, B.L. (2007) *J. Org. Chem.*, **72**, 6635.
103. van den Heuvel, M.G.L. and Dekker, C. (2007) *Science*, **317**, 333.
104. Kay, E.R., Leigh, D.A. and Zerbetto, F. (2007) *Angew. Chem. Int. Ed.*, **46**, 72–191.
105. For an excellent review on molecular rotors, see: Kottas, G.S., Clarke, L.I., Horinek, D. and Michl, J. (2005) *Chem. Rev.*, **105**, 1281.
106. Astumian, R.D. (1997) *Science*, **276**, 917.
107. Koumura, N., Zijlstra, R.W.J., van Delden, R.A., Harada, N. and Feringa, B.L. (1999) *Nature*, **401**, 152.
108. (a) Koumura, N., Geertsema, E.M., Meetsma, A. and Feringa, B.L. (2000) *J. Am. Chem. Soc.*, **122**, 12005; (b) Koumura, N., Geertsema, E.M., van Gelder, M.B., Meetsma, A. and Feringa, B.L. (2002) *J. Am. Chem. Soc.* **124**, 5037.
109. ter Wiel, M.K.J., van Delden, R.A., Meetsma, A. and Feringa, B.L. (2005) *J. Am. Chem. Soc.*, **127**, 4208.
110. Harada, N., Koumura, N. and Feringa, B.L. (1997) *J. Am. Chem. Soc.*, **119**, 7256.
111. Groenhof, G. (2003) *Understanding Light Induced Conformational Changes in Molecular Systems from first Principles*, PhD Thesis, University of Groningen.
112. Grimm, S., Brauchle, C. and Frank, I. (2005) *ChemPhysChem*, **6**, 1943.
113. (a) ter Wiel, M.K.J., van Delden, R.A., Meetsma, A. and Feringa, B.L. (2005) *J. Am. Chem. Soc.*, **127**, 14208; (b) see also: Klok, M., Walko, M., Geertsema, E.M., Ruangsupapichat, N., Kistemaker, J.C.M., Meetsma, A. and Feringa, B.L. (2008) *Chem. Eur. J.*, **14**, 11183.
114. Overcrowded alkenes, see also: Biedermann, P.U., Stezowski, J.J. and Agranat, I. (2001) *Eur. J. Org. Chem.*, 15.
115. For stereochemical definitions, see: Eliel, E.L. and Wilen, S.H. (1994) *Stereochemistry of Organic Compounds*, John Wiley & Sons, Inc., New York.
116. Qu, D.-H. and Feringa, B.L. (2010) *Angew. Chem. Int. Ed.*, **49**, 1107–1109.
117. Berg, H.C. (2003) *Ann. Rev. Biochem.*, **72**, 19.
118. Ruangsupapichat, N., Pollard, M.M., Harutyunyan, S.R. and Feringa, B.L. (2011) *Nature Chem.*, **3**, 153.
119. Pollard, M.M., Klok, M., Pijper, D. and Feringa, B.L. (2007) *Adv. Funct. Mater.*, **17**, 718.
120. (a) Zijlstra, R.W.J., van Duijnen, P.T., Feringa, B.L., Steffen, T., Duppen, K. and Wiersma, D.A. (1997) *J. Phys. Chem. A*, **101**, 9828; (b) Schudeeboom, W., Jonker, S.A., Warman, J.M., de Haas, M.P., Vermeulen, M.J.W., Jager, W.F., de Lange, B., Feringa, B.L. Fessenden, R.W. (1993) *J. Am. Chem. Soc.*, **115**, 3286.
121. ter Wiel, M.K.J., van Delden, R.A., Meetsma, A. and Feringa, B.L. (2003) *J. Am. Chem. Soc.*, **125**, 15076.
122. Vicario, J., Meetsma, A. and Feringa, B.L. (2005) *Chem. Commun.*, 5910.
123. Vicario, J., Walko, M., Meetsma, B.L. and Feringa, B.L. (2006) *J. Am. Chem. Soc.*, **128**, 5127.
124. Pijper, D., van Delden, R.A., Meetsma, A. and Feringa, B.L. (2005) *J. Am. Chem. Soc.*, **127**, 17612.
125. (a) Pollard, M.M., Meetsma, A. and Feringa, B.L. (2008) *Org. Biomol. Chem.*, **6**, 507; (b) Klok, M., Boyle, N., Pryce, M.T., Meetsma, A., Browne, W.R. and Feringa, B.L. (2008) *J. Am. Chem. Soc.*, **130**, 10484; (c) Augulis, R., Klok, M., Feringa, B.L. and van Loosdrecht, P.H.M. (2008) *Phys. Status*

Solidi C., **6**, 181; (d) Kulago, A.A., Klok, M., Mes, E.M., Brouwer, A.M. and Feringa, B.L. (2010) *J. Org. Chem.*, **75**, 666–679; (e) Landaluce, T.F., London, G., Pollard, M.M., Rudolf, P. and Feringa, B.L. (2010) *J. Org. Chem.*, **75**, 5323; (f) Klok, M., Browne, W.R. and Feringa, B.L. (2009) *Phys. Chem. Chem. Phys.*, **11** (40), 9124–9131.

126. Geertsema, E.M., van der Molen, S.J., Martens, M. and Feringa, B.L. (2009) *Proc. Natl. Acad. Sci. USA*, **106**, 16919.

127. (a) Kronemeijer, A.J., Akkerman, H.B., Kudernac, T., van Wees, B., Feringa, B.L., Blom, P. and de Boer, B. (2008) *Adv. Mater.*, **20**, 1467; (b) van der Molen, S., Liao, J., Kudernac, T., Agustsson, J.S., Bernard, L., Calame, M., van Wees, B.J., Feringa, B.L. and Schonenberger, C. (2009) *Nano Lett.*, **9**, 76; (c) Browne, W.R. and Feringa, B.L. (2009) *Ann. Rev. Phys. Chem.*, **60**, 407.

128. van Delden, R.A., ter Wiel, M.K.J., Pollard, M.M., Vicaro, J., Koumura, N. and Feringa, B.L. (2005) *Nature*, **437**, 1337.

129. Carroll, G., Pollard, M.M., van Delden, R. and Feringa, B.L. (2010) *Chem. Sci.*, **1**, 97.

130. Pollard, M.M., Lubomska, M., Rudolf, P. and Feringa, B.L. (2007) *Angew. Chem. Int. Ed.*, **46**, 1278.

131. London, G., Carroll, G.T., Fernandez Landaluce, T., Pollard, M.M., Rudolf, P. and Feringa, B.L. (2009) *Chem. Commun.*, 1712.

132. Cnossen, A., Pijper, D., Kudernac, T., Pollard, M.M., Katsonis, N. and Feringa, B.L. (2009) *Chem. Eur. J.*, **15**, 2768.

133. Morin, J.-F., Shirai, Y. and Tour, J.M. (2006) *Org. Lett.*, **8**, 1713.

134. (a) Muraoka, T., Kinbara, K., Kobayashi, Y. and Aida, T. (2003) *J. Am. Chem. Soc.*, **125**, 5612–5613; (b) Muraoka, T., Kinbara, K. and Aida, T. (2006) *Nature*, **440**, 512–515; (c) Muraoka, T., Kinbara, K. and Aida, T. (2007) *Chem. Commun.*, 1441–1443.

135. Muraoka, T., Kinbara, K. and Aida, T. (2006) *Nature*, **440**, 512–515.

136. (a) Maxein, G. and Zentel, R. (1995) *Macromolecules*, **28**, 8438; (b) Muller, M. and Zentel, R. (1996) *Macromolecules*, **29**, 1609; (c) Pieroni, O., Fissi, A., Angelini, N. and Lenci, F. (2001) *Acc. Chem. Res.*, **34**, 9; (d) Ciardelli, F. and Pieroni, O. (2001) *Molecular Switches*, Chapter 13 (ed. B.L. Feringa), Wiley-VCH Verlag GmbH, Weinheim; (e) Ueno, A., Takahashi, K., Anzai, J.I. and Osa, T. (1981) *J. Am. Chem. Soc.*, **103**, 6410; (f) Kim, M.-J., Yoo, S.-Y. and Kim, D.-Y. (2006) *Adv. Funct. Mater.*, **16**, 2089; (g) Kickova, A., Donovalova, J., Kasak, P. and Putala, M. (2010) *New. J. Chem.*, **34**, 1109.

137. (a) Browne, W.R. and Feringa, B.L. (2010) *Chimia*, **64**, 398; (b) Oosterling, M.L.C.M., Schoevaars, A.M., Haitjema, H.J. and Feringa, B.L. (1996) *Israel J. Chem.*, **36**, 341.

138. Pijper, D. and Feringa, B.L. (2007) *Angew. Chem. Int. Ed.*, **20**, 3693.

139. (a) Lifson, S., Green, M.M., Andreola, C. and Peterson, N.C. (1989) *J. Am. Chem. Soc.*, **111**, 8850; (b) Green, M.M., Peterson, N.C., Sato, T., Teramoto, A., Cook, R. and Lifson, S. (1995) *Science*, **268**, 1860.

140. Green, M.M., Garetz, B.A., Munoz, B., Chang, H., Hoke, S. and Cooks, G.J. (1995) *J. Am. Chem. Soc.*, **117**, 4181.

141. Pijper, D., Jongejan, M.G.M., Meetsma, A. and Feringa, B.L. (2008) *J. Am. Chem. Soc.*, **130**, 4541.

142. Carroll, G.T., Jongejan, M.G.M., Pijper, D. and Feringa, B.L. (2010) *Chem. Sci.*, **1**, 469.

143. (a) Ikeda, T. and Kanazawa, A. (2001) in *Molecular Switches*, Chapter 12 (ed. B.L. Feringa), Wiley-VCH Verlag GmbH, Weinheim; (b) Bahadur, B. (ed.) (1991) *Liquid Crystals: Applications and Uses*, vol I-III, World Scientific, Singapore; (c) Ikeda, T. and Tsutsumi, O. (1995) *Science*, **268**, 1873; (d) Ichimura, K. (2000) *Chem. Rev.*, **100**, 1874.

144. (a) Eelkema, R. and Feringa, B.L. (2006) *Org. Biomol. Chem.*, **4**, 3729; (b) For other recent studies on the use of chiroptical switches in liquid crystals, see: Tejedor, R.M., Oriol, L.,

Serrano, J.L. and Sierra, T.J. (2008) *Mater. Chem.*, **18**, 2899; (c) Mathews, M. and Tamaoki, N. (2008) *J. Am. Chem. Soc.*, **130**, 11409; (d) Goh, M. and Akagi, K. (2008) *Liq. Cryst.*, **35**, 953.

145. (a) Feringa, B.L., Huck, N.P.M. and van Doren, H.A. (1995) *J. Am. Chem. Soc.*, **117**, 9929; (b) van Delden, R.A., van Gelder, M.B., Huck, N.M.P. and Feringa, B.L. (2003) *Adv. Func. Mater.*, **13**, 319.

146. Eelkema, R. and Feringa, B.L. (2006) *Chem., Asian J.*, **1**, 367.

147. Eelkema, R., Pollard, M.M., Vicario, J., Katsonis, N., Ramon, B.S., Batiaansen, C.W.M., Broer, D.J. and Feringa, B.L. (2006) *Nature*, **440**, 163.

148. (a) Eelkema, R., Pollard, M.M., Katsonis, N., Vicario, J., Broer, D.J. and Feringa, B.L. (2006) *J. Am. Chem. Soc.*, **128**, 14397; (b) Bosco, A., Jongejan, M.G.M., Eelkema, R., Katsonis, N., Lacaze, E., Ferrarini, A. and Feringa, B.L. (2008) *J. Am. Chem. Soc.*, **130**, 14615.

149. (a) de Loos, M., Feringa, B.L. and van Esch, J.H. (2005) *Eur. J. Org. Chem.*, 3615; (b) Weiss, R.G. and Terech, P. (eds) (2006) *Molecular Gels; Materials with Self-Assembled Fibrillar Networks*, Springer-Verlag, Berlin.

150. de Jong, J.J.D., Lucas, L.N., Kellogg, R.M., van Esch, J.H. and Feringa, B.L. (2004) *Science*, **304**, 278.

151. (a) de Jong, J.J.D., Hania, P.R., Pugzlys, A., Lucas, L.N., de Loos, M., Kellogg, R.M., Feringa, B.L., Duppen, K. and van Esch, J.H. (2005) *Angew. Chem. Int. Ed.*, **44**, 2373; (b) for a recent switchable organogel with near IR CD modulation, see: Zheng, J., Qiao, W., Wan, X., Gao, J.P. and Wang, Z.Y. (2008) *Chem. Mater.*, **20**, 6163; (c) Chiral amplification using photoresponsive organogelators, see: de Jong, J.J.D., Tiemersma-Wegman, T.D., van Esch, J.H. and Feringa, B.L. (2005) *J. Am. Chem. Soc.*, **127**, 13804.

6
Multistate/Multifunctional Molecular-Level Systems: Photochromic Flavylium Compounds

Fernando Pina, A. Jorge Parola, Raquel Gomes, Mauro Maestri, and Vincenzo Balzani

6.1
Introduction

The expression 'molecular-level switch' has usually two distinct meanings [1–11]. The first definition refers to a molecular-level device, incorporated in a molecular-level wire, that can reversibly interrupt the movement of electrons or electronic energy across it in response to an external stimulus. These systems are of interest in the fields of molecular electronics and molecular photonics. The second definition describes any molecular-level system that can be reversibly interconverted between two or more different states by use of an external stimulus. These systems are extensively employed for analytical purposes and have connections with molecular-level computation and information processing.

Any molecular-level system that can be reversibly switched between two different states by use of an external stimulus can be taken as a basis for storing information, that is for memory purposes [12]. An ideal molecular-level memory should be stable and easy to write, and its switched form should be stable, easy to read and erasable when necessary. Systems that undergo an irreversible change can be used as permanent memories (e.g. photography and dosimetry). In more complex systems, switching can be performed among more than two states. This possibility can be exploited for obtaining memories that are permanent unless they are erased on purpose, or for performing logic operations.

Switching processes may be carried out under thermodynamic or kinetic control [2]. Systems under complete thermodynamic control can be stable, easy to write, easy to read and erasable, but for some applications they are useless, because they need a permanent stimulation – as soon as the stimulus is removed the molecule reverts to its initial state. An example is fluorescent sensors and indicators, the emission intensity of which is modulated by the presence of a particular substrate. The most interesting systems for memory purposes are those in which at least one stage is under kinetic control, with the metastable species separated from the thermodynamically stable state by an energy barrier high enough to delay, or, in a limiting case, to prevent the back thermal reaction. Kinetic control can operate for extremely diverse time periods, from picoseconds (for some electronic excited

Molecular Switches, Second Edition. Edited by Ben L. Feringa and Wesley R. Browne.
© 2011 Wiley-VCH Verlag GmbH & Co. KGaA. Published 2011 by Wiley-VCH Verlag GmbH & Co. KGaA.

states) to years. In kinetically controlled systems a second stimulus can be used to detrap the metastable state and to reverse the process. It should be noted that in a system under thermodynamic control it is not possible to address single molecules because of rapid equilibration between them. In contrast, for a system under kinetic control a single molecule of the assembly can be addressed, maintained in the switched state and interrogated (e.g. by single-molecule spectroscopy) [13].

6.2
Energy Stimulation

The three most important types of stimuli that can be used to switch molecular or supramolecular species are light energy (photons), electrical energy (electrons/holes) and chemical energy (in the form of protons, metal ions, etc.) [14]. In photochemical stimulation the most common switching processes are related to photoisomerization or photoinduced redox reactions; for electrochemical inputs, the induced processes are, of course, redox reactions. Interestingly, molecular switching has also been achieved by manipulation of individual molecules by scanning tunnelling microscopy [15].

Compared to chemical stimulation, photochemical and, to some extent, electrochemical stimuli can be switched on and off easily and rapidly. A further advantage of the use of photochemical and electrochemical techniques is that photons and electrons, besides supplying the stimulus to make a switch work (i.e. 'to write' the information bit), can also be useful 'to read' the state of the system and thus to control and monitor its operation. It should also be noted that photochemical and electrochemical inputs and outputs are among the easiest to interface to macroscopic systems, making them amenable to the multiscale engineering required for the eventual creation of real devices.

Another important distinction can be made [3, 11, 16]. The external stimulus causes, of course, both electronic and nuclear rearrangements in the molecular-level system to which it is applied. Usually, one of the two types of rearrangements prevails or is more relevant to the performed function. When switching involves large nuclear movements, particularly in supramolecular systems, the mechanical aspect might become more interesting than the switching process itself [9, 11, 16–20].

6.3
Photochromic Systems

The term *'photochromic'* is applied to molecules that can be reversibly interconverted, with at least one of the reactions being induced by light excitation, between two forms with different absorption spectra. The two forms, of course, differ not only in their absorption spectra, but also in several other properties such as redox potentials, acid/base strength, dielectric constant, and so on.

Photochromism (once called *phototropism*) has a long history in the scientific literature: the first example was reported in 1867 by Fritzsche [21]. In his famous paper entitled 'The Photochemistry of the Future' (1912), Giacomo Ciamician [22] discussed the importance of photochromic substances and also mentioned the possibility of using such compounds for an unexpected purpose: *'Phototropic substances, which often assume very intense colours in the light, and afterwards return in the darkness to their primitive colour, might be used very effectively. Such substances might well attract the attention of fashion ... The dress of a lady, so prepared, would change its colour according to the intensity of light. Passing from darkness to light the colours would brighten up, thus conforming automatically with the environment: the last word of fashion for the future'*. Photochromism is also a phenomenon characteristic of most biological photoreceptors such as rhodopsin (vision) and phytochrome (photomorphogenesis) [23]. Until the middle of the last century research on photochromic compounds was performed mainly in academic centres. Around 1960, however, it was recognized that this phenomenon has considerable commercial interest (e.g. photochromic glasses); since then, most of the research has been carried out in industrial laboratories.

The classical applications of photochromic materials can been classified as follows [24]:

1) applications depending upon sensitivity to irradiation (e.g. self-developing photography, protective materials, camouflage and decoration);
2) applications depending upon reversibility (e.g. protection against sunlight, smart windows, protection against intense flashes of light, data storage and retrieval);
3) applications depending upon thermal, chemical or physical properties (e.g. temperature indicators, photoresist technology, photocontractile polymers, Q-switches and security printing).

Currently, much interest is devoted to the possibility of using photochromic compounds for information processing at the molecular level. The first scientist who carried out systematic investigations on photochromic compounds as computer memory elements was Hirshberg [25]. In the following years, there has been strong development of research on photochromic molecular memories, with a great number of patents granted, particularly in Japan [24].

In photochromic systems the interconverting species are isomers, because the photoreaction simply causes rearrangement of the electronic and nuclear structure of the molecule, with or without reversible bond breaking. Light excitation causes switching from a stable isomer **X** to a higher-energy isomer **Y** that is expected to reconvert to **X** on overcoming a more or less high energy barrier (Figure 6.1). Photochromic systems, therefore, are under kinetic control. After photochemical conversion (a process that, by use of lasers, can be performed in a few femtoseconds) [7], a spontaneous back reaction is expected to occur. This reaction, however, can be fast or slow depending on the system. Sometimes, the photoproduct might be kinetically inert and the process can be reverted only by use of a second light stimulus. Depending on the thermal stability of the

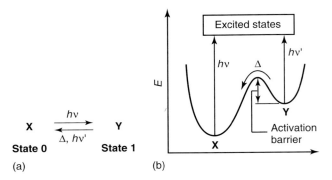

Figure 6.1 Schematic representation of a simple photochromic reaction (a) and of its energy profile (b).

photogenerated isomers photochromic systems can thus be classified into two categories:

- T-type (thermally reversible type), when the photogenerated isomers revert thermally to their initial forms;
- P-type (photochemically reversible type), when they do not revert to the initial isomers even at elevated temperatures.

For the latter compounds it is particularly important that the absorption bands of the two forms do not overlap if selective excitations are to lead to pure species. It should be noted that, with a few exceptions [26, 27], the photogenerated isomer **Y** cannot be converted, by excitation with light of the same or different wavelength, into a third isomer **Z**.[1] The best-studied families of photochromic compounds, as illustrated in the various chapters of this volume, are diarylethenes, fulgides, spiropyrans, azobenzenes, dihydroazulenes and chalcones derived from flavylium compounds.

Photochromic reactions are always accompanied by undesirable side reactions that might compromise practical applications. If a photochromic reaction occurring with almost unitary quantum yield is accompanied by a side reaction with quantum yield 0.001, 63% of the initial concentration will be decomposed after 1000 cycles.

6.4
Bistable and Multistable Systems

The simplest photochromic compounds are bistable species that can be interconverted between two forms (**X** and **Y**) exhibiting different colours [24, 28]. As mentioned above, most photochromic compounds change their colour by photoexcitation and revert more or less slowly to their initial state when kept in the dark

---

1) Of course, this restriction does not apply to interconversions between excited states.

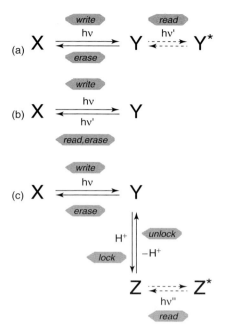

Figure 6.2 Schematic representation of the behaviour of three types of photochromic systems. (a) The photochemical reaction of the form **X** is thermally reverted in the dark. (b) The photochemical reaction of the form **X** can be reverted only by light excitation of the form **Y**. (c) The form **Y**, which goes back to **X** under light excitation, can be transformed by a second stimulus (e.g. an acid/base reaction) into another form **Z** that is stable towards light excitation and, when necessary, can be reconverted to **Y**. For more details, see text.

(Figure 6.2a). Compounds exhibiting this behaviour are useless for information storage (or switching purposes) since the written information (switching state) is spontaneously erased (back converted) after a relatively short time.

Other photochromic compounds do not return to the initial state thermally, but can undergo reversible photoisomerization (Figure 6.2b). Such compounds can be used for optoelectronic devices. However, they present a severe problem. The light used for reading the written data (detecting the switching state) causes the back conversion of the sampled molecules and therefore the gradual loss of information (state definition). Several attempts have been made to overcome this difficulty, including the use of photochemically inactive infrared light to read the status of the system [29, 30].

A general approach to avoiding destructive reading is to combine two reversible processes that can be addressed by means of two different stimuli (dual-mode systems) [31–37]. The additional stimulus can be another photon [31], heat [32], an electron [33–35], a proton [36], or even something more subtle, such as formation of a hydrogen bond [37]. In such systems (Figure 6.2c), light is used to convert **X** to **Y** (*write*); a second stimulus (e.g. a proton in Figure 6.2c) is then used to transform

Y (which would be reconverted back to X by a direct photon-reading process) into Z, another stable state of the system (*lock*) that can be optically detected without being destroyed (*read*). By use of this process the change caused by the writing photon is protected. When the written information must be erased, Z is reverted to Y (*unlock*) by an opposite stimulus (e.g. a base in Figure 6.2c) and Y is then converted back to X (*erase*). Such a *write-lock-read-unlock-erase* cycle can constitute the basis for optical memory systems with multiple storage and nondestructive readout capacity.

The concept of dual-mode stimulation can be further expanded. Systems can be devised that are capable of existing in several forms (*multistate*) that can be interconverted by different external stimuli (*multifunctional*). In this chapter, we will discuss the multistate/multifunctional character of the chemistry of synthetic photochromic flavylium compounds and show that the examination of complex chemical systems from the viewpoints of 'molecular-level device' [10, 11] and 'molecular-level logic function' [6, 8] may reveal very interesting aspects and may be useful to introduce new concepts in the field of chemical research [14].

6.5
Nature of the Species Involved in the Chemistry of Flavylium Compounds

As with anthocyanins [38], which are one of the most important sources of colour in flowers and fruits, synthetic flavylium salts (2-phenyl-1-benzopyrylium salts) in aqueous solutions undergo various structural transformations [39] that can be driven by pH changes and light excitation [40]. Such transformations are often accompanied by quite dramatic colour changes or colour disappearance. The thermal and photochemical reactions of several synthetic flavylium salts have been investigated in much detail [41–44]. It has been shown that some of these compounds can perform *write-lock-read-unlock-erase* cycles and can also exhibit a multistate/multifunctional behaviour.

The basic scheme to discuss the structural transformations of flavylium-type compounds is that shown in Figure 6.3 [39, 40]. As we will see below, other forms may also be involved, depending on the nature of the substituents. The flavylium cation AH^+, which is the stable form in strongly acidic solution, can be easily prepared by acid condensation of salicylaldehyde and acetophenone derivatives, as well as by other routes [45–47]. In moderately acidic or neutral solution, the thermodynamically stable form is generally the neutral *trans*-2-hydroxychalcone species **Ct**, which is formed from AH^+ through the two intermediate compounds **B2** and **Cc**. **B2** is a hemiketal species, obtained by hydration in the 2-position of the flavylium cation, and **Cc** is a *cis*-2-hydroxychalcone, formed from the hemiketal **B2** through a tautomeric ring-opening process. The interesting feature of these systems is that the AH^+ and **B2** forms can be reversibly interconverted by pH changes [38, 39], whereas **Cc** and **Ct** can be interconverted by photoexcitation [40, 41, 48]. Since the **B2** and **Cc** forms are in tautomeric equilibrium, it follows that pH and light stimulation can be used to cause interconversion between the four fundamental forms (Figure 6.3). Furthermore, the AH^+ form can be hydrated not

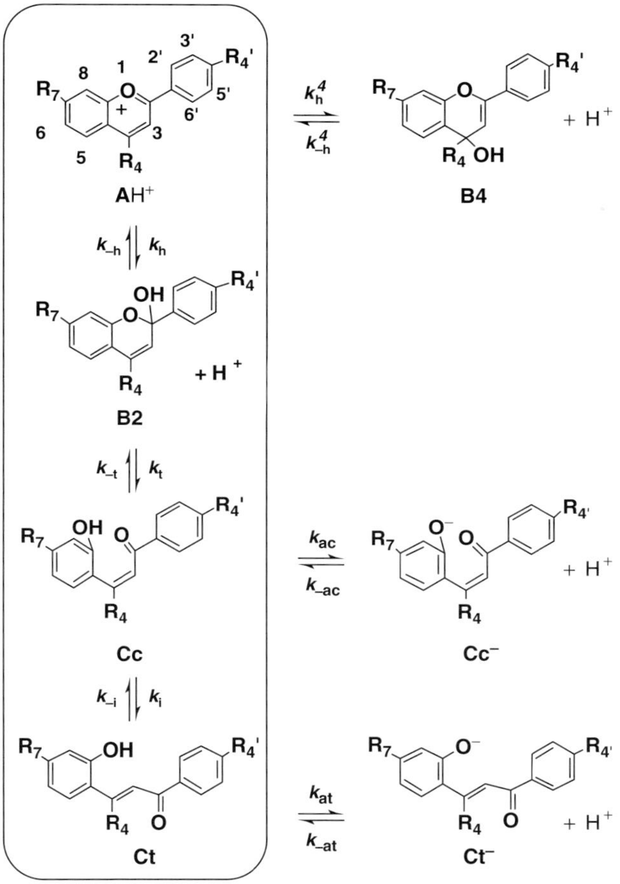

Figure 6.3 Structural transformations of the flavylium-type compounds. Only the most important forms are shown.

only in the 2-, but also in the 4-position, to give the **B4** species. The **Cc** and **Ct** can undergo deprotonation to give the **Cc**$^-$ and **Ct**$^-$ monoanions, respectively, which, being *cis/trans* isomers, can in principle be interconverted by light excitation. As we will see later, depending on the nature of the substituents, other acid/base equilibria and *cis/trans* couples may be present. It is therefore clear that in these systems pH changes coupled with light excitation may provide for very intricate series of chemical reactions, with dramatic changes in the absorption spectra (i.e. in the colour of the system). A further interesting aspect is that some of the species exhibit fluorescence, that is not only another analytical 'handle' to control the behaviour of the system, but also a very interesting signal for the purpose of information processing. Several studies concerning the thermodynamic as well as the kinetic aspects of the thermal reactions of flavylium-type compounds have been reported in the literature [38, 39], with the photochemical and photophysical aspects being systematically examined only more recently [49–59]. As we shall see below, pH jump, temperature jump and flash photolysis experiments allow for the measurement of the rate constants of some of the reactions involved, and steady-state titration experiments (by using UV-VIS and NMR techniques) allow for the measurement of equilibrium constants.

6.5.1
Thermodynamics of Flavylium Compounds

The studies performed so far on the chemical behaviour of synthetic flavylium compounds in acidic or neutral aqueous solutions showed that almost all the compounds behave in a manner that can be summarized in the following equations:

$$AH^+ + 2H_2O \rightleftharpoons B4 + H_3O^+ \quad K_h^4 \tag{6.1}$$

$$AH^+ + 2H_2O \rightleftharpoons B2 + H_3O^+ \quad K_h \tag{6.2}$$

$$B2 \rightleftharpoons Cc \quad K_t \tag{6.3}$$

$$Cc \rightleftharpoons Ct \quad K_i \tag{6.4}$$

The species involved in Equations 6.1–6.4 are the same species shown in Figure 6.3, that is the flavylium cation AH^+, two hemiketal forms $B2$ and $B4$ obtained by hydration of the flavylium cation in the 2 or 4 position, respectively, the *cis*-chalcone species Cc formed from hemiketal $B2$ through a tautomeric process, and the *trans*-chalcone form Ct resulting from the isomerization of *cis*-chalcone.

As shown previously [50, 60] the mole fraction of the acidic form AH^+ can be obtained from Equation 6.5:

$$\frac{[AH^+]}{C_0} = \alpha = \frac{[H^+]}{[H^+] + K'_a} \tag{6.5}$$

where C_0 is the total concentration and K'_a is given by Equation 6.6:

$$K'_a = K_h + K_h K_t + K_h K_t K_i \tag{6.6}$$

Equation 6.5 accounts for the complex equilibria described by Equations 6.2–6.4 in terms of a single acid–base equilibrium (Equation 6.7) between the acidic species AH^+ and a conjugated base CB having a concentration equal to the sum of the concentrations of all the species present at the equilibrium (i.e. $B2$, Cc and Ct)[2] and mole fraction given by Equation 6.8 [50, 60]:

$$AH^+ + 2H_2O \rightleftharpoons CB + H_3O^+ \quad K'_a \tag{6.7}$$

$$\frac{[CB]}{C_0} = \frac{[B2] + [Cc] + [Ct]}{C_0} = \beta = \frac{K'_a}{[H^+] + K'_a} = 1 - \alpha \tag{6.8}$$

The individual expressions of the mole fractions of each component of CB can be easily calculated as shown in Refs. [50, 60]. If the pH is sufficiently high to

---

2) Hemiketal B4 was not included since it is usually formed in percentages below 1% [39c,d].

consider all the compounds in the form of **CB** ($\beta = 1$), the relationships shown in Equation 6.9 apply:

$$\frac{[\mathbf{B2}]}{C_0} = \frac{K_h}{K'_a}; \quad \frac{[\mathbf{Cc}]}{C_0} = \frac{K_h K_t}{K'_a}; \quad \frac{[\mathbf{Ct}]}{C_0} = \frac{K_h K_t K_i}{K'_a} \tag{6.9}$$

The global constant K'_a (defined in Equation 6.6) can be experimentally determined by following the decrease in the absorbance of the flavylium cation (e.g. the mole fraction of $\mathbf{AH^+}$) as a function of pH. The remaining constants can be calculated by measuring the mole fraction of each component of **CB** at the equilibrium, when $\mathbf{AH^+}$ is no longer present. This is not a trivial task, and in most cases NMR techniques must be used together with pH jumps. In particular, the value of K'_a can be measured by using a UV/Vis spectrophotometer if the hydration reaction takes place on a timescale of seconds [48]. If the hydration is faster, stopped flow apparatus are required. When the rate of the thermal *cis–trans* isomerization reaction (Equation 6.4) is much slower than the rates of the reactions involved in the other equilibriums, a pseudoequilibrium can be obtained and Equations 6.10–6.13 can be used to evaluate the pseudoequilibrium constant ($K^\wedge_a$) and the mole fractions of the various forms:

$$\frac{[\mathbf{AH^+}]}{C_0} = \alpha = \frac{[H^+]}{[H^+] + K^\wedge_a} \tag{6.10}$$

$$K^\wedge_a = K_h + K_h K_t \tag{6.11}$$

$$\frac{[\mathbf{CB}]}{C_0} = \frac{[\mathbf{B2}] + [\mathbf{Cc}]}{C_0} = \beta = \frac{K^\wedge_a}{[H^+] + K^\wedge_a} = 1 - \alpha \tag{6.12}$$

$$\frac{[\mathbf{B2}]}{C_0} = \frac{K_h}{K^\wedge_a}; \quad \frac{[\mathbf{Cc}]}{C_0} = \frac{K_h K_t}{K^\wedge_a} \tag{6.13}$$

6.6
Thermal Reactions of the 4'-Methoxyflavylium Ion

In order to illustrate the complex reaction network of these systems, we will now focus on the behaviour of the 4'-methoxyflavylium ion (Figure 6.3; $R_4 = R_7 = H$, $R_{4'} = OCH_3$) [53]. A spectroscopic and kinetic investigation of the transformations undergone by the 4'-methoxyflavylium ion was originally performed by McClelland and Gedge [39c]. By using the pH-jump technique, they found that seven different species are involved, as transient or equilibrium compounds, depending on the experimental conditions (Figure 6.3). The absorption spectra of the strongly coloured 4'-methoxyflavylium cation $\mathbf{AH^+}$ ($\lambda_{max} = 435$ nm, $\varepsilon = 42\,000$ M^{-1} cm^{-1}), the colourless *trans*-2'-hydroxy-4'-methoxychalcone **Ct** ($\lambda_{max} = 350$ nm, $\varepsilon = 18\,000$ M^{-1} cm^{-1}), and the **B2** and **Cc** mixture are shown in Figure 6.4.

The mole-fraction distribution [39c, 53] of the various species in aqueous solution at 25 °C as a function of pH, obtained from the equilibrium constants (see Equations 6.8 and 6.9), is shown in Figure 6.5. The thermodynamically stable form in the pH range 2–8 is the *trans*-2'-hydroxy-4'-methoxychalcone, **Ct**, that, at higher pH, is transformed into its anion, **Ct**$^-$ (Figure 6.5, solid lines). In strongly acidic solutions,

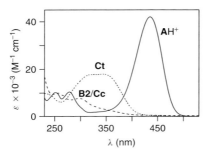

Figure 6.4 Absorption spectra in aqueous solution at 25 °C of the 4′-methoxyflavylium compound: **AH**$^+$ at pH = 1.0, **Ct** at pH = 4.0 and **B2/Cc** mixture at pH = 7.0 [53]. Reproduced from reference [53]. Copyright ACS (1997).

AH$^+$ becomes thermodynamically stable; however, **Ct** cannot be converted to **AH**$^+$ because of the large activation barrier that involves isomerization of **Ct** to the intermediate compound **Cc**. Furthermore, a solution of **AH**$^+$ is almost indefinitely stable at room temperature below pH 3, since under such conditions a very large kinetic barrier prevents conversion of **AH**$^+$ to the thermodynamically stable **Ct** form via the hydrated species **B2** and the **Cc** isomer (Figure 6.3). At higher pH, however, **AH**$^+$ is very reactive [38c]. For example, starting from an aqueous solution of **AH**$^+$ at 25 °C and pH = 1, a pH jump to pH = 4.29 leads within a few seconds to a pseudoequilibrium consisting of 50% **AH**$^+$, 33.2% **B2**, 0.3% **B4** and 16.5% **Cc** (Figure 6.5). A much slower reaction follows (half-life 19.7 h), resulting in complete conversion to the thermodynamically stable form **Ct**.

At pH 8, **AH**$^+$ reacts mainly with solvent water (half-life 0.44 s) to produce 64% **B4**, 24% **B2** and 12% **Cc**, the last two being in equilibrium with each other (half-life of the equilibration, 7×10^{-5} s^{-1}) [38c]. This is followed by another fast reaction (half-life 66 s) in which **B4**, a product of kinetic control of the initial hydration of **AH**$^+$, is converted via **AH**$^+$ to **B2** and **Cc**, yielding a pseudoequilibrated mixture of 66.3% **B2**, 33.1% **Cc** and 0.6% **B4** (Figure 6.5, dashed lines). A much slower reaction (half-life 9.9 h) then occurs, resulting in complete conversion to **Ct**.

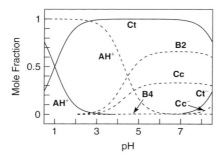

Figure 6.5 Mole-fraction distribution in aqueous solution at 25 °C as a function of pH for the 4′-methoxyflavylium compound. Solid lines refer to the species obtained at the thermodynamic equilibrium. Dashed lines refer to species obtained bringing **AH**$^+$ solutions from pH = 1 to higher pH values by the pH-jump technique or by exciting **Ct** solutions by flash light. Such species reach a pseudoequilibrium in the second timescale and then undergo a very slow thermal reaction to **Ct** [53]. Reproduced from reference [53]. Copyright ACS (1997).

6.7
Photochemical Behaviour of the 4′-Methoxyflavylium Ion

As described above, in the pH range 2–8 the colourless *trans*-2′-hydroxy-4′-methoxychalcone **Ct** is the thermodynamically stable species and therefore it is the final product of the transformations of the strongly coloured 4′-methoxyflavylium ion AH^+. Even at pH = 1, when AH^+ is the thermally stable species, **Ct** can be kinetically stable because of the high energy barrier of its transformation to **Cc**. **Ct**, however, can be converted into AH^+ by a photochemical reaction [53]. As expected from the thermal behaviour of the system, the photoreaction causes a transient or an almost permanent effect depending on temperature and pH of the irradiated solution.

6.7.1
Continuous Irradiation

Continuous irradiation of 2.3×10^{-5} M aqueous solutions of **Ct** at pH = 1.0 with 365-nm light causes strong spectral changes, with five isosbestic points and formation of a very intense band in the visible region with a maximum at 435 nm (Figure 6.6a) [53]. Analysis of the spectral changes shows that the photoreaction converts **Ct** into AH^+, without formation of sizeable amounts of other products. The quantum yield of the photoreaction is 0.04, independent of the presence of dioxygen in solution. At pH = 1.0, no back reaction takes place and irradiation with 434-nm light, corresponding to the maximum of the absorption band of AH^+ (Figure 6.4), does not cause any effect.

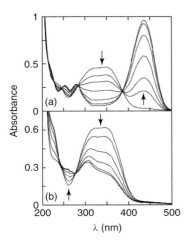

Figure 6.6 Spectral changes caused by continuous irradiation of an aqueous solutions of the **Ct** form of 4′-methoxyflavylium ion with 365 nm light: (a) pH = 1.0, [**Ct**] = 2.5×10^{-5} M; the curves correspond to the following irradiation times: 0; 0.5; 1; 2; 4; 7; 12 min. (b) pH 7.0, [**Ct**] = 3.2×10^{-5} M; the curves correspond to the following irradiation times: 0; 0.25; 1.5; 3; 6; 10 min [53]. Reproduced from reference [53]. Copyright ACS (1997).

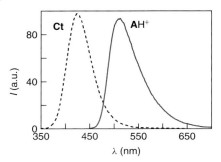

Figure 6.7 Fluorescence spectra in aqueous solution at 25 °C of the **AH$^+$** (pH = 1.0) and **Ct** (pH = 4.0) forms of the 4′-methoxyflavylium ion [53]. Reproduced from reference [53]. Copyright ACS (1997).

When irradiation of **Ct** is carried out at pH = 4.0, the quantum yield of the photoreaction leading from **Ct** to **AH$^+$** does not change, but the expected thermal back reaction of **AH$^+$** to **Ct** is observed. The rate of the back reaction increases with temperature (activation energy 93 kJ mol^{-1} at pH = 4.0). Irradiation at pH = 7.0 causes the spectral changes shown in Figure 6.6b. At this pH the disappearance of **Ct** does not cause any increase of absorbance in the visible spectral region, showing that **AH$^+$** is not formed. Furthermore, the back reaction is very fast so that complete disappearance of **Ct** cannot be observed. This is in full agreement with the expectations based on the data shown in Figure 6.5, which indicate that at pH = 7.0 the pseudoequilibrated mixture of products is constituted essentially by the open **Cc** and closed **B2** *cis* forms. As is always the case for aromatic derivatives of ethylene [61], the absorption spectrum of this mixture of *cis* species is less intense and slightly blue-shifted compared to the spectrum of the *trans* form (Figure 6.4). Under such conditions, irradiation of the mixture with 313-nm light causes the reverse *cis* → *trans* photoisomerization reaction with an apparent quantum yield of about 0.5 (based on the total light absorbed by **Cc** and **B2**).

Interestingly, **Ct** and **AH$^+$** exhibit intense fluorescence bands with λ_{max} at 430 and 530 nm, respectively (Figure 6.7) [53]. The fluorescence lifetime is shorter than 1 ns in both cases. It is worth noting that the occurrence of the above-described thermal and photochemical reactions can also be followed by fluorescence measurements.

6.7.2
Pulsed Irradiation

Flash photolysis is a powerful technique to investigate the kinetics of conversion of the various forms of flavylium ions [48]. Even by a simple flash-photolysis apparatus with a time resolution of about 0.2 s it is possible to obtain kinetic data that can complement and/or replace those obtainable by the pH-jump technique.

Flash excitation [53] of a 6.0 × 10^{-5} M aqueous solutions of **Ct** at 25 °C and pH = 3.0 or 7.0 causes a bleaching in the 300–400 nm region, that can be assigned to the

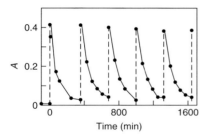

Figure 6.8 Behaviour of a 1.0×10^{-5} M aqueous solutions of the **Ct** form of the 4′-methoxyflavylium ion at pH = 3.0 and 60 °C under 365-nm light excitation (dashed lines) followed by dark periods (full lines) [53]. Reproduced from reference [53]. Copyright ACS (1997).

disappearance of **Ct**. At pH 3, a strong increase in absorbance in the 400–500 nm region is observed, as expected for the formation of AH^+. The absorbance *vs.* time traces show that **Ct** disappears within the timescale of the flash, but its disappearance does not lead directly to AH^+. One or more intermediate products are formed (**Cc** and **B2** according to the scheme of Figure 6.3), which then convert completely to AH^+ in a few seconds. At pH = 7.0, the decrease of absorbance in the 300–400 nm region, corresponding to the disappearance of **Ct**, is not accompanied by an increase in absorbance in the visible region because in neutral solution AH^+ is not stable and the main products of the photoreaction are **B2** and **Cc** (Figure 6.4). None of the thermal and photochemical processes observed are affected by the presence of dioxygen in the solution.

In order to check the degree of reversibility of the observed reactions, a 1.0×10^{-5} M aqueous solution of **Ct** at pH = 3.0 and 60 °C was irradiated at 365 nm. After 20 min of irradiation, which causes the formation of the coloured form AH^+, the solution was kept in the dark, at 60 °C, until a practically complete bleaching of the visible absorption of AH^+ had occurred. Then, light excitation was again performed. The changes in absorbance at 435 nm obtained repeating five times these light/dark cycles are shown in Figure 6.8. As one can see, the degree of reversibility of the system is satisfactory [53].

In conclusion, the photochemical behaviour is in agreement with the behaviour observed by pH jump experiments. Although **Cc** is obviously the primary product of flash excitation, the observed species and their survival time (*from seconds to years*) before going back to the thermodynamically stable form **Ct** depend on temperature and pH.

6.8
Flavylium Ions with OH Substituents

In flavylium compounds that carry OH substituents, other forms, not present in the above-discussed 4′-methoxyflavylium compound, can be obtained because of the deprotonation of the OH group, as illustrated in Figure 6.9 for the

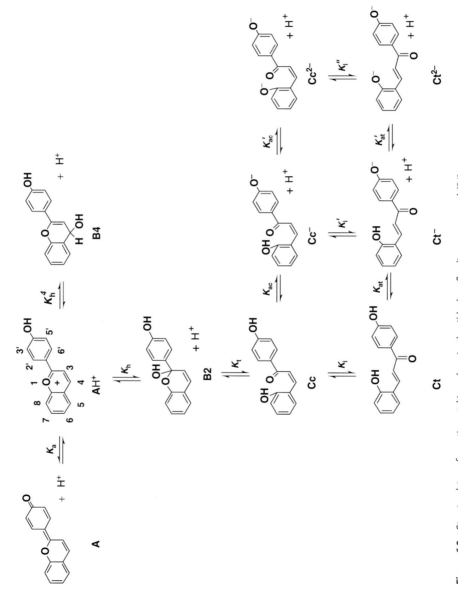

Figure 6.9 Structural transformations taking place in the 4′-hydroxyflavylium compound [54].

4′-hydroxyflavylium ion [54]. The new species are the quinoidal base **A**, obtained by simple deprotonation of the **AH$^+$** flavylium cation, and the dianionic **Cc^{2-}** and **Ct^{2-}** forms, obtained by second deprotonation of **Cc** and **Ct**. The roles played by these forms depend on the specific compound and pH conditions. For example, in the case of 4′-hydroxyflavylium ion the **Ct^{2-}** species exhibits fluorescence and the **Cc^{2-}** one undergoes photoisomerization to **Ct^{2-}** (for more details, *vide infra*). Interestingly, for the 4-methyl-7-hydroxy and 7,4′-dihydroxyflavylium compounds both the **AH$^+$** cation and the **A** quinoidal base exhibit fluorescence. Moreover, in the former compound, for which only two forms (**AH$^+$** and **A**) are observed, the pK_a of the ground state (4.4) is higher than the apparent pK_a of the excited state (0.7) and a very efficient adiabatic excited-state proton-transfer reaction (yield = 0.95) transforms *AH$^+$ into *A. These results show that the 4-methyl-7-hydroxyflavylium compound behaves as a four-level system and suggests that it could be used, in principle, to obtain a laser effect [55].

When a hydroxyl group is introduced in position 2′ of the flavylium core, the respective 2,2′-dihydroxychalcones can cyclize to form flavanones in basic media or cyclize back to the flavylium in acidic media. Although this new flavanone species is not photoreactive it allows the number of possible states available for the system to be enlarged [62].

6.9
Flavylium Ions with Other Substituents

Besides methoxy and hydroxyl substituents, it is possible to synthesize flavylium ions bearing more strongly electrodonating groups, such as amino groups, or strongly withdrawing groups. The inclusion of amino functional groups is of great interest since it permits considerable red shifts in the absorption spectra, providing a great variety of colours. Furthermore, the presence of an amino substituent allows for the formation of protonated species such as **AH$_2^{2+}$**, **Ct$^+$** (see Figure 6.10) and, being a strong electron donor, makes more difficult the hydration step extending to higher values the pH range where the flavylium ion (**AH$^+$**) predominates [42]. For instance, for the 4′-dimethylamino compound the pK_a **AH$_2^{2+}$/AH$^+$** is −0.6 and the pK_a **Ct$^+$/Ct** is about 2. The value of pK'_a for this compound is 6.9, much higher than those usually observed with flavylium ions containing oxygen donor groups. The charge injected by the amino group into the benzopyrylium core increases the double bond character of the C–N bond [42, 63, 64]. The 4′-dimethylamino compound has a high barrier for *cis–trans* isomerization, similar to that of the 4′-hydroxy and 4′-methoxy derivatives, while 7-diethylamino-4′-hydroxy has a low barrier. In both cases, **Ct$^+$** can be formed upon a pH jump from basic solutions (**Ct$^-$** or **Ct^{2-}**) and gives rise to the formation of **AH$^+$** with a rate that depends strongly upon the specific nature of the substituents – taking days (4′-dimethylamino) [42] or just a few seconds (7-diethylamino-4′-hydroxy) [63]. In aqueous solutions (with no added organic solvent) [51a, 65], photochemical reactions of neutral alkylamino

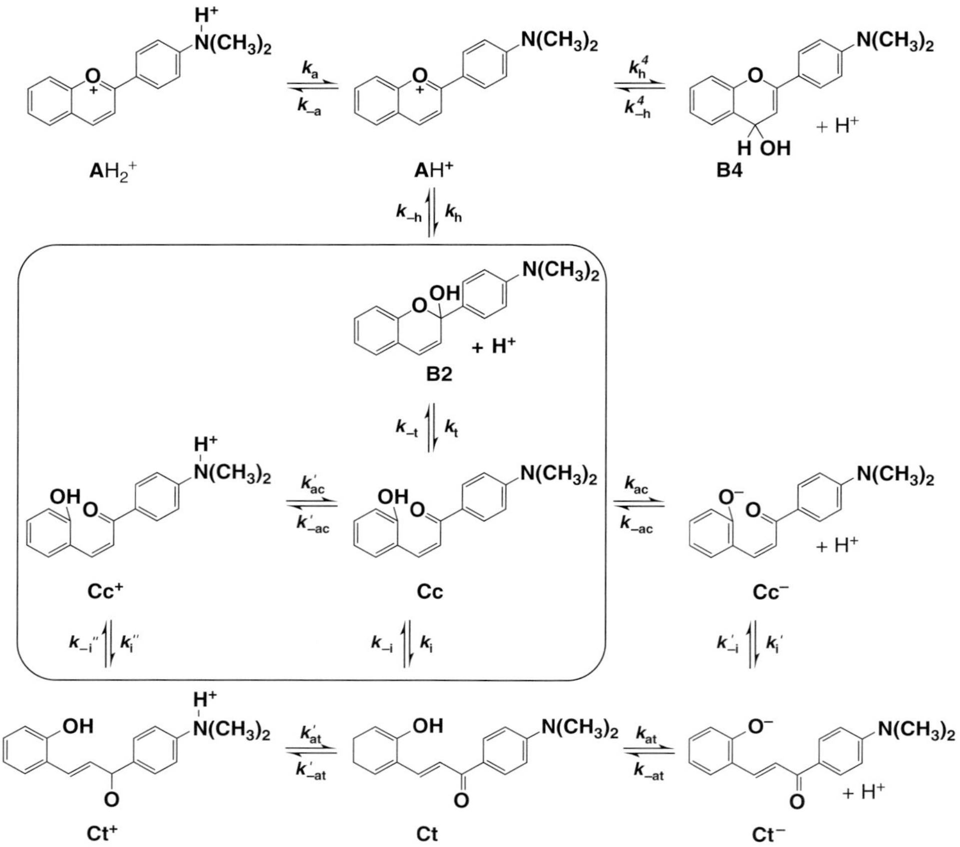

Figure 6.10 Structural transformations taking place in the 4′-dimethylaminoflavylium compound. The area within the square highlights the species that cannot be identified from the changes in absorption spectra [42].

substituted *trans*-chalcone species have never been observed; however, ionized chalcones can exhibit photochemistry [42].

By contrast, the introduction of nitro groups facilitates the hydration step, providing very low pK'_a values (Figure 6.11). For instance, the 4′-hydroxy-6-nitroflavylium, has a pK'_a of −0.6 and also a very high isomerization barrier (90.4 kJ mol^{-1}) [66]. The network of chemical reactions of this compound has some peculiarities, namely, a new reaction channel that leads to **Cc⁻** from **B2⁻** and the formation of **B2** and **Cc** upon a pH jump from 1 to ~4. The behaviour of this system was clarified by stopped-flow experiments. Upon a pH jump from 0.7 to ~4, the species observable immediately after the dead time of the stopped flow apparatus is the flavylium cation (λ_{max} = 450 nm), that disappears (9 s^{-1}) to give **Cc** in equilibrium with **B2** (λ_{max} = 308 nm). By contrast, a pH jump to 13.2 immediately leads to

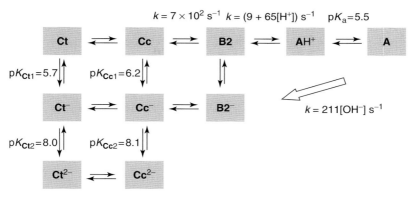

Figure 6.11 Scheme of the network of chemical reactions for 4′-hydroxy-6-nitroflavylium and their kinetic or thermodynamic constants [66].

the quinoidal base **A** ($\lambda_{max} = 500$ nm) that evolves to the **Cc^{2-}** ($\lambda_{max} = 361$ nm) species according to a first-order process with a rate constant of 38 s^{-1}. After the pH jumps to basic pH values, the rate constant for the disappearance of **A** depends linearly on the hydroxyl concentration; this suggests that the quinoidal base **A**, formed during the dead time of the stopped-flow apparatus, undergoes a hydroxyl attack, most probably leading to **B2$^-$** and then, depending on pH, to **Cc$^-$** or **Cc^{2-}** through a fast ring opening. A freshly prepared solution at pH 0.7 (**AH$^+$** form) was submitted to a pH jump to 3.2, attaining the pseudoequilibrium, and subsequently to a pH jump to pH 0. The last step, leading again to **AH$^+$** formation, was followed by stopped flow. The decrease in absorbance at 320 nm shows two consecutive processes; a fast process (masked by the dead time of the apparatus) and a second one, monitored either at 320 nm (decrease of **Cc** absorption) or at 470 nm (increase of **AH$^+$** absorption) that follows first-order kinetics with a rate constant of 65 s^{-1}. This behaviour can be interpreted by considering that, at pH 3.2, **Cc** is the main species and the faster process is the tautomerization reaction that leads to **B2**. The second and slower process can thus be identified as the dehydration reaction that forms the flavylium cation **AH$^+$** [66].

Some pseudoflavylium compounds where the 1-benzopyrylium moiety is preserved and other groups are introduced were described. 4′-Hydroxynaphthoflavylium, where the benzopyrylium unit was replaced by a naphthopyrylium one, shows photochromism, rate and thermodynamic constants very similar to those previously reported for the parent 4′-hydroxyflavylium [54] except the rate constant of the *cis–trans* isomerization that is much higher for the naphthoflavylium derivative [67]. The low *cis–trans* isomerization barrier was explained by the higher delocalization of the C3–C4 double through the naphthalene ring. Several 2-styryl-1-benzopyrylium salts were studied and are characterized by red-shifted absorption maxima. This family of compounds showed no evidence for the existence of an extra isomerization at the styryl moiety, following the same pH- and light-dependent reaction network of flavylium salts [68]. However, due to higher charge delocalization of the positive charge, the observed rate constants are about threefold lower compared with the respective flavylium parent salts.

Chromenes are known for their photochromism based on electrocyclic ring opening at the pyran C—O bond. Included in the flavylium reaction network there are hemiketals that correspond to hydroxyphenylchromenes that opens up the possibility of enlarging the flavylium reaction network to include a second type of photochromic system. However, the **B2** species are elusive states of the network, equilibrating in the subsecond timescale with **Cc** and reverting back to the flavylium with pH-dependent rates. Substituting the 2-phenyl ring by a 2-(4-pyridine) would contribute to the stabilization of the hydroxypyridinechromene since protonation of the pyridine moiety would prevent flavylium formation even under very acidic conditions. The pseudochalcone (*E*)-3-(2-hydroxyphenyl)-1-(pyridin-4-yl)prop-2-en-1-one presents a reaction network that includes **B2**, **Cc**, **Ct**, **B2$^+$**, **Cc$^+$** and **Ct$^+$** and exhibits *cis–trans* photoisomerization, as well as photochromism between **B2** and **Ct** [69]. The irradiation of **Ct** in MeOH/H$_2$O (1 : 1) at 365 nm produces **B2** almost quantitatively through two consecutive photochemical reactions: **Ct** → **Cc** photoisomerization followed by **Cc** → **B2** photo ring closure with a global quantum yield of 0.02. On the other hand, irradiation of **B2** at 254 nm leads to a photostationary state composed by 80% **Ct** and 20% **B2**, with a quantum yield of 0.21.

6.10
Energy-Level Diagrams

As discussed above for the 4′-methoxyflavylium compound, pH jump, temperature jump and flash-photolysis experiments allow for the measurement of the rate constants of some of the reactions involved, and steady-state titration experiments (by using UV-Vis and NMR techniques) allow the measurement of equilibrium constants. The values obtained for the most important processes of five flavylium compounds are gathered in Table 6.1.

As we have seen before, the complex equilibria involving the species present at moderately acid pH (Figures 6.9–6.11) can be described in terms of a single acid–base equilibrium (Equation 6.7) between the acid species **AH$^+$** and a conjugated base **CB** that represents all the species present at equilibrium; therefore its concentration is now equal to the sum of the concentrations of the species **A**, **B2**, **Cc** and **Ct** and $K'_a = K_a + K_h + K_h K_t + K_h K_t K_i$ [53, 60]. The equilibrium constant of such an overall process is also given in Table 6.1. By using the thermodynamic data shown in the table, an energy-level diagram can be constructed for each compound. Simplified versions of such diagrams (Figures 6.12–6.16) can then be used to illustrate the behaviour of the various compounds [42, 53–55, 57–59] and to discuss the effect of the substituents.

Table 6.1 Thermodynamic and kinetic constants for some structural transformations of synthetic flavylium compounds.[a]

Substituent	7-OH[b]	4′,7-diOH[c]	None[d]	4′-OH[e]	4′-OMe[f]
K'_a	2.0×10^{-3}	8.9×10^{-4}	2.3×10^{-2}	1.26×10^{-2}[g]	8.0×10^{-2}
K_a	2.8×10^{-4}	1.0×10^{-4}	–	3.16×10^{-6}	–
K_h	8.0×10^{-6}	1.4×10^{-6}	9.8×10^{-4}	3.6×10^{-6}	3.4×10^{-5}
K_t	–	–	0.06	1	0.50
K_i	500	1.4×10^3	400	3500	About 100
k_h (s^{-1})	0.48^a	1.8×10^{-2}	4.6	8.9×10^{-2}	0.47
k_{-h} (s^{-1} M^{-1})	$3 \times 10^{4\,h}$	$1.3 \times 10^{4\,h}$	4.7×10^3	$2.5 \times 10^{4\,h}$	1.38×10^4
k_i (s^{-1})	0.57^h	0.26^h	4.1×10^{-4}	3.7×10^{-5}	5.8×10^{-5}
k_{-i} (s^{-1})	8.3×10^{-4}	1.8×10^{-4}	1.1×10^{-6}	$<10^{-7}$	$<10^{-6}$

[a] Measured by means of pH-jump techniques at 25 °C, unless otherwise noted.
[b] Reference [58].
[c] Reference [48].
[d] Reference [57].
[e] References [53, 39d].
[f] References [39c, 52].
[g] At 60 °C.
[h] Measured by flash photolysis.

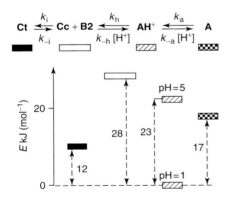

Figure 6.12 Energy-level diagram for the species involved in the equilibria of the 7,4′-dihydroxyflavylium compound [59].

An intuitive way to present the interconversion processes in flavylium-type compounds is their description by a hydraulic analogy [59]. Using such an analogy, the behaviour of aqueous solution of flavylium ions upon a pH jump from 1.0 to 4.2 can be schematically represented as in Figure 6.17. In the case of 4′-hydroxyflavylium, **Cc** converts very slowly to **Ct** and thus **B2** and **Cc** accumulate, whereas for 7,4′-dihydroxyflavylium and 7-hydroxyflavylium **Cc** converts very rapidly to **Ct** so that **Cc** and **B2** disappear as soon as they are formed.

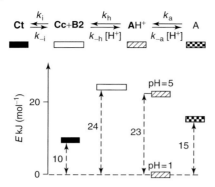

Figure 6.13 Energy-level diagram for the species involved in the equilibria of the 7-hydroxyflavylium compound [59].

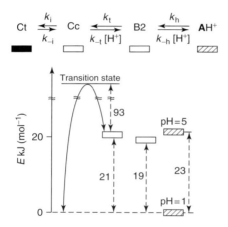

Figure 6.14 Energy-level diagram for the species involved in the equilibria of the 4′-methoxyflavylium compound [53].

The hydraulic analogy can also be used to illustrate the photochemical behaviour of these compounds, the light playing the role of a pump. The scheme shown in Figure 6.18 is appropriate for 4′-hydroxyflavylium.

6.11
Chemical Process Networks

As mentioned above, molecular or supramolecular systems capable of existing in different forms (*multistate*) that can be interconverted by different external stimuli (*multifunctional*) are interesting for both basic and applicative reasons. As we have seen above (Figures 6.3, 6.9, 6.10 and 6.11), the flavylium compounds can be interconverted into a number of different transient and stable forms by

6.11 Chemical Process Networks

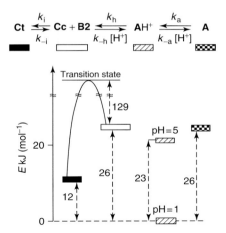

Figure 6.15 Energy-level diagram for the species involved in the equilibria of the 4′-hydroxyflavylium compound [54].

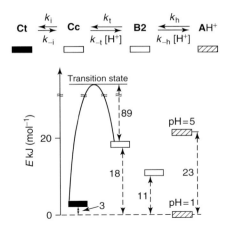

Figure 6.16 Energy-level diagram for the species involved in the equilibria of the unsubstituted flavylium compound [58].

using two different inputs, namely, light and changes in pH. Several interesting aspects emerge when the resulting networks of chemical processes are analysed in terms of 'molecular-level device' and 'molecular-level logic function', as illustrated below. To illustrate these aspects, we will mainly discuss the cases of the 4′-methoxy-, 4′-hydroxy- and unsubstituted flavylium compounds where the high activation energy of the **Cc** → **Ct** reaction provides considerable kinetic stability to **Cc** and related structures. For the sake of simplicity, in the following discussion the **A**, **B4** and **B2** transient species have been neglected since in the compounds under examination they are always in rapid equilibrium with either AH^+ or **Cc**.

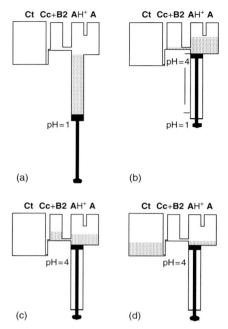

Figure 6.17 Hydraulic analogy for the description of the behaviour of the flavylium compounds upon a pH jump from pH = 1.0 to 4.0 [59]. (a) The system is equilibrated at pH = 1.0; (b) the pH is changed to 4.0; the pH jump has an effect comparable to raising the piston and the figure represents the situation immediately after the proton transfer process; (c) when the $cis \rightarrow trans$ isomerization is very slow, it is possible to obtain an intermediate (pseudoequilibrium) state involving the species AH^+, **A**, **B** and **Cc** and (d) thermodynamic equilibrium at pH = 4.0 [59]. Reproduced from reference [59]. Copyright Wiley (2001).

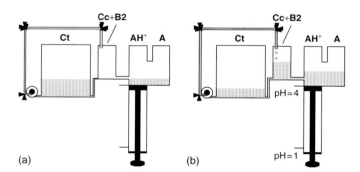

Figure 6.18 Hydraulic analogy for the photochemical reaction of **Ct** form of the flavylium compounds. Light behaves like a pump that increases (in a transient mode as represented, or in steady state) the quantity of liquid in the reservoir **Cc**. (a) Before irradiation and (b) immediately after the light flash [59]. Reproduced from reference [59]. Copyright Wiley (2001).

6.11.1
Write-Lock-Read-Unlock-Erase Cycles

As discussed at the beginning of this chapter, photochromic systems represent potential molecular-level memory devices. A number of problems, however, must be solved for practical applications. A most challenging one is to find systems with multiple storage and nondestructive readout capacity, that is where the record can be erased when necessary, but is not destroyed by the readout. The 4′-hydroxyflavylium [54] and 4′-methoxyflavylium [53] ions can operate through the *write-lock-read-unlock-erase* cycle illustrated in Figure 6.2c and therefore they can be taken as a basis for optical memory systems with multiple storage and nondestructive readout capacity.

The behaviour of the 4′-methoxyflavylium ion can be described making reference to Figure 6.19 (which refers to a solution at pH = 3.0), where **Ct**, **Cc** and **AH$^+$** play the role of the generic species **X**, **Y** and **Z** of Figure 6.2c (**B2** is not shown because it is always in equilibrium with **Cc**):

1) stable form (**Ct** ≡ **X**) can be photochemically converted by irradiation with 365-nm light (*write*) into a form (**Cc** ≡ **Y**) that can be reconverted back either thermally or on optical readout;
2) by a second stimulus (addition of acid, which can also be present from the beginning without perturbing the behaviour of the system), **Cc** ≡ **Y** can be converted into a kinetically inert form **AH$^+$** ≡ **Z** (*lock*);
3) the **AH$^+$** ≡ **Z** form shows a spectrum (Figure 6.4) clearly distinct from that of **Ct** ≡ **X** and is photochemically inactive, so that it can be optically detected (*read*) without being erased;

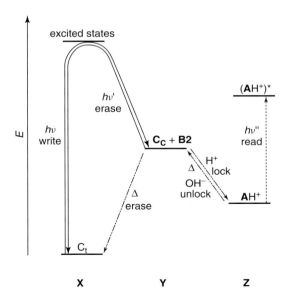

Figure 6.19 Schematic energy-level diagram for the species involved in the *write-lock-read-unlock-erase* cycle in the case of the 4′-methoxyflavylium ion [53].

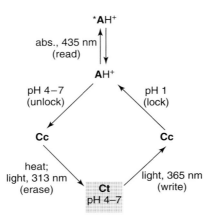

Figure 6.20 Write-lock-read-unlock-erase cycle starting from the **Ct** form of the 4′-hydroxyflavylium compound [54].

4) by addition of base, $AH^+ \equiv Z$ can be reconverted into $Cc \equiv Y$ (*unlock*);
5) $Cc \equiv Y$ can be thermally or photochemically reconverted into the initial $Ct \equiv X$, form (*erase*).

It should be noted that the locking time of the written information bit is not indefinite (at 25 °C and pH = 3.0, the half-life of the back reaction from AH^+ to **Ct** is about eight days).

In the case of the 4′-hydroxyflavylium ion [54], a similar behaviour is observed that can be described on the basis of the energy-level diagram of Figure 6.15 or the simpler scheme shown in Figure 6.20:

1) in the range $4 < pH < 7$, the stable or kinetically inert (depending on pH) colourless **Ct** species can be photochemically converted (365-nm light) into the thermodynamically unstable, but relatively inert, **Cc** form (*write*);
2) by a second stimulus (addition of acid), **Cc** can be converted into the kinetically inert or thermodynamically stable (depending on pH) AH^+ form (*lock*); if the initial pH is 1, the **Cc** species autolocks as AH^+;
3) the AH^+ species is photochemically inactive and shows an absorption spectrum clearly distinct from that of **Ct**, so that it can be optically detected (*read*);
4) by addition of base, AH^+ can be reconverted into **Cc** (*unlock*);
5) **Cc** can be reconverted to the initial **Ct** form by a thermal or a photochemical reaction (*erase*).

A cycle equal to that just described can be designed with 4′-methylflavylium ion. Among the 4′-substituted flavylium ions, this compound exhibits the lowest energy barrier for the *cis–trans* isomerization reaction. Lowering the activation energy of the *cis–trans* isomerization reaction favours autoerasing, which discourages its use in an optical memory system [70].

The 4′-hydroxy-3-methylflavylium compound can also be used to operate a write-lock-read-unlock-erase cycle [71].

Due to its isomerization barrier, also 4′-hydroxy-6-nitroflavylium [66] can be efficiently used in *write-lock-read-unlock-erase* cycles. This compound performs like

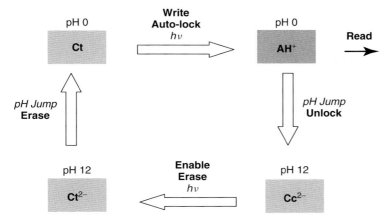

Figure 6.21 Optical memory based on 4′-hydroxy-6-nitroflavylium, write/lock-read-unlock-enable/erase–erase cycle [71].

an optical memory by means of a slightly different cyclic process that involves the following steps: *write/lock-read-unlock-enable/erase-erase* (Figure 6.21):

1) starting with the **Ct** form at pH 0, light excitation causes the isomerization to **Cc** (write) that spontaneously converts (autolock) to **AH$^+$** because of the acidity of the solution;
2) **AH$^+$** is photostable and can thus be examined by UV/Vis absorption spectroscopy (read) without being erased;
3) when the information has to be erased, **AH$^+$** can be converted into **Cc^{2-}** by a pH jump to pH 12 (unlock);
4) **Cc^{2-}** can then be isomerized by light excitation to **Ct^{2-}**, a process (enable-erase) that finally allows formation of the original **Ct** species by a pH jump to pH 0 (erase).

The 3′,4′-(methylenedioxy)flavylium compound, whose network of thermal and photochemical reactions is shown in Figure 6.22, offers the possibility of conceiving two coupled photochromic systems [72]. In the pH range 1–6, a cycle capable of *write-read-erase* can be designed.

1) The starting point is the metastable **Ct** species at pH 1.0. Using near-UV light, **Ct** is converted into **AH$^+$** (write and autolock).
2) Once again **AH$^+$** can be easily readout.
3) The system can be erased by a pH jump to pH 6.0 in order produce **Cc** (in equilibrium with **B2**), which is thermally reconverted into **Ct**.
4) To prepare the system for a new cycle (enable), a second pH jump to 1 should be performed.

The other photochromic system is even more interesting because there is the possibility of two consecutive *write* steps [73].

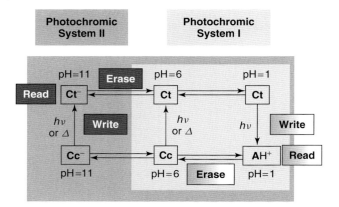

Figure 6.22 *Write-read-erase* cycles for 3′,4′-(methylenedioxy)flavylium covering acid (photochromic system i, light grey) and basic (photochromic system II) pH regions [73]. For details, see text.

1) Starting, for example with **Cc⁻** at pH 11, the first *write* step consists of the irradiation of this species that totally converts into **Ct⁻**.
2) **Ct⁻** can be examined by UV/Vis absorption spectroscopy (read)
3) The system can be erased through a pH jump to 1 leading to the metastable **Ct** species.
4) At this point a second *write* step (the same as the previous cycle) is possible, converting the system into **AH⁺**. This last reaction can also be carried out thermally.
5) Finally, the system should be submitted to a second pH jump to the starting pH value, so that **Cc⁻** species is recovered.

For the 4′-dimethylamino compound, it is also possible to perform a *write-lock-read-unlock-erase* cycle, operating however in a different way [42].

6.11.2
Reading without Writing in a *Write-Lock-Read-Unlock-Erase* Cycle

A generally overlooked difficulty with photochromic systems is that the starting form (**Ct** in the above discussion) is the photoreactive one, so that it cannot be read by absorption spectroscopy without writing. With the 4′-hydroxyflavylium ion, this difficulty can be overcome starting from **AH⁺**, which is the thermodynamically stable form at pH = 1, and performing a *write-lock-read-unlock-erase* cycle as illustrated in Figures 6.23 and 6.24 [54]. Since **AH⁺** is not photosensitive, it can be read by light excitation (i.e. by recording its absorption spectrum) without writing. Then it can be unlocked by a pH jump to 12, which yields the metastable **Cc²⁻** form. At this stage, one can write the optical information obtaining the stable (locked) **Ct²⁻** form that can then be read. When necessary, the information

6.11 Chemical Process Networks | 207

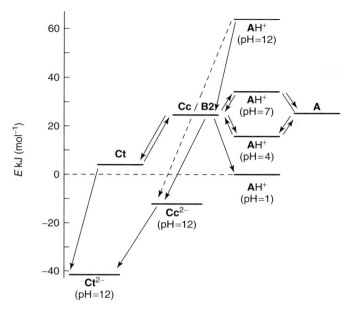

Figure 6.23 Energy-level diagram for the species involved in the structural transformations of the 4'-hydroxyflavylium compound in the pH range 1–12 [54].

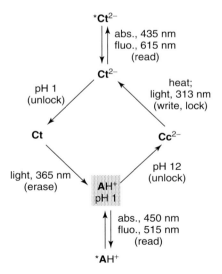

Figure 6.24 Read-write-lock-read-unlock-erase cycle starting from the AH^+ form of the 4'-hydroxyflavylium compound [54].

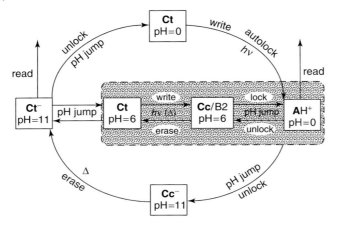

Figure 6.25 *Write-lock-read-unlock-erase* cycles for the unsubstituted flavylium ion [58]. For more detail, see text.

stored into Ct^{2-} can be unlocked by a pH jump yielding **Ct** and can then be erased by light excitation. The same functionality can be obtained starting from Ct^{2-}.

In the case of the unsubstituted flavylium cation [58], in addition to a cycle such as that schematized in Figure 6.20, a cycle based on the anionic species can also be performed (Figure 6.25). This second cycle starts at pH = 11 with the Ct^- form that, being not photosensitive, can be read without writing. Then, two different paths can be followed. The first one begins with a pH jump to pH = 6 that leads to **Ct** and goes on as described above (shaded area in Figure 6.25). The second path starts with a pH jump from 11 to 0, leading to **Ct** that can be photochemically written (and locked because of the low pH) to AH^+. In this form, the information can be stored permanently and read without erasing since AH^+ is thermally and photochemically stable. When necessary, AH^+ can be unlocked by a pH jump to 11 and thermally erased to give back Ct^-. An advantage of this cycle lies in the possibility of reading the system in both the initial (nonwritten) and final (written) states without writing or erasing. Moreover, Ct^- is more stable than **Ct**, so that the durability of the system could increase. A disadvantage is given by the fact that in this cycle autounlocking cannot occur so that two pH changes per cycle are needed [58].

The pH-driven cycle of compound 7-(*N*,*N*-diethylamino)-4'-hydroxyflavylium [63] is an example of unidirectional cycle because the reverse reaction follows a different pathway from the forward reaction.

6.11.3
Micelle Effect on the Write-Lock-Read-Unlock-Erase Cycle

Interaction of micelles with the flavylium network introduces significant modifications on the mole-fraction distribution of the various species as well as

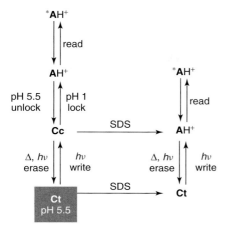

Figure 6.26 *Write-lock-read-unlock-erase* cycles for the 4′-hydroxyflavylium ion, starting from the **Ct** form at pH 5.5. Left-hand side: light and pH jump inputs in the absence of micelles. This part is equivalent to the cycle shown in Figure 6.20. Right-hand side: light input at the autolocking pH in the presence of SDS micelles [63].

on the photochemical response of the network. The flavylium cation, due to its positive charge, is stabilized by the negatively charged sodium dodecyl sulfate (SDS) micelles, whereas the contrary occurs in the case of the positively charged cetyltrimethylammonium bromide (CTAB). On the other hand, neutral polyoxyethylene(10)-isooctylphenylether (Triton X-100) micelles tend to stabilize the uncharged (basic) forms [57]. As a general behaviour the neutral chalcones have a propensity to be stabilized in the presence of micelles most probably because they are fairly soluble in water and prefer the hydrophobic core of the micelles. In addition to the effect on the mole-fraction distribution of the various species in solution, the presence of micelles also influences their interconversion rates. Addition of micelles can therefore be considered as a third external stimulus (together with light excitation and pH jump) capable of changing the state of this multistate/multifunctional molecular-level system. Particularly interesting is the possibility of changing the autolock pH of a photochromic reaction by addition of micelles. For example, (Figure 6.26) in the case of the compound 4′-hydroxyflavylium, starting from a **Ct** solution at pH 5.5, it is possible to design an alternative *write-lock-read-unlock-erase* cycle based on addition of SDS micelles and exploit autolocking and thermal erasing processes of this system [57]. It was demonstrated that the effect of SDS is mainly on the hydration process, K_h, leaving the other equilibrium and rate constants essentially unaffected. More specifically, the largest effect observed on the stabilization of the flavylium cation in the presence of SDS micelles is a result of the increasing of its dehydration rate constant, k_{-h} [74].

A necessary requirement to carry out write-lock-read-unlock cycles based on flavylium salts is the existence of a high barrier for the *trans–cis* isomerization. However, compounds such as luteolinidin (a natural deoxyanthocyanidin, 5,7,3',4'-tetrahydroxyflavylium) which have a low barrier, can also be used to store information if the following sequence of steps is carried out: [75]

1) *Write* – light irradiation at 365 m, that converts the equilibrium mixture **Ct/A** at pH 6 (70/30) into **A**, which can be read (at 500 nm).
2) *Lock* – as **A** tends to give back the thermodynamically stable species **Ct**, returning to the initial equilibrium, a lock step (pH jump to pH ≤ 2) is needed.
3) *Read* – the flavylium cation thus formed can be easily read out.
4) *Unlock* – pH jump to 6 and a thermal reaction (erase) that gives back the initial state.

Addition of CTAB to aqueous solutions of compounds exhibiting a small thermal barrier for the *cis–trans* isomerization allows the operation of very efficient photochromic systems [73, 76]. As an example, in the case of 7,4'-dihydroxyflavylium at pH 1.5 the main species is the **Ct**-CTAB adduct (**Ct$_m$**), while in water **AH$^+$** predominates at this pH. After addition of CTAB micelles to an aqueous solution of **AH$^+$**, the system evolves to **Ct$_m$** with concomitant disappearance of the yellow flavylium cation.

The reverse processes (appearance of **AH$^+$**) can be achieved by irradiation at 365 nm (the photoreaction takes about 4 min to be complete) [73]. This kind of systems present several advantages over those in water. In particular, they allow the observation of a better colour contrast upon irradiation.

Photochromic systems based on the interaction of CTAB micelles with flavylium derived 2-hydroxychalcones, can also be designed for aminoflavylium ions. These compounds give rise to a variety of beautiful colours, but totally lack photochemistry in pure water. While in water (2 < pH < 5) **AH$^+$** (red-pink) is more stable than **Ct** (yellow), in the presence of CTAB micelles the species **Ct$_m$** is even more stable than **AH$^+$**. Interesting enough is the photochemical production of **AH$^+$** upon irradiation of **Ct$_m$**, followed by its very fast migration to the bulk water and slow conversion back to **Ct$_m$** [76]. The system cannot be considered a permanent memory, but it can be viewed as a temporary one. Similar results were observed with 7-hydroxy-2-(4-N,N-dimethylaminostyryl)-1-benzopyrylium and 7-hydroxy-2-(4-hydroxystyryl)-1-benzopyrylium in CTAB micelles [68].

Virtually, flavylium ions exhibiting all the colours of the rainbow can be synthesized and their photochromism explored in this way, with the additional advantage of keeping the overall mass of the system constant, because no pH jumps are needed.

6.11.4
Permanent and Temporary Memories

The human brain contains shallow and deep memory forms [77]. The network of processes interconverting the various species of the 4'-hydroxyflavylium ion

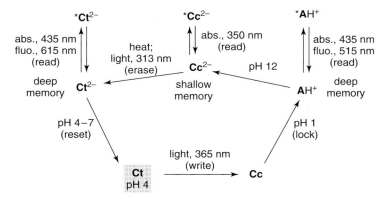

Figure 6.27 A write-lock-read-unlock-erase cycle with two memory levels based on the 4′-hydroxyflavylium compound [54].

allows the presence of different levels of memory [54]. Once the permanent (deep) AH^+ form of memory has been obtained (*write* and *lock*, Figure 6.27), a jump to pH 12 leads to the formation of a temporary (shallow) memory state, Cc^{2-}, whose spontaneous slow erasure to give the deep Ct^{2-} memory can be accelerated by light. Reset can then be accomplished by a back pH jump to pH 4.

6.11.5
Oscillating Absorbance Patterns

Another feature of the 4′-hydroxyflavylium ion should be emphasized [54]. Starting from AH^+, alternation of pH jump and light excitation causes oscillation patterns of absorbance at different wavelengths, as shown in Figure 6.28. Such patterns may be interesting for signal generation and information processing [78].

6.11.6
Colour-Tap Effect

Because of the competition between the pH-dependent rate of the reaction leading from the uncoloured **Cc** form to the coloured AH^+ and **A** species and the pH-independent back *cis* → *trans* isomerization, the amount of coloured species formed upon light excitation of the **Ct** solution depends on pH. In other words, the pH plays the role of a tap for the colour intensity generated by light excitation [59]. This also means that this system can be viewed as a light-switchable pH indicator. In the case of 7,4′-dihydroxyflavylium the colour-tap effect is larger than for 7-hydroxyflavylium.

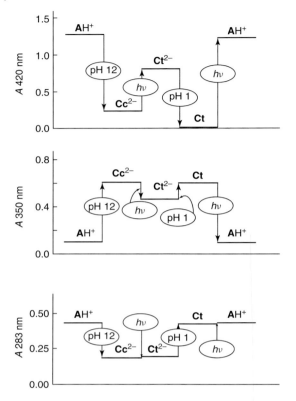

Figure 6.28 Absorbance oscillations caused by alternate pH jump and light excitation on a 3.3×10^{-5} M aqueous solution starting from the **AH$^+$** form of the 4'-hydroxyflavylium compound at pH = 1 [54].

6.11.7
Logic Operations

From the viewpoint of logic operations [8, 79, 80], simple (bistable) photochromic systems perform YES/NO functions. Multistate/multifunctional molecular-level systems can be taken as bases for more complex logic operations. Chemical systems capable to perform AND [79], OR [79], eXclusive OR (XOR) [81] and XNOR [82] logic operations and integration of logic functions and sequential operation of gates at the molecular scale have been reported [83]. (A complete survey of chemical systems capable of performing as logic gates is given in Chapter 18.) With the 4'-hydroxyflavylium compound, light excitation and pH jumps can be taken as inputs, and absorbance or fluorescence as outputs. Starting from the nonemitting **Ct** species (Figure 6.7) and taking the emission of **AH$^+$** at 515 nm as output signal, jump to pH = 1 alone or light excitation alone are not able to generate the output,

Table 6.2 Truth table for the AND logic behaviour of the 4′-hydroxyflavylium compound starting from **Ct** at pH = 5.5.

Input 1[a]	Input 2[b]	Output[c]
0	0	0
1	0	0
0	1	0
1	1	1

[a]pH jump to pH 1.
[b]Light excitation at 365 nm.
[c]Absorbance at 435 nm or emission at 515 nm of the AH^+ form.

whereas when these two inputs are applied in series, the output is obtained (AND logic function, Table 6.2) [54].

For the 4′-hydroxyflavylium compound in the presence of micelles [57], starting from **Ct** at pH 5.0 and taking the formation of the AH^+ absorption at 436 or 450 nm as an output, the truth table for the effect of the three inputs (pH jump to 1.0, addition of SDS, light excitation) shows a peculiar pattern (Table 6.3) corresponding to an OR function that is activated only in the presence of the third input (enabled OR).

An interesting system, based on the 4′-methoxyflavylium ion, capable of behaving according to an XOR logic has been reported [84]. The system consists of an aqueous solution containing the **Ct** form of the 4′-methoxyflavylium ion (AH^+), and the $[Co(CN)_6]^{3-}$ complex ion (as a potassium salt). The absorption spectra of **Ct** and $[Co(CN)_6]^{3-}$ are shown in Figure 6.29. For the present discussion, the most relevant

Table 6.3 Truth table for the enabled OR logic behaviour of the 4′-hydroxyflavylium compound starting from **Ct** at pH = 5.5.

Input 1[a]	Input 2[b]	Input 3[c]	Output[d]
1	0	0	0
0	1	0	0
1	1	0	0
0	0	0	0
1	0	1	1
0	1	1	1
1	1	1	1
0	0	1	0

[a]pH jump to pH 1.
[b]SDS micelle.
[c]Light excitation at 365 nm.
[d]Absorbance at 435 nm or emission at 515 nm of the AH^+ form.

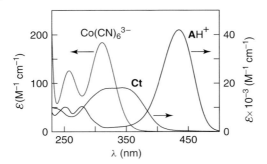

Figure 6.29 Absorption spectra of $[Co(CN)_6]^{3-}$ and of the **Ct** and **AH**$^+$ forms of the 4'-methoxyflavylium compound in aqueous solution [84]. Reproduced from reference [84]. Copyright ACS (2000).

aspects of the thermal and photochemical reactions of these two compounds, are as follows [53, 85]. Excitation by 365-nm light of **Ct**, which is the thermodynamically stable form of the flavylium species in the pH range 3–7, causes the already discussed *trans* → *cis* photoisomerization reaction (quantum yield: 0.04): [53]

$$\text{Ct} \xrightleftharpoons{h\nu} \text{Cc} \xrightleftharpoons[\text{OH}^-]{\text{H}^+} \text{AH}^+ \tag{6.14}$$

As we have seen above, if the solution is sufficiently acid (pH < 4), the **Cc** isomer is rapidly protonated with conversion to the 4'-methoxyflavylium ion **AH**$^+$, which is kinetically stable under such pH conditions and exhibits an intense absorption band with maximum at 434 nm (Figure 6.29) and an emission band with a maximum at 530 nm (Figure 6.7). At higher pH values, however, protonation does not occur and the **Cc** photoproduct is back converted to **Ct**. As far as $[Co(CN)_6]^{3-}$ is concerned, excitation by 254- or 365-nm light in acid or neutral aqueous solution causes the dissociation of a CN^- ligand from the metal coordination sphere (quantum yield = 0.31), with a consequent increase in pH [85]:

$$[Co(CN)_6]^{3-} + H_3O^+ + h\nu \rightarrow [Co(CN)_5(H_2O)]^{2-} + HCN \tag{6.15}$$

When an acid solution (pH = 3.6) containing 2.5×10^{-5} M **Ct** and 2.0×10^{-2} M $[Co(CN)_6]^{3-}$ is irradiated at 365 nm, most of the incident light is absorbed by **Ct** (Figure 6.29), which undergoes photoisomerization to **Cc**. Since the pH of the solution is sufficiently acidic, **Cc** is rapidly protonated (Reaction 6.14), with the consequent appearance of the absorption band with maximum at 434 nm (Figure 6.29) and of the emission band with maximum at 530 nm (Figure 6.7) characteristic of the **AH**$^+$ species. On continuing irradiation, it can be observed that the absorption and emission bands increase in intensity, reach a maximum value and then decrease to complete disappearance. These results show that **AH**$^+$ first forms and then disappears with increasing irradiation time. The reason for

Table 6.4 Truth table for the XOR (eXclusive OR) logic behaviour of the 4'-methoxyflavylium-$[Co(CN)_6]^{3-}$ system starting from pH = 3.6.

Input 1[a]	Input 2[a]	Output[b]
0	0	0
1	0	1
0	1	1
1	1	0

[a]The two inputs are identical and consist of the amount of photons necessary to achieve the absorbance value corresponding to the top of the curve of Figure 6.25.
[b]Absorbance at 435 nm or emission at 515 nm of the AH^+ form.

the off-on-off behaviour of AH^+ under continuous light excitation is related to the effect of Reaction 6.15 on Reaction 6.14. As **Ct** is consumed by Reaction 6.14 with formation of AH^+, an increasing fraction of the incident light is absorbed by $[Co(CN)_6]^{3-}$, which undergoes Reaction 6.15. Such a photoreaction causes an increase in the pH of the solution. This not only prevents further formation of AH^+, which would imply protonation of the **Cc** molecules that continue to be formed by light excitation of **Ct**, but also causes the back reaction to **Cc** (and, then, to **Ct**) of the previously formed AH^+ molecules. Clearly, the examined solution performs like a threshold device as far as the input (light)/output (spectroscopic properties of AH^+) relationship is concerned. Instead of a continuous light source, pulsed (flash) irradiation can be used. Under the input of only one flash, a strong change in absorbance at 434 nm is observed, due to the formation of AH^+. After two flashes, however, the change in absorbance practically disappears. In other words, an output (434 nm absorption) can be obtained only when *either* input 1 (flash I) *or* input 2 (flash II) are used, whereas there is no output under the action of *none* or *both* inputs. This shows (Table 6.4) that the above-described system behaves according to an XOR logic, under control of an intrinsic threshold mechanism. It is noteworthy that the input and output signals have the same nature (light) [84].

6.11.8
Multiple Reaction Patterns

In the case of the unsubstituted flavylium cation we have seen that different paths can be followed to obtain the same result (Figure 6.25). For the 4'-hydroxyflavylium compound the network of processes is even more intricate and several species are interconnected by multiple reaction patterns [54]. For example, in order to go from AH^+ to **Ct** three different routes can be chosen, as is pictorially represented in Figure 6.30:

1) jump to pH 12, with formation of Cc^{2-}, followed by excitation with 313-nm light to obtain Ct^{2-}, and by a jump to pH = 6;

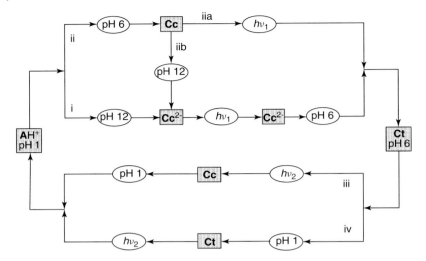

Figure 6.30 The network of processes caused by pH jumps and light excitations interconnecting the **AH$^+$** and **Ct** forms of the 4′-hydroxyflavylium compound [54].

2) jump to pH = 6 to form **Cc**; at this stage, one can choose between two subroutes, that is (i) light excitation with at 313-nm or (ii) jump to pH 12 to merge into the preceding path which leads to **Ct** via **Cc^{2-}** and **Ct^{2-}**.
3) Once **Ct** has been obtained, one can go back to **AH$^+$** by two different routes:
4) light excitation at 365 nm to obtain **Cc** and subsequent jump to pH = 1;
5) jump to pH = 1 and subsequent light excitation at 365 nm.

Interestingly, in some cases, that is starting from **Ct** at pH 6, one obtains the same result (**AH$^+$**) regardless of the order in which light excitation (365 nm) and pH jump (pH = 1) are applied. In other cases, however, this is not true because the system exhibits a 'memory effect'. For example (not shown in Figure 6.30), starting from **AH$^+$** at pH = 1, light excitation followed by pH jump to 12 leads to **Cc^{2-}**, whereas when the two inputs are applied in the reverse order one gets **Ct^{2-}**. Since **Cc^{2-}** and **Ct^{2-}** exhibit very different spectroscopic properties (e.g. **Ct^{2-}** exhibits fluorescence, whereas **Cc^{2-}** does not), from the state of the system after the two inputs one can establish in which sequence the two inputs have been applied.

A network of processes governed by external inputs can be used as a model system [76, 86, 87] for understanding the chemical basis of complex biological systems [88, 89].

6.11.9
Upper-Level Multistate Cycles

While several new multistate systems can be designed, it is desirable to aim at higher levels of operation, for instance, through the assembly of several multistate systems together, each system operating by its own kinetic and thermodynamic rules.

A new level of complexity could be introduced by connecting two multistate networks based on 4′-acetamidoflavylium and 4′-aminoflavylium, possessing six- and sevenfold multistates, respectively, and that can be reversibly interchanged by inputs of light, pH jumps and heat. The two networks are connected irreversibly, the 4′-acetamidoflavylium being transformable into the 4′-aminoflavylium by means of a heat input in extremely acidic solutions (Figure 6.31).

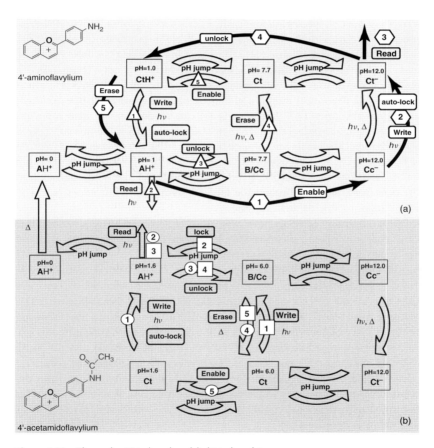

Figure 6.31 Thermal, pH-induced and light-induced interconverting pathways between all possible forms in 4′-acetamidoflavylium (b) and 4′-aminoflavylium (a) systems, allowing several *write-lock-read-unlock-erase* cycles involving different species [44]. For details, see text. Reproduced from reference [44]. Copyright ACS (2000).

In the case of 4′-acetamidoflavylium [44] (Figure 6.31b, numbers inside squares), one possible cycle starts with the **Ct** species in the equilibrium at pH = 6.0:

1) The photochemical *write* step gives a mixture of **B/Cc** (the system has the drawback of not being completely converted into **B/Cc** state).
2) The *lock* step results from the pH jump 6–1.6, forming **AH**$^+$, which is stable and can be read.
3) The *unlock* step corresponds to the pH jump 1.6–6 to give **B/Cc**.
4) The erasing can be achieved by an heat input. For this flavylium, yet another cycle can be conceived, by starting with metastable **Ct** species at acidic pH (Figure 6.31b, numbers inside circles).

In the case of the compound 4′-aminoflavylium (Figure 6.31a) [44], cycles starting from the **Ct** or the **Ct**$^-$ species are not feasible as they are not photoactive. However, it is possible to start with the **CtH**$^+$ species available in pseudoequilibrium at pH 1 as a metastable species (Figure 6.31a, numbers inside triangles). Because of the low pH value of the system, the autolock step is spontaneous upon irradiation (1), the *cis*-chalcone evolving to **B** (both possibly protonated) and finally to **AH**$^+$, which can be read (2). The *unlock* step (3), giving **Cc** (in this compound **B** is residual) is made by a pH jump 1–7. The erasing (4) can be done by an heat input or, better, by using light to allow the formation of **Ct**. At this pH value this compound is thermodynamically stable and photoinactive and the system can be stored. To allow the next cycle, it is necessary to use the *enable* step (5) through a pH jump from 7 to 1. Other cycles involving the basic species of the 4′-aminoflavylium can be conceived, passing through the ionized *cis* and *trans* chalcones (see, e.g. Figure 6.31a, numbers inside hexagons).

While each memory cycle operates over states of a given flavylium system, this upper level of information treatment would operate over memory cycles within a system and over crossing from one system to another.

The sequence of instructions necessary to carry out an optical memory cycle may be defined as its algorithm. On this basis, the *write-lock-read-unlock-erase* cycle starting with the **Ct** species of 4′-acetamidoflavylium at pH = 6.0 (Figure 6.31b, numbers in squares) is described by an algorithm consisting of the following sequence of instructions: (i) irradiation at 365 nm, (ii) pH jump from 6.0 to 1.0, (iii) measurement of the absorption at 438 nm (**AH**$^+$ absorption), (iv) pH jump 1.0–6.0 and (v) thermal input. Indeed, each memory cycle operates according to a defined and unique algorithm. Moreover, it is possible to change from the 4′-acetamidoflavylium network to the 4′-aminoflavylium one by another algorithm containing two instructions: (i) pH jump to pH = 0 (or even pH = 1.0) or (ii) thermal input (Figure 6.31).

A careful analysis of the mole-fraction distribution (Figure 6.32) of species at the pseudo- and at the final equilibrium for each flavylium system provides a means for building up flow diagrams to allow operation over memory cycles and system crossing. Overlapping the information on Figure 6.31 with the 4′-acetamidoflavylium distribution of species in Figure 6.32 defines an operational pH range for each memory cycle. A similar exercise can be done for the 4′-aminoflavylium [44].

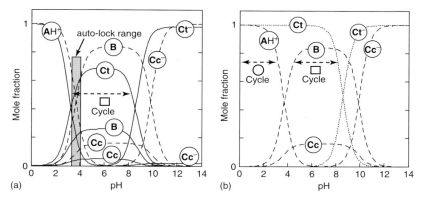

Figure 6.32 Mole fraction distribution of 4′-acetamidoflavylium: (a) solid line thermodynamic equilibrium, dashed line – pseudoequilibrium **AH⁺/B/Cc/Cc⁻** and (b) traced line – pseudoequilibrium **AH⁺/B/Cc/Cc⁻**, dotted line – pseudoequilibrium **Ct/Ct⁻** [44]. Reproduced from reference [44]. Copyright Wiley (2005).

With this information, a flow diagram such as the one shown in Figure 6.33 can be constructed. As an example, if the **Ct⁻** species of 4′-acetamidoflavylium is regarded as the starting point of the upper-level cycle, the first step would consist of a pH input to acid. If pH > 8, the **Ct⁻** form remains unchanged and a further pH input is necessary (return). At 4.5 < pH < 8, the squares cycle can operate, while for pH < 4.5 the autolock pH cycle (circles) is available. After operation of the squares cycle, the system is in the **Ct** multistate, and other inputs of pH can be made (return). On the other hand, after operation of the circles cycle, a heat input permits passage to the 4′-aminoflavylium network. At this point, a new pH input is needed to continue: (i) if pH < 7 the hexagons cycle can be performed and (ii) if pH > 7, the system is in its pseudoequilibrium and a heat input is necessary to produce the *trans*-chalcone species. The next step is a further pH input: (i) if pH > 2, it is necessary to return in order to continue and (ii) if pH < 2, the triangles cycle can operate, and finally stop.

6.11.10
Multiswitchable System Operated by Proton, Electron and Photon Inputs

The substitution of the flavylium compound in position 6 by a hydroxyl group can be compared with the same substitution in position 4′ in terms of the existence of a kinetic barrier for the *cis–trans* isomerization [90]. In the case of the 6-hydroxyflavylium, the quinone/hydroquinone character of the respective chalcones (see Reaction 6.16) allows us to obtain quasireversible cyclic voltammetric waves. It is well known that flavylium salts and anthocyanins can be involved in several electrochemical processes, but no reversible electrochemistry was previously

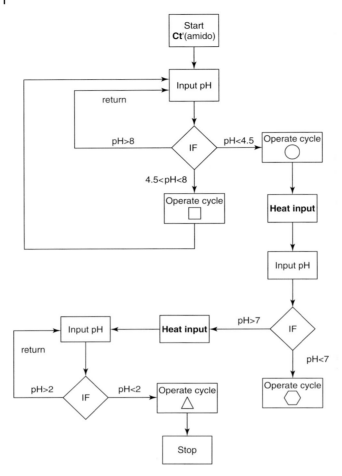

Figure 6.33 Example of a flow diagram over memory cycles, defining a sequence of algorithms to be operated in both 4′-acetamidoflavylium and 4′-aminoflavylium chemical networks [44].

reported on flavylium species, not even on flavylium-derived chalcones [91].

$$\text{Ct} \underset{}{\overset{-2H^+,\ -2e^-}{\rightleftarrows}} \text{Ct}_{ox} \qquad (6.16)$$

This compound serves to illustrate for the first time in the flavylium network of chemical reactions the possibility of using electrochemistry as a new input to reach other states of the network.

6.11.11
Nonaqueous Media and Steps towards Solid-State Devices

As discussed in the previous sections, for the operation of an optical memory, the existence of a *cis–trans* thermal barrier that prevents the back reaction is required, otherwise the written information is erased before it can be read. The kinetic barrier for *cis–trans* isomerization is much higher in ionic liquids, which can be explained by a more rigid and stable distribution of **Ct** in the ionic liquid that needs to be disrupted during the thermal isomerization owing to the formation of less-planar intermediates, as well as by the large viscosity of the ionic liquid [92]. This desired property allows to operate a *write-read-erase* cycle, for a compound that *per se*, cannot be used as a memory due to the lack of thermal barrier, such as 7,4′-dihydroxyflavylium [48].

A very similar cycle was reported using the 4′-hydroxyflavylium compound [93]. Moreover, 7-(N,N-diethylamino)-4′-hydroxyflavylium exhibits photochromism when used in a water–ionic liquid ([bmim][PF$_6$]) biphasic systems [94].

The performance of flavylium-based photochromic systems is increased by their incorporation into Pluronic® F-127 matrices, which switch from polymeric solutions to micelles to gels with changes in temperature depending on copolymer concentration [95]. Two compounds 7,4′-dihydroxyflavylium and 7-(N,N-diethylamino)-4-hydroxyflavylium, both exhibiting a small thermal *cis–trans* isomerization barrier in water were investigated. In the first system, the flavylium in the gel photoswitches from the colourless **Ct** species to the yellow **AH$^+$** with $\Phi = 0.04$ at pH 2.2 or to the orange quinoidal base (**A**) with quantum yield $\Phi = 0.015$ (25 °C) at pH 5.2. The second system does not exhibit photochemistry in water but, when incorporated into the gel matrix, switches from yellow to red with $\Phi = 0.01$ at pH 4.9.

In order to achieve progress towards real practical applications, encapsulation of synthetic flavylium salts in water-permeable crosslinked poly(2-hydroxyethyl methacrylate) (PHEMA) polymer matrices was carried out and a solid multistate/multifunctional system was obtained upon application of light and pH stimuli [96]. This system represents a step towards real applications, but it does present some drawbacks, such as the existence of some leaching during the immersion in acidic and basic solutions (that are used to operate the cycles) from the more accessible sites. This leaching is reduced in the next cycles probably occurring in more inaccessible sites [97].

Pursuing the same aim of trying to achieve solid-state systems based on flavylium ions [98], some photochromic systems in solid hydrogel matrices have been reported, for instance, agar-gel and sol-gel matrices. It was possible to conclude that in the sol-gel matrix the colouration/decolouration could be repeated over 50 cycles without significant fatigue. In a silica sol-gel matrix, a fairly good reversibility was obtained up to about 30 cycles; in the later cycles, extensive degradation took place. This promising result could, however, be obtained only with specific derivatives of flavylium ions, 4′-(N,N-dimethylamino) flavylium in the case of agar-gel and 4′-(N-octylamino) flavylium in the case of a sol-gel matrix [98].

Encapsulation of flavylium cations in zeolites has also been pursued [97, 99]. In general, the positively charged flavylium cations are stabilized by the aluminosilicate framework of the zeolites. More interestingly, it was possible to encapsulate the neutral *trans*-chalcone of 7,4′-dihydydroxyflavylium in the one-dimensional channels of zeolite L [99]. The encapsulated **Ct** was shown to generate AH^+ when the **Ct**-loaded crystals were suspended in water, which proves that isomerization, tautomerization and dehydration reactions take place inside the zeolite L.

A phase-change thermochromic system was described based on a 4-substituted flavylium dye 4-(2-carboxyphenyl)-7-diethylamino-4′-dimethylamino-1-benzopyrylium, that is able to form a leuco lactone form in the presence of a developer (ethyldiisopropylamine) and a suitable solvent (e.g. acetonitrile, *n*-pentadecanonitrile) [100]. The leuco form is a spirolactone species, which ring opens at low temperature (below the solvent melting point) to form the blue flavylium cation. Decarboxylation of the lactone to give 4-phenyl-7-diethylamino-4′-dimethylamino-1-benzopyrylium was observed upon irradiation of the system with UV light, erasing the thermochromic effect.

6.12
Conclusions

Synthetic flavylium compounds can exist in several forms (*multistate*) that can be interconverted by more than one type of external stimuli (*multifunctional*). The intricate network of their reactions, when examined from the viewpoints of 'molecular-level device' and 'molecular-level logic function', reveals that these systems exhibit very interesting properties.

In our brain, neurons store, exchange and retrieve information via extremely complicated chemical processes [88, 89]. Synthetic multistate/multifunctional systems may play the role of models to begin understanding the chemical basis of complex biological processes [84]. It is not at all clear whether 'wet' artificial systems can find real applications, for example in molecular-scale computers [101, 102]. In any case, the study of molecular or supramolecular species capable of existing in different forms that can be interconverted by external stimuli is a topic of great interest since it introduces new concepts in the field of chemistry and stimulates the ingenuity of research workers engaged in the 'bottom up' approach to nanotechnology.

Acknowledgements

This work was supported in Portugal by the Fundação para a Ciência e Tecnologia and FEDER (projects PTDC/QUI/67786/2006 and PTDC/QUI-QUI/104129/2008), and in Italy by MURST (PRIN 2006034123 and PRIN 2006030320) and the University of Bologna (Funds for Selected Research Topics). R. G. thanks a PhD grant from FCT (Portugal SFRH/BD/27282/2006).

References

1. Balzani, V. and Scandola, F. (1991) *Supramolecular Photochemistry*, Horwood, Chichester.
2. Ward, M.D. (1997) *Chem. Ind.*, (16), 640.
3. Balzani, V., Credi, A. and Venturi, M. (1999) in *Supramolecular Science: Where it is and Where it is Going* (eds R. Ungaro and E. Dalcanale), Kluwer, Dordrecht, p. 1.
4. Ward, M.D. (2001) *J. Chem. Ed.*, **78**, 323.
5. Pease, A.R. and Stoddart, J.F. (2001) *Struct. Bond.*, **99**, 189.
6. Tomasulo, M., Giordani, S. and Raymo, F.M. (2005) *Adv. Funct. Mater*, **15**, 787.
7. Gust, D., Moore, T.A. and Moore, A.L. (2006) *Chem. Commun.*, 1169.
8. de Silva, A.P., Uchiyama, S., Vance, T.P. and Wannalerse, B. (2007) *Coord. Chem. Rev.*, **251**, 1623.
9. Kay, E.R., Leigh, D.A. and Zerbetto, F. (2007) *Angew. Chem. Int. Ed.*, **46**, 72.
10. Ballardini, R., Ceroni, P., Credi, A., Gandolfi, M.T., Maestri, M., Semeraro, M., Venturi, M. and Balzani, V. (2007) *Adv. Funct. Mater.*, **17**, 740.
11. Balzani, V., Credi, A. and Venturi, M. (2008) *Molecular Devices and Machines*, 2nd edn, Wiley-VCH Verlag GmbH, Weinheim.
12. Irie M. (2000) *Chem. Rev.*, **100** (5), 1683–1684 (Photochromism: memories and switches).
13. (a) Zander, Ch., Enderlein, J. and Keller, R.A. (2002) *Single Molecule Detection in Solution*, Wiley-VCH Verlag GmbH, Weinheim; (b) Samorì, P. (2006) *Scanning Probe Microscopies beyond Imaging: Manipulation of Molecules and Nanostructures*, Wiley-VCH Verlag GmbH, Weinheim.
14. Balzani, V., Credi, A. and Venturi, M. (2008) *Chem. Eur. J.*, **14**, 26.
15. (a) Moresco, F., Meyer, G. and Rieder, K.-H. (2001) *Phys. Rev. Lett.*, **86**, 672; (b) Loppacher, C., Guggisberg, M., Pfeiffer, O., Meyer, E., Bammerlin, M., Lüthi, R., Schlittler, R., Gimzewski, J.K., Tang, H. and Joachim, C. (2003) *Phys. Rev. Lett.*, **90**, 066107; (c) Qiu, X.H., Nazin, G.V. and Ho, W. (2004) *Phys. Rev. Lett.*, **93**, 196806.
16. Balzani, V., Credi, A. and Venturi, M. (2000) in *Stimulating Concepts in Chemistry* (eds M. Shibasaki, J.F. Stoddart and F. Vögtle), Wiley-VCH Verlag GmbH, Weinheim, p. 255.
17. (a) Balzani, V., Credi, A., Raymo, F.M. and Stoddart, J.F. (2000) *Angew. Chem. Int. Ed.*, **39**, 3348; (b) Balzani, V., Credi, A., Silvi, S. and Venturi, M. (2006) *Chem. Soc. Rev.*, **35**, 1135.
18. Sauvage, J.-P. (2001) *Struct. Bond.*, **99** (Special volume on Molecular Machines and Motors).
19. Kelly, T.R. (2005) *Top. Curr. Chem.*, **262** (Special volume on Molecular Machines).
20. Balzani, V., Credi, A. and Venturi, M. (2007) *Nanotoday*, **2**, 18.
21. Fritzsche, J. (1867) *C. R. Acad. Sci.*, **69**, 1035.
22. Ciamician, G. (1912) *Science*, **36**, 385.
23. Horspool, W.M. and Song, P.-S. (1995) *Handbook of Organic Photochemistry and Photobiology*, CRC, Boca Raton.
24. Brown, G.H. (1971) *Photochromics*, Wiley-Interscience, New York.
25. (a) Hirshberg, Y. (1956) *J. Am. Chem. Soc.*, **78**, 2303; (b) Hirshberg, Y. (1960) *New Sci.*, **7**, 1243.
26. (a) Michell, R.H., Iyer, V.S., Mahadevan, R., Venugopalan, S. and Zhou, P. (1996) *J. Org. Chem.*, **61**, 5116; (b) Michell, R.H., Ward, T.R., Wang, Y. and Dibble, P.W. (1999) *J. Am. Chem. Soc.*, **121**, 2601.
27. Zhao, W. and Carreira, E.M. (2002) *J. Am. Chem. Soc.*, **124**, 1582.
28. Dürr, H. and Bouas-Laurent, H. (1990) *Photochromism – Molecules and Systems*, Elsevier, Amsterdam.
29. Dvornikov, A.S. and Rentzepis, P.M. (1994) *Mol. Cryst. Liq. Cryst.*, **246**, 379.
30. Seibold, M. and Port, H. (1996) *Chem. Phys. Lett.*, **252**, 135.
31. Uchida, K. and Irie, M. (1993) *J. Am. Chem. Soc.*, **115**, 6442.
32. Irie, M. (1993) *Mol. Cryst. Liq. Cryst.*, **227**, 263.
33. Daub, J., Salbeck, J., Knöchel, T., Fischer, C., Kunkely, H. and

Rapp, K.M. (1989) *Angew Chem. Int. Ed. Eng.*, **28**, 1494.

34. Iyoda, T., Saika, T., Honda, K. and Shimidzu, T. (1989) *Tetrahedron Lett.*, **30**, 5429.
35. Daub, J., Fischer, J., Salbeck, J. and Ulrich, K. (1990) *Adv. Mater.*, **8**, 366.
36. Yokoyama, Y., Ymamane, T. and Kurita, Y. (1991) *J. Chem. Soc., Chem. Commun.*, 1722.
37. Irie, M., Miyatake, O., Uchida, K. and Eriguchi, T. (1994) *J. Am. Chem. Soc.*, **116**, 9894.
38. (a) Brouillard, R. (1988) in *The Flavonoids, Advances in Research* (ed. J. Harborne), Chapman & Hall, London, p. 525; (b) Brouillard, R. (1982) in *Anthocyanins as Food Colors*, Chapter 2 (ed. P. Markakis), Academic Press, New York.
39. (a) Brouillard, R. and Dubois, J.E. (1997) *J. Am. Chem. Soc.*, **99**, 1359; (b) Brouillard, R. and Delaporte, J. (1997) *J. Am. Chem. Soc*, **99**, 8461; (c) McClelland, R.A. and Gedge, S. (1980) *J. Am. Chem. Soc.*, **102**, 5838; (d) McClelland, R.A. and McGall, G.H. (1982) *J. Org. Chem.*, **47**, 3730.
40. Pina, F., Maestri, M. and Balzani, V. (1999) *Chem. Commun.*, (2), 107.
41. Maestri, M., Pina, F. and Balzani, V. (2001) in *Multistate/Multifunctional Molecular Level Systems. Photochromic Flavylium Compounds* (ed. B.L. Feringa), Wiley-VCH Verlag GmbH, Weinheim, p. 339.
42. Roque, A., Lodeiro, C., Pina, F., Maestri, M., Dumas, S., Passaniti, P. and Balzani, V. (2003) *J. Am. Chem. Soc.*, **125**, 987.
43. (a) Matsushima, R., Fujimoto, S. and Tokumura, K. (2001) *Bull. Chem. Soc. Jpn.*, **74**, 827; (b) Horiuchia, H., Tsukamotoa, A., Okajimaa, T., Shirasea, H., Okutsua, T., Matsushima, R. and Hiratsuka, H. (2009) *J. Photochem. Photobiol., A: Chem*, **205**, 203.
44. Giestas, L., Folgosa, F., Lima, J.C., Parola, A.J. and Pina, F. (2005) *Eur. J. Org. Chem.*, (19), 4187.
45. Perkin, W.H., Robinson, R. and Tuner, M.R. (1908) *J. Chem. Soc., Trans.*, **93**, 1085.
46. (a) Michaelidis, C. and Wizinger, R. (1951) *Helv. Chim. Acta*, **34**, 1761; (b) Johnson, A.W. and Melhuish, R.R. (1947) *J. Chem. Soc.*, 346.
47. Katritzky, A.R., Czerney, P., Levell, J.R. and Du, W.H. (1998) *Eur. J. Org. Chem.*, (11), 2623.
48. (a) Pina, F., Melo, M.J., Flamigni, L., Ballardini, R. and Maestri, M. (1997) *New. J. Chem.*, **21**, 969; (b) Maestri, M., Ballardini, R., Pina, F. and Melo, M.J. (1997) *J. Chem. Educ.*, **74**, 1314.
49. (a) Haucke, G., Czerney, P., Igney, C. and Hartmann, H. (1989) *Ber. Bunsenges. Phys. Chem.*, **93**, 805; (b) Haucke, G., Czerney, P., Steen, D., Rettig, W. and Hartmann, H. (1993) *Ber. Bunsenges. Phys. Chem.*, **97**, 561.
50. (a) Figueiredo, P., Lima, J.C., Santos, H., Wigand, M.-C., Brouillard, R., Maçanita, A.L. and Pina, F. (1994) *J. Am. Chem. Soc.*, **116**, 1249; (b) Pina, F., Benedito, L., Melo, M.J., Parola, A.J. and Bernardo, M.A. (1996) *J. Chem. Soc., Faraday Trans.*, **92**, 1693.
51. (a) Matsushima, R., Mizuno, H. and Itoh, H. (1995) *J. Photochem. Photobiol., A: Chem.*, **89**, 251; (b) Matsushima, R., Mizuno, H. and Kajiura, A. (1994) *Bull. Chem. Soc. Jpn.*, **67**, 1762; (c) Matsushima, R. and Suzuki, M. (1992) *Bull. Chem. Soc. Jpn.*, **65**, 39.
52. (a) Paulo, L., Freitas, A.A., da Silva, P.F., Shimizu, K., Quina, F.H. and Maçanita, A.L. (2006) *J. Phys. Chem. A*, **110**, 2089; (b) Fernandes, A.C., Romão, C.C., Rosa, C.P., Vieira, V.P., Lopes, A., da Silva, P.F. and Maçanita, A.L. (2004) *Eur. J. Org. Chem.*, (23), 4877; (c) da Silva, P.F., Lima, J.C., Quina, F.H. and Maçanita, A.L. (2004) *J. Phys. Chem. A*, **108**, 10133; (d) Maçanita, A.L., Moreira, P.F., Lima, J.C., Quina, F.H., Yahwa, C. and Vautier-Giongo, C. (2002) *J. Phys. Chem. A*, **106**, 1248.
53. Pina, F., Melo, M.J., Maestri, M., Ballardini, R. and Balzani, V. (1997) *J. Am. Chem. Soc.*, **119**, 5556.
54. Pina, F., Roque, A., Melo, M.J., Maestri, M., Belladelli, L. and Balzani, V. (1998) *Chem. Eur. J.*, **4**, 1184.
55. Pina, F., Melo, M.J., Santos, M.H., Lima, J.C., Abreu, I., Ballardini, R. and

Maestri, M. (1998) *New J. Chem.*, (22), 1093.
56. Pina, F. (1998) *J. Chem. Soc., Faraday Trans.*, **94**, 2109.
57. Roque, A., Pina, F., Alves, S., Ballardini, R., Maestri, M. and Balzani, V. (1999) *J. Mater. Chem.*, **9**, 2265.
58. Pina, F., Melo, M.J., Passaniti, P., Camaioni, N., Maestri, M. and Balzani, V. (1999) *Eur. J. Org. Chem*, **11**, 3139.
59. Pina, F., Melo, M.J., Parola, A.J., Maestri, M. and Balzani, V. (1998) *Chem. Eur. J.*, **4**, 2001.
60. Pina, F., Maestri, M. and Balzani, V. (2003) in *Handbook of Photochemistry and Photobiology*, Chapter 3 (ed. H.S. Nalwa), American Scientific Publishers, Stevenson Ranch.
61. Saltiel, J. and Sun, Y.-P. (1990) in *Photochromic Systems-Molecules and Systems*, Chapter 4 (eds H. Dürr and H. Bouas-Laurent), Elsevier, Amsterdam.
62. Petrov, V., Gomes, R., Parola, A.J., Jesus, A., Laia, C.A.T. and Pina, F. (2008) *Tetrahedron*, **64**, 714.
63. Moncada, M.C., Fernandez, D., Lima, J.C., Parola, A.J., Lodeiro, C., Folgosa, F., Melo, M.J. and Pina, F. (2004) *Org. Biomol. Chem.*, **2**, 2802.
64. Laia, C.A.T., Parola, A.J., Folgosa, F. and Pina, F. (2007) *Org. Biomol. Chem.*, **5**, 69.
65. Matsushima, R., Fujimoto, S. and Tokumura, K. (2001) *Bull. Chem. Soc. Jpn.*, **74**, 827.
66. Moncada, M.C., Parola, A.J., Lodeiro, C., Pina, F., Maestri, M. and Balzani, V. (2004) *Chem. Eur. J.*, **10**, 1519.
67. Gavara, R., Petrov, V. and Pina, F. (2010) *Photochem. Photobiol. Sci.*, **9**, 298.
68. (a) Gomes, R., Diniz, A.M., Jesus, A., Parola, A.J. and Pina, F. (2009) *Dyes Pigm.*, **81**, 69; (b) Diniz, A.M., Gomes, R., Parola, A.J., Laia, C.A.T. and Pina, F. (2009) *J. Phys. Chem. B*, **113**, 719.
69. Leydet, Y., Parola, A.J. and Pina, F. (2010) *Chem. Eur. J.*, **16**, 545.
70. Maestri, M., Pina, F., Roque, A. and Passaniti, P. (2000) *J. Photochem. Photobiol. A: Chem.*, **137**, 21.

71. Roque, A., Lodeiro, C., Pina, F., Maestri, M., Ballardini, R. and Balzani, V. (2002) *Eur. J. Org. Chem.*, (16), 2699.
72. Fernandez, D., Folgosa, F., Parola, A.J. and Pina, F. (2004) *New J. Chem.*, **28**, 1221.
73. Gomes, R., Parola, A.J., Laia, C.A.T. and Pina, F. (2007) *Photochem. Photobiol. Sci.*, (6), 1003.
74. Parola, A.J., Pereira, P., Pina, F. and Maestri, M. (2007) *J. Photochem. Photobiol. A: Chem.*, **185**, 383.
75. Melo, M.J., Moura, S., Maestri, M. and Pina, F. (2002) *J. Mol. Struct.*, **612**, 245.
76. Gomes, R., Parola, A.J., Laia, C.A.T. and Pina, F. (2007) *J. Phys. Chem. B*, **111**, 12059.
77. (a) Eichenbaum, H. (1997) *Science*, **277**, 330; (b) Iyoda, T., Saika, T., Honda, K. and Shimidzu, T. (1989) *Tetrahedron Lett.*, **30**, 5429.
78. (a) Cartwright, H.M. (1993) *Application of Artificial Intelligence in Chemistry*, Oxford University Press Inc., New York; (b) Nilson, N.J. (1998) *Artificial Intelligence: A New Synthesis*, M. Kaufmann Publishers, San Francisco.
79. de Silva, A.P., Gunaratne, H.Q.N. and McCoy, C.P. (1993) *Nature*, **364**, 42.
80. Balzani, V., Credi, A. and Venturi, M. (2003) *ChemPhysChem*, **3**, 49.
81. Credi, A., Balzani, V., Langford, S.J. and Stoddart, J.F. (1997) *J. Am. Chem. Soc.*, **119**, 2679.
82. Asakawa, M., Ashton, P.R., Balzani, V., Credi, A., Mattersteig, G., Matthews, O.A., Montalti, M., Spencer, N., Stoddart, J.F. and Venturi, M. (1997) *Chem. Eur. J.*, **3**, 1992.
83. De Silva, A.P. and Uchiyama, S. (2007) *Nature Nanotechnol.*, **2**, 399.
84. Pina, F., Melo, M.J., Maestri, M., Passaniti, P. and Balzani, V. (2000) *J. Am. Chem. Soc.*, **122**, 4496.
85. Moggi, L., Bolletta, F., Balzani, V. and Scandola, F. (1966) *J. Inorg. Nucl. Chem.*, **28**, 2589.
86. Amatore, C., Thouin, L. and Warkocz, J.-S. (1999) *Chem. Eur. J.*, **5**, 456.
87. Faulkner, S., Parker, D. and Williams, J.A.G. (1999) in *Supramolecular Science: Where it is and Where it is Going* (eds

88. Beale, R. and Jackson, T. (1990) *Neural Computing: An Introduction*, Adam Hilger, Bristol.
89. Hubel, D.H. (1995) *Eye, Brain and Vision*, Scientific American Library, New York.
90. Roque, A., Lima, J.C., Parola, A.J. and Pina, F. (2007) *Photochem. Photobiol. Sci.*, **6**, 381.
91. Jimenez, A., Pinheiro, C., Parola, A.J., Maestri, M. and Pina, F. (2007) *Photochem. Photobiol. Sci.*, **6**, 372.
92. Pina, F., Lima, J.C., Parola, A.J. and Afonso, C.A.M. (2004) *Angew. Chem. Int. Ed.*, **43**, 1525.
93. Fernandez, D., Parola, A.J., Branco, L.C., Afonso, C.A.M. and Pina, F. (2004) *J. Photochem. Photobiol. A: Chem.*, **168**, 185.
94. Pina, F., Parola, A.J., Melo, M.J., Laia, C.A.T. and Afonso, C.A.M. (2007) *Chem. Commun.*, **16**, 1608.
95. (a) Pina, F. and Hatton, A.T. (2008) *Langmuir*, **24**, 2356; (b) Gomes, R., Laia, C.A.T. and Pina, F. (2009) *J. Phys. Chem. B*, **113**, 11134.
96. Galindo, F., Lima, J.C., Luis, S.V., Parola, A.J. and Pina, F. (2005) *Adv. Funct. Mater.*, **15**, 541.
97. (a) Kohno, Y., Shibata, Y., Oyaizu, N., Yoda, K., Shibata, M. and Matsushima, R. (2008) *Micropor. Mesopor. Mater.*, **1143**, 373; (b) Kohno, Y., Tsubota, S., Shibata, Y., Nozawa, K., Yoda, K., Shibata, M. and Matsushima, R. (2008) *Micropor. Mesopor. Mater.*, **116**, 70; (c) Kohno, Y., Ito, M., Kurata, M., Ikoma, S., Shibata, M., Matsushima, R., Tomita, Y., Maeda, Y. and Kobayashi, K. (2011) *J. Photochem. Photobiol. A: Chem.*, **218**, 87.
98. Matsushima, R., Kato, K. and Ishigai, S. (2002) *Bull. Chem. Soc. Jpn.*, **75**, 2079.
99. Gomes, R., Albuquerque, R.Q., Pina, F., Parola, A.J. and De Cola, L. (2010) *Photochem. Photobiol. Sci.*, **9**, 916.
100. Gavara, R., Laia, C.A.T., Parola, A.J. and Pina, F. (2010) *Chem. Eur. J.*, **16**, 7760.
101. (a) Clarkson, M.A. (1989) *Byte*, **14**, 268; (b) Rouvray, D. (1998) *Chem. Br.*, **34** (2), 26; (c) Rouvray, D. (2000) *Chem. Br.*, **36** (12), 46; (d) Ball, P. (2000) *Nature*, **406**, 118.
102. (a) (2001) *Sci. Am.*, **285** (3) (Special issue on Nanotechnology); (b) Ballardini, R., Ceroni, P., Credi, A., Gandolfi, M.T., Maestri, M., Semeraro, M., Venturi, M. and Balzani, V. (2007) *Adv. Funct. Mater.*, **17**, 740.

7
Nucleic-Acid-Based Switches

Eike Friedrichs and Friedrich C. Simmel

7.1
Molecular Switches Made from DNA and RNA

Among the many molecule types from which molecular switches have been derived, nucleic acids play a special role. In nature, nucleic acids are responsible for the storage of genetic information, its transmission and regulation. Whereas deoxyribonucleic acid (DNA) is the material of the genetic code, RNA assumes a wider range of roles, for example as messenger RNA (mRNA), as a structural component of ribosomes (rRNA) or as a transcriptional regulator (e.g. microRNAs).

Chemically, DNA and RNA are heteropolymers, which are composed of a sequence of units called '*nucleotides*'. The backbone of nucleic-acid polymers consists of (deoxy)ribose sugars joined by phosphodiester bonds. The sugars are linked to one of four types of molecules called '*bases*' via a glycosidic bond. The bases are the purines adenine and guanine, and the pyrimidines cytosine and thymine (for DNA) or uracil (for RNA), respectively. Sugar plus phosphate plus base make up one nucleotide. Importantly, the bases may bind to each other via hydrogen-bond interactions. In the standard (Watson–Crick) base-pairing scheme, guanine binds to cytosine via three hydrogen bonds, whereas adenine binds to thymine (or uracil) via two hydrogen bonds. Hence, the bases participating in such interactions are called '*complementary*'. The same term is used for two oligonucleotides, which contain sequences of corresponding bases. Under favourable reaction conditions, two such complementary strands may 'hybridize' with each other to form a double-stranded molecule – the famous double helix.

The all-important informational character of nucleic acids is based on the combinatorial variability of the sequences of the nucleotides (there are, e.g. $4^{20} = 1.1 \times 10^{12}$ different possible oligonucleotides of a length of 20 units), combined with molecular recognition between DNA or RNA molecules with sequence complementarity.

The fact that two nucleic acid oligomers bind to each other strongly only when their base sequences are complementary has inspired many researchers in nanotechnology to use DNA molecules as a 'smart glue' for nanomaterials or as a

Molecular Switches, Second Edition. Edited by Ben L. Feringa and Wesley R. Browne.
© 2011 Wiley-VCH Verlag GmbH & Co. KGaA. Published 2011 by Wiley-VCH Verlag GmbH & Co. KGaA.

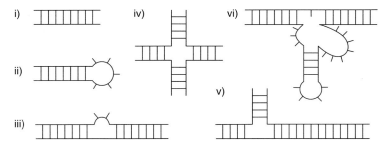

Figure 7.1 Typical nucleic-acid secondary structure elements: (i) Double-stranded conformation; (ii) stem-loop structure (hairpin); (iii) bulge; (iv) Holliday junction (double helix 'crossing' with strand exchange); (v) three-strand junction and (vi) structure of a hammerhead ribozyme composed of double-stranded, hairpin and bulge elements.

construction material itself. In these applications, nucleic acids play a hybrid role – they are both information carriers and structural material.

In the last decade, the properties of DNA or RNA molecules have also been exploited to realize a large variety of molecular switches, ranging from simple fluorescent sensors to complex reaction networks with computational functions.

Most of these switches are based on the fact that – depending on the sequence – nucleic acid structures can assume more than only one conformation. Transitions between the different conformations can be brought about by hybridization or removal of 'effector' oligonucleotides, by changes in buffer composition or by addition or removal of ligands stabilizing one of the alternative conformations.

In contrast to the synthetic molecular devices previously realized by organic chemists, however, nucleic-acid-based switches can often simply be 'designed' by the choice of appropriate nucleotide sequences. Molecular switch design is facilitated by the many biophysical and computational tools [2–4], which are available for investigation and prediction of DNA or RNA secondary structure, typical elements of which are shown in Figure 7.1. Accurate prediction has been made possible by extensive biochemical studies of how base sequence and ambient conditions affect secondary structure, and the theoretical knowledge derived from it [5–12]. Promising candidates for nucleic-acid molecular switches are easily transferred from computational models into real-world molecular structures by automated synthesis methods and standard biochemical protocols.

The emphasis in this field is therefore usually put on design and physical characterization, not on synthesis of the molecular switches. This has the advantage that nucleic-acid-based switches can be very complex structures, even though the synthesis effort is comparatively low. Consequently, artificial nucleic-acid switches are somewhat closer to naturally occurring biological switches than to purely synthetic devices. Interestingly, one can even recognize a convergence in the research efforts in nucleic-acid nanoscience and RNA biochemistry, blurring the boundary between 'natural' and 'artificial' molecular switches. In particular, in recent years it has been recognized that RNA has a wider spectrum of biological functions than has been previously known. For example, riboregulators play an important role in gene regulation and are in many cases similar in structure and function to artificially constructed nucleic-acid switches.

In the following paragraphs we will introduce the basic design principles of molecular switches made from nucleic acids and will then survey a variety of applications of these switches in nanoscience and biology. Section 7.2 deals with *allosteric ribozymes* – also called *aptazymes*, indicating that they derive from the combination of an *aptamer* and a *ribozyme* – as examples of artificial allosteric switches based on nucleic acids. Section 7.3 will cover *regulatory RNA* molecules, which are strikingly similar to artificially designed nucleic-acid switches, but also occur naturally. Simple *sensor applications*, as they can be easily derived from nucleic-acid switches will be addressed in Section 7.4. Switchable structures with a dedicated information-processing function are covered in the context of *DNA computing* in Section 7.5. Prototype *nanomachines, switchable aptamers* and *DNA walkers* will be surveyed in Section 7.6. We will turn to some recent developments in *DNA-based switchable materials* in Section 7.7 and conclude with a short outlook in Section 7.8.

7.2
Switchable Ribozymes

It was discovered only in the 1980s that certain RNA species in a ciliate organism exhibit catalytic activity. This came as quite a surprise, because at that time only proteins were thought capable of catalytic action in biological systems [13, 14].

As in biological systems many enzymes are activated or deactivated by undergoing allosteric changes triggered by an external stimulus, researchers in the field wondered if such allosteric behaviour could also be found in ribozymes. A number of (deoxy)ribozymes – we will use the term ribozyme loosely for both catalytically active RNA and DNA molecules – have been discovered and examined in detail since, including several DNAzymes, the self-cleaving hammerhead and hairpin ribozyme [15–23]. Yet, no allosterically triggered ribozymes had been found to occur naturally for a long time. Inspired by this, such allosteric behaviour was incorporated artificially into a large number of ribozymes by use of aptamers. Aptamers are oligonucleotides that bind strongly and specifically to a ligand (a small molecule or a protein), and can thus be regarded as nucleic acid analogues of antibodies. Aptamers can be evolved to bind to virtually any given ligand utilizing a method called systematic evolution of ligands by exponential enrichment (SELEX), in which they are selected from a random pool of sequences according to their binding properties [24–28]. Aptamers have turned out to be an important component of many molecular switches (such as aptazymes) and will play a role in most of the later sections.

Switchable ribozymes can be categorized by the stimulus they react with – ligand molecules or nucleic-acid molecules, for example – or by the kind of reaction, that is catalysed – cleavage of nucleic acids or peroxidase activity, for instance. Yet, driven by the notion that it is preferable to emphasize the mechanism by which switching can be induced, we will here focus on highlighting these.

7.2.1
Ribozyme Switching by Antisense Interaction

As it is the three-dimensional structure of a ribozyme domain that determines its catalytic capability, forcing the catalytically active domain into a double-stranded conformation by adding a nucleic-acid strand, that is complementary to the ribozyme sequence is the most straightforward way of controlling a ribozyme's activity. Based on this antisense interaction approach, Porta and Lizardi [29] were the first to construct an artificial switchable ribozyme, a modified version of the hammerhead ribozyme. Addition of an effector recognition and a ribozyme recognition sequence to the 5′-end of the ribozyme, forces the ribozyme domain into a double-stranded conformation by hybridization to the ribozyme recognition region. With this modification, the ribozyme domain cannot form properly, unless a DNA effector strand is added that hybridizes to the effector-recognition region of the ribozyme, releasing the ribozyme sequence. This folds into its active formation hereon and initiates cleavage of a suitable substrate.

A second example of an antisense activated artificial ribozyme [30] demonstrates the feasibility to generate even completely new allosteric ribozymes – not by rational design, but by *in vitro* selection – in this case a switchable ribozyme ligase ('L1'). In its inactive conformation, the L1 ribozyme is folded back on itself, preventing the RNA substrate from binding. A DNA effector strand, however, can hybridize to L1. This affects the secondary structure of L1, and after rearrangement its RNA substrate binding region is exposed, allowing an increase of the ligase activation by a factor of 10 000 [30].

Masking parts of the RNA substrate binding site of the DNAzyme effectively inhibits hybridization between the ribozyme and a substrate strand. The inhibiting effector strand can be removed from the DNAzyme by addition of a removal strand. Upon binding to an overhang region of the effector strand, the removal strand displaces the DNAzyme in a branch migration process (see also DNA machines, Section 7.6), thus releasing and activating it [31].

An example of an antisense controlled ribozyme with possible medical applications is given by the 'maxizyme'. This is a nucleic-acid structure that can recognize nucleic acid target sequences and – upon binding to it – forms a cavity that can capture Mg^{2+} ions, which are catalytically indispensable for a number of enzymatic reactions in living cells. Using mRNA coding for a leukaemia-related protein as the maxizyme target sequence, apoptosis was induced in cells with a chromosome including a critical gene [32, 33].

7.2.2
Ribozyme Deactivation by Steric Hindrance

In biological systems, an enzyme's activity is often dependent on the presence of coenzymes. Therefore, triggering a ribozyme by a small ligand molecule – instead of a nucleic-acid molecule – was an obvious goal to achieve. To this end, Breaker and coworkers constructed a self-cleaving ribozyme fused to an ATP binding aptamer, for which the self-cleaving rate is reduced 180-fold upon addition of ATP. Catalytic inhibition in this case is induced by a steric hindrance mechanism: The ribozyme domain is not destroyed in the presence of ATP – as in the case of the antisense interactions described before. Instead, the aptamer domain – remaining conformationally heterogeneous in the absence of ATP – is stabilized after addition of ATP and due to steric interference between the aptamer and ribozyme tertiary structures, the ribozyme activity is strongly reduced [34]. An allosteric OFF-ribozyme – one that is activated in the presence of ATP – was created as well; as was another ribozyme that could be controlled by theophylline [34, 35].

7.2.3
Ribozyme Activation by Complex Stabilization

Activation of ribozymes as a response to either an antisense strand or the presence of a ligand molecule can also be mediated by stabilization of the ribozyme–substrate complex. In the case of an antisense mediator strand, hybridization of the ribozyme and the substrate is not favoured without any assistance. The mediator strand, however, can hybridize to a catalytically inactive domain of the ribozyme as well as to the substrate strand and thus the mediator strand provides additional complementary bases for the substrate to bind. Therefore, a three-arm junction is formed, increasing the activity of two specific DNAzymes as well as that of a variant of the hammerhead ribozyme by a factor of 20–30 (Figure 7.2b) [36]. An increase of cleavage activity by a factor of 35 is reported for two DNAzymes and a hammerhead variant in a similar substrate-ribozyme stabilization approach, where activation is triggered by either ATP or (flavin mononucleotide) (FMN), compare Figure 7.2a [37].

7.2.4
Ligand-Induced Stabilization of the Ribozyme Domain

Araki *et al.* [38] constructed an allosteric ribozyme switch consisting of a hammerhead ribozyme and a FMN domain, in which binding of FMN to the aptamer domain induced the proper formation of the hammerhead domain needed for catalytic activity.

To demonstrate the feasibility of the stabilization approach for another ribozyme, FMN, ATP or theophylline aptamers, respectively, were fused to a ribozyme ligase. An aptazyme resulted with an activity increase by a factor of 260–1600-fold in the presence of the ligand, depending on the particular ligand molecule [39].

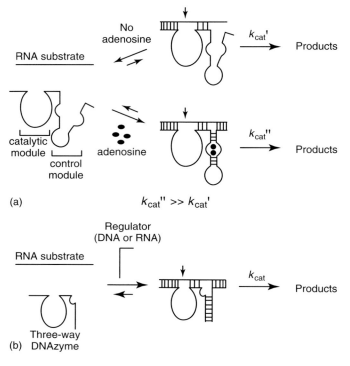

Figure 7.2 Switching principles for switchable ribozymes. (a) Complex stabilization by a ligand. Without the presence of adenosine, hybridization of the RNA substrate to the ribozyme – comprising a catalytic and a control module – is not favoured and ribozyme activity is low. Presence of adenosine stabilizes substrate hybridization to the ribozyme and cleavage activity highly enhanced. (b) Complex stabilization by antisense interaction. By hybridizing to both ribozyme and substrate, the presence of the regulator DNA or RNA activates the ribozyme. Figure reproduced from Ref. [37] with kind permission by Elsevier Limited.

The number of ligand molecules can be easily expanded utilizing *in vitro* selection methods. For example, ribozymes have been engineered that are activated by the presence of divalent metal ions [40].

7.3
Regulatory RNA Molecules

7.3.1
Riboswitches

The focus of this chapter is on *artificial* molecular switches based on nucleic acids. Nonetheless, naturally occurring RNA regulatory elements in gene-expression control – such as the recently discovered 'riboswitches', give insight into the

Riboswitch Regulatory Pathways

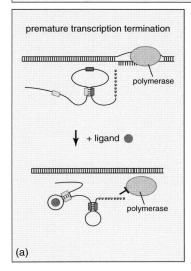
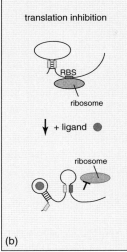
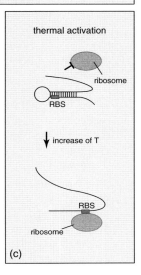

Figure 7.3 Three pathways of riboswitch regulation of genetic expression. (a) Premature transcription-termination pathway: Binding of a ligand molecule initiates formation of the terminator domain. The mRNA produced lacks the coding region and cannot serve as a template for protein production. (b) Translational inhibition pathway: A ligand induced conformational change leads to masking of the ribosomal binding site (RBS). This inhibits binding of the mRNA to the ribosome and thus translation. (c) Thermal translation activation pathway. An increase in temperature opens the RBS-masking stem region, exposing the RBS and initiating translation.

mechanisms by which switching of nucleic-acid molecules can be achieved in general. Due to the great interest they generate from a biochemical and biological point of view, many of them have been examined in considerable detail. Incidentally, some riboswitches have been re-engineered by life scientists in the meantime, adding an artificial tinge to the subject. Furthermore, in biological systems – apart from riboswitches – there is an abundance of other pathways by which RNA molecules interfere with gene expression – like short interfering RNA (siRNA) or microRNA – that are not covered by this section. For a detailed discussion of riboswitches and other natural regulatory RNA the reader is referred to more specialized reviews [18, 41–51].

Riboswitches have been under intense investigation since their discovery in 2001 [52]. Riboswitch-mediated gene regulation is triggered by ligand molecules in most of the cases, while some switches are triggered thermally. The latter are also referred to as *RNA thermometers* [53]. In the thermodynamic pathway (Figure 7.3c), translation at low temperatures is often inhibited by a stem structure in the 5′-upstream uncoding region of the mRNA masking the ribosome binding site (RBS). When the temperature is increased, the stem melts, which allows the ribosome to bind to the RBS and to begin translation.

Riboswitches that are triggered by a ligand molecule usually code for a protein that is involved in the metabolism of this particular molecule. Upon binding of the ligand to an aptamer region of the riboswitch mRNA 5′-noncoding region, many mRNA riboswitches alter their conformational structure in such a way that they can no longer serve as a template for protein synthesis. This is mediated either by premature termination of mRNA transcription or by translational inhibition.

In the premature-termination pathway (Figure 7.3a), binding of the ligand leads to the formation of a hairpin terminator structure in the noncoding upstream region, resulting in mRNA molecules lacking the coding region. In the absence of the ligand, however, the hairpin terminator structure is not favoured energetically and does not form, so the full mRNA is transcribed [54–59].

If translation is inhibited via the translational pathway (Figure 7.3b), a full length mRNA is transcribed either way. Yet, a competing domain mechanism regulates the translational activity of the mRNA: In the presence of the ligand molecule a stem structure is stabilized, in which the RBS is masked, while in the absence of the ligand the ribosome can bind to the RBS and initiate translation [54, 60–62].

Remarkably, there is a strong bias in nature towards switches that turn off gene expression in the presence of the ligand molecule, as the ones described before. Only a small number of switches is reported to work the opposite way [63–65].

Riboswitches have so far been identified in bacteria, fungi and plants. Thus, riboswitches are reckoned to be an archaic pathway of gene regulation [66, 67]. Among the ligands for riboswitches are amino acids (lysine [68, 69]), nucleotides (adenine [57, 64], guanine [57]), sugar derivates (glucosamine-6-phosphate [70]), vitamins (cobalamine [61], thiamine [71, 72]) and coenzymes (FMN [54, 55], thiamine pyrophosphate (TPP) [60], S-adenosyl-methionine (SAM [59, 73]).

The glmS riboswitch in certain Gram-positive bacteria is the only mRNA riboswitch to self-cleave by a hitherto unknown mechanism upon binding of its ligand (glucosamine-6-phosphate). Displaying catalytic activity, it can also be regarded as a naturally occurring allosteric ribozyme [70].

Recently, a riboswitch with two aptamer domains was identified in B. Clausii, responding to the coenzyme B12 and S-adenosylmethionine, respectively, just as a Boolean NOR gate. Thus, translation is inhibited whenever both coenzyme B12 and S-adenosylmethionine are present at the same time [74].

7.3.2
Synthetic RNA Regulatory Switches

An artificial riboswitch has been constructed recently in *E. coli*: A gene coding for green fluorescent protein (GFP) was modified by the addition of an RBS recognition region upstream of the RBS. By this, the mRNA that is transcribed cannot be translated, because the RBS recognition region folds back, forms a hairpin and sequesters the RBS. This is why this particular mRNA is referred to as cis-repression RNA (crRNA). Translation can be activated by a noncoding so-called trans-activating RNA (taRNA), produced from a second promoter. This taRNA can

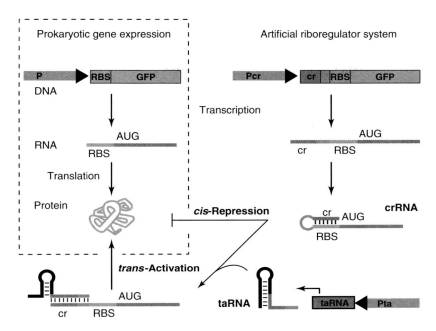

Figure 7.4 An artificial riboswitch. Inside the box: Native prokaryotic gene expression: The ribosome binding site (RBS) is unmasked; mRNA transcribed from a gene is translated into a protein. Outside the box: An RBS recognizing cis-repressing sequence (cr) has been artificially inserted into the gene. Translation of a protein is inhibited by cr masking the RBS. A second promoter allows transcription of a trans-activating RNA (taRNA) that unmasks the mRNA's RBS by hybridizing to it and thus activating translation of the protein. Figure reproduced from Ref. [75] with kind permission by Nature Publishing Group.

hybridize to the crRNA, which unmasks the RBS. The activation can be monitored fluorescently by measuring the GFP expression level (Figure 7.4) [75].

Trans-acting artificial regulatory RNA has also been demonstrated in a eukaryotic system by so-called antiswitches. Binding of an effector molecule – tetracycline or theophylline – induces a conformational change of the antiswitch, revealing an antisense domain, that is designed to pair with a 15-base region around the start codon of an mRNA coding for GFP or yellow fluorescent protein (YFP). Repression of protein expression can be monitored fluorescently as described before [76]. A different method for artificial gene regulation was demonstrated in yeast that was based on aptazyme-mediated self-cleavage of mRNA. By using several aptazymes in series, this approach was utilized to employ logic control of gene expression [77].

For transcription, a double-stranded promoter region is crucial. Without the promoter being double stranded, transcription is virtually turned off. This is exploited by an unorthodox method of trans-acting gene repression, which has been demonstrated *in vitro* recently, resulting in a system with a bistable expression behaviour [78].

7.4
Sensor Applications

7.4.1
Switches are Sensors

Nucleic-acid switches react on external stimuli by changing their conformational structure. Thus, in a trivial way, *any* nucleic-acid switch can be regarded as a sensor of some sort for whatever causes the switching event. Considering nucleic acids in aqueous solutions, this includes changes in temperature [79], pH value [80], ion concentration, hybridization to other nucleic acid sequences, binding of ligand molecules or even interaction with light [82].

7.4.2
Sensor-Construction Requirements

For sensing applications a highly specific and sensitive switching behaviour as well as a quick and simple readout mechanism are crucial. This, of course, drastically impairs the practical usefulness of many switches for sensing purposes on the one hand, while on the other the number of readout mechanisms is narrowed. As the structural change of the molecule will alter its diffusional properties, the state of the switch can be readout by the help of gel electrophoresis [83] or fluorescence correlation spectroscopy (FCS) measurements [84], for example. While the first is time consuming, the latter is too sophisticated to serve as a fast and simple sensing method. Therefore, these detection mechanisms do not qualify for typical sensing applications.

Nevertheless, sensing methods are still manifold. For instance, conformational changes of an aptamer when binding to a ligand can be accompanied by a change in its electrical conductance properties, which can be measured [85].

Most nucleic acid sensors, however, make use of fluorescent dyes, whose emission characteristics are affected by conformational changes of the nucleic-acid sensor molecules. This includes a phenomenon called fluorescence resonance energy transfer (FRET), where energy can be transferred from one dye (the energy donor) to another dye (the energy acceptor) without photon emission from the donor. The efficiency of this process is strongly distance dependent ($\sim R^{-6}$). Dependent on the particular donor–acceptor pair, the so-called Förster radius – which is the distance of 50% FRET efficiency – is typically in the range of 20–100 Å. Therefore, covalently attaching a FRET dye pair in a distance close to the Förster radius at positions of the nucleic-acid switch that are expected to alter their distance upon switching significantly, will also lead to a significant change in the fluorescence emission of the donor dye. For a thorough discussion of FRET phenomena there is an abundance of literature available [86–90].

FRET-related assays come in a variety of forms. Among the most widely used are molecular beacons [91, 92], short DNA oligonucleotides with one FRET dye at their

5'-end, the other at their 3'-end. Except for a sequence in its centre, a molecular beacon is self-complementary, so that it folds into a hairpin structure, with the dyes in the vicinity at the stem's end. If the hairpin structure is chosen to be complementary to a target sequence, binding to this target sequence opens up the stem and thus reduces the FRET efficiency, which results in an increased donor fluorescence intensity [93]. Incorporation of an aptamer sequence into the hairpin region of a molecular beacon can be utilized for creating sensors for literally any molecule that can be recognized by an aptamer: It just has to be ensured that formation of the hairpin-stem and the ligand-bound aptamer structures are mutually exclusive and that the latter is favoured in the presence of the ligand only.

As long as the emission of an ensemble of fluorophores is detected, fluctuations in the distance of single FRET pairs – that affect the FRET efficiency and thus the fluorescence intensity – are averaged out. This is why it is not even essential for a beacon to form a rigid structure like a hairpin in the ligand unbinding state, as long as it can be distinguished from the aptamer binding conformation by its mean fluorescence level. For instance, in the case of an aptamer-based sensor for the protein thrombin, binding of the protein triggers a conformational change of the aptamer strand from a loose random coil to a compact stacked quadruplex structure, which is accompanied by a significant change in the fluorescence intensity of a FRET pair; concentrations in the range of 100 pM can be detected with high sensitivity and specificity [94].

This kind of aptamer secondary structure stabilization is also used for a DNA cocaine sensor. Here also, no stable secondary structure is believed to form in the absence of the ligand molecule. However, formation of the aptamer domain after binding of a cocaine molecule is accompanied by a significant change in the mean dye–quencher distance and is thus causing a change in the fluorescence signal [95].

Even aptamers consisting of two separate strands have been designed in such a way that actually forming the ligand binding aptamer structure by hybridization of the two strands is only favoured in the presence of the ligand molecule due to a number of sequence mismatches. A pair of FRET dyes – with one dye attached to each strand – serves to signal whether the strands are hybridized (low fluorescence intensity) or not (high intensity). This could be shown to be feasible for a cocaine as well as for an ATP sensor [96].

Competing formation of two aptamer domains can also be used to build a sensor. Two exclusive structures can be adopted reversibly by a switch that acts as a sensor for the specific cooling pathway. After heating up, fast cooling traps the switch in a local energy minimum that is retained for several weeks at room temperature, while it ends up in its lowest-energy conformation when it is cooled down slowly. A change in fluorescence intensity is also caused by a different mean distance of a FRET pair in two exclusive conformations [79].

Dyes do not have to be covalently bound to the switch; using dye aptamer sequences, no quenching dye is required, either. Allosteric changes of noncovalent interactions of the fluorophore malachite green with the sensor upon binding of one of the ligands ATP, FMN and theophylline increases the fluorescence signal significantly [97].

A ligand-binding aptamer domain competing with a dye-binding aptamer domain can also be used for sensor construction, if the dye-binding domain is stabilized only in the absence of the ligand. If this is the case, upon binding of the ligand, the dye is set free, resulting in absorption attenuation of the system [98], fluorescence intensity decrease of the dye or even its precipitation [98].

Sensors can also make use of luminescence instead of fluorescence reactions. A DNAzyme that enhances the catalytic peroxidase activity of hemin is used to foster the chemoluminescent reaction of luminol. As described in references [99, 100], triggering of the activation of this particular DNAzyme is due to the presence of the M13 phage DNA, for which it serves as a detector.

The same DNAzyme also catalyses the oxidation of 2,2'-azino-bis3-ethylbenzthiaz oline-6-sulfonic acid (ABTS). As this changes the molar absorptivity of this molecule, absorption measurements can be easily utilized for detection [100].

7.4.3
Signal Amplification for Lowering Detection Limit

Without signal amplification, for every molecule that is triggering one particular molecular switch there will be the stoichiometrically equivalent change in intensity of one dye molecule that can be exploited for sensing. If a small number of molecules has to be detected, this signal might be too low and an additional amplificational step is needed.

Discussing nucleic-acid molecular switches, generally all mechanisms by which nucleic acids can be amplified can be considered for signal amplification. For instance, polymerase chain reaction (PCR)-based signal amplification has been utilized for the purpose of sensing a specific DNA sequence in a copy number as low as 40 molecules [100].

For practical reasons, however, an isothermal method like rolling circle amplification (RCA) is often favourable over a method involving thermocycles like PCR. RCA has actually been used for sensing a target DNA single strand. If it is present, an RCA product is generated that comprises a DNAzyme sequence repeatedly that catalyses the oxidation of ABTS, which can be detected by changes in absorption. The detection limit for this method is reported to be 1 pM [101]. To demonstrate a practical application of this method, it could be shown to be successful in sensing M13 phage viral ssDNA [102].

Of course, signal amplification is not restricted to sensors for DNA sequences. For example, a cocaine sensor has been developed in which formation of a cocaine aptamer domain upon binding activates the elongation of a mediator strand by a DNA polymerase. After elongation, a restriction endonuclease produces a nick, giving rise to a second round of elongation by the polymerase. Mediator strands hybridize to doubly labelled molecular beacons, increasing their fluorescence emission [103].

Signal amplification can also be generated without the assistance of proteins at all. This can be approached by means of a technique called hybridization chain reaction (HCR). In this technique, mutual hybridization of hairpin loops

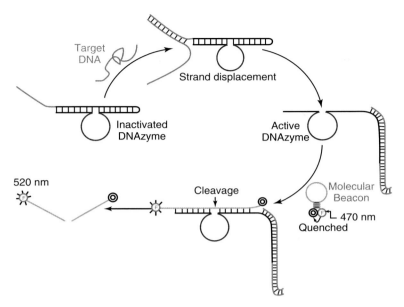

Figure 7.5 The inactivated DNAzyme (i) exhibits a toehold to which the target DNA can hybridize. By a strand displacement reaction (ii) the DNAzyme is activated (iii). A molecular beacon – doubly labelled with a FRET-pair – hybridizes to the DNAzyme domain and is cleaved (iv). This leads to an increased fluorescence level (v). As the DNAzyme can cleave molecular beacons repeatedly, the recognition of a single target DNA molecule is amplified by numerous molecular beacon cleaving events. Figure reproduced from Ref. [108] with kind permission by Elsevier Limited.

is kinetically hindered, but triggered by hybridization to a target strand or the binding event of a ligand – in this case ATP – to an aptamer region contained in the hairpin loops. By this binding event, a cascade of hybridization events is generated. The emerging long nicked double-stranded DNA built from the hairpin loop monomers can be monitored by gel electrophoresis or by fluorescence [104]. Recently, the HCR technique was demonstrated to also work with RNA molecules, and even *in vivo!* [105].

Generally, allosteric ribozymes – which have been discussed in greater detail in a previous section of this chapter – that are activated or deactivated by hybridizing to a target sequence or binding of a ligand can be used for signal amplification in sensing of the target sequence or the ligand, respectively.

Allosteric nucleic-acid-cleaving ribozymes that have been already modified to work as functional sensors include an allosteric deoxyribozyme that recognizes a target nucleic-acid sequence. Binding to this sequence activates the deoxyribozyme, which then cleaves a doubly labelled fluorescent substrate or molecular beacon repeatedly, resulting in a strong increase of fluorescence intensity (Figure 7.5) [106–108]. Involving a hammerhead ribozyme, a theophylline-dependent sensor was realized [109]. For further reading on nucleic acid-based sensor applications, the reader is referred to [45, 66, 110–112].

7.5
DNA Computing

In analogy with electronics, molecular switches can also be regarded as components of molecular 'computers'. A conventional switch, which is triggered by the presence of a particular 'input' molecule simply reports the presence of the input (it is a sensor for that molecule – see above). If the conformational change is dependent on several input molecules (or other parameters), the switch can report a logical function of the inputs (e.g. an 'AND' gate). In recent years, a variety of such computational elements based on DNA have been reported, and these may find application as advanced sensors, which can analyse complex mixtures of molecules, for example for medical diagnosis. As such, the development of 'autonomous' DNA-based molecular computers has developed into a sub-branch of DNA computing [113].

The DNA computing field was founded in 1994, starting with a seminal work by Leonard Adleman [114], in which biochemical techniques were ingeniously applied to solve certain computational problems such as the 'Hamiltonian path' problem or related satisfiability problems. Later, it was shown that the information-processing potential of DNA can also be exploited to evaluate simple logical expressions with 'autonomous' DNA devices.

Utilizing the catalytic properties of (deoxy)ribozymes [21–23], Stojanovic and coworkers reported several DNA-based logic gates and circuits [115–120]. These gates are based on DNA constructs, in which folding of a deoxyribozyme into an enzymatically active conformation is inhibited in the absence of certain input molecules. Recognition of molecular input signals restores the catalytic activity of the deoxyribozymes. Using this principle, logical NOT, AND, XOR gates and others were realized.

In a different approach, Seelig *et al*. constructed logical networks based on strand displacement by DNA branch migration, and inhibition of hybridization of DNA hairpins [12, 121, 122]. In Ref. [123], hybridization of a DNA 'output sequence' with another strand (e.g. a downstream gate) was inhibited by hybridization with protecting strands. DNA 'input strands' could remove the protecting strands by branch migration, releasing the output strand. With this concept, AND, OR and NOT gates were constructed, and several of these gates could be linked together to evaluate a complex logical expression. To be able to link together several computational stages, an elaborate signal restoration concept [121] was developed. This computational problem – and its solution – is analogous to the fanout concept in electronics, where the output of a transistor has to be able to drive several downstream transistors within a circuit. Similar restoration or amplification techniques should be of interest for any application, in which cascades or networks of interacting molecular switches are required.

Other examples for information processing with DNA or RNA molecules have been reported, among others, by Benenson *et al*. [124, 125], Soreni *et al*. [126], Weizmann *et al*. [127, 128] or Miyoshi *et al*. [129]. These examples demonstrate that nucleic-acid-based switches can be used to function as sensors, which do not merely report simple YES/NO information about the presence of a single molecule species, but that are capable of analysing a mixture of biomolecules in accordance with a certain 'diagnostic rule'.

7.6
DNA Machines

Utilization of DNA conformational transitions to induce nanoscale motion is a highly promising approach towards the realization of nanoscale machines, motors and even robots for nanotechnology [130–136].

Quite a number of prototypes at different levels of complexity have been demonstrated in recent years. Some of the simpler nanomechanical switches make use of buffer-induced conformational changes, while others rely on sequence-dependent switching caused by hybridization reactions.

7.6.1
Prototype Machines Based on the i-Motif Transition

At low pH values, four strands of DNA containing cytosine triples can associate with each other via noncanonical base pairs between protonated and nonprotonated cytosine bases, resulting in a four-stranded structure termed '*i-motif*'. If a single oligonucleotide contains four sequences of cytosine triples, the i-motif transition can also occur intramolecularly, resulting in a chair-like structure. The formation of the intramolecular i-motif from a random coil structure is characterized by a sharp transition at pH 6.5. At low proton concentrations, the i-motif strand is able to form a double-stranded structure with a complementary strand, while at high proton concentrations it adopts an i-motif conformation, which leaves the counterstrand unhybridized. Thus, by switching the pH between a value of 5 and 8, the oligonucleotide can be switched between a double-stranded and an i-motif conformation [80]. In an exciting recent development it could be shown that an i-motif based molecular switch can actually be employed as a pH sensor inside living cells [81].

Instead of adding hydrochloric acid and sodium hydroxide alternately by hand to induce the conformational switching, the i-motif can be driven autonomously by an oscillating chemical reaction – for example by a variant of the Landolt reaction, that is periodically changing the pH value of the reaction solution [137]. The i-motif strands can also be attached to a surface [138]; this can be utilized to fix them on the bottom of a continuous-stirred-tank reactor (CSTR), in which chemical oscillations can be maintained for arbitrarily long periods of time [139]. Attaching i-motif oligonucleotides on a silicon cantilever tip causes a reversible, substantial downward bending of the cantilever, which shows that the i-motif conformational changes can be even harnessed to perform micromechanical work [140].

7.6.2
Tweezers – a Prototype System for Reversible Switching Devices

DNA tweezers were one of the first artificial DNA switches that could be reversibly switched between an open and a closed conformation. The switching mechanism employed for the tweezers system – strand displacement by branch migration – has become a standard technique used in various other DNA-based switches since [141]. Tweezers comprise of three DNA strands A, B and C, with the first half of strand A hybridized to strand B and the second half hybridized to C. These two halves form rigid double-helical arms, but – as a four-base sequence in the very centre of A remains unhybridized to either B or C – can move quite freely with respect to each other. Single-stranded overhangs of B and C protrude A on both sides, dangling floppily around in the open conformation. Addition of a fuel strand F, that is complementary to the single-stranded overhang regions of B and C is able to close the tweezer by hybridization; by this, both double-helical tweezer arms formed from A, B and C are pulled together, closing the tweezers. An unhybridized overhang 'toehold' section of F allows its removal from the tweezer: Upon addition of a release strand R, that is complementary to F, hybridization of R to the toehold section of F is followed by strand displacement by branch migration. This results in a duplex DNA waste product – formed from F and R – and a reopening of the tweezer arms [142].

With a slight modification – by ligating the free dangling ends of B and C to a joint BC strand – a nanoactuator results that can be reversibly switched between a relaxed and a straightened conformation, as it is illustrated in Figure 7.6 [143]. It is even possible to construct such a nanoactuator with two distinct straightened conformations, resulting in a three-state switchable device, with the adopted state depending on which of the two fuel strands is used [144].

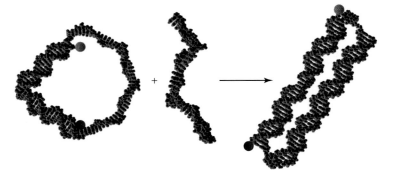

Figure 7.6 A DNA nanoactuator. The doubly labelled actuator in its relaxed state (left) consists of two strands of DNA hybridized together, exposing a long single-stranded region. Hybridization of an effector strand to the single-stranded loop region stretches the actuator, resulting in a spatial separation of the fluorophores.

Instead of adding opening and closing strands by hand, tweezers can be opened or closed by mRNA, that is transcribed from an artificial gene, encoding the fuel or release strand sequence. Transcription of the mRNA then can be suppressed in response to the presence of regulatory proteins, mimicking natural gene-regulation processes [145].

Abandoning the fuel-release strand approach, the tweezers-based nanoactuator described in Ref. [116] could be turned into a system continuously switching between an open and a closed conformation. The tweezer arms described in Ref. [146], are joined by a deoxyribozymes with endonuclease activity. When a substrate strand hybridizes to the deoxyribozyme, the tweezer arms are stretched apart from each other. Cleavage of the substrate strand by the deoxyribozyme returns the tweezers into the closed state. The cleaved substrate strand falls off and is replaced by an intact one, opening the tweezer again. The system can also be halted and restarted by adding a brake or a removal strand, respectively, following the toehold approach of the original system described before [147].

7.6.3
Switchable Aptamers

To demonstrate the versatility of the fuel-release strand approach, it was also used to force aptamers from their ligand-binding aptamer conformation into a double-stranded conformation, depriving the aptamer of its ligand-binding capability. For example, a thrombin-binding aptamer could be triggered to grab and release thrombin upon alternately adding fuel and release strands [148, 149].

The same approach was applied to an aptamer for Taq DNA polymerase that inhibits the polymerase activity when folded into its proper polymerase-binding conformation. Addition of an effector strand renders the polymerase active, while it is deactivated by addition of a removal strand. In this way, the fuel–antifuel strand approach even allows us to control an enzyme's activity (Figure 7.7) [150].

7.6.4
Devices Based on Double-Crossover Motifs

Based on Holliday junction structures [151–153] – 'intersections' of two DNA double helices that exchange strands with each other – and other DNA crossover motifs [154–156] it has been shown that it is possible to create sophisticated supramolecular structures like a DNA 'cage' [157], DNA Borromean rings [158] or even DNA 'origami' [133, 134].

In fact, one of the first supramolecular structures that could be switched between two distinct states was formed from two rigid DNA 'double-crossover' molecules – structures composed of two Holliday junctions joined together. In Ref. [159], two of such double-helical twin domains were connected by a 4.5-turn double-helical bridge that contained the particular sequence $d(CG)_{10}$. This sequence is known to switch from the right-handed B-form of DNA to the left-handed Z-form in the presence of cobalt hexamine. This transition twists the double-crossover parts of

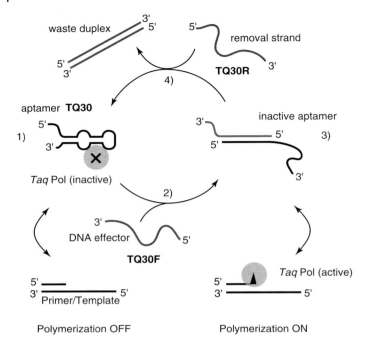

Figure 7.7 A switchable aptamer controlling the activity of Taq Pol (Thermus acquaticus polymerase). (1) Aptamer TQ 30 in its Taq Pol binding conformation, inactivating the polymerase. (2) Addition of effector strand TQ30F inactivates the aptamer (3), and Taq Pol is activated. (4) Addition of removal strand TQ30R displaces TQ30 from the effector strand by branch migration and Aptamer TQ30 folds back into its Taq Pol inhibiting conformation (1). With each cycle a TQ30F-TQ30R double-strand waste product is generated. Figure reproduced from Ref. [150] with kind permission.

the device against each other, resulting in a measurable distance change of several nanometers between the DNA double-crossover domains [159].

Following a different strategy, the pH-dependent transition from duplex to triplex DNA was utilized to develop a nanoactuator capable of similar movements as the tweezers in the previous section [160]. It is also possible to create large lattices from crossover structures, and by incorporation of DNA switchable elements, a DNA supramolecular lattice was realized with a switchable lattice constant [161].

A more complex switch based on multiple crossover structures has been demonstrated in Ref. [162]. In this work, four interwoven strands of DNA can form two distinct multiple crossover motifs – PX (paranemic crossover) and JX_2 (paranemic crossover with two juxtaposed sites). Which of the two motifs is formed in particular, is dependent on the addition of two more strands, A1 and A2 or B1 and B2, respectively. Again applying the fuel and removal strand approach, fuel strands A1 and A2 can be removed by the corresponding removal strands. By this, the PX motif is switched to an intermediate state, but can be switched into the JX_2 motif by addition of B1 and B2, which, can be removed by removal strands, leaving the device in the intermediate state again. Addition of A1 and A2 switch it back into the PX conformation.

Extending the system via cover strands – additional double-helical arms on both sides of the device – makes switching addressable with the help of RNA strands instead of DNA as before. DNA trapezoids that work as bridges between two adjacent PX-JX$_2$-devices in a device array are flipped by the switching of devices, which can be visualized by atomic force microscopy (AFM) imaging [162].

An array of PX-JX$_2$ devices could also be successfully integrated into a 2D DNA network, with each of the devices addressable separately by set–unset strands. A hairpin 'robot' arm, that is connected to each of the devices switches from an 'up' to a 'down'-configuration with respect to the substrate (Figure 7.8) [163]. An real "molecular assembly line" based on this concept was later demonstrated using a DNA origami substrate [164].

7.6.5
Walkers – towards DNA-Based Motors

Protein-based molecular motors like kinesin or myosin exemplify that conformational switching on the molecular level can be turned into active, directed motion on a scale much larger than the molecular dimensions they arise from. Analogously, converting the mechanical movements exhibited in DNA molecules that are switched between different conformations into directed movement of the molecule along a suitable substrate is addressed by a number of recent publications in the field of artificial DNA systems. Though the achievements described seem rather humble compared to natural molecular motors, they constitute interesting proof-of-principle systems, employing a completely different class of molecules for nanoscale movement than have been developed by evolution.

In Ref. [124], with the help of triple-crossover (TX) DNA molecules a footpath for a DNA walker was created, imitating microtubules, on which natural kinesin motors walk. The walker itself consists of two double-helical domains connected by three flexible, nine-nucleotide unpaired DNA linker strands. Each of the two helical domains has a distinct single-stranded overhang, called a *'foot'* that can hybridize to a set strand. A set strand on its part can hybridize not only to a 'foot', but also to a 'foothold' strand, which is incorporated equidistantly in the footpath strand. In addition, the set strands do exhibit a toehold section that serves as a point of attack for an unset strand, which a set strand can hybridize to, releasing the walker's foot by strand displacement. Addition of set strands, followed by the addition of the respective unset strands makes the walker move on the footpath. The set–unset waste products can be removed after each step, which is facilitated by attachments of biotin to all 5′-ends of unset strands: These allow removal by use of streptavidin-coated magnetic beads and a magnet. To prevent the walker from walking backwards, three different foothold strands are used [165].

A simplified version of this walker, in which it is reduced to a dsDNA with two noncomplementary, single-stranded overhangs serving as feet, was demonstrated in Ref. [125]. Using four foothold sequences, and correspondingly, four set and

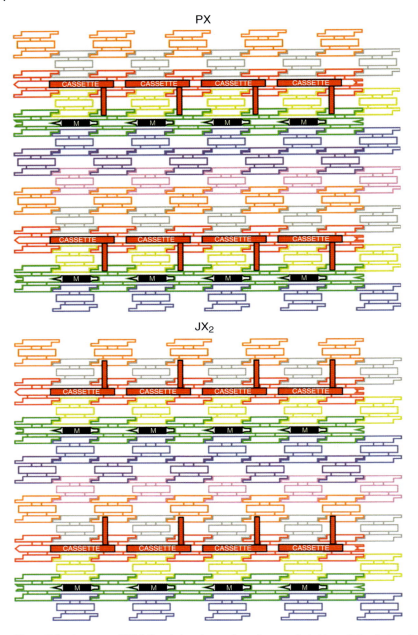

Figure 7.8 An array of PX-JX2 switches integrated into a 2D DNA network. A hairpin 'robot' arm connected to each of the devices is switched with the device from a "down" to an "up"-configuration with respect to the substrate. Switching of the devices between the two states is facilitated with set and unset strands. Figure reproduced from Ref. [163] with kind permission by AAAS.

unset strands, five steps with this system could be demonstrated [166]. Later on, a variety of more sophisticated walker schemes were developed that enabled autonomous walking with coordinated motion of the feet of the walkers [167, 168].

As we have seen, regression of a walker can be circumvented by the use of different foothold strands. But there is another way of preventing the walker from back motion that can be regarded as a 'burning bridges' strategy, involving ribozymes or enzymes: A DNAzyme that contains a catalytic core and two recognition arms is hybridized to a foothold strand by means of its two recognition arms, while the catalytic core forms a bulge between them. The DNAzyme's catalytic core can cleave the upper end of the foothold strand, which is hybridized to the upper recognition arm of the DNAzyme. By this, the upper recognition arm of the walker can hybridize to the next foothold's upper end. This is followed by a strand displacement process, in which the lower end of the DNAzyme eventually hybridizes to the next foothold's lower end, releasing the previous foothold completely. Because the upper end of the previous foothold strand is missing after each step, no backwards movement is possible along the track [169]. DNAzymes are also used in "molecular spiders". Here several DNAzyme "legs" are attached to a "body" comprised of a strepatvidin protein [170]. The spiders can diffusively move along substrate tracks, e.g. defined on a DNA origami structure [171].

A nicking enzyme to cut off the upper end of a foothold strand – instead of a DNAzyme – has been successfully used for a walker on a DNA substrate as well [172], while a combination of two endonuclease nicking enzymes and a ligase could be utilized to hand over a small DNA 'package' along a DNA track [173].

It has to be noted that in all of the systems realized so far, there is only a very superficial similarity to naturally occurring molecular motors like myosin or kinesin. In particular, there is no power stroke involved in the motion of the DNA systems. Rather, the Brownian motion of the 'feet' searching for a binding site is rectified by an irreversible chemical step. In that sense – from a physical point of view – DNA walkers are more closely related to motors such as DNA or RNA polymerase, or possibly biological 'polymerization motors'.

In fact, polymerization motors can also be directly employed with DNA. Very recently, utilizing the 'hybridization chain reaction' introduced in section 7.4.3, Pierce and coworkers [174] could demonstrate a DNA-based version of the polymerization motor employed by the bacterium Rickettsia rickettsii.

7.7
Switchable Molecular Networks and Materials

Incorporation of switchable elements into larger molecular networks can result in materials with switchable properties. DNA networks with switchable lattice constants [161] and Seeman's arrays of robotic arms [163] introduced in the previous section can be regarded as examples of such materials.

DNA-based molecular structures have also been suggested as switchable containers for nano-objects. In Ref. [157], it was shown that a protein can be placed into the cavity of a supramolecular tetrahedron assembled from four DNA single strands. It has been suggested to switch the shape of the tetrahedron using DNA or RNA effector molecules in order to release the captured protein, resulting in a controlled delivery system for the protein. The same application is envisioned for a molecular "box" made from DNA origami that can be opened and closed using "key" and "lock" DNA strands [1].

Switchable bioinorganic hybrid systems have been realized with gold nanoparticles. The particles can be reversibly aggregated and redispersed by exploiting the pH-driven conformational changes of i-motif oligonucleotides (cf. Section 7.6.1) [175] or by utilizing the common fuel-release strand approach [176]. In both cases, switching is accompanied by a change in plasmonic communication between the Au nanoparticles, resulting in a visible colour change of the solution containing the hybrid structure.

DNA molecules can even be utilized to switch the macroscopic rheological properties of a material such as a polyacrylamide gel. Chemically modified oligonucleotides can be copolymerized with acrylamide, resulting in linear polyacrylamide with oligonucleotide side chains. Adding oligonucleotides complementary to the side chains crosslinks the polyacrylamide and causes gelation. Removal of the crosslinking strands utilizing the strand displacement approach redissolves the gel into a sol. It could be shown that this allows us to reversibly switch the viscoelastic properties of the gel [177]. The switchable gel may find application as controlled release system for nanoparticles [178] or as a DNA-adjustable nanosieve.

7.8
Conclusion and Outlook

Nucleic acids are extremely versatile molecules for the rational design and construction of complex molecular switches. Their unique, sequence-dependent molecular-recognition properties enable programmable self-assembly of these molecular structures, and also programmability of the switching events themselves – that is DNA or RNA-based switches can be made to react to the presence of specific 'input' or 'effector' molecules.

As has been surveyed in this chapter, nucleic acids have been used to realize simple conformational switches, motor-like molecular walkers, autonomous information-processing devices and also switchable materials. Potential applications of these structures can be found in nanotechnology – as components of active materials or molecular 'assembly lines' – but also, and predominantly, in biomedicine. Here, they can serve as transducers in biosensors, as autonomous molecular diagnosis units or as components of drug-delivery devices.

There is also a strong similarity – in structure, mechanism and function – between artificially constructed or evolved DNA and RNA switches and naturally occurring riboswitches. Riboswitches may inspire future developments of artificial RNA switches, and implementation of such artificial nucleic-acid structures *in vivo* will eventually make them important components for synthetic biology.

Acknowledgements

The authors acknowledge financial support by the Nanosystems Initiative Munich (NIM) and the Human Frontier Science Program (HFSP).

References

1. Andersen, E.S. et al. (2009) Self-assembly of a nanoscale DNA box with a controllable lid. *Nature*, **459** (7243), 73–76.
2. Lu, Z.J., Turner, D.H. and Mathews, D.H. (2006) A set of nearest neighbor parameters for predicting the enthalpy change of RNA secondary structure formation. *Nucleic Acids Res.*, **34** (17), 4912–4924.
3. Xayaphoummine, A., Bucher, T. and Isambert, H. (2005) Kinefold web server for RNA/DNA folding path and structure prediction including pseudoknots and knots. *Nucleic Acids Res.*, **33**, W605–W610.
4. Hofacker, I.L. (2003) Vienna RNA secondary structure server. *Nucleic Acids Res.*, **31** (13), 3429–3431.
5. Serra, M.J. and Turner, D.H. (1995) Predicting thermodynamic properties of RNA. *Energ. Biol. Macromol.*, **259**, 242–261.
6. Turner, D.H., Sugimoto, N. and Freier, S.M. (1988) RNA structure prediction. *Ann. Rev. Biophys. Biophys. Chem.*, **17**, 167–192.
7. Mathews, D.H. et al. (1999) Expanded sequence dependence of thermodynamic parameters improves prediction of RNA secondary structure. *J. Mol. Biol.*, **288** (5), 911–940.
8. SantaLucia, J. and Hicks, D. (2004) The thermodynamics of DNA structural motifs. *Ann. Rev. Biophys. Biomol. Struct.*, **33**, 415–440.
9. Dimitrov, R.A. and Zuker, M. (2004) Prediction of hybridization and melting for double-stranded nucleic acids. *Biophys. J.*, **87** (1), 215–226.
10. Mathews, D.H. (2006) Revolutions in RNA secondary structure prediction. *J. Mol. Biol.*, **359** (3), 526–532.
11. Batey, R.T. and Doudna, J.A. (1998) The parallel universe of RNA folding. *Nature Struct. Biol.*, **5** (5), 337–340.
12. Green, S.J., Lubrich, D. and Turberfield, A.J. (2006) DNA hairpins: fuel for autonomous DNA devices. *Biophys. J.*, **91** (8), 2966–2975.
13. Guerriertakada, C. et al. (1983) The RNA moiety of ribonuclease-P is the catalytic subunit of the enzyme. *Cell*, **35** (3), 849–857.
14. Kruger, K. et al. (1982) Self-splicing RNA – auto excision and auto-cyclization of the ribosomal-RNA intervening sequence of tetrahymena. *Cell*, **31** (1), 147–157.
15. Chen, X., Li, N. and Ellington, A.D. (2007) Ribozyme catalysis of metabolism in the RNA world. *Chem. Biodivers.*, **4** (4), 633–655.
16. Doudna, J.A. and Cech, T.R. (2002) The chemical repertoire of natural ribozymes. *Nature*, **418** (6894), 222–228.
17. Emilsson, G.M. and Breaker, R.R. (2002) Deoxyribozymes: new activities and new applications. *Cell. Mol. Life Sci.*, **59** (4), 596–607.
18. Hammann, C. and Westhof, E. (2007) Searching genomes for ribozymes and riboswitches. *Genome Biol.*, **8** (4), Art. No. 210, 1–11.
19. Lilley, D.M.J. (2005) Structure, folding and mechanisms of ribozymes. *Curr. Opin. Struct. Biol.*, **15** (3), 313–323.
20. Link, K.H. et al. (2007) Engineering high-speed allosteric hammerhead ribozymes. *Biol. Chem.*, **388** (8), 779–786.
21. Breaker, R.R. (1997) DNA enzymes. *Nature Biotechnol.*, **15** (5), 427–431.
22. Bartel, D.P. and Szostak, J.W. (1993) Isolation of new ribozymes from a large pool of random sequences. *Science*, **261** (5127), 1411–1418.

23. Santoro, S.W. and Joyce, G.F. (1997) A general purpose RNA-cleaving DNA enzyme. *Proc. Natl. Acad. Sci. USA*, **94** (9), 4262–4266.
24. Marshall, K.A. and Ellington, A.D. (2000) In vitro selection of RNA aptamers. *RNA-Ligand Interact. Part B*, **318**, 193–214.
25. Ellington, A.D. and Szostak, J.W. (1990) In vitro selection of RNA molecules that bind specific ligands. *Nature*, **346** (6287), 818–822.
26. Burgstaller, P. and Famulok, M. (1994) Isolation of RNA aptamers for biological cofactors by in-vitro selection. *Angew. Chem. Int. Ed. Engl.*, **33** (10), 1084–1087.
27. Wilson, D.S. and Szostak, J.W. (1999) In vitro selection of functional nucleic acids. *Ann. Rev. Biochem.*, **68**, 611–647.
28. Potyrailo, R.A. *et al.* (1998) Adapting selected nucleic acid ligands (aptamers) to biosensors. *Anal. Chem.*, **70** (16), 3419–3425.
29. Porta, H. and Lizardi, P.M. (1995) An allosteric hammerhead ribozyme. *Biotechnology*, **13** (2), 161–164.
30. Robertson, M.P. and Ellington, A.D. (1999) In vitro selection of an allosteric ribozyme that transduces analytes to amplicons. *Nature Biotechnol.*, **17** (1), 62–66.
31. Chen, Y. and Mao, C.D. (2005) Regulating enzyme activities in a multiple-enzyme complex. *Chembiochem*, **6** (6), 999–1002.
32. Kuwabara, T. *et al.* (1998) A novel allosterically trans-activated ribozyme, the maxizyme, with exceptional specificity in vitro and in vivo. *Mol. Cell*, **2** (5), 617–627.
33. Soda, Y. *et al.* (2004) A novel maxizyme vector targeting a bcr-abl fusion gene induced specific cell death in Philadelphia chromosome-positive acute lymphoblastic leukemia. *Blood*, **104** (2), 356–363.
34. Tang, J. and Breaker, R.R. (1998) Mechanism for allosteric inhibition of an ATP-sensitive ribozyme. *Nucleic Acids Res.*, **26** (18), 4214–4221.
35. Tang, J. and Breaker, R.R. (1997) Rational design of allosteric ribozymes. *Chem. Biol.*, **4** (6), 453–459.
36. Wang, D.Y. *et al.* (2002) A general approach for the use of oligonucleotide effectors to regulate the catalysis of RNA-cleaving ribozymes and DNAzymes. *Nucleic Acids Res.*, **30** (8), 1735–1742.
37. Wang, D.Y., Lai, B.H.Y. and Sen, D. (2002) A general strategy for effector-mediated control of RNA-cleaving ribozymes and DNA enzymes. *J. Mol. Biol.*, **318** (1), 33–43.
38. Araki, M. *et al.* (1998) Allosteric regulation of a ribozyme activity through ligand-induced conformational change. *Nucleic Acids Res.*, **26** (14), 3379–3384.
39. Robertson, M.P. and Ellington, A.D. (2000) Design and optimization of effector-activated ribozyme ligases. *Nucleic Acids Res.*, **28** (8), 1751–1759.
40. Zivarts, M., Liu, Y. and Breaker, R.R. (2005) Engineered allosteric ribozymes that respond to specific divalent metal ions. *Nucleic Acids Res.*, **33** (2), 622–631.
41. Batey, R.T. (2006) Structures of regulatory elements in mRNAs. *Curr. Opin. Struct. Biol.*, **16** (3), 299–306.
42. Isaacs, F.J., Dwyer, D.J. and Collins, J.J. (2006) RNA synthetic biology. *Nature Biotechnol.*, **24** (5), 545–554.
43. Breaker, R.R. (2004) Natural and engineered nucleic acids as tools to explore biology. *Nature*, **432** (7019), 838–845.
44. Schwalbe, H. *et al.* (2007) Structures of RNA switches: insight into molecular recognition and tertiary structure. *Angew. Chem. Int. Ed.*, **46** (8), 1212–1219.
45. Lai, E.C. (2003) RNA sensors and riboswitches: self-regulating messages. *Curr. Biol.*, **13** (7), R285–R291.
46. Vitreschak, A.G. *et al.* (2004) Riboswitches: the oldest mechanism for the regulation of gene expression? *Trends Genet.*, **20** (1), 44–50.
47. Winkler, W.C. and Breaker, R.R. (2005) Regulation of bacterial gene expression by riboswitches. *Ann. Rev. Microbiol.*, **59**, 487–517.
48. Tucker, B.J. and Breaker, R.R. (2005) Riboswitches as versatile gene control

elements. *Curr. Opin. Struct. Biol.*, **15** (3), 342–348.
49. Nudler, E. (2006) Flipping riboswitches. *Cell*, **126** (1), 19–22.
50. Coppins, R.L., Hall, K.B. and Groisman, E.A. (2007) The intricate world of riboswitches. *Curr. Opin. Microbiol.*, **10** (2), 176–181.
51. Mandal, M. and Breaker, R.R. (2004) Gene regulation by riboswitches. *Nature Rev. Mol. Cell Biol.*, **5** (6), 451–463.
52. Stormo, G.D. and Ji, Y.M. (2001) Do mRNAs act as direct sensors of small molecules to control their expression? *Proc. Natl. Acad. Sci. USA*, **98** (17), 9465–9467.
53. Narberhaus, F., Waldminighaus, T. and Chowdhury, S. (2006) RNA thermometers. *FEMS Microbiol. Rev.*, **30** (1), 3–16.
54. Winkler, W.C., Cohen-Chalamish, S. and Breaker, R.R. (2002) An mRNA structure that controls gene expression by binding FMN. *Proc. Natl. Acad. Sci. USA*, **99** (25), 15908–15913.
55. Mironov, A.S. *et al.* (2002) Sensing small molecules by nascent RNA: a mechanism to control transcription in bacteria. *Cell*, **111** (5), 747–756.
56. McDaniel, B.A.M. *et al.* (2003) Transcription termination control of the S box system: direct measurement of S-adenosylmethionine by the leader RNA. *Proc. Natl. Acad. Sci. USA*, **100** (6), 3083–3088.
57. Mandal, M. and Breaker, R. (2004) An mRNA structure that differentially controls gene expression by binding guanine or adenine. *Biophys. J.*, **86** (1), 32a.
58. Grundy, F.J., Lehman, S.C. and Henkin, T.M. (2003) The L box regulon: lysine sensing by leader RNAs of bacterial lysine biosynthesis genes. *Proc. Natl. Acad. Sci. USA*, **100** (21), 12057–12062.
59. Epshtein, V., Mironov, A.S. and Nudler, E. (2003) The riboswitch-mediated control of sulfur metabolism in bacteria. *Proc. Natl. Acad. Sci. USA*, **100** (9), 5052–5056.
60. Sudarsan, N., Barrick, J.E. and Breaker, R.R. (2003) Metabolite-binding RNA domains are present in the genes of eukaryotes. *RNA-a Publ. RNA Soc.*, **9** (6), 644–647.
61. Nou, X.W. and Kadner, R.J. (2000) Adenosylcobalamin inhibits ribosome binding to btuB RNA. *Proc. Natl. Acad. Sci. USA*, **97** (13), 7190–7195.
62. Nahvi, A. *et al.* (2002) Genetic control by a metabolite binding mRNA. *Chem. Biol.*, **9** (9), 1043–1049.
63. Mandal, M. and Breaker, R.R. (2004) Adenine riboswitches and gene activation by disruption of a transcription terminator. *Nature Struct. Mol. Biol.*, **11** (1), 29–35.
64. Serganov, A. *et al.* (2004) Structural basis for discriminative regulation of gene expression by adenine- and guanine-sensing mRNAs. *Chem. Biol.*, **11** (12), 1729–1741.
65. Mandal, M. *et al.* (2004) A glycine-dependent riboswitch that uses cooperative binding to control gene expression. *Science*, **306** (5694), 275–279 & **306** (5701), 1477.
66. Kaempfer, R. (2003) RNA sensors: novel regulators of gene expression. *EMBO Rep.*, **4** (11), 1043–1047.
67. Kubodera, T. *et al.* (2003) Thiamine-regulated gene expression of aspergillus oryzae thiA requires splicing of the intron containing a riboswitch-like domain in the 5′-UTR. *FEBS Lett.*, **555** (3), 516–520.
68. Sudarsan, N. *et al.* (2003) An mRNA structure in bacteria that controls gene expression by binding lysine. *Genes Dev.*, **17** (21), 2688–2697.
69. Rodionov, D.A. *et al.* (2003) Regulation of lysine biosynthesis and transport genes in bacteria: yet another RNA riboswitch? *Nucleic Acids Res.*, **31** (23), 6748–6757.
70. Winkler, W.C. *et al.* (2004) Control of gene expression by a natural metabolite-responsive ribozyme. *Nature*, **428** (6980), 281–286.
71. Winkler, W., Nahvi, A. and Breaker, R.R. (2002) Thiamine derivatives bind messenger RNAs directly to regulate bacterial gene expression. *Nature*, **419** (6910), 952–956.
72. Miranda-Rios, J., Navarro, M. and Soberon, M. (2001) A conserved RNA

structure (thi box) is involved in regulation of thiamin biosynthetic gene expression in bacteria. *Proc. Natl. Acad. Sci. USA*, **98** (17), 9736–9741.
73. Winkler, W.C. et al. (2003) An mRNA structure that controls gene expression by binding S-adenosylmethionine. *Nature Struct. Biol.*, **10** (9), 701–707.
74. Sudarsan, N. et al. (2006) Tandem riboswitch architectures exhibit complex gene control functions. *Science*, **314** (5797), 300–304.
75. Isaacs, F.J. et al. (2004) Engineered riboregulators enable post-transcriptional control of gene expression. *Nature Biotechnol.*, **22** (7), 841–847.
76. Bayer, T.S. and Smolke, C.D. (2005) Programmable ligand-controlled riboregulators of eukaryotic gene expression. *Nature Biotechnol.*, **23** (3), 337–343.
77. Win, M.N. and Smolke, C.D. (2008) Higher-order cellular information processing with synthetic RNA devices. *Science*, **322** (3900), 456–460.
78. Kim, J., White, K.S. and Winfree, E. Construction of an in vitro bistable circuit from synthetic transcriptional switches. (2006) *Mol. Syst. Biol.* 2, Art No. 68, 1–12.
79. Viasnoff, V., Meller, A. and Isambert, H. (2006) DNA nanomechanical switches under folding kinetics control. *Nano Lett.*, **6** (1), 101–104.
80. Liu, D.S. and Balasubramanian, S. (2003) A proton-fuelled DNA nanomachine. *Angew. Chem. Int. Ed.*, **42** (46), 5734–5736.
81. Modi, S. et al. (2009) A DNA nanomachine that maps spatial and temporal pH changes inside living cells. *Nat Nanotechnol*, **4** (5), 325–330.
82. Liu, H.J. et al. (2007) Light-driven conformational switch of i-motif DNA. *Angew. Chem. Int. Ed.*, **46** (14), 2515–2517.
83. Westermeier, R. et al. (1997) *Electrophoresis in Practise*, Wiley-VCH Verlag GmbH.
84. Schwille, P. (2001) Fluorescence correlation spectroscopy and its potential for intracellular applications. *Cell Biochem. Biophys.*, **34** (3), 383–408.
85. Fahlman, R.P. and Sen, D. (2002) DNA conformational switches as sensitive electronic sensors of analytes. *J. Am. Chem. Soc.*, **124** (17), 4610–4616.
86. Stryer, L. and Haugland, R.P. (1967) Energy transfer – a spectroscopic ruler. *Proc. Natl. Acad. Sci. USA*, **58** (2), 719–771.
87. Selvin, P.R. (2000) The renaissance of fluorescence resonance energy transfer. *Nature Struct. Biol.*, **7** (9), 730–734.
88. Stryer, L. (1978) Fluorescence energy-transfer as a spectroscopic ruler. *Ann. Rev. Biochem.*, **47**, 819–846.
89. Lakowicz, J.R. (2006) *Principles of Fluorescence Spectroscopy*, Springer Science and Business Media, LLC, New York.
90. Weiss, S. (1999) Fluorescence spectroscopy of single biomolecules. *Science*, **283** (5408), 1676–1683.
91. Tyagi, S. and Kramer, F.R. (1996) Molecular beacons: Probes that fluoresce upon hybridization. *Nat. Biotech.*, **14** (3), 303.
92. Wang, K. et al. (2009) Molecular Engineering of DNA: Molecular Beacons. *Angew. Chem. Int. Ed.*, **48** (5), 856–870.
93. Tan, W.H., Wang, K.M. and Drake, T.J. (2004) Molecular beacons. *Curr. Opin. Chem. Biol.*, **8** (5), 547–553.
94. Li, J.W.J., Fang, X.H. and Tan, W.H. (2002) Molecular aptamer beacons for real-time protein recognition. *Biochem. Biophys. Res. Commun.*, **292** (1), 31–40.
95. Stojanovic, M.N., de Prada, P. and Landry, D.W. (2001) Aptamer-based folding fluorescent sensor for cocaine. *J. Am. Chem. Soc.*, **123** (21), 4928–4931.
96. Stojanovic, M.N., de Prada, P. and Landry, D.W. (2000) Fluorescent sensors based on aptamer self-assembly. *J. Am. Chem. Soc.*, **122** (46), 11547–11548.
97. Stojanovic, M.N. and Kolpashchikov, D.M. (2004) Modular aptameric sensors. *J. Am. Chem. Soc.*, **126** (30), 9266–9270.
98. Stojanovic, M.N. and Landry, D.W. (2002) Aptamer-based colorimetric probe for cocaine. *J. Am. Chem. Soc.*, **124** (33), 9678–9679.

99. Travascio, P. et al. (2001) The peroxidase activity of a hemin-DNA oligonucleotide complex: free radical damage to specific guanine bases of the DNA. *J. Am. Chem. Soc.*, **123** (7), 1337–1348.
100. Cheglakov, Z. et al. (2006) Ultrasensitive detection of DNA by the PCR-induced generation of DNAzymes: The DNAzyme primer approach. *Chem. Commun.*, (30), 3205–3207.
101. Tian, Y., He, Y. and Mao, C.D. (2006) Cascade signal amplification for DNA detection. *Chembiochem*, **7** (12), 1862–1864.
102. Cheglakov, Z. et al. (2007) Diagnosing viruses by the rolling circle amplified synthesis of DNAzymes. *Org. Biomol. Chem.*, **5** (2), 223–225.
103. Shlyahovsky, B. et al. (2007) Spotlighting of cocaine by an autonomous aptamer-based machine. *J. Am. Chem. Soc.*, **129** (13), 3814–3815.
104. Dirks, R.M. and Pierce, N.A. (2004) Triggered amplification by hybridization chain reaction. *Proc. Natl. Acad. Sci. USA*, **101** (43), 15275–15278.
105. Choi, H.M.T. et al. (2010) Programmable in situ amplification for multiplexed imaging of mRNA expression. *Nat. Biotechnol.*, **28** (11), 1208–1212.
106. Hartig, J.S. et al. (2004) Sequence-specific detection of microRNAs by signal-amplifying ribozymes. *J. Am. Chem. Soc.*, **126** (3), 722–723.
107. Stojanovic, M.N., de Prada, P. and Landry, D.W. (2001) Catalytic molecular beacons. *Chembiochem*, **2** (6), 411–415.
108. Tian, Y. and Mao, C.D. (2005) DNAzyme amplification of molecular beacon signal. *Talanta*, **67** (3), 532–537.
109. Frauendorf, C. and Jaschke, A. (2001) Detection of small organic analytes by fluorescing molecular switches. *Bioorg. Med. Chem.*, **9** (10), 2521–2524.
110. Muller, S., Strohbach, D. and Wolf, J. (2006) Sensors made of RNA: tailored ribozymes for detection of small organic molecules, metals, nucleic acids and proteins. *IEEE Proc. Nanobiotechnol.*, **153** (2), 31–40.
111. Silverman, S.K. (2003) Rube Goldberg goes (ribo)nuclear? Molecular switches and sensors made from RNA. *RNA-a Publ. RNA Soc.*, **9** (4), 377–383.
112. Cho, E.J. Lee, J.-W. and Ellington, A.D. (2009) Applications of Aptamers as Sensors. *Annu Rev Anal Chem*, **2**, 241–264.
113. Jonoska, N. (2004) Trends in computing with DNA. *J. Comput. Sci. Technol.*, **19** (1), 98–113.
114. Adleman, L.M. (1994) Molecular computation of solutions to combinatorial problems. *Science*, **266** (5187), 1021–1024.
115. Kolpashchikov, D.M. and Stojanovic, M.N. (2005) Boolean control of aptamer binding states. *J. Am. Chem. Soc.*, **127** (32), 11348–11351.
116. Stojanovic, M.N., Mitchell, T.E. and Stefanovic, D. (2002) Deoxyribozyme-based logic gates. *J. Am. Chem. Soc.*, **124** (14), 3555–3561.
117. Stojanovic, M.N., Nikic, D.B. and Stefanovic, D. (2003) Implicit-OR tiling of deoxyribozymes: construction of molecular scale OR, NAND, and four-input logic gates. *J. Serb. Chem. Soc.*, **68** (4-5), 321–326.
118. Stojanovic, M.N. and Stefanovic, D. (2003) Deoxyribozyme-based half-adder. *J. Am. Chem. Soc.*, **125** (22), 6673–6676.
119. Stojanovic, M.N. and Stefanovic, D. (2003) A deoxyribozyme-based molecular automaton. *Nature Biotechnol.*, **21** (9), 1069–1074.
120. Macdonald, J. et al. (2006) Medium scale integration of molecular logic gates in an automaton. *Nano Lett.*, **6** (11), 2598–2603.
121. Seelig, G., Yurke, B. and Winfree, E. (2006) Catalyzed relaxation of a metastable DNA fuel. *J. Am. Chem. Soc.*, **128** (37), 12211–12220.
122. Turberfield, A.J. et al. (2003) DNA fuel for free-running nanomachines. *Phys. Rev. Lett.*, **90** (11), Art. No. 118102, 1–4.
123. Seelig, G. et al. (2006) Enzyme-free nucleic acid logic circuits. *Science*, **314** (5805), 1585–1588.
124. Adar, R. et al. (2004) Stochastic computing with biomolecular automata.

125. Benenson, Y. et al. (2001) Programmable and autonomous computing machine made of biomolecules. *Nature*, **414** (6862), 430–434.
126. Soreni, M. et al. (2005) Parallel biomolecular computation on surfaces with advanced finite automata. *J. Am. Chem. Soc.*, **127** (11), 3935–3943.
127. Niazov, T. et al. (2006) Concatenated logic gates using four coupled biocatalysts operating in series. *Proc. Natl. Acad. Sci. USA*, **103** (46), 17160–17163.
128. Weizmann, Y. et al. (2005) Endonuclease-based logic gates and sensors using magnetic force-amplified readout of DNA scission on cantilevers. *J. Am. Chem. Soc.*, **127** (36), 12666–12672.
129. Miyoshi, D. Inoue, M. and Sugimoto, N. (2006) DNA logic gates based on structural polymorphism of telomere DNA molecules responding to chemical input signals. *Angew. Chem. Int. Ed.*, **45** (46), 7716–7719.
130. Deng, Z.X., Lee, S.H. and Mao, C.D. (2005) DNA as nanoscale building blocks. *J. Nanosci. Nanotechnol.*, **5** (12), 1954–1963.
131. Feldkamp, U. and Niemeyer, C.M. (2006) Rational design of DNA nanoarchitectures. *Angew. Chem. Int. Ed.*, **45** (12), 1856–1876.
132. Seeman, N.C. (2003) DNA in a material world. *Nature*, **421** (6921), 427–431.
133. Rothemund, P.W.K. (2006) Folding DNA to create nanoscale shapes and patterns. *Nature*, **440** (7082), 297–302.
134. Douglas, S.M. et al. (2009) Self-assembly of DNA into nanoscale three-dimensional shapes. *Nature*, **459** (7245), 414–418.
135. Lu, Y. and Liu, J.W. (2006) Functional DNA nanotechnology: emerging applications of DNAzymes and aptamers. *Curr. Opin. Biotechnol.*, **17** (6), 580–588.
136. Bath, J. and Turberfield, A.J. (2007) DNA nanomachines. *Nature Nanotechnol.*, **2** (5), 275–284.

137. Liedl, T. and Simmel, F.C. (2005) Switching the conformation of a DNA molecule with a chemical oscillator. *Nano Lett.*, **5** (10), 1894–1898.
138. Liu, D.S. et al. (2006) A reversible pH-driven DNA nanoswitch array. *J. Am. Chem. Soc.*, **128** (6), 2067–2071.
139. Liedl, T., Olapinski, M. and Simmel, F.C. (2006) A surface-bound DNA switch driven by a chemical oscillator. *Angew. Chem. Int. Ed.*, **45** (30), 5007–5010.
140. Shu, W.M. et al. (2005) DNA molecular motor driven micromechanical cantilever arrays. *J. Am. Chem. Soc.*, **127** (48), 17054–17060.
141. Zhang, D.Y. and Seelig, G. (2011) Dynamic DNA nanotechnology using strand-displacement reactions. *Nature Chemistry*, **3** (2), 103–113.
142. Yurke, B. et al. (2000) A DNA-fuelled molecular machine made of DNA. *Nature*, **406** (6796), 605–608.
143. Simmel, F.C. and Yurke, B. (2002) A DNA-based molecular device switchable between three distinct mechanical states. *Appl. Phys. Lett.*, **80** (5), 883–885.
144. Simmel, F.C. and Yurke, B. (2001) Using DNA to construct and power a nanoactuator. *Phys. Rev. E*, **6304** (4), Art. No. 041913, 1–5.
145. Dittmer, W.U. et al. (2005) Using gene regulation to program DNA-based molecular devices. *Small*, **1** (7), 709–712.
146. Chen, Y., Wang, M.S. and Mao, C.D. (2004) An autonomous DNA nanomotor powered by a DNA enzyme. *Angew. Chem. Int. Ed.*, **43** (27), 3554–3557.
147. Chen, Y. and Mao, C.D. (2004) Putting a brake on an autonomous DNA nanomotor. *J. Am. Chem. Soc.*, **126** (28), 8626–8627.
148. Reuter, A., Dittmer, W.U. and Simmel, F.C. (2007) Kinetics of protein-release by an aptamer-based DNA nanodevice. *Eur. Phys. J. E*, **22** (1), 33–40.
149. Dittmer, W.U., Reuter, A. and Simmel, F.C. (2004) A DNA-based machine that can cyclically bind and release thrombin. *Angew. Chem. Int. Ed.*, **43** (27), 3550–3553.

150. Friedrichs, E. and Simmel, F.C. (2007) Controlling DNA polymerization with a switchable aptamer. *Chembiochem*, **8** (14), 1662–1666.
151. Holliday, R. (1964) Mechanism for gene conversion in fungi. *Genet. Res.*, **5** (2), 282–304.
152. Seeman, N.C. and Kallenbach, N.R. (1983) Design of immobile nucleic-acid junctions. *Biophys. J.*, **44** (2), 201–209.
153. Mao, C.D., Sun, W.Q. and Seeman, N.C. (1999) Designed two-dimensional DNA Holliday junction arrays visualized by atomic force microscopy. *J. Am. Chem. Soc.*, **121** (23), 5437–5443.
154. Seeman, N.C. (1982) Nucleic-acid junctions and lattices. *J. Theor. Biol.*, **99** (2), 237–247.
155. Fu, T.J. and Seeman, N.C. (1993) DNA double-crossover molecules. *Biochemistry*, **32** (13), 3211–3220.
156. LaBean, T.H. et al. (2000) Construction, analysis, ligation, and self-assembly of DNA triple crossover complexes. *J. Am. Chem. Soc.*, **122** (9), 1848–1860.
157. Erben, C.M., Goodman, R.P. and Turberfield, A.J. (2006) Single-molecule protein encapsulation in a rigid DNA cage. *Angew. Chem. Int. Ed.*, **45** (44), 7414–7417.
158. Mao, C.D., Sun, W.Q. and Seeman, N.C. (1997) Assembly of Borromean rings from DNA. *Nature*, **386** (6621), 137–138.
159. Mao, C.D. et al. (1999) A nanomechanical device based on the B-Z transition of DNA. *Nature*, **397** (6715), 144–146.
160. Chen, Y., Lee, S.H. and Mao, C. (2004) A DNA nanomachine based on a duplex-triplex transition. *Angew. Chem. Int. Ed.*, **43** (40), 5335–5338.
161. Feng, L.P. et al. (2003) A two-state DNA lattice switched by DNA nanoactuator. *Angew. Chem. Int. Ed.*, **42** (36), 4342–4346.
162. Zhong, H. and Seeman, N.C. (2006) RNA used to control a DNA rotary nanomachine. *Nano Lett.*, **6** (12), 2899–2903.
163. Ding, B. and Seeman, N.C. (2006) Operation of a DNA robot arm inserted into a 2D DNA crystalline substrate. *Science*, **314** (5805), 1583–1585.
164. Gu, H. et al. (2010) A proximity-based programmable DNA nanoscale assembly line. *Nature*, **465** (7295), 202–205.
165. Sherman, W.B. and Seeman, N.C. (2004) A precisely controlled DNA biped walking device. *Nano Lett.*, **4** (7), 1203–1207.
166. Shin, J.S. and Pierce, N.A. (2004) A synthetic DNA walker for molecular transport. *J. Am. Chem. Soc.*, **126** (35), 10834–10835.
167. Omabegho, T. Sha, R. and Seeman, N.C. (2009) A Bipedal DNA Brownian Motor with Coordinated Legs. *Science*, **324** (5923), 67–71.
168. Green, S. Bath, J. and Turberfield, A. (2008) Coordinated Chemomechanical Cycles: A Mechanism for Autonomous Molecular Motion. *Phys. Rev. Lett.*, **101** (23), Art No. 238101.
169. Tian, Y. et al. (2005) Molecular devices – a DNAzyme that walks processively and autonomously along a one-dimensional track. *Angew. Chem. Int. Ed.*, **44** (28), 4355–4358.
170. Pei, R. et al. (2006) Behavior of polycatalytic assemblies in a substrate-displaying matrix. *J. Am. Chem. Soc.*, **128** (39), 12693.
171. Lund, K. et al. (2010) Molecular robots guided by prescriptive landscapes. *Nature*, **465** (7295), 206–210.
172. Bath, J., Green, S.J. and Turberfield, A.J. (2005) A free-running DNA motor powered by a nicking enzyme. *Angew. Chem. Int. Ed.*, **44** (28), 4358–4361.
173. Yin, P. et al. (2004) A unidirectional DNA walker that moves autonomously along a track. *Angew. Chem. Int. Ed.*, **43** (37), 4906–4911.
174. Venkataraman, S. et al. (2007) An autonomous polymerization motor powered by DNA hybridization. *Nature Nanotechnol.*, **2** (8), 490–494.
175. Sharma, J. et al. (2007) pH-driven conformational switch of "i-motif" DNA for the reversible assembly of gold nanoparticles. *Chem. Commun.*, (5), 477–479.
176. Niemeyer, C.M., Ceyhan, B. and Hazarika, P. (2003) Oligofunctional DNA-gold nanoparticle conjugates.

Angew. Chem. Int. Ed., **42** (46), 5766–5770.

177. Lin, D.C., Yurke, B. and Langrana, N.A. (2004) Mechanical properties of a reversible, DNA-crosslinked polyacrylamide hydrogel. *J. Biomech. Eng. Trans. ASME*, **126** (1), 104–110.

178. Liedl, T. *et al.* (2007) Controlled trapping and release of quantum dots in a DNA-switchable hydrogel. *Small*, **3** (10), 1688–1693.

Part II
Switching in Containers, Polymers and Channels

8
Switching Processes in Cavitands, Containers and Capsules

Vladimir A. Azov and François Diederich

8.1
Introduction

In recent years, chemists have demonstrated considerable imagination in the construction of molecular machines and molecular switches [1]. Much effort has been applied to the development of molecular architectures, displaying various types of molecular motion: *cis–trans* isomerization of double bonds [2, 3], rotation around single bonds [4], shuffling motion in rotaxanes [5], ring rotation in catenanes [6], coiling/decoiling motion in supramolecular systems [7] and several others [1]. On the other hand, controllable switching movements of large host molecules with cavities – cavitands, molecular containers and capsules – have been less exploited. While most molecular switches and machines demonstrate spatial displacement of molecular or supramolecular components, switching of molecular hosts not only induces similar molecular motions but additionally provides a way to control guest binding and release. This chapter reviews construction and functioning of switchable molecular and supramolecular hosts with preorganized cavitary binding sites and discusses application of these systems for induction of large molecular motions and reversible alteration of molecular encapsulation properties.

Receptors with well-defined interior binding sites, usually spherical or cylindrical in shape, and enabling fully constricted guest complexation are called *container molecules* or *capsules*. Covalently constructed hosts are usually referred to as *container molecules*, whereas reversibly self-assembled architectures are called *capsules*. However, naming in the literature is not consistent and preorganized receptors with more open, accessible cavity binding sites, defined by Cram [8] as cavitands, are frequently also called *container molecules*. Several other terms were coined to define molecular containers of specific shape (for example molecular baskets), or the method of assembly (coordination cages).

Following the introduction of the first covalent capsular receptors by Collet and coworkers [9] and Cram and coworkers [10] and the first supramolecularly assembled counterparts by Rebek and coworkers [11], research on molecular encapsulation has greatly expanded, producing fascinating molecular architectures with guest-hosting sites of various shape and sizes. Covalent molecular containers

Molecular Switches, Second Edition. Edited by Ben L. Feringa and Wesley R. Browne.
© 2011 Wiley-VCH Verlag GmbH & Co. KGaA. Published 2011 by Wiley-VCH Verlag GmbH & Co. KGaA.

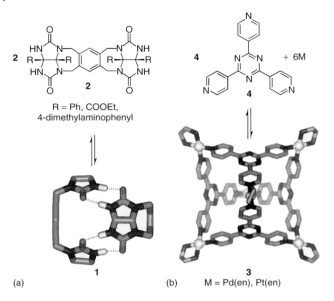

Figure 8.1 Representative examples of self-assembled molecular capsules. (a) 'Tennis ball' **1** is held together by a seam of 8 H-bonds connecting two identical components **2** [11]. (b) The metal-coordinated octahedral cage **3** is assembled from ligand **4** [20a,b]. Some substituents are omitted for clarity. en = ethylenediamine.

were investigated to explore template effects during their formation [12] and to prepare and stabilize in their shielded inner phase delicate unstable molecules and reactive intermediates such as cyclobutadiene and benzyne [13]. Among the more open covalent molecular containers, resorcin[4]arene-based cavitands [14] and cucurbit[n]urils [15], which are discussed in greater detail in this chapter, have attracted particular interest. Reversibly self-assembling container and capsule architectures are stabilized by multiple H-bonding networks between self-complementary building blocks [16], as illustrated by the first such structure, the Rebek 'tennis ball' **1** [11] (Figure 8.1a). Popular components for such self-assembly are glycoluril derivatives (as in **2**), calix[4]arene tetraurea derivatives [17], imide-substituted resorcin[4]arene cavitands [18] or resorcin[4]- or pyrogallo[4]arenes, with the latter assembling to giant hexameric capsules [19]. The second important principle for self-assembly leading to formation of containers and capsules relies on metal coordination, as illustrated for the octahedral cage **3** [20a,b] that is obtained by tetrahedral assembly of four 2,4,6-tris(4′-pyridyl)-1,3,5-triazenes (**4**) around four Pd(II) or Pt(II) centres (Figure 8.1b). Since the first reports on spherical assemblies by the groups of Saalfrank and coworkers [21], Fujita and coworkers [20], Stang and coworkers [22], Dalcanale and coworkers [23] and Raymond and coworkers [24], metal-ion coordination has been elegantly and vigorously exploited [25] to construct new cavities for single or multiple guest encapsulation or to create new reaction [26] and catalytic [27] sites.

The understanding of the principles governing the occupancy of capsular cavities has been greatly enhanced [28], leading to the 55% rule by Mecozzi and Rebek that states that inclusion complexes are favoured when a guest occupies 55 ± 9% of the space available within a host [29]. On the other hand, proper control over the process of host–guest complex formation and the kinetics of guest uptake and release [30] is less developed; very few studies were aimed directly at controlling the structures of capsules, containers and cavitands by stimulus-induced molecular switching.

Several mechanisms exist to control molecular motion and to switch on/off the guest-binding affinity of molecular and supramolecular container molecules. Supramolecular H-bonded capsules generally disassemble readily into the individual components upon changing from noncompetitive apolar to polar solvents that compete for H-bonding. Similarly, changes in temperature, competition with stronger chelating ligands for the metal ions or alteration of their redox states can lead to the dissociation of metal coordination capsules. Less-destructive methods to change reversibly the geometry of containers and cavitands as well as the affinity of the inner space for guest binding are of course very desirable. Molecular containers and cavitands with sidewalls and portals that can be reversibly opened and closed under various stimuli are indeed at the centre of this chapter.

8.2
Switchable Covalently Constructed Cavitands and Container Molecules

One of the most fascinating classes of dynamic receptors with switchable guest-hosting properties comprises the resorcin[4]arene cavitand family – vase-shaped molecules with walls formed by four aromatic moieties, – initially introduced by Cram and coworkers in the early 1980s [14, 31, 32]. These compounds, for example quinoxaline-bridged cavitand **5** and pyrazine-bridged cavitand **6** (Figure 8.2), are readily prepared in two steps by condensation of resorcinol and an appropriate

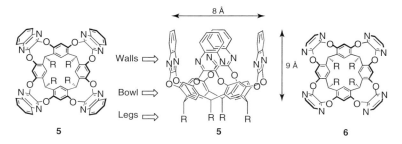

Figure 8.2 Structures of quinoxaline-bridged (**5**) and pyrazine-bridged (**6**) resorcin[4]arene cavitands and side view of the *vase* conformation of **5** showing approximate dimensions and domain names. R-groups may vary.

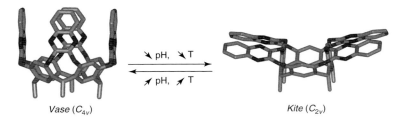

Vase (C_{4v})   Kite (C_{2v})

Figure 8.3 Molecular models of the *vase* and *kite* conformers of quinoxaline-bridged resorcin[4]arene cavitand **5**. The equilibrium is temperature and pH dependent. H-atoms and legs are omitted for clarity.

aldehyde to afford a so-called octol [33], followed by fourfold bridging of four pairs of hydroxyl groups by 2,3-dichloroquinoxaline or 2,3-dibromopyrazine.

A remarkable property of these systems is the reversible, temperature-dependent switching between a closed *vase* and an open *kite* conformations (Figure 8.3). The *vase* conformation (C_{4v} symmetry) of **5**, with a hollow, hydrophobic cavity of about 320 Å^3 volume and suitable for guest encapsulation [34], is prevalent in solution at room temperature and above, as well as in most solid-state structures. The *kite* conformer (C_{2v} symmetry), with a flat, extended surface about 19.3 Å × 15.6 Å in size (estimated using an available crystal structure) [35], is predominant at $T \leq 213$ K in nonpolar solvents, such as CDCl$_3$. Cavitand **6** also experiences the same equilibrium, although in a different temperature window [36]: the *vase* → *kite* transition occurs at much higher temperatures, compared with **5**, and the *vase* conformation becomes favourable in CDCl$_3$ only at temperatures ≥ 350 K.

Cavitands **5** and **6** belong to the family of switchable molecular receptors that display an intramolecular switching mechanism: the geometry is altered exclusively by conformational interconversion. The switching (Figure 8.3) resembles the movement of a mechanical gripper and changes the overall geometry of the cavitand molecule dramatically. *Vase* → *kite* switching not only destroys the inner cavity but also the wall flaps open up in opposite directions, and this offers the opportunity to induce large-scale, geometrically well-defined movements.

8.2.1
Characterization of *Vase* and *Kite* Conformations in the Solid State and in Solution

Both *vase* and *kite* conformations of bridged resorcin[4]arene cavitands are now structurally characterized in the solid state; their contrasting geometries are depicted in Figure 8.4. Several crystal structures of *vase* conformers (Figure 8.4a) of quinoxaline-bridged cavitands [31a, 34, 37, 38], such as **5a** [38] (R = C$_5$H$_{11}$) and pyrazine-bridged cavitand **6a** [36] (R = C$_6$H$_{13}$) have been reported. *Kite* conformers were first observed in the crystal structures of so-called velcrands [31b, 35, 39, 40], cavitands with four methyl substituents located in the *ortho*-positions to the O-atoms of the resorcinol moieties, which prevent for steric reasons the transition to the *vase* geometry. Recently, the crystal structure of octanitro derivative **7** [41]

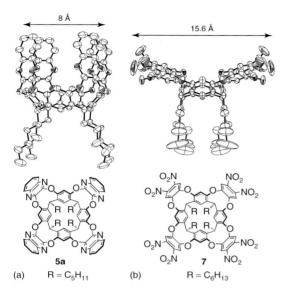

Figure 8.4 ORTEP representations of (a) quinoxaline-bridged cavitand **5a** in the closed *vase* conformation [38] and (b) octanitro-cavitand **7** in the extended *kite* conformation [38].

($R = C_6H_{13}$), which exists exclusively in the *kite* conformation in solution and in solid state, was solved [38] (Figure 8.4b). Velcrands tend to form *face*-to-*face* dimers (so-called 'velcraplexes') in solution and in the solid state [31b, 39, 40], whereas the crystal lattice of **7** shows infinite 'head-to-tail' columns with voids filled by Me_2CO molecules [38]. Cavitand **7** most probably prefers exclusively the *kite* conformation due to intramolecular dipolar repulsions of the eight nitro groups in the *vase* form. The avoidance of similar, intermolecular repulsions would also explain the absence of 'velcraplex-type' dimer formation.

Cram and coworkers [14, 31] showed that the *vase–kite* equilibration is conveniently monitored by ^1H NMR spectroscopy. In the *vase* form in $CDCl_3$, the methine proton (located on a bridging carbon atom between two resorcinarene rings) appears at $\delta \approx 5.5$ ppm, whereas upon switching to the *kite* form, this resonance shifts upfield to $\delta \approx 3.7$ ppm. Later, it was shown that the *vase* → *kite* interconversion can also be monitored by characteristic changes in the UV-Vis absorption spectrum [37].

For nearly two decades, temperature-induced switching was the only way to achieve isomerization between *vase* (prevalent at room temperature and above) and *kite* (below 213 K) forms. In recent years, other stimuli were introduced to induce this switching. Acidification of a solution of **5b** ($R = C_6H_{13}$) in $CDCl_3$ or CD_2Cl_2 with deuterated trifluoroacetic acid (TFA, CF_3COOD, 0.3–0.5 M) leads to complete conversion from the *vase* into the *kite* conformation at room temperature [37]. This protonation-induced switching is fully reversible and, upon addition of a base such as K_2CO_3 or a tertiary amine, the *vase* form is fully recovered. The driving force for *vase* → *kite* interconversion at low pH is, presumably, the protonation of the mildly basic quinoxaline N-atoms. This leads to repulsive *Coulombic* interactions in the *vase* form that vanish upon transition to the *kite* conformation. Weaker acids, such as acetic acid, do not induce *vase* → *kite* interconversion.

The *vase* → *kite* transition of quinoxaline-bridged resorcin[4]arene cavitands in $CDCl_3$ solution and at air/water interfaces can also be triggered by the addition of Zn(II) ions [42] (see Section 8.2.2 below). Other examples for metal-ion-induced conformational *vase* → *kite* switching were realized by decorating the upper rim of the cavitand walls with ligating groups such as phosphonates: upon addition of La(OTf)$_3$ (OTf = triflate), velcraplex-like dimers form, with the metal ions coordinating to the rim functionalities of the two cavitands involved. At the same time, guests previously bound in the cavity of the *vase* form, such as quinuclidinium cation, adamantane derivatives or the drug ibuprofen, are released. Addition of better ligands, such as nitrate, regenerates the *vase* form under restitution of the initial complexes ('caviplexes') [43].

Variable-temperature nuclear magnetic resonance (VT-NMR) studies of cavitand **5b** in acidic CD_2Cl_2 led to the discovery of a second, *kite 1-kite 2*, switching process [44] (Scheme 8.1) – an exchange proceeding between two degenerate *kite* conformations. This equilibrium is fast on the NMR timescale at room temperature, and the ^{1}H NMR spectrum of the protonated cavitand at 298 K is dynamically averaged showing an apparent fourfold symmetry of the *vase* conformer. Upon cooling to circa 250 K, the number of signals doubles and the

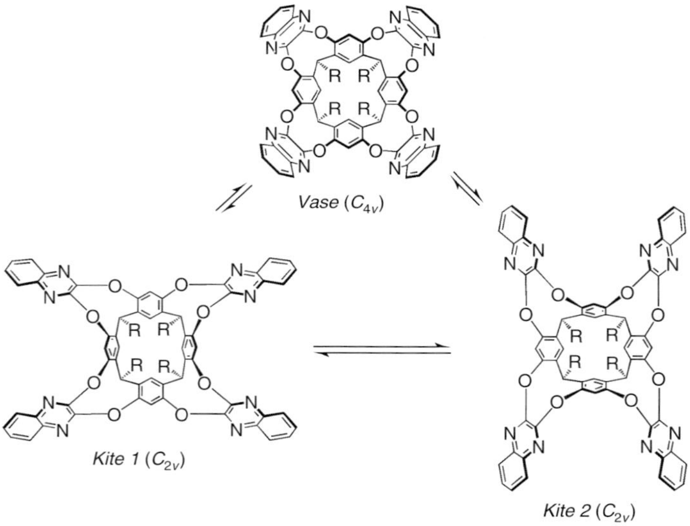

Vase (C_{4v})

Kite 1 (C_{2v})

Kite 2 (C_{2v})

Scheme 8.1 Interconversions between *vase*, *kite 1* and *kite 2* conformers of resorcin[4]arene cavitand **5b** (R = C_6H_{13}) with four quinoxaline flaps [44].

spectrum becomes consistent with the presence of two slowly interconverting degenerate *kite* conformers with C_{2v}-symmetry and similar to the spectrum of the *kite* conformation of unprotonated **5b** in CD_2Cl_2 at 193 K.

Already in the early 1980s, Cram and coworkers [14, 31] postulated that the temperature dependence of the *vase–kite* equilibrium is caused by solvation effects. At low temperatures, solvation (enthalpy of solvation ΔH_{solv}) of the larger, solvent-exposed surface favours the *kite* conformer, whereas at higher temperature, the entropic term $T\Delta S_{solv}$ for solvation of the larger *kite* surface becomes unfavourable and the *vase* conformation starts to predominate. Detailed VT-NMR studies of the quinoxaline-bridged cavitand **5b**, pyrazine-bridged cavitand **6a** (R = C_6H_{13}), and octanitro-cavitand **7** as well as a solvent scan of the conformational behaviour of **5b** and **6a** provided a comprehensive insight into the driving forces that govern *vase–kite* interconversion of the resorcin[4]arene cavitands [36, 44]. While the earlier hypothesis by Cram and coworkers were confirmed, additional influences of solvation were revealed.

From VT-NMR scans, the activation parameters for the *kite 1–kite 2* equilibration of **5b**, **6a** and **7**, as well as the kinetic and thermodynamic parameters for the *vase–kite* transition of **5b**, were determined (Table 8.1). The similarity of the activation parameters for the *kite 1* → *kite 2* switching of **7** to those determined for the *kite* → *vase* transition of **5b** suggested that both switching processes proceed through similar, presumably partial-*vase*-like transition states. The values of ΔH and ΔS for the *vase* → *kite* transition of **5b** evidenced that it is a ΔH_{solv}-driven process, whereas the entropic term favours the reverse process at elevated temperature. Still, this hypothesis needed to be further extended to explain the higher relative preference for the *kite* conformation of pyrazine-bridged cavitand **6a** with respect to quinoxaline-bridged **5b**, despite the smaller *kite* surface available for solvation.

The solvent dependence clearly showed that not only higher *kite* stabilization, but also weaker *vase* stabilization are important solvent-mediated forces driving the *vase* → *kite* transition [36]. The deeper cavity of **5b** is efficiently stabilized by the inclusion of nonpolar solvent molecules, such as $CHCl_3$, CH_2Cl_2 and especially aromatic solvents, leading to a predominance of the *vase* conformation already at room temperature. In comparison, the shallower cavity of **6** with the

Table 8.1 Thermodynamic and kinetic parameters for different switching modes of cavitands **5b**, **6a** and **7** (R = C_6H_{13}) [36, 44].

Cavitand	Conditions	Switching mode	ΔH (kcal·mol^{-1})	ΔS (cal·mol^{-1}·K^{-1})	$\Delta H^{\neq}$ (kcal·mol^{-1})	$\Delta S^{\neq}$ (cal·mol^{-1}·K^{-1})
5b	CD_2Cl_2	Vase → kite	-5.8 ± 0.5	-25.2 ± 1.5	3.8 ± 2	-33.2 ± 4
5b	CD_2Cl_2	Kite → vase	5.8 ± 0.5	25.2 ± 1.5	9.7 ± 2	-7.8 ± 4
6a	$CDCl_3$	Kite 1-kite 2	a	a	12.4 ± 1.5	0.1 ± 3
7	$CDCl_3$	Kite 1-kite 2	a	a	10.9 ± 1.5	-14.0 ± 3

[a]Degenerate equilibrium.

less-polarizable pyrazine walls benefits less from solvation and, therefore, the *vase* conformation is only observed at higher temperatures (in comparison with **5b**). The importance of cavity solvation is vividly illustrated by the destabilizing effect of 1,3,5-trimethylbenzene (mesitylene), in which both **5b** and **6** switch to the *vase* form only at $T > 350-370$ K. A bulky mesitylene molecule does not fit into the cavity of the *vase*-shaped receptors [45].

Quite unexpectedly, solute-solvent H-bonding interactions were also identified as forces stabilizing the *kite* conformation, thereby affecting the equilibrium. Solvents belonging to the chlorinated hydrocarbon series (CH_2Cl_2, $CHCl_3$, $CHCl_2CHCl_2$, $CHBr_3$), with substantial H-bonding acidity $\Sigma\alpha$ [46], can participate in H-bonding interactions with the slightly basic nitrogen atoms of the pyrazine- and quinoxaline bridged cavitands. In the open *kite* geometry, the cavitand nitrogen atoms are much more accessible for participation in such weak H-bonding interactions with solvent molecules than in the closed *vase* geometry, which explains why such interactions shift the equilibrium towards the *kite* conformation. Accordingly, the *kite* geometry is also more favourable for the pyrazine- than for the quinoxaline-bridged cavitands, due to the slightly higher basicity of the pyrazine nitrogen atoms.

Thus, *vase–kite* conformational switching is governed by the efficiency of the solvation of the *vase* cavity and the *kite* surface, as well as by weak H-bonding interactions with solvent in the *kite* form. It is induced by stimuli such as changes in temperature and pH or upon addition of metal ions (Zn(II)).

8.2.2
Cavitand Immobilization on Surfaces and Switching at Interfaces

The construction of practical molecular devices requires the immobilization of the cavitands on solid supports. Accordingly, various modifications of the cavitand legs (Figure 8.2) with functional groups for fixation on surfaces have been pursued. Functional cavitand legs not only serve for surface immobilization [47, 48], but also enhance the solubility in aqueous media [49] and enable formation of supramolecular coordination polymers [40].

Cavitands with dialkyl thioether legs were used for the formation of self-assembled monolayers (SAMs) on gold surfaces [48]. SAMs formed by **5c** on Au(111) were prepared by solution deposition and imaged by ultrahigh-vacuum scanning tunnelling microscopy (UHV-STM) at the molecular level, showing a well-ordered monolayer (Figure 8.5) with the individual, tightly packed cavitand molecules being fixed in the *vase* conformation. Cavitands with shorter legs afforded only poorly ordered SAMs. The tightly packed monolayers prevented any induction and observation of *vase–kite* switching.

The amphiphilic cavitand **5d**, with polar carboxylic ester legs [42], was prepared for *Langmuir* monolayer formation on an aqueous subphase (Figure 8.6). Since the molecular-area requirements estimated from X-ray crystal structures for the *vase* (120 Å^2) and *kite* (270 Å^2) forms significantly differ, it was possible to detect conformational switching in monolayers at the air/water interface by evaluating the pressure–area ($\Pi - A$) isotherms [42]. Upon lowering the pH from 7 to 1 by

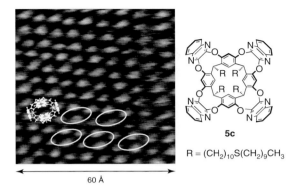

Figure 8.5 STM image of a SAM of cavitand **5c** adsorbed on Au(111) showing the structural model [48]. The ellipses outline individual molecules of **5c**. Reproduced from Ref. [48] with permission. Copyright RSC (2001).

addition of CF$_3$COOH to the aqueous subphase, the area per molecule indeed increased from 125 to 225 Å^2, suggesting formation of a monolayer of **5d** in the *kite* conformation. The fact that the theoretical value of 270 Å^2 for the *kite* form was not reached, can be explained by still incomplete protonation of the cavitand at pH 1. It was subsequently discovered that addition of Zn(OAc)$_2$ to the water subphase (Figure 8.6) also induced the *vase* → *kite* transition. Actually, in this experiment the molecular-area requirement changed as predicted from 125 Å^2 (*vase*) to 270 Å^2 (*kite*) (Figure 8.6). Zn(II) ion-induced *vase* → *kite* conformational switching was subsequently also observed by ^{1}H NMR spectroscopy in CDCl$_3$ by monitoring the chemical shift of the methine proton resonance as described above. *Job* plot analysis suggested that the switching results from specific complexation between

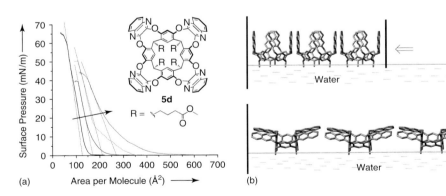

Figure 8.6 (a) Pressure–area (Π–A) isotherms of cavitand **5d** on a pure water (black line) subphase and in the presence of increasing amounts of Zn(OAc)$_2$ (grey lines) in the aqueous subphase [42]. The arrow denotes the increment of Zn(OAc)$_2$ concentration. (b) The molecular-area requirements of **5d** in *Langmuir* monolayers differ strongly in the *vase* and *kite* forms. Reproduced from Ref. [42] with permission. Copyright RSC (2004).

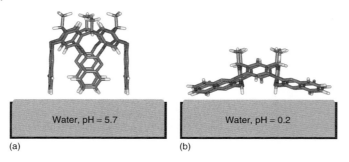

Figure 8.7 Illustration of the proposed orientation of quinoxaline-bridged cavitand **5e** with four $C_{11}H_{23}$ alkyl chain legs at the lower rim on (a) neutral and (b) acidic water subphases [51]. Alkyl chains are omitted for clarity.

cavitand **5d** in the *kite* form and Zn(II) ions, with a 1 : 2 binding stoichiometry. This implies coordination of each Zn(II) ion to two neighbouring quinoxaline N-atoms, which is geometrically only possible in the C_{2v}-symmetric *kite* conformation. Upon addition of larger amounts of methanol, decomplexation occurs and the *vase* form is recovered.

Conformational transitions of quinoxaline-bridged cavitand **5e**, with lipophilic *n*-decyl legs, in *Langmuir* films were also studied using surface second-harmonic generation (SHG) at different pH values (between 5.7 and 0.1) of the aqueous subphase [50]. It was shown that a monolayer of the quinoxaline-bridged cavitand on water switched from the *vase* to the *kite* conformation at sufficiently low (<0.6) pH. Further investigations using sum-frequency vibrational spectroscopy (SFVS) provided direct structural information about conformational and orientational preferences of **5e** in the film (Figure 8.7) [51]. The cavitands prefer the upside-down orientation [50b] on the water/air interface with hydrophobic alkyl chains pointing away from water, and undergo *vase* → *kite* switching in the monolayer upon reducing the pH of the water subphase from 5.7 to 0.2. In the *vase* form, the quinoxaline flaps orient almost perpendicularly to the water surface, whereas in the *kite* form, they lie nearly flat on this surface.

8.2.3
Synthetic Modifications of the Upper Rim

Upper-rim modification of bridged resorcin[4]arene cavitands provides straight access to a diversity of switchable molecular architectures featuring different functions, such as dynamic molecular receptors and sensors, redox-switchable cavitands and systems undergoing observable multinanometer-sized contraction/expansion motions. Modifications of the cavity walls are popular to tune the size and the inner properties of the cavitands. While most of this work involves the 'symmetric' replacement of all four bridging quinoxaline moieties by four new, identical wall components [18, 41, 43, 47, 49, 52], some 'asymmetric' systems with one [34b, 38, 47c, 53–55] or two [38, 55] flaps differing from the residual ones have also been reported.

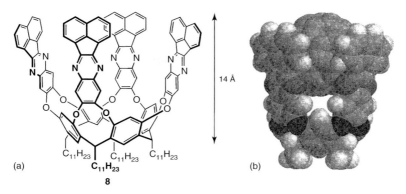

Figure 8.8 (a) Structure of the deep cavitand **8** showing approximate dimensions [41a]. (b) Side view of the energy-minimized CPK models of the $C_{60} \cdot \mathbf{8}$ complex. (Adapted from Ref. [41a] with permission. Copyright ACS (1999).)

Most of the modifications were aimed at stiffening the cavity-featuring *vase* form [18, 41b, 47, 49, 52] and, thus, eliminating conformational switchings. Here, we only discuss modifications that left the ability of cavitands to switch fully functional.

The symmetrically substituted deep cavitand **8** [41a] (Figure 8.8a) seems to prefer a *vase*-like structure in various solvents, although peak broadening indicated some conformational dynamics on the NMR timescale already at 295 K. When the samples were heated above 330 K, sharp spectra characteristic of a C_{4v} symmetric *vase* arose. Cavitand **8** has a higher preference for the *kite* conformation than quinoxaline-bridged **5** due to its extended flaps that provide a large surface for favourable solvation in the *kite* form. The cavity of **8** is complementary in size to fullerene C_{60} (for a molecular model, see Figure 8.8b) and 1:1 complexation of the carbon sphere was indeed detected by UV-Vis spectroscopy in toluene, yielding an association constant of $K_a = 900 \pm 250 \, \text{M}^{-1}$ at 293 K.

Self-folding cavitands **9** and **10** [52a], (Figure 8.9) displaying in the *kite* form a dimerization mode similar to velcrands [31b, 35, 39, 40], were prepared by Rebek and coworkers. D_{2d}-symmetric dimers of **9** and **10** ('velcraplexes'), composed of two *kite* conformers facially bound to each other, displayed remarkable solution stability with dimerization constants $K_{\text{dimer}} \geq 10^5 \, \text{M}^{-1}$ in *p*-xylene-d_{10}. The dimers reversibly dissociate into the C_{2v}-symmetric monomers and switch into the C_{4v}-symmetric vase with a deep cavity upon changing the solvent polarity, for example by adding $(CD_3)_2SO$ to the *p*-xylene-d_{10} solution of **10b**. The high kinetic and thermodynamic stability of the D_{2d}-symmetric dimers is due to intermolecular hydrogen bonds between the amide groups on the upper rim of the two associating cavitands. This stabilizing H-bonding network is disrupted upon addition of the competing solvent $(CD_3)_2SO$.

For the preparation of asymmetrically bridged cavitands, an efficient access to partially bridged cavitands such as **11–13** ($R = C_6H_{13}$) (Scheme 8.2), is of

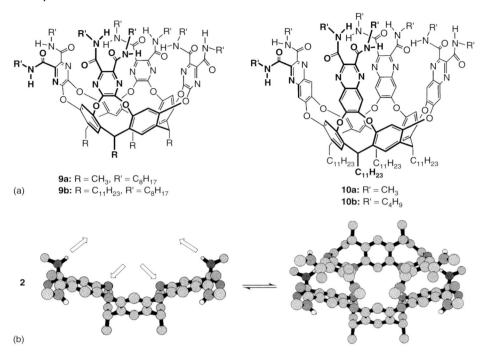

Figure 8.9 (a) Structures of the cavitands **9** and **10** [52a]. (b) Energy-minimized structures of monomeric and dimeric **9**. Alkyl chains are omitted for clarity. (Adapted from Ref. [52a] with permission. Copyright Wiley (2000).)

paramount importance. Cavitand **11** with three quinoxaline walls was first prepared in 55% yield by Dalcanale and coworkers [34b] by bridging octol **14** with 3 equiv. of 2,3-dichloroquinoxaline. Later, variation of the amounts of 2,3-dichloroquinoxaline, fine tuning of the reaction conditions and the use of advanced chromatographic separation methods allowed the preparation of *anti-* and *syn-*doubly bridged cavitands **12** and **13** with yields of 4 and 20%, respectively (Scheme 8.2) [38, 55]. Shortly after, a much more efficient (55% yield), selective synthesis of **12a** (with R = C_5H_{11}) by controlled scission of two quinoxaline bridges of **5a** by catechol was reported by Castro *et al.* [56]. Presumably, transfer of the quinoxaline moiety from two resorcinol HO-groups of **5a** to the two HO-groups of catechol is driven by strain release due to cleavage of two nine-membered diether rings and this strain release seems to be the largest if two cavity walls are removed in the *anti* position. Bridging of **11–13** with the substituted dichlorodiazaphthalimide derivatives **15** provided the upper-rim-modified cavitands **16–18** [38, 55].

The fully bridged cavitands **15–18** all undergo reversible temperature- and pH-triggered *vase–kite* conformational switching [44, 55], although in the thermal process they showed a slightly higher preference for the *kite* conformation than the parent cavitand **5b**. Remarkably, even the partially bridged resorcin[4]arenes **11–13**

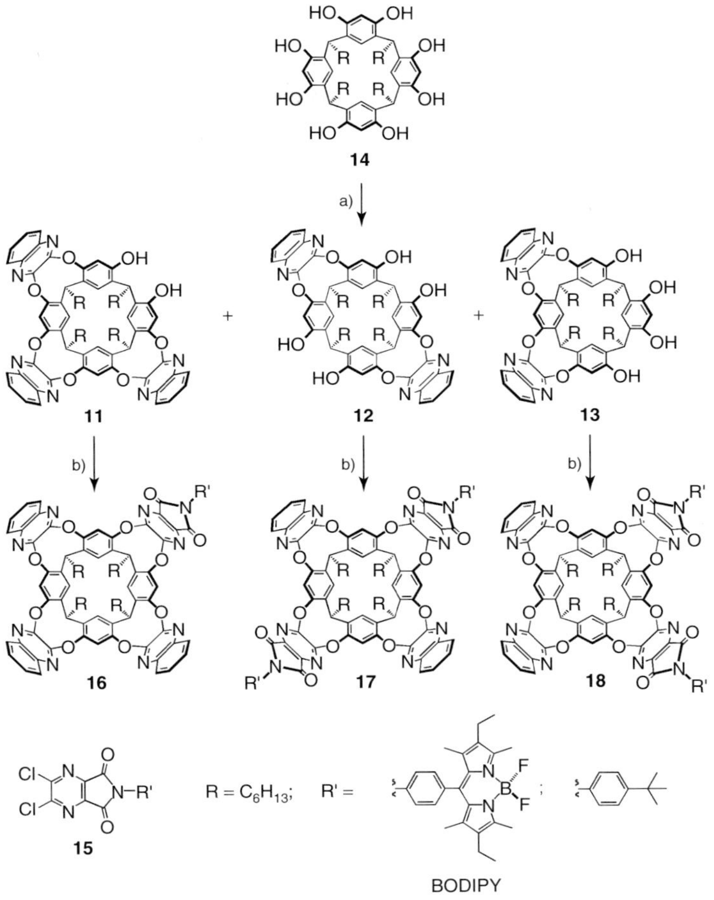

Scheme 8.2 Synthesis of partially and differentially bridged cavitands [38]. (a) 2,3-Dichloroquinoxaline, K$_2$CO$_3$, Me$_2$SO and (b) **15**, K$_2$CO$_3$, Me$_2$SO.

all undergo this switching. For diol **11**, low-temperature experiments revealed that the switching was not complete, that is both *kite* and *vase* conformers were present in a circa 1 : 1 ratio at 193 K (CD$_2$Cl$_2$). For tetrols **12** and **13**, low-temperature scans could not be performed since they started to precipitate from the solution below 273 K. On the other hand, complete (reversible) conversion of **11**–**13** from the *vase* to the *kite* form can be induced upon addition of CF$_3$COOD at room temperature.

Partially bridged cavitands, such as **11**–**13**, feature an interesting receptor site, characterized by a hydrophilic region comprising an array of phenolic HO groups for H-bonding, and hydrophobic domains shaped by the quinoxaline flaps. Thus, cleft-type receptors such as **12** should be capable of sandwiching lipophilic substrates between their two aromatic flaps (Figure 8.10a). Indeed, ^{1}H NMR

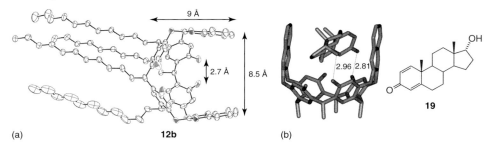

Figure 8.10 (a) Dimensions of the cleft-type binding site of receptor **12b** as revealed by X-ray crystal structure analysis [57]. (b) Results of the computer-assisted docking studies for the **12b·19** complex, showing distances between O-atoms involved in H-bonding.

binding titrations (298 K) with the receptor **12b** (in which the *n*-hexyl legs of **12** are replaced by *n*-dodecyl legs, Figure 8.10a) in CDCl$_3$ (298 K) revealed the formation of stable complexes (K_a values up to 10^3 M^{-1}) with selected steroids [57] (Figure 8.10b). In these complexes, the central steroidal ABCD-ring core is sandwiched between the two quinoxaline moieties, whereas functional groups at the A and D ring termini interact with the phenolic HO groups. This binding mode is supported by molecular modelling, as illustrated in Figure 8.10b for the 1 : 1 complex of steroid **19** ($K_a = 536$ M^{-1}). Steroid recognition is diastereoselective and the diastereoisomer of **19** with a β-OH group only binds with $K_a = 104$ M^{-1}, since this hydroxyl group is sterically unable to participate in H-bonding interactions.

8.2.4
Modular Construction of Extended Switches with Giant Expansion–Contraction Cycles

The interest in molecular-scale devices capable of controlled mechanical movements on the nanoscale and mimicking the corresponding natural phenomena such as muscle expansion/contraction has been growing steadily over the past decade [1]. The resorcinarene cavitand backbone is the ideal scaffold for the construction of molecular switches undergoing geometrically precisely defined, multinanometer-sized movements between two stable states. Extension of the wall flaps increases the distance between their termini in the *kite*, but not in the *vase* conformation, leading to molecular motion of increasing amplitude. Cavitands with extended, rigid oligo(phenylene ethynylene) arms attached to one (as in **16**) or two *N*-arylated diazaphthalimide cavity wall flaps in *anti* (as in **17**) and *syn* (as in **18**) positions can be prepared using the iodinated cavitand derivatives as key intermediates [58]. The functionalized cavity walls can be enlarged to the desirable size by Sonogashira cross-couplings [59] with rigid, terminally ethynylated or iodinated oligo(phenylene ethynylene) oligomers. The rigid attachment of the rod-like arms ensures that the separation of their termini (R^1 and R^2 in Figure 8.11)

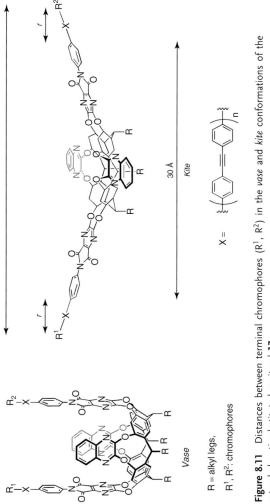

Figure 8.11 Distances between terminal chromophores (R^1, R^2) in the *vase* and *kite* conformations of the extended *anti*-substituted cavitand **17**.

in the extended *kite* conformation is well defined and conformational freedom is substantially restricted. Thus, it should be possible to switch the interplanar distance of terminal chromophores between circa 8 Å (*vase*) and circa (30 + 2r) Å (*kite*, with r the distance between the termini and the arylimide anchor).

The synthesis of the donor–acceptor boron dipyrromethene (4,4-difluoro-4-bora-3a,4a-diaza-sindacene, BODIPY) dye-labelled cavitand **20** for fluorescence resonance energy transfer (FRET) studies vividly illustrates the highly convergent, modular assembly of extended cavitands from readily accessible building blocks: partially bridged cavitand **12**, an *N*-arylated dichlorodiazaphthalimide as modified flap unit (such as **15**), oligo(phenylene ethynylene) spacers and two modified donor and acceptor BODIPY dyes [58a, b] (Figure 8.12). Computer modelling, based on X-ray crystal structures of bridged resorcin[4]arene cavitands in both *vase* and *kite* conformations [35, 38], suggested a ≈1000% difference in the distance between the dye pair in the contracted (≈7 Å) and expanded (≈7 nm, distance between the two B-atoms of the BODIPY dyes) states. Very recently, a family of the dye-labeled cavitands bearing the same BODIPY dyes with different lengths of oligo(phenylene-ethynylene) spacers was prepared using the same modular methodology [60].

The large-scale expansion/contraction movement of **20**, resulting from reversible *vase*–*kite* switching, induced by changes in temperature or pH, was clearly proven by both ^{1}H NMR and FRET measurements. Theoretical considerations [61] suggested that a large decrease in the intramolecular FRET efficiency would be measurable at distances >5 nm between donor and acceptor dyes. In accordance, experiments showed that in the expanded *kite* state, with a spatial separation of donor and acceptor of ≈7 nm, the FRET efficiency became dramatically reduced and the emission intensity of the acceptor was reduced to >5% of the value measured for the *vase* state. This result could only be obtained with a rigid structure such as **20**, with limited available conformational space.

The recently reported study using fluorophore-appended resorcin[4]arene cavitands provided evidence for a hitherto poorly recognized conformational flexibility of both *vase* and *kite* conformations [62], giving an explanation for the relatively low FRET efficiency observed for the cavitand **20** in the *vase* state [58a, b]. Unlike in the solid state, where cavitands usually show a perfect *vase* conformation with opposite walls approximately 8 Å apart, much longer distances between the wall tips and nonplanar orientation of the opposite walls can be expected in solution.

8.2.5
Electrochemically Triggered Switching

Resorcin[4]arene-based cavitands featuring two extended bridges consisting of quinoxaline-fused tetrathiafulvalene (TTF) moieties were constructed in an ambitious synthesis [63] to realize for the first time reversible *vase*–*kite* interconversion under the stimulus of electrochemical electron-transfer processes (Figure 8.13). In the neutral form, cavitand **21** was expected to adopt the *vase* form, whereas, upon oxidation of both TTF moieties to the mono- or dicationic states, the open

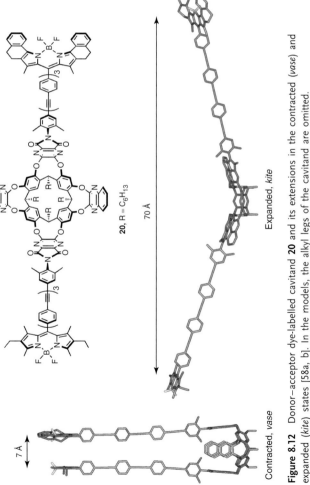

Figure 8.12 Donor–acceptor dye-labelled cavitand **20** and its extensions in the contracted (*vase*) and expanded (*kite*) states [58a, b]. In the models, the alkyl legs of the cavitand are omitted.

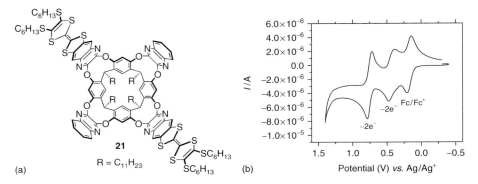

Figure 8.13 (a) Structure of the bis-TTF cavitand **21** (R = C$_{11}$H$_{23}$) [63]. (b) CV of **21** (0.5 mM) in CH$_2$Cl$_2$ (0.1 M Bu$_4$NPF$_6$) at 293 K; scan rate: 100 mV/s. Reproduced from Ref. [63] with permission. Copyright Swiss Chemical Society (2006).

kite geometry should be preferred due to Coulombic repulsion between the two cationic wall flaps in the *vase* form.

VT-NMR experiments confirmed that **21** undergoes temperature-triggered *vase–kite* conversion, similar to the parent quinoxaline-bridged cavitands **5**. Cyclic voltammetry (CV) studies disclosed that compound **21** expectedly undergoes two reversible 2e$^-$ oxidation steps, the first one leading to the bis(TTF radical cation) and the second one to the bis(TTF dication) (Figure 8.13). The first redox couple showed strong broadening of the oxidation/reduction peaks implying close proximity of the two TTF moieties leading to some degree of electronic coupling. The narrow second 2e$^-$ oxidation step suggested that upon further oxidation the two chromophores are at larger distance and no longer in electronic communication. These results present first evidence for an electrochemically induced *vase–kite* conformational switching of **21**. Future investigations will address how the redox properties affect the host–guest binding abilities of the bis-TTF cavitands to obtain further support for the postulated switching.

8.2.6
Switching Molecular Containers

Although a variety of closed-shell host molecules with fascinating inner-phase binding and reactivity properties have been prepared [9, 10, 12, 13], the construction of container molecules with switchable portals that allow controlled uptake and release of a guest molecule at ambient temperature has been barely addressed. Molecular baskets **22a,b** and tube **23** (Scheme 8.3) represent rare examples of molecular containers with apertures that can be opened or closed in a controlled way [64]. In their closed conformation in neutral solution, these compounds are highly selective hosts for the encapsulation of suitably sized cycloalkanes. However, addition of CF$_3$COOD converts these hosts into open-portal conformers

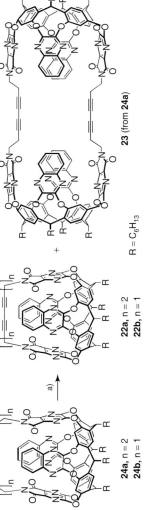

Scheme 8.3 Synthesis of container molecules **22a,b** and **23** (R = C$_6$H$_{13}$) [64]. (a) CuCl, CuCl$_2$, air, N,N-dimethylformamide, 20 °C, 16 h.

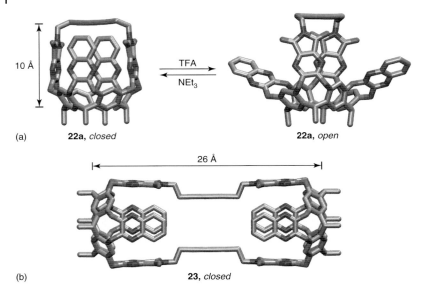

Figure 8.14 Molecular models (Spartan '04, AM1) of basket **22a** (a) and tube **23** (b) showing approximate dimensions for the closed forms of the two containers [64]. The hexyl chains on the lower rims as well as the hydrogen atoms are omitted for clarity.

with rapid release of encapsulated guest molecules, whereas reneutralization with base induces reuptake of the latter.

The two container molecules **22a** and **23** are formed in a 10 : 1 ratio by oxidative acetylenic coupling of **24a** (Scheme 8.3). The ratio is independent of the concentration of starting material. Isolable quantities of a dimerization product could not be obtained using precursor **24b**, probably because the resulting tube would have been too strained. Molecular modelling studies show that the closed form of container **22a** is a roughly spherical basket ($10 \times 8 \times 10$ Å), whereas **23** is a cylindrical tube ($26 \times 8 \times 10$ Å) (Figure 8.14). At room temperature in solvents such as $(CD_3)_2CO$, $CDCl_3$ or mesitylene-d_{12}, the container molecules are exclusively present in their closed forms, as indicated by NMR spectroscopy. Both **22a** and **23** undergo pH-dependent switching upon addition of CF_3COOD due to protonation of the quinoxaline N-atoms and subsequently induced Coulombic repulsion between the two ionized flaps. The ^{1}H NMR spectrum of tube **23** shows switching-induced upfield shifts of around 2 ppm for all resorcinarene methine proton resonances in two cavitand halves, indicative of complete opening to fully extended *kite* forms. In sharp contrast, the opening movement in basket **22a** is strongly reduced. The methine protons under the two flexible quinoxaline flaps shift upfield by circa 1 ppm (at $[CF_3COOD] = 8$ M), indicative of partial opening, whereas the methine protons below the bridged imide flaps hardly shift at all.

Molecular baskets **22a,b** have one binding site and tube **23** two identical ones. Experiments in $(CD_3)_2CO$ showed that **22** and **23** have high affinity to cycloalkanes; the host–guest exchange was slow on the ^1H NMR timescale. Interestingly, in the 1:1 complex of cyclohexane with tube **23**, the guest does not exchange between the two host sites on the NMR timescale. Binding to baskets **22** was strongly enhanced when mesitylene is used as a solvent. Whereas acetone molecules solvate the interior of the container, mesitylene solvent molecules are too large and do not compete for the binding site [36, 44, 45].

Binding of cyclohexane by **22a** is favoured in relation to other cycloalkanes since its molecular volume complies best with the 55% rule by Mecozzi and Rebek [29]: it occupies 56% of the available space, whereas cyclopentane (47%) and cycloheptane (65%) are not optimal for efficient binding. Cyclohexane binds in acetone with K_a around $6\,M^{-1}$, whereas the association constant in mesitylene amounts to $K_a \approx 4 \times 10^4\,M^{-1}$ (298 K). The slightly smaller molecular basket **22b** shows best selectivity for cyclopentane, whereas cycloheptane is not bound. Both molecular baskets displayed much higher binding efficiency with alicyclic ethers, such as oxacyclopentane and oxacyclohexane, due to additional polar stabilization, such as dipolar $C_2O \cdots C{=}O$ interaction between a guest and the imide moieties [64c].

Acid-induced conformational switching of **22** and **23** turns off their hosting ability. Complete release of cyclohexane occurred rapidly at $[CF_3COOD] = 1.4\,M$ for **22a** and at $[CF_3COOD] = 0.2\,M$ for **23**. The switching mechanism is fully reversible: upon neutralization of acidic solutions with appropriate amounts of NEt_3, guest binding was restored rapidly and completely for both **22a** and **23**. While the half-life for the cyclohexane complex of **22a** in mesitylene is in the order of hours, with a first-order rate decomplexation constant $k_{out} = 2.5 \times 10^{-3}\,s^{-1}$, switching enables nearly instantaneous decomplexation. As already mentioned above, capsule **22a** does not fully open to release the guest: the bridged imide flaps retain the same orientation as in a fully formed *vase* and only the two quinoxaline moieties bend out enabling guest escape. These findings of partially opened conformers are of relevance for future developments of switchable containers based on bridged resorcin[4]arene cavitands.

The 'hinged' molecular capsules **25a–c** [65] (Figure 8.15) are comprised of two unequal bridged resorcin[4]arene cavitand halves, covalently connected by a joint 1,4,5,8-tetraazaanthracene-2,3,6,7-tetrayl bridge. Variation in the length of the three alkyl bridges in one of the two cavitand halves from $-CH_2-$ (in **25a**) to $-(CH_2)_3-$ (in **25c**) reduces the depth of the capsule cavity and changes the overall cavity dimensions from circa $15 \times 8\,Å$ to circa $14 \times 8\,Å$, respectively. The second half in **25a–c**, the quinoxaline-bridged cavitand, undergoes reversible temperature- or acid-induced switching to an open conformer that strongly affects the entire capsule geometry. Investigations into how these changes affect guest binding properties are underway.

Switching the entrance to molecular containers can also be performed by means of gates or 'revolving doors' [66, 67] – bulky groups attached to the open end of the cavity that can close upon a bound guest or rotate away to open the cavity. By regulating the rate of opening and closing of these gates it may be possible to

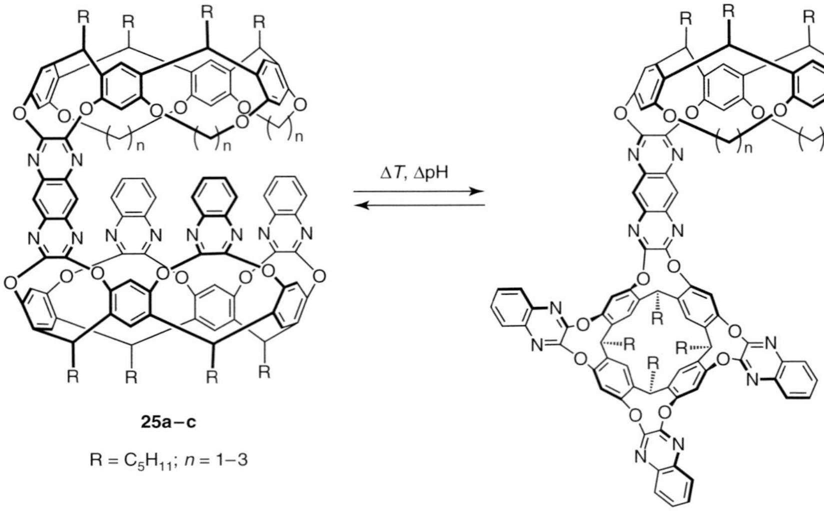

Figure 8.15 Conformational 'open-close' switching of the 'hinged' molecular capsules **25a–c** [65].

25a–c

R = C$_5$H$_{11}$; n = 1–3

influence the kinetics of the guest exchange. Trifold-gated molecular basket **26a** (Figure 8.16) with three phenolic gates controlling the entrance into the cavity is an example of such a system [67]. A network of H-bonds between the three phenolic HO groups on top of the molecular basket is holding the gates together, thereby closing the cavity in CDCl$_3$. This network, however, can be disrupted by the addition of base (Et$_3$N) leading to the opening of the cavity. Neutralization with CF$_3$COOH reinstalls the closed container. Mobility of the gates of the modified molecular basket **26b**, which has three metal coordination sites attached to its upper rim, could be controlled by addition of Ag(I) [68a,b] or Cu(I) ions [68c].

Dynamic ^{1}H NMR spectroscopic measurements of molecular basket **26c** with the self-folding upper rim showed that the thermodynamic stability of its CCl$_4$ complex in CD$_2$Cl$_2$ was −2.7 kcal/mol ($\Delta G°$, 298 K), whereas the activation free energy for CCl$_4$ departing the basket was found to be 13.1 kcal/mol($\Delta G^{\neq}$, 298 K) [69a]. A guest-exchange process, proceeding through the basket gates rotating about their axis, was considered to be the most probable one. Estimation of kinetic parameters for the translocation of *t*-BuBr (CD$_2$Cl$_2$, 226 K) in molecular basket **26c** gave values $k_{in} \approx 520$ M^{-1} s^{-1} and $k_{out} \approx 4.7$ s^{-1}. It was also possible to slow down the rate of exchange by replacement of three acetyl residues in **26c** with other groups, with benzoates being the most efficient [69b]. Additional kinetic studies showed that the rate of guest exchange also strongly depends on the guest structure and nature [69c].

Although the release of large guests encapsulated in covalent molecular containers is usually very slow [12b, 70], incorporation of photoswitches provides control over this process. Photoswitchable covalent capsules were reported by Deshayes and coworkers [71], who introduced 3-nitro-*o*-xylylene moieties to bridge

8.2 Switchable Covalently Constructed Cavitands and Container Molecules | 281

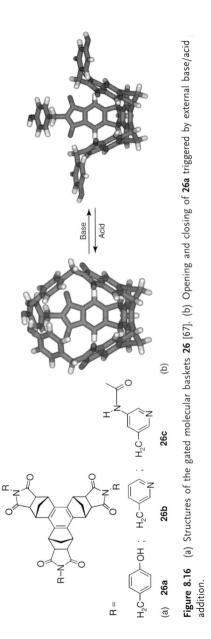

Figure 8.16 (a) Structures of the gated molecular baskets **26** [67]. (b) Opening and closing of **26a** triggered by external base/acid addition.

phenolic HO groups of the two resorcin[4]arene cavitands forming the closed container (a 'hemicarcerand'). Upon irradiation, the benzylic ether bond between cavitand and nitroxylylene moiety is cleaved, leading to the release of encapsulated guests such as N-methylpyrrolidine or N,N-dimethylacetamide. The rate of this irreversible photocleavage process can be controlled by modifying either the absorption properties of the host or the intensity of the irradiation.

8.2.7
Cucurbit[n]urils

The cucurbit[n]uril (CB[n]) family comprises several macrocyclic, barrel-shaped molecular receptors with a cavity 9.1 Å deep and 4.4–8.8 Å ($n = 5$–8) wide (Figure 8.17) [72]. With two open apertures (similar to many cyclophane hosts and the cyclodextrins), uptake and release of complimentarily sized guests into the cavity usually is a rather easy process. The two apertures are lined each with a row of convergent urea carbonyl groups, and binding properties are often governed by H-bonding between the guest and these strongly H-bond accepting C=O groups. Protonated ammonium ions are particularly well bound by CB[n] receptors since strong ionic H-bonds are formed with the C=O groups at the cavity entrances.

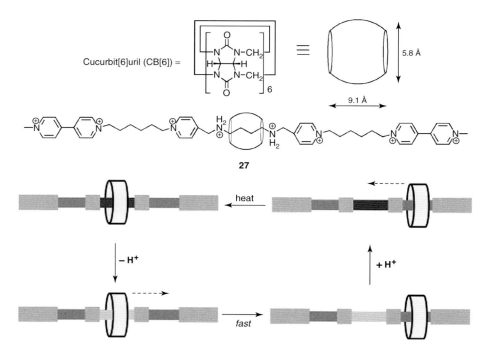

Figure 8.17 Switching cycle of the bistable cucurbit[6]uril-based [2]pseudorotaxane **27** [73c]. (Adapted from Ref. [73c].)

Complexation is further assisted by apolar (hydrophobic binding and desolvation) effects as well as by ion–dipole interactions with charged centre.

The structure of cucurbit[n]urils favours construction of switching molecular devices with the rotaxane architecture. Several reports describe the synthesis of molecular shuttles, where a cucurbituril wheel is shuffled along the pseudorotaxane axle upon protonation/deprotonation of functional groups, such as secondary ammonium ions, located along it [73]. The bistable [2]pseudorotaxane **27** [73c] represents a remarkable example of a molecular device behaving as a kinetically controlled molecular switch. In neutral D$_2$O, the CB[6] wheel preferentially binds to the central 1,4-diammoniumbutane moiety of the axle. Upon addition of Hünig's base, causing deprotonation, the wheel shifts onto the hexyl chain, bridging the terminal bipyridinium and a pyridinium ring. In this position, CB[6] undergoes ion–dipole interactions and efficient C=O$\cdots$H–C interactions. However, simply adding back DCl does not induce immediate migration back to the centre. Rather, additional heating is necessary to overcome the high activation barrier ($\Delta G^{\ddagger} = 26$ kcal mol^{-1}) on the way back onto the 1,4-diammoniumbutane moiety.

Cucurbiturils were also used for designs of other types of switchable supramolecular systems. Nau and coworkers [74] have developed various cucurbituril-based fluorescent switches, in which fluorescence of a dye undergoes a large and predictable change upon complexation of its anchor with a cucurbit[n]uril macrocycle. Functionality of a complex three-way supramolecular switch developed by Kim and coworkers [75] is based on the ability of the large CB[8] to encapsulate two identical or different guest molecules, such as electron-poor methyl viologen (A) or electron-rich TTF derivatives (A). Chemically or electrochemically controllable three-way interconversion between hetero-guest-pair (D–A) and homo-guest-pair (D$_2$ or A$_2$) inclusion complexes of CB[8] is a key feature of this switching system.

8.3
H-Bonded Molecular Capsules

Hydrogen bonding is one of the favourite weak forces used for supramolecular construction owing to its specificity, directionality and high relevance to biological processes. As an example, supramolecular assembly in crystals ('crystal engineering') relies heavily on the repetitive association of specifically designed H-bonding 'synthons' [76]. Rebek and coworkers [11] were the first to show that well-designed self-complementary building blocks, capable of forming multiple H-bonds, also provide access to self-assembled molecular containers and capsules possessing different geometries, sizes, stabilities and symmetries [16–19]. The kinetic stability of such capsules depends on the number of the formed H-bonds and their pattern, and half-lives can range from milliseconds to hours. All H-bonded supramolecular capsules are extremely susceptible to solvent polarity, and solvents competing effectively for H-bonds usually lead to rapid disassembly and release of encapsulated guests.

8.3.1
Glycoluril-Derived H-Bonded Capsules

A significant number of H-bonded supramolecular capsules is assembled from 'C-shaped' glycoluril building blocks. The first one, the so-called 'tennis ball' (**1**, Figure 8.1) has a cavity capable of hosting guests with volumes of about 50 Å^3, such as methane, ethane and noble gases. By expansion of the spacer size between the two glycoluril units, while retaining shape and terminal H-bonding sites to allow for dimerization, self-assembling dimers with cavity volumes up to 320 Å^3 were obtained. They are capable of recognizing and binding larger guest molecules [77]. The spacers usually contain additional H-bonding sites that contribute to the stability of the capsule as shown in Figure 8.18 for the so-called 'softball' **28**, made from two moieties of **29**.

The mechanism of guest exchange is dependent on the internal flexibility of the dimeric host. Whereas for small capsules, derived from two rigid monomeric units, the complete dissociation of the two complementary halves is considered to be necessary [78], guest uptake and release by the larger and more flexible containers can proceed by the opening of one or two flaps [79, 80]. In the case of the small 'tennis ball', held together by eight H-bonds, complete dissociation and guest exchange takes place on the timescale of seconds [78]. Complete dissociation of the larger 'softball' **28** into two halves **29** would require rupture of all 16 H-bonds, which is a much more energetically unfavourable process. The guest exchange proceeds in a two-step sequence (Figure 8.18) through opening of one of the flaps [79]. First, the initial guest is exchanged for 'smaller' solvent molecules and subsequently, complexation of the second guest occurs under desolvation. As a consequence, guest exchange is faster than dissociation [81]. Indeed, the rate of the exchange process depends on the concentration of incoming guests. Exchange proceeds with half-lives of several minutes at millimolar guest concentrations [79] (3–7 mM, taken in $\geq$10-fold excess to host concentration).

8.3.2
Calix[4]arene and Resorcin[4]arene Capsules

With their bowl-like shapes, calix[4]arenes and resorcin[4]arenes are appropriate, widely used subunits for the construction of self-assembled molecular capsules. The concave face of these bowl-shaped molecules represents one half of a closed molecular container, and a variety of functional groups can mediate dimerization. Calix[4]arenes **30** substituted at their wide rim by urea groups (Figure 8.19) are self-complementary components and form dimers **31** in noncompetitive solvents, which are held together by a directional seam of 16 H-bonds around the equator of a capsule [17], provided a suitable guest (often a solvent molecule) is present. A variety of aromatic, aliphatic and cationic guests can be hosted within the twisted, bipyramidal cavity with an approximate volume of 180 Å^3.

The thermodynamic stability of dimers **31** in apolar solvents is so high that equilibration between monomer and dimer is virtually nonexistent. Measured

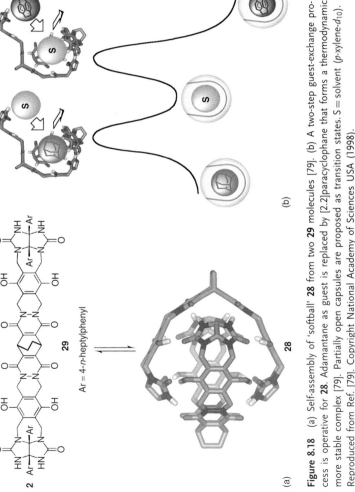

Figure 8.18 (a) Self-assembly of 'softball' **28** from two **29** molecules [79]. (b) A two-step guest-exchange process is operative for **28**. Adamantane as guest is replaced by [2.2]paracyclophane that forms a thermodynamically more stable complex [79]. Partially open capsules are proposed as transition states. S = solvent (p-xylene-d_{10}). Reproduced from Ref. [79]. Copyright National Academy of Sciences USA (1998).

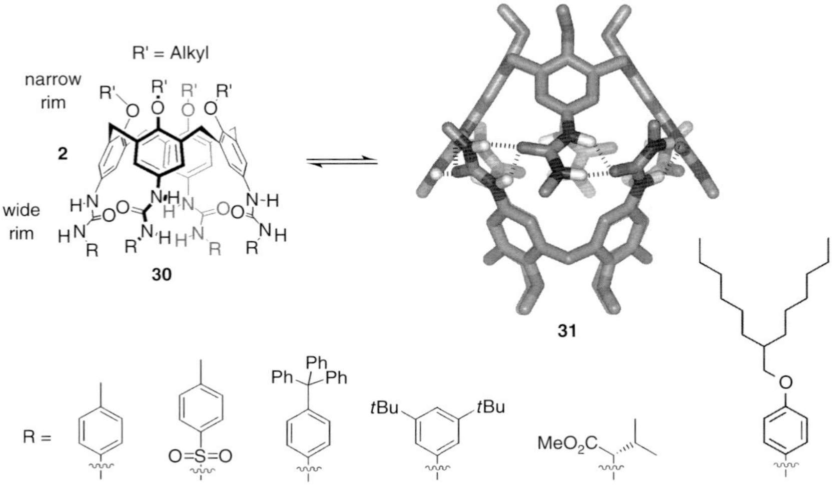

Figure 8.19 Self-assembly of the calix[4]arene tetraurea capsules **31** from monomers **30** [17].

dissociation constants and guest exchange rates depend strongly on the nature of the solvent and guest, the calixarene substitution pattern and often on the experimental method used. First nuclear Overhauser effect spectroscopy nuclear magnetic resonance (NOESY-NMR) studies of calixarene capsules with decreased symmetry in benzene solution (c = 3–5 mM) led to values of 0.26 s^{-1} for the overall rate constant of the capsule dissociation/recombination process and 0.47 s^{-1} for the rate constant of guest (benzene) in/out exchange [82]. FRET studies performed on differently substituted tetraurea calix[4]arenes labelled at the narrow rim by appropriate coumarine dyes, gave values in the range of 10^6–6×10^8 M^{-1} for the thermodynamic self-association constants and of 5×10^{-3}–5×10^{-6} s^{-1} for the dissociation rate constant [83]; measurements were performed under high dilution conditions (50–500 nM). Substitution of the narrow rim of the calixarene is mostly responsible for such a dramatic difference in rates: bulkier or H-bonding substituents inhibit the calixarene monomer from forming a 'pinched cone' conformation as it dissociates, due to steric hindrance at the lower rim [83]. This assumption was proven by a systematic study using capsules **31** with different substituents on the lower rims of calixarene hemispheres [84]. Half-lifetimes ($t_{1/2}$) for guest exchange (C$_6$H$_6$) against a solvent molecule (C$_6$D$_{12}$) varied from $t_{1/2} \approx 1.5$ h (R' = Me) to $t_{1/2} \approx 2700$ h (R' = Bn) with the tetra-OH derivative lying in between them with $t_{1/2} \approx 180$ h. Further kinetic studies showed that the rate of guest release/exchange depends not only on the structure of the calixarene, but also on the guest and the solvent [85]: half-lifetimes for guest exchange varied between $t_{1/2} \approx 3$ h (CHCl$_3$) and $t_{1/2} \approx 2000$ h (C$_6$H$_{12}$) for the exchange of the indicated guest against a solvent molecule (C$_6$D$_{12}$) in capsule **31** (R' = C$_{10}$H$_{21}$).

Resorcin[4]arene cavitand **32** with four imide functions around its upper rim self-assembles in most apolar solvents into the cylindrical dimer **33** [18, 28], held

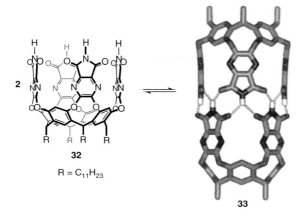

Figure 8.20 Self-assembly of the tetra-diazaphthalimide-bridged cavitand **32** into the cylindrical capsule **33** [18].

together by eight bifurcated hydrogen bonds (Figure 8.20). Its inner space [18e] of circa 425 Å^3 is usually filled by solvent, but when mesitylene is used, the solvent molecules are too large to be accommodated by the capsule and do not compete with guest binding. The lifetimes of complexes formed by **33** can vary from milliseconds to days depending on solvent and guest nature [18], making guest exchange slow on the NMR timescale and allowing observation of separate resonances for free and encapsulated guests.

^{1}H NMR magnetization transfer studies performed in mesitylene revealed that the exchange of small guests, such as benzene, proceeds within seconds ($k_{obs} = 0.3–1.4\,s^{-1}$ for $[C_6H_6] = 16–160\,mM$) and without full dissociation of the capsule [86]. Since the cavity accommodates two small guest molecules best (for example one benzene and one *p*-xylene molecule), the exchange occurs presumably through opening of two imide flaps of one cavitand half (Figure 8.21), followed by a displacement of one guest molecule by another one. Similar to the findings for container **22** (Section 8.2.6), full opening of two flaps (as suggested in Figure 8.21) might actually not be required for guest release and exchange. The exchange of larger guests [86], such as replacing 4,4′-dimethylbiphenyl with 4,4′-dimethylstilbene, proceeds much slower. The most probable mechanism is analogous to that for capsule **28** (Figure 8.18), with an intermediate capsule filled by solvent molecules.

Capsules such as **33**, with the two cavitand halves labelled with pyrene and perylene as donor and acceptor fluorophores, respectively, were used for FRET dissociation studies at nanomolar (500 nM) concentrations [87]. The rate constant of $1.9 \times 10^{-3}\,s^{-1}$ for the exchange of the capsule subunits was determine in toluene, corresponding to a half-life ($t_{1/2}$) of 6 min, whereas in mesitylene the exchange rate constant is reduced to $6.0 \times 10^{-6}\,s^{-1}$ ($t_{1/2} = 32\,h$). The 3 orders of magnitude faster exchange in toluene can presumably be attributed to the ability of toluene to solvate the inner cavity of the capsule, filling the necessary space as it

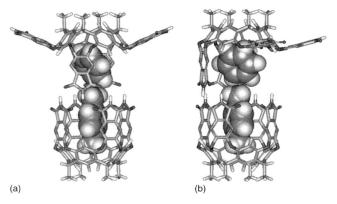

(a) (b)

Figure 8.21 Proposed intermediate structures of **33** involved in benzene guest exchange with (a) two opposite walls opened and (b) two adjacent walls opened [86]. Recent studies of guest exchange in basket **22** (Figure 8.14) suggest that the flap portals may not have to open as widely as indicated here [64].

dissociates and recombines. Guests influence the exchange process significantly: the thermodynamically favoured guests slow down capsule exchange rates (for example $t_{1/2} > 100$ h for undecane), whereas the thermodynamically unfavoured ones increase them by 3–4 orders of magnitude (measurements were performed with 1000 equiv. of a guest). Addition of protic solvents, such as methanol, increases the rate of exchange and finally leads to destruction of the capsules [87, 88]. In the presence of small amounts of protic solvents, guest exchange proceeds by complete dissociation of the capsules [89].

Although capsules **33** are photochemically inert, an elegant method was developed to induce photochemically controlled reversible encapsulation. The switching principle is based on the light-induced *trans* → *cis* isomerization of 4,4′-dimethylazobenzene, the *trans* isomer of which is tightly encapsulated in **33**, whereas the *cis* isomer is not, thus affording control over encapsulation of a second guest [89]. Heating of a solution induces reverse *cis* → *trans* isomerization of 4,4′-dimethylazobenzene, as a consequence restoring the starting state. The functionality of the system was effectively demonstrated by a supramolecular on/off switch, in which fluorescence of a *trans*-stilbene derivative was reversibly commutated by external light/heat stimuli [88].

Self-assembly of resorcin[4]arenes **34** and of the closely related pyrogallol[4]arenes leads to the largest of the H-bonded supramolecular capsules [19] such as **35** with a cavity volume of about 1400 Å^3 (Figure 8.22). The hexameric resorcinarene capsule **35** is held together by a network of 60 hydrogen bonds, including eight water molecules (therefore, wet solvents are required for their self-assembly), whereas pyrogallolarene capsules do not require water for their formation and stabilization. The very large cavity permits encapsulation of several small or even large guests, for example eight benzene molecules, three biphenyl molecules and

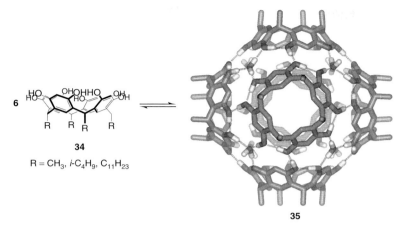

Figure 8.22 Self-assembly of the hexameric resorcin[4]arene capsule **35** [19].

even bulky tetraalkylammonium salts. However, what is gained in size and in synthetic simplicity is lost in the stability of the capsules. The rates of guest exchange are usually high. Rate constants for release of the guest, measured for a series of tetraalkylammonium salts, were found to be in the range of $0.04–20\,s^{-1}$ [90] and $0.36\,s^{-1}$ for 1,2-*cis*-cyclohexanediol [91]. Exchange rates are dependent on guest size: larger guests require more energy for their release from the capsule [90].

Although several possible mechanisms of guest exchange can be envisaged, due to the presence of multiple components in the assembly, it is extremely difficult to experimentally elucidate the dominant one. The most probable mechanism would include dissociation of one resorcinarene unit to give a pentameric intermediate with one open portal. This process should be energetically not too costly, since a minimum number of H-bonds is broken, which would reflect the moderate activation free enthalpy $\Delta G^{\ddagger}$ of circa 16 kcal/mol measured for the release of tetra(*n*-butyl)ammonium bromide [90].

Hexameric capsules incorporating structural subunits labelled with pyrene and perylene as donor and acceptor fluorophores, respectively, were used for FRET assembly and exchange studies [92]. Experiments performed at nanomolar (500 nM) concentrations in various wet solvents allowed estimation of the rates for the exchange of resorcin[4]arene monomers. The exchange was relatively fast with half-lives spanning from 3 min (C_6D_6), to 10 min (CD_2Cl_2) and to 46 min ($CDCl_3$). Addition of an appropriate guest provided additional stabilization for the capsules, increasing half-lives in CD_2Cl_2 up to 16 min for tetra(*n*-butyl)antimony bromide. Furthermore, it was shown that hexameric resorcin[4]arene assemblies could be fully destroyed by addition of 1.5% *v*/*v* of methanol – a solvent known to disrupt the H-bonds.

Several other types of self-assembling H-bonding capsules based on calixarene or resorcinarene building blocks have been reported [93–95]. Information related to kinetics of the assembly of these dynamic structures and the mechanisms of guest exchange, however, is scarce.

8.3.3
Multicomponent Self-Assembled Molecular Containers

Analysis of the mechanism for guest exchange and estimation of kinetic constants for assembly/disassembly becomes even more difficult if capsules are composed of several different components and if two or more isomeric assemblies can form.

Multicomponent rosette-shaped supramolecular containers composed upon self-assembly between components containing melamine moieties and barbituric or cyanuric acid groups as H-bonding counterparts are among the most fascinating ones [96, 97]. Kinetic stabilities for several hexarosette assemblies of different size were measured using both chiral amplification and racemization experiments [97b]. It was shown that solvent temperature and polarity have tremendous effects on the kinetic stability. In addition, stability is strongly dependent on the number of H-bonds holding the structures together: half-lives for double (36 H-bonds), tetra- (72 H-bonds) and hexarosette (108 H-bonds) assemblies were found to be 8.4 min, 5.5 h and 150 h in chloroform at 50 °C, respectively.

8.4
Assembly and Disassembly of Metal-Ion-Coordination Cages

The assembly of multitopic ligands around metal ions has provided access to a great diversity of molecular containers of various size and shape. Whereas H-bonds prefer rectilinear arrangement of components across the seam connecting curved subunits of the molecular containers, metal-ion joints are usually the sources of curvature themselves, interconnecting linear or flat polydentate ligands into three-dimensional structures [25c]. Due to the relatively high strength of metal-to-ligand bonds, metal coordination cages are generally quite stable species and can self-assemble also in protic solvents such as methanol and water. Still, they can usually be decomposed by the addition of strong nucleophiles and are susceptible to elevated temperatures.

Resorcin[4]arenes functionalized at the wider upper rim with four nitrile groups, such as **36**, form molecular capsules, such as **37**, upon mixing in a 1:2 ratio with suitable Pd(II) or Pt(II) complexes [23, 98] (Figure 8.23). The resulting supramolecular cage, bound by a seam of four coordinated Pd(II) or Pt(II) ions, bears a +8 charge, and its cavity holds one of the eight counterions ($CF_3SO_3^-$ in **37**). Studies of the kinetic stability of capsule **37a** by ^{1}H NOESY and ^{19}F NOESY experiments provided formation/dissociation rate constant $k_{obs} \approx 0.30\,s^{-1}$ in $CDCl_3$ and $k_{obs} \approx 5.2\,s^{-1}$ in $CD_3NO_2/CDCl_3$ (7:1) mixture [98b]. The kinetic stability of the capsules is higher if the assembly involves square-planar Pt(II) rather than Pd(II) coordination. Thus, switching guest binding and other properties of the capsule can be fine tuned: use

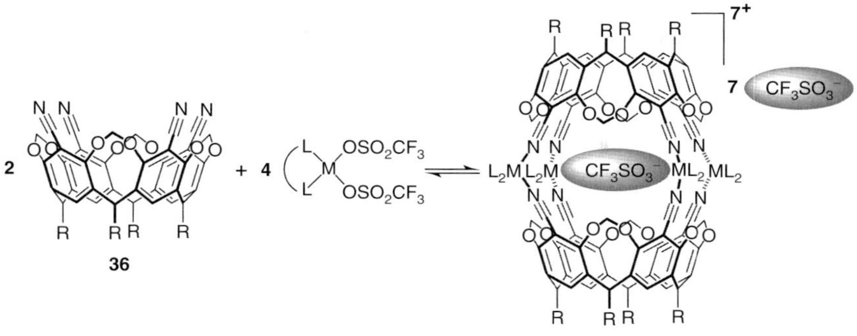

Figure 8.23 Self-assembly of the resorcin[4]arene-based metal-coordination cage **37** [23, 98]. dppp = 1,2-bis(diphenylphosphino)propane.

37a: R=$C_{11}H_{23}$, M=Pd, L_2=dppp
37b: R=C_6H_{13}, M=Pt, L_2=dppp
37c: R=$C_{11}H_{23}$, M=Pt, L_2=dppp

of Pt(II) ions makes them kinetically stable within the timeframe of the NMR experiment (no guest exchange is observed), whereas reversible assembly/disassembly and equilibration with oligomeric species occurs with Pd(II) ions. Additionally, mild increase of the solvent polarity substantially decreases the kinetic stability of the coordination capsules. Presumably, polar solvents provide better stabilization of a polar dissociative transition state, generated during the exchange process.

Harrison and coworkers [99] devised a similar family of molecular containers comprising two resorcin[4]arene hemispheres modified each with four iminodiacetate groups which coordinate to a seam of Co(II), Cu(II) or Fe(II) ions. The capsules were reported to be stable in water and a variety of organic solvents, but could be destroyed by pH manipulation: acidification leads to disassembly, whereas neutralization reinstitutes the capsules [99a].

The host–guest chemistry of the tetrahedral cages **38** [M_4L_6] (M = Fe(III), Al(III), Ga(III), In(III), Ti(IV) or Ge(IV), composed of four coordinating N,N'-bis(2,3-dihydroxybenzoyl)-1,5-diaminonaphthalene (**39**) ligands, was extensively studied by the group of Raymond [24, 25a, b, 26c, d, e, 27]. These supramolecular containers have an interior confined by six bis(catechol) ligands spanning along the edges between four metal atoms at the corners of a tetrahedron and possess a total charge of −12 (Figure 8.24). The hosts have high propensity for encapsulation of cationic species (such as tetraalkylammonium ions) inside their relatively hydrophobic cavities. The highly charged host–guest assemblies are quite soluble in polar solvents, such as H_2O, MeOH, Me_2SO and N,N-dimethylformamide (DMF).

Extensive studies using NMR spin saturation transfer measurements were performed with **38** to determine the mechanism of guest exchange [100]. Two possible exchange mechanisms were considered: *via* aperture expansion and *via* rupture of a metal–ligand bond. Experimental evidence, that is (i) similar rates for

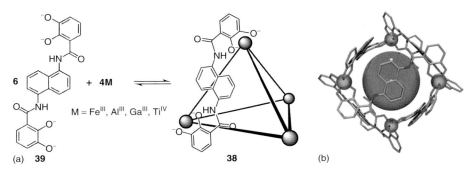

Figure 8.24 (a) The tetrahedral M_4L_6 molecular container **38** is assembled from six bis-catecholate ligands **39** and four metal cations [100]. (b) A spherical guest (such as tetraalkylammonium cation) encapsulated in the tetrahedral cage **38**. (Adapted from Ref. [30].)

guest exchange observed for labile ($[Ga_4L_6]^{12-}$) and inert ($[Ti_4L_6]^{8-}$ and $[Ge_4L_6]^{8-}$) capsules and (ii) rate sensitivity to size and conformation of a guest, as well as molecular modelling studies suggested a strong case for the nondissociative guest-exchange mechanism involving aperture expansion. Nevertheless, the rates of guest exchange are moderate and comparable with those for processes proceeding *via* the dissociation/association mechanism. The rate constants for the self-exchange process lie in the range of $0.003-4.4\,s^{-1}$ for a series of tetraalkylammonium/phosphonium derivatives. Only the exchange of very large guests is strongly retarded and may require partial ligand dissociation. In a recent study with D-substituted molecular guests, inverse kinetic isotope effects of up to 11% were observed in the guest-exchange process from ($[Ga_4L_6]^{12-}$) coordination capsule [101]. Although the faster exchange rates for D-substituted guests can be simply rationalized in terms of the smaller steric requirements of deuterium, the more comprehensive explanation presumes the coupling of the host aperture expansion with the guest C-H/D vibrational modes at the sterically strained transition state.

The group of Fujita *et al.* [20, 25e,f, 26a, b] use heterocyclic ligands in combination with *cis*-enforced square-planar Pd(II) and Pt(II) complexes for the construction of highly symmetric positively charged supramolecular capsules (for an example of such a capsule, see **3** in Figure 8.1b). The capsules are soluble in water, and their hydrophobic cavity aptly binds a variety of organic guest molecules, preferentially aromatic compounds and organic anions. Examples are *o*- and *m*-terphenyls, tetrabenzylsilane, trialkoxysilanes and adamantane carboxylate. Similar to **37**, Pt(II)-coordinated **3** is kinetically very stable and tolerant to acids (pH < 1), bases (pH > 11) or nucleophiles (such as NEt_3) due to the inertness of a Pt(II)–pyridine coordinative bond, whereas its Pd(II) counterpart immediately decomposes upon addition of acid or base [20c]. Unfortunately, no detailed studies related to kinetics of guest exchange were performed with these remarkable molecular host systems.

Thus, in the case of molecular containers held together by strong metal–ligand bonds and often possessing relatively large apertures between the walls units, switching in a sense of reversible partial or full dissociation is often difficult to achieve and to study. Guest uptake/release processes are more likely to proceed *via* skeletal deformation and aperture expansion.

8.5 Conclusions

Molecular containers, capsules and cavitands are remarkable architectures capable of holding other molecules in closed cavities inside their own chests. As was shown in this chapter, the properties and functions of some of them can, to some extent, be controlled by external stimuli. However, only relatively few detailed studies were specifically aiming for an improved understanding of switching properties such as reversible guest exchange and, in the case of supramolecular capsules, reversible assembly and disassembly. A more general knowledge about such processes, in particular, in closed-shell hosts, is still under development.

The future research directions are likely to include the following: (i) extension of the scope and accuracy of stimuli (such as temperature and pH changes, light or redox processes) used for switching capsular shape and guest-hosting capacity, (ii) tuning the selectivity of dynamically (and 'dynamically combinatorially') assembled molecular vessels, (iii) construction of devices capable of large-scale molecular movement within supramolecular assemblies on the way to switchable sensors and molecular machines and (iv) development of functional systems for supramolecular catalysis. The unique inner phase of container molecules has indeed been shown to greatly facilitate and accelerate stoichiometric uni- and bimolecular transformations. For such processes to become catalytic and show turnover, the containers must possess addressable built-in portals for product release. Other practical devices in reach through this research are controllable nanoreactors, molecular motors and smart drug-delivery vessels. Molecular-gripper-type systems could be envisioned for the construction of smart cantilever tips for atomic force microscopy (AFM) and other manipulations on the nanoscopic scale. All these possible technological advances, however, rely on a detailed understanding of the basic physical-organic principles governing the involved switching processes.

Acknowledgements

This work was supported by the Swiss National Science Foundation and the NCCR 'Nanoscale Science', Basel. The authors are grateful to Prof. J. Rebek, Jr, Prof. E. Dalcanale, Prof. L. M. Gutierrez-Tunstad and Prof. J. D. Badjić for the structures of their receptors.

References

1. For reviews on molecular switches and molecular machines see: (a) Kay, E.R., Leigh, D.A. and Zerbetto, F. (2007) *Angew. Chem. Int. Ed.*, **119**, 72–196; (b) Kay, E.R., Leigh, D.A. and Zerbetto, F. (2007) *Angew. Chem. Int. Ed.*, **46**, 72–191; (c) Kay, E.R., Leigh, D.A. and Zerbetto, F. (2001) *Acc. Chem. Res.*, **34**, 410–522 (special issue on molecular machines); (d) Balzani, V., Credi, A., Raymo, F.M. and Stoddart, J.F. (2000) *Angew. Chem. Int. Ed.*, **112**, 3484–3530; (e) Balzani, V., Credi, A., Raymo, F.M. and Stoddart, J.F. (2000) *Angew. Chem. Int. Ed.*, **39**, 3348–3391; (f) Balzani, V., Venturi, M. and Credi, A. (2003) *Molecular Devices and Machines: A Journey into the Nano World*, Wiley-VCH Verlag GmbH, Weinheim.
2. Dugave, C. and Demange, L. (2003) *Chem. Rev.*, **103**, 2475–2532.
3. (a) Shinkai, S., Nakaji, T., Ogawa, T., Shigematsu, K. and Manabe, O. (1981) *J. Am. Chem. Soc.*, **103**, 111–115; (b) Muraoka, T., Kinbara, K., Kobayashi, Y. and Aida, T. (2003) *J. Am. Chem. Soc.*, **125**, 5612–5613; (c) Norikane, Y. and Tamaoki, N. (2004) *Org. Lett.*, **6**, 2595–2598.
4. (a) Kelly, T.R., De Silva, H. and Silva, R.A. (1999) *Nature*, **401**, 150–152; (b) Kelly, T.R., Silva, R.A., De Silva, H., Jasmin, S. and Zhao, Y. (2000) *J. Am. Chem. Soc.*, **122**, 6935–6949; (c) Koumura, N., Zijlstra, R.W.J., van Delden, R.A., Harada, N. and Feringa, B.L. (1999) *Nature*, **401**, 152–155; (d) Koumura, N., Geertsema, E.M., Meetsma, A. and Feringa, B.L. (2000) *J. Am. Chem. Soc.*, **122**, 12005–12006; (e) ter Wiel, M.K.J., van Delden, R.A., Meetsma, A. and Feringa, B.L. (2003) *J. Am. Chem. Soc.*, **125**, 15076–15086; (f) Fujita, T., Kuwahara, S. and Harada, N. (2005) *Eur. J. Org. Chem.*, 4533–4543; (g) Fletcher, S.P., Dumur, F., Pollard, M.M. and Feringa, B.L. (2005) *Science*, **310**, 80–82.
5. (a) Jiménez, M.C., Dietrich-Buchecker, C. and Sauvage, J.-P. (2000) *Angew. Chem.*, **112**, 3422–3425; (b) Jiménez, M.C., Dietrich-Buchecker, C. and Sauvage, J.-P. (2000) *Angew. Chem. Int. Ed.*, **39**, 3284–3287; (c) Jimenez-Molero, M.C., Dietrich-Buchecker, C. and Sauvage, J.-P. (2003) *Chem. Commun.*, 1613–1616; (d) Tseng, H.-R., Vignon, S.A. and Stoddart, J.F. (2003) *Angew. Chem. Int. Ed.*, **115**, 1529–1533; (e) Tseng, H.-R., Vignon, S.A. and Stoddart, J.F. (2003) *Angew. Chem. Int. Ed.*, **42**, 1491–1495; (f) Bottari, G., Leigh, D.A. and Pérez, E.M. (2003) *J. Am. Chem. Soc.*, **125**, 13360–13361; (g) Kang, S., Vignon, S.A., Tseng, H.-R. and Stoddart, J.F. (2004) *Chem. Eur. J.*, **10**, 2555–2564; (h) Liu, Y., Flood, A.H., Bonvallet, P.A., Vignon, S.A., Northrop, B.H., Tseng, H.-R., Jeppesen, J.O., Huang, T.J., Brough, B., Baller, M., Magonov, S., Solares, S.D., Goddard, W.A., Ho, C.-M. and Stoddart, J.F. (2005) *J. Am. Chem. Soc.*, **127**, 9745–9759.
6. (a) Amabilino, D.B., Dietrich-Buchecker, C.O., Livoreil, A., Pérez-García, L., Sauvage, J.-P. and Stoddart, J.F. (1996) *J. Am. Chem. Soc.*, **118**, 3905–3913; (b) Asakawa, M., Ashton, P.R., Balzani, V., Credi, A., Hamers, C., Mattersteig, G., Montalti, M., Shipway, A.N., Spencer, N., Stoddart, J.F., Tolley, M.S., Venturi, M., White, A.J.P. and Williams, D.J. (1998) *Angew. Chem. Int. Ed.*, **110**, 357–361; (c) Asakawa, M., Ashton, P.R., Balzani, V., Credi, A., Hamers, C., Mattersteig, G., Montalti, M., Shipway, A.N., Spencer, N., Stoddart, J.F., Tolley, M.S., Venturi, M., White, A.J.P. and Williams, D.J. (1998) *Angew. Chem. Int. Ed.*, **37**, 333–337; (d) Korybut-Daszkiewicz, B., Więckowska, A., Bilewicz, R., Domagała, S. and WoYniak, K. (2004) *Angew. Chem. Ind. Ed.*, **116**, 1700–1704; (e) Korybut-Daszkiewicz, B., Więckowska, A., Bilewicz, R., Domagała, S. and WoYniak, K. (2004) *Angew. Chem. Int. Ed.*, **43**, 1668–1672; (f) Mobian, P., Kern, J.-M. and Sauvage, J.-P. (2004) *Angew. Chem. Ind. Ed.*, **116**, 2446–2449; (g) Mobian,

P., Kern, J.-M. and Sauvage, J.-P. (2004) *Angew. Chem. Int. Ed.*, **43**, 2392–2395.
7. (a) Berl, V., Huc, I., Khoury, R.G., Krische, M.J. and Lehn, J.-M. (2000) *Nature*, **407**, 720–723; (b) Berl, V., Huc, I., Khoury, R.G. and Lehn, J.-M. (2001) *Chem. Eur. J.*, **7**, 2798–2809; (c) Berl, V., Huc, I., Khoury, R.G. and Lehn, J.-M. (2001) *Chem. Eur. J.*, **7**, 2810–2820; (d) Barboiu, M. and Lehn, J.-M. (2002) *Proc. Natl. Acad. Sci. USA*, **99**, 5201–5206; (e) Dolain, C., Maurizot, V. and Huc, I. (2003) *Angew. Chem. Ind. Ed.*, **115**, 2844–2846; (f) Dolain, C., Maurizot, V. and Huc, I. (2003) *Angew. Chem. Int. Ed.*, **42**, 2738–2740.
8. Cram, D.J. (1983) *Science*, **219**, 1177–1183.
9. (a) Gabard, J. and Collet, A. (1981) *J. Chem. Soc., Chem. Commun.*, 1137–1139; (b) Canceill, J., Cesario, M., Collet, A., Guilhem, J. and Pascard, C. (1985) *J. Chem. Soc., Chem. Commun.*, 361–363; (c) Canceill, J., Lacombe, L. and Collet, A. (1985) *J. Am. Chem. Soc.*, **107**, 6993–6996.
10. (a) Cram, D.J., Karbach, S., Kim, Y.H., Baczynskyj, L. and Kalleymeyn, G.W. (1985) *J. Am. Chem. Soc.*, **107**, 2575–2576; (b) Cram, D.J., Karbach, S., Kim, Y.H., Baczynskyj, L., Marti, K., Sampson, R.M. and Kalleymeyn, G.W. (1988) *J. Am. Chem. Soc.*, **110**, 2554–2560.
11. (a) Wyler, R., de Mendoza, J. and Rebek, J. Jr (1993) *Angew. Chem. Ind. Ed. Engl.*, **105**, 1820–1821; (b) Wyler, R., de Mendoza, J. and Rebek, J. Jr (1993) *Angew. Chem. Int. Ed.*, **32**, 1699–1701; (c) Branda, N., Wyler, R. and Rebek, J. Jr (1994) *Science*, **263**, 1267–1268.
12. (a) Sherman, J.C. (1995) *Tetrahedron*, **51**, 3395–3422; (b) Jasat, A. and Sherman, J.C. (1999) *Chem. Rev.*, **99**, 931–967; (c) Sherman, J. (2003) *Chem. Commun.*, 1617–1623.
13. (a) Cram, D.J., Tanner, M.E. and Thomas, R. (1991) *Angew. Chem. Ind. Ed. Engl.*, **103**, 1048–1051; (b) Cram, D.J., Tanner, M.E. and Thomas, R. (1991) *Angew. Chem. Int. Ed.*, **30**, 1024–1027; (c) Warmuth, R. (1997) *Angew. Chem. Ind. Ed. Engl.*, **109**, 1406–1409; (d) Warmuth, R. (1997) *Angew. Chem. Int. Ed.*, **36**, 1347–1350; (e) Warmuth, R. (2001) *Eur. J. Org. Chem.*, 423–437.
14. Moran, J.R., Karbach, S. and Cram, D.J. (1982) *J. Am. Chem. Soc.*, **104**, 5826–5828.
15. Freeman, W.A., Mock, W.L. and Shih, N.-Y. (1981) *J. Am. Chem. Soc.*, **103**, 7367–7368.
16. (a) Conn, M.M. and Rebek, J. Jr (1997) *Chem. Rev.*, **97**, 1647–1668; (b) Rebek, J. Jr (1999) *Acc. Chem. Res.*, **32**, 278–286; (c) Böhmer, V. and Vysotsky, M.O. (2001) *Aust. J. Chem.*, **54**, 671–677.
17. (a) Shimizu, K.D. and Rebek, J. Jr (1995) *Proc. Natl. Acad. Sci. USA*, **92**, 12403–12407; (b) Rebek, J. Jr (2000) *Chem. Commun.*, 637–643; (c) Bogdan, A., Rudzevich, Y., Vysotsky, M.O. and Böhmer, V. (2006) *Chem. Commun.*, 2941–2952.
18. (a) Heinz, T., Rudkevich, D.M. and Rebek, J. Jr (1998) *Nature*, **394**, 764–766; (b) Heinz, T., Rudkevich, D.M. and Rebek, J. Jr (1999) *Angew. Chem. Ind. Ed.*, **111**, 1206–1209; (c) Heinz, T., Rudkevich, D.M. and Rebek, J. Jr (1999) *Angew. Chem. Int. Ed.*, **38**, 1136–1139; (d) Körner, S.K., Tucci, F.C., Rudkevich, D.M., Heinz, T. and Rebek, J. Jr (2000) *Chem. Eur. J.*, **6**, 187–195; (e) Ajami, D., Iwasawa, T. and Rebek, J. Jr (2006) *Proc. Natl. Acad. Sci. USA*, **103**, 8934–8936.
19. (a) MacGillivray, L.R. and Atwood, J.L. (1997) *Nature*, **389**, 469–472; (b) Yamanaka, M., Shivanyuk, A. and Rebek, J. Jr (2004) *J. Am. Chem. Soc.*, **126**, 2939–2943; (c) Avram, L. and Cohen, Y. (2004) *J. Am. Chem. Soc.*, **126**, 11556–11563; (d) Rissanen, K. (2005) *Angew. Chem. Ind. Ed.*, **117**, 3718–3720; (e) Rissanen, K. (2005) *Angew. Chem. Int. Ed.*, **44**, 3652–3654; (f) Evan-Salem, T., Baruch, I., Avram, L., Cohen, Y., Palmer, L.C. and Rebek, J. Jr (2006) *Proc. Natl. Acad. Sci. USA*, **103**, 12296–12300.
20. (a) Fujita, M., Oguro, D., Miyazava, M., Oka, H., Yamaguchi, K. and Ogura, K. (1995) *Nature*, **378**, 469–471;

(b) Ibukuro, F., Kusukawa, T. and Fujita, M. (1998) *J. Am. Chem. Soc.*, **120**, 8561–8562; (c) Fujita, M., Nagao, S. and Ogura, K. (1995) *J. Am. Chem. Soc.*, **117**, 1649–1650.

21. (a) Saalfrank, R.W., Stark, A., Peters, K. and von Schnering, H.G. (1988) *Angew. Chem. Ind. Ed. Engl.*, **100**, 878–880; (b) Saalfrank, R.W., Stark, A., Peters, K. and von Schnering, H.G. (1988) *Angew. Chem. Int. Ed.*, **27**, 851–853; (c) Saalfrank, R.W., Stark, A., Bremer, M. and Hummel, H.-U. (1990) *Angew. Chem. Ind. Ed. Engl.*, **102**, 292–295; (d) Saalfrank, R.W., Stark, A., Bremer, M. and Hummel, H.-U. (1990) *Angew. Chem. Int. Ed.*, **29**, 311–314; (e) Saalfrank, R.W., Burak, R., Breit, A., Stalke, D., Herbst-Irmer, R., Daub, J., Porsch, M., Bill, E., Müther, M. and Trautwein, A.X. (1994) *Angew. Chem. Ind. Ed. Engl.*, **106**, 1697–1699; (f) Saalfrank, R.W., Burak, R., Breit, A., Stalke, D., Herbst-Irmer, R., Daub, J., Porsch, M., Bill, E., Müther, M. and Trautwein, A.X. (1994) *Angew. Chem. Int. Ed.*, **33**, 1621–1623.

22. (a) Stang, P.J., Olenyuk, B., Muddiman, D.C. and Smith, R.D. (1997) *Organometallics*, **16**, 3094–3096; (b) Stang, P.J. and Olenyuk, B. (1997) *Acc. Chem. Res.*, **30**, 502–518.

23. (a) Jacopozzi, P. and Dalcanale, E. (1997) *Angew. Chem. Ind. Ed. Engl.*, **109**, 665–667; (b) Jacopozzi, P. and Dalcanale, E. (1997) *Angew. Chem. Int. Ed.*, **36**, 613–615.

24. (a) Beissel, T., Powers, R.E. and Raymond, K.N. (1996) *Angew. Chem. Ind. Ed. Engl.*, **108**, 1166–1168; (b) Beissel, T., Powers, R.E. and Raymond, K.N. (1996) *Angew. Chem. Int. Ed. Engl.*, **35**, 1084–1086; (c) Caulder, D.L., Powers, R.E., Parac, T.N. and Raymond, K.N. (1998) *Angew. Chem. Ind. Ed. Engl.*, **110**, 1940–1943; (d) Caulder, D.L., Powers, R.E., Parac, T.N. and Raymond, K.N. (1998) *Angew. Chem. Int. Ed.*, **37**, 1840–1843; (e) Parac, T.N., Caulder, D.L. and Raymond, K.N. (1998) *J. Am. Chem. Soc.*, **120**, 8003–8004.

25. (a) Caulder, D.L. and Raymond, K.N. (1999) *Acc. Chem. Res.*, **32**, 975–982; (b) Fiedler, D., Leung, D.H., Bergman, R.G. and Raymond, K.N. (2005) *Acc. Chem. Res.*, **38**, 351–360; (c) Leininger, S., Olenyuk, B. and Stang, P.J. (2000) *Chem. Rev.*, **100**, 853–908; (d) Seidel, S.R. and Stang, P.J. (2002) *Acc. Chem. Res.*, **35**, 972–983; (e) Fujita, M., Umemoto, K., Yoshizawa, M., Fujita, N., Kusukawa, T. and Biradha, K. (2001) *Chem. Commun.*, 509–518; (f) Fujita, M., Tominaga, M., Hori, A. and Therrien, B. (2005) *Acc. Chem. Res.*, **38**, 371–380.

26. (a) Yoshizawa, M., Takeyama, Y., Kusukawa, T. and Fujita, M. (2002) *Angew. Chem. Ind. Ed.*, **114**, 1403–1405; (b) Yoshizawa, M., Takeyama, Y., Kusukawa, T. and Fujita, M. (2002) *Angew. Chem. Int. Ed.*, **41**, 1347–1349; (c) Fiedler, D., Bergman, R.G. and Raymond, K.N. (2004) *Angew. Chem. Ind. Ed.*, **116**, 6916–6919; (d) Fiedler, D., Bergman, R.G. and Raymond, K.N. (2004) *Angew. Chem. Int. Ed.*, **43**, 6748–6751; (e) Leung, D.H., Bergman, R.G. and Raymond, K.N. (2006) *J. Am. Chem. Soc.*, **128**, 9781–9797.

27. (a) Leung, D.H., Fiedler, D., Bergman, R.G. and Raymond, K.N. (2004) *Angew. Chem. Ind. Ed.*, **116**, 981–984; (b) Leung, D.H., Fiedler, D., Bergman, R.G. and Raymond, K.N. (2004) *Angew. Chem. Int. Ed.*, **43**, 963–966; (c) Fiedler, D., van Halbeek, H., Bergman, R.G. and Raymond, K.N. (2006) *J. Am. Chem. Soc.*, **128**, 10240–10252.

28. (a) Hof, F., Craig, S.L., Nuckolls, C. and Rebek, J. Jr (2002) *Angew. Chem. Ind. Ed.*, **114**, 1556–1578; (b) Hof, F., Craig, S.L., Nuckolls, C. and Rebek, J. Jr (2002) *Angew. Chem. Int. Ed.*, **41**, 1488–1508; (c) Rebek, J. Jr (2005) *Angew. Chem. Ind. Ed.*, **117**, 2104–2115; (d) Rebek, J. Jr (2005) *Angew. Chem. Int. Ed.*, **44**, 2068–2078; (e) Hof, F. and Rebek, J. Jr (2002) *Proc. Natl. Acad. Sci. USA*, **99**, 4775–4777.

29. Mecozzi, S. and Rebek, J. Jr (1998) *Chem. Eur. J.*, **4**, 1016–1022.

30. Pluth, M.D. and Raymond, K.N. (2007) *Chem. Soc. Rev.*, **36**, 161–171.

31. (a) Moran, J.R., Ericson, J.L., Dalcanale, E., Bryant, J.A., Knobler, C.B. and Cram, D.J. (1991) *J. Am. Chem. Soc.*, **113**, 5707–5714; (b) Cram, D.J., Choi, H.-J., Bryant, J.A. and Knobler, C.B. (1992) *J. Am. Chem. Soc.*, **114**, 7748–7765; (c) Cram, D.J. and Cram, J.M. (1994) *Container Molecules and Their Guests*, Royal Society of Chemistry, Cambridge, pp. 107–130.
32. Azov, V.A., Beeby, A., Cacciarini, M., Cheetham, A.G., Diederich, F., Frei, M., Gimzewski, J.K., Gramlich, V., Hecht, B., Jaun, B., Latychevskaia, T., Lieb, A., Lill, Y., Marotti, F., Schlegel, A., Schlittler, R.R., Skinner, P.J., Seiler, P. and Yamakoshi, Y. (2006) *Adv. Funct. Mater.*, **16**, 147–156.
33. (a) Högberg, A.G.S. (1980) *J. Org. Chem.*, **45**, 4498–4500; (b) Högberg, A.G.S. (1980) *J. Am. Chem. Soc.*, **102**, 6046–6050; (c) Tunstad, L.M., Tucker, J.A., Dalcanale, E., Weiser, J., Bryant, J.A., Sherman, J.C., Helgeson, R.C., Knobler, C.B. and Cram, D.J. (1989) *J. Org. Chem.*, **54**, 1305–1312.
34. (a) Dalcanale, E., Soncini, P., Bacchilega, G. and Ugozzoli, F. (1989) *J. Chem. Soc., Chem. Commun.*, 500–502; (b) Soncini, P., Bonsignore, S., Dalcanale, E. and Ugozzoli, F. (1992) *J. Org. Chem.*, **57**, 4608–4612.
35. Ihm, H., Ahn, J.-S., Lah, M.S., Ko, Y.H. and Paek, K. (2004) *Org. Lett.*, **6**, 3893–3896.
36. Roncucci, P., Pirondini, L., Paderni, G., Massera, C., Dalcanale, E., Azov, V.A. and Diederich, F. (2006) *Chem. Eur. J.*, **12**, 4775–4784.
37. Skinner, P.J., Cheetham, A.G., Beeby, A., Gramlich, V. and Diederich, F. (2001) *Helv. Chim. Acta*, **84**, 2146–2153.
38. Azov, V.A., Skinner, P.J., Yamakoshi, Y., Seiler, P., Gramlich, V. and Diederich, F. (2003) *Helv. Chim. Acta*, **86**, 3648–3670.
39. Bryant, J.A., Knobler, C.B. and Cram, D.J. (1990) *J. Am. Chem. Soc.*, **112**, 1254–1255.
40. (a) Pirondini, L., Stendardo, A.G., Geremia, S., Campagnolo, M., Samorì, P., Rabe, J.P., Fokkens, R. and Dalcanale, E. (2003) *Angew. Chem. Ind. Ed.*, **115**, 1422–1425; (b) Pirondini, L., Stendardo, A.G., Geremia, S., Campagnolo, M., Samorì, P., Rabe, J.P., Fokkens, R. and Dalcanale, E. (2003) *Angew. Chem. Int. Ed.*, **42**, 1384–1387.
41. (a) Tucci, F.C., Rudkevich, D.M. and Rebek, J. Jr (1999) *J. Org. Chem.*, **64**, 4555–4559; (b) Rudkevich, D.M., Hilmersson, G. and Rebek, J. Jr (1998) *J. Am. Chem. Soc.*, **120**, 12216–12225.
42. Frei, M., Marotti, F. and Diederich, F. (2004) *Chem. Commun.*, 1362–1363.
43. (a) Amrhein, P., Wash, P.L., Shivanyuk, A. and Rebek, J. Jr (2002) *Org. Lett.*, **4**, 319–321; (b) Amrhein, P., Shivanyuk, A., Johnson, D.W. and Rebek, J. Jr (2002) *J. Am. Chem. Soc.*, **124**, 10349–10358.
44. Azov, V.A., Jaun, B. and Diederich, F. (2004) *Helv. Chim. Acta*, **87**, 449–462.
45. Chapman, K.T. and Still, W.C. (1989) *J. Am. Chem. Soc.*, **111**, 3075–3077.
46. (a) Abraham, M.H. (1993) *Chem. Soc. Rev.*, **22**, 73–83; (b) Martin, S.D., Poole, C.F. and Abraham, M.H. (1998) *J. Chromatogr., A*, **805**, 217–235; (c) Ishihama, Y. and Asakawa, N. (1999) *J. Pharm. Sci.*, **88**, 1305–1312.
47. (a) Far, A.R., Rudkevich, D.M., Haino, T. and Rebek, J. Jr (2000) *Org. Lett.*, **2**, 3465–3468; (b) Saito, S. and Rebek, J. Jr (2001) *Bioorg. Med. Chem. Lett.*, **11**, 1497–1499; (c) Far, A.R., Cho, Y.L., Rang, A., Rudkevich, D.M. and Rebek, J. Jr (2002) *Tetrahedron*, **58**, 741–755.
48. Yamakoshi, Y., Schlittler, R.R., Gimzewski, J.K. and Diederich, F. (2001) *J. Mater. Chem.*, **11**, 2895–2897.
49. Haino, T., Rudkevich, D.M., Shivanyuk, A., Rissanen, K. and Rebek, J. Jr (2000) *Chem. Eur. J.*, **6**, 3797–3805.
50. (a) Lagugné-Labarthet, F., An, Y.Q., Yu, T., Shen, Y.R., Dalcanale, E. and Shenoy, D.K. (2005) *Langmuir*, **21**, 7066–7070; (b) Lagugné-Labarthet, F., Yu, T., Barger, W.R., Shenoy, D.K., Dalcanale, E. and Shen, Y.R. (2003) *Chem. Phys. Lett.*, **381**, 322–328.
51. Pagliusi, P., Lagugné-Labarthet, F., Shenoy, D.K., Dalcanale, E. and Shen, Y.R. (2006) *J. Am. Chem. Soc.*, **128**, 12610–12611.

52. (a) Tucci, F.C., Rudkevich, D.M. and Rebek, J. Jr (2000) *Chem. Eur. J.*, **6**, 1007–1016; (b) Shivanyuk, A., Rissanen, K., Körner, S.K., Rudkevich, D.M. and Rebek, J. Jr (2000) *Helv. Chim. Acta*, **83**, 1778–1790; (c) Hof, F., Trembleau, L., Ullrich, E.C. and Rebek, J. Jr (2003) *Angew. Chem. Ind. Ed.*, **115**, 3258–3261; (d) Hof, F., Trembleau, L., Ullrich, E.C. and Rebek, J. Jr (2003) *Angew. Chem. Int. Ed.*, **42**, 3150–3153; (e) Biros, S.M., Ullrich, E.C., Hof, F., Trembleau, L. and Rebek, J. Jr (2004) *J. Am. Chem. Soc.*, **126**, 2870–2876.
53. Vincenti, M., Dalcanale, E., Soncini, P. and Guglielmetti, G. (1990) *J. Am. Chem. Soc.*, **112**, 445–447.
54. (a) Renslo, A.R., Tucci, F.C., Rudkevich, D.M. and Rebek, J. Jr (2000) *J. Am. Chem. Soc.*, **122**, 4573–4582; (b) Lücking, U., Tucci, F.C., Rudkevich, D.M. and Rebek, J. Jr (2000) *J. Am. Chem. Soc.*, **122**, 8880–8889; (c) Tucci, F.C., Renslo, A.R., Rudkevich, D.M. and Rebek, J. Jr (2000) *Angew. Chem. Ind. Ed.*, **112**, 1118–1121; (d) Tucci, F.C., Renslo, A.R., Rudkevich, D.M. and Rebek, J. Jr (2000) *Angew. Chem. Int. Ed.*, **39**, 1076–1079; (e) Starnes, S.D., Rudkevich, D.M. and Rebek, J. Jr (2001) *J. Am. Chem. Soc.*, **123**, 4659–4669; (f) Lücking, U., Chen, J., Rudkevich, D.M. and Rebek, J. Jr (2001) *J. Am. Chem. Soc.*, **123**, 9929–9934.
55. Azov, V.A., Diederich, F., Lill, Y. and Hecht, B. (2003) *Helv. Chim. Acta*, **86**, 2149–2155.
56. Castro, P.P., Zhao, G., Masangkay, G.A., Hernandez, C. and Gutierrez-Tunstad, L.M. (2004) *Org. Lett.*, **6**, 333–336.
57. Cacciarini, M., Azov, V.A., Seiler, P., Künzer, H. and Diederich, F. (2005) *Chem. Commun.*, 5269–5271.
58. (a) Azov, V.A., Schlegel, A. and Diederich, F. (2005) *Angew. Chem. Ind. Ed.*, **117**, 4711–4715; (b) Azov, V.A., Schlegel, A. and Diederich, F. (2005) *Angew. Chem. Int. Ed.*, **44**, 4635–4638; (c) Azov, V.A., Schlegel, A. and Diederich, F. (2006) *Bull. Chem. Soc. Jpn.*, **79**, 1926–1940.
59. Marsden, J.A. and Hailey, M.M. (2004) in *Metal-Catalyzed Cross-Coupling Reactions*, vol. **1** (eds. A. de Meijere and F. Diederich), 2nd edn, Wiley-VCH Verlag GmbH, Weinheim, pp. 317–394.
60. Pochorovski, I., Breiten, B., Schweizer, W.B. and Diederich, F. (2010) *Chem. Eur. J.*, **16**, 12590–12602.
61. (a) Förster, T. (1948) *Ann. Phys.*, **437**, 55–75; (b) Selvin, P.R. (1995) *Methods Enzymol.*, **246**, 300–334; (c) Sapsford, K.E., Berti, L. and Medintz, I.L. (2006) *Angew. Chem. Ind. Ed.*, **118**, 4676–4704; (d) Sapsford, K.E., Berti, L. and Medintz, I.L. (2006) *Angew. Chem. Int. Ed.*, **45**, 4562–4588.
62. Shirtcliff, L.D., Xu, H. and Diederich, F. (2010) *Eur. J. Org. Chem.*, 846–855.
63. Frei, M., Diederich, F., Tremont, R., Rodriguez, T. and Echegoyen, L. (2006) *Helv. Chim. Acta*, **89**, 2040–2057.
64. (a) Gottschalk, T., Jaun, B. and Diederich, F. (2007) *Angew. Chem. Ind. Ed.*, **119**, 264–268; (b) Gottschalk, T., Jaun, B. and Diederich, F. (2007) *Angew. Chem. Int. Ed.*, **46**, 260–264; (c) Gottschalk, T., Jarowski, P.D. and Diederich, F. (2008) *Tetrahedron*, **64**, 8307–8317.
65. Kang, S.-W., Castro, P.P., Zhao, G., Nuñez, J.E., Godinez, C.E. and Gutierrez-Tunstad, L.M. (2006) *J. Org. Chem.*, **71**, 1240–1243.
66. Hooley, R.J., Van Anda, H.J. and Rebek, J. Jr (2006) *J. Am. Chem. Soc.*, **128**, 3894–3895.
67. Maslak, V., Yan, Z., Xia, S., Gallucci, J., Hadad, C.M. and Badjić, J.D. (2006) *J. Am. Chem. Soc.*, **128**, 5887–5894.
68. (a) Yan, Z., Xia, S., Gardlik, M., Seo, W., Maslak, V., Gallucci, J., Hadad, C.M. and Badjić, J.D. (2007) *Org. Lett.*, **9**, 2301–2304; (b) Gardlik, M., Yan, Z., Xia, S., Rieth, S., Gallucci, J., Hadad, C.M. and Badjić, J.D. (2009) *Tetrahedron*, **65**, 7213–7219; (c) Rieth, S., Yan, Z., Xia, S., Gardlik, M., Chow, A., Fraenkel, G., Hadad, C.M. and Badjić, J.D. (2008) *J. Org. Chem.*, **73**, 5100–5109.

69. (a) Wang, B.-Y., Bao, X., Yan, Z., Maslak, V., Hadad, C.M. and Badjić, J.D. (2008) *J. Am. Chem. Soc.*, **130**, 15127–15133; (b) Wang, B.-Y., Rieth, S. and Badjić, J.D. (2009) *J. Am. Chem. Soc.*, **131**, 7250–7252; (c) Rieth, S., Bao, X., Wang, B.-Y., Hadad, C.M. and Badjić, J.D. (2010) *J. Am. Chem. Soc.*, **132**, 773–776.
70. (a) Robbins, T.A. and Cram, D.J. (1995) *J. Chem. Soc., Chem. Commun.*, 1515–1516; (b) Garcia, C., Humilière, D., Riva, N., Collet, A. and Dutasta, J.-P. (2003) *Org. Biomol. Chem.*, **1**, 2207–2216.
71. (a) Piatnitski, E.L. and Deshayes, K.D. (1998) *Angew.Chem. Ind. Ed. Engl.*, **110**, 1022–1024; (b) Piatnitski, E.L. and Deshayes, K.D. (1998) *Angew. Chem. Int. Ed. Engl.*, **37**, 970–972.
72. (a) Lagona, J., Mukhopadhyay, P., Chakrabarti, S. and Isaacs, L. (2005) *Angew. Chem. Ind. Ed.*, **117**, 4922–4949; (b) Lagona, J., Mukhopadhyay, P., Chakrabarti, S. and Isaacs, L. (2005) *Angew. Chem. Int. Ed.*, **44**, 4844–4870; (c) Kim, K. (2002) *Chem. Soc. Rev.*, **31**, 96–107; (d) Lee, J.W., Samal, S., Selvapalam, N., Kim, H.-J. and Kim, K. (2003) *Acc. Chem. Res.*, **36**, 621–630; (e) Kim, K., Selvapalam, N., Ko, Y.H., Park, K.M., Kim, D. and Kim, J. (2007) *Chem. Soc. Rev.*, **36**, 267–279; (f) Geras'ko, O.A., Samsonenko, D.G. and Fedin, V.P. (2002) *Russ. Chem. Rev.*, **71**, 741–760.
73. (a) Mock, W.L. and Pierpont, J. (1990) *J. Chem. Soc., Chem. Commun.*, 1509–1511; (b) Jun, S.I., Lee, J.W., Sakamoto, S., Yamaguchi, K. and Kim, K. (2000) *Tetrahedron Lett.*, **41**, 471–475; (c) Lee, J.W., Kim, K. and Kim, K. (2001) *Chem. Commun.*, 1042–1043; (d) Tuncel, D. and Katterle, M. (2008) *Chem. Eur. J.*, **14**, 4110–4116.
74. (a) Praetorius, A., Bailey, D.M., Schwarzlose, T. and Nau, W.M. (2008) *Org. Lett.*, **10**, 4089–4092; (b) Pischel, U., Uzunova, V.D., Remón, P. and Nau, W.M. (2010) *Chem. Commun.*, 2635–2637.
75. Hwang, I., Ziganshina, A.Y., Ko, Y.H., Yun, G. and Kim, K. (2009) *Chem. Commun.*, 416–418.
76. (a) Desiraju, G.R. (1997) *Chem. Commun.*, 1475–1482; (b) Desiraju, G.R. (2002) *Acc. Chem. Res.*, **35**, 565–573; (c) Gavezzotti, A. (2002) *J. Mol. Struct.*, **615**, 5–12; (d) Hollingsworth, M.D. (2002) *Science*, **295**, 2410–2413.
77. (a) Meissner, R.S., Rebek, J. and de Mendoza, J. Jr (1995) *Science*, **270**, 1485–1488; (b) Rivera, J.M., Martín, T. and Rebek, J. Jr (1998) *J. Am. Chem. Soc.*, **120**, 819–820.
78. Szabo, T., Hilmersson, G. and Rebek, J. Jr (1998) *J. Am. Chem. Soc.*, **120**, 6193–6194.
79. Santamaría, J., Martín, T., Hilmersson, G., Craig, S.L. and Rebek, J. Jr (1999) *Proc. Natl. Acad. Sci. USA*, **96**, 8344–8347.
80. Wang, X. and Houk, K.N. (1999) *Org. Lett.*, **1**, 591–594.
81. (a) Rivera, J.M., Craig, S.L., Martín, T. and Rebek, J. Jr (2000) *Angew. Chem. Int. Ed.*, **112**, 2214–2216; (b) Rivera, J.M., Craig, S.L., Martín, T. and Rebek, J. Jr (2000) *Angew. Chem. Int. Ed.*, **39**, 2130–2132.
82. Mogck, O., Pons, M., Böhmer, V. and Vogt, W. (1997) *J. Am. Chem. Soc.*, **119**, 5706–5712.
83. Castellano, R.K., Craig, S.L., Nuckolls, C. and Rebek, J. Jr (2000) *J. Am. Chem. Soc.*, **122**, 7876–7882.
84. Vatsouro, I., Alt, E., Vysotsky, M. and Böhmer, V. (2008) *Org. Biomol. Chem.*, 998–1003.
85. Vysotsky, M.O. and Böhmer, V. (2000) *Org. Lett.*, **2**, 3571–3574.
86. Craig, S.L., Lin, S., Chen, J. and Rebek, J. Jr (2002) *J. Am. Chem. Soc.*, **124**, 8780–8781.
87. Barret, E.S., Dale, T.J. and Rebek, J. Jr (2007) *J. Am. Chem. Soc.*, **129**, 8818–8824.
88. Amaya, T. and Rebek, J. Jr (2004) *J. Am. Chem. Soc.*, **126**, 14149–14156.
89. (a) Dube, H., Ajami, D. and Rebek, J. Jr (2010) *Angew. Chem. Int. Ed.*, **122**, 3260–3263; (b) Dube, H., Ajami, D. and Rebek, J. Jr (2010) *Angew. Chem. Int. Ed.*, **49**, 3192–3195; (c) Dube,

H., Ams, M.R. and Rebek, J. Jr (2010) *J. Am. Chem. Soc.*, **132**, 9984–9985.
90. Yamanaka, M., Shivanyuk, A. and Rebek, J. Jr (2004) *J. Am. Chem. Soc.*, **126**, 2939–2943.
91. Palmer, L.C., Shivanyuk, A., Yamanaka, M. and Rebek, J. Jr (2005) *Chem. Commun.*, 857–858.
92. (a) Barrett, E.S., Dale, T.J. and Rebek, J. Jr (2007) *J. Am. Chem. Soc.*, **129**, 3818–3819; (b) Barrett, E.S., Dale, T.J. and Rebek, J. Jr (2008) *J. Am. Chem. Soc.*, **130**, 2344–2350.
93. (a) Chapman, R.G., Olovsson, G., Trotter, J. and Sherman, J.C. (1998) *J. Am. Chem. Soc.*, **120**, 6252–6260; (b) Chapman, R.G. and Sherman, J.C. (1998) *J. Am. Chem. Soc.*, **120**, 9818–9826.
94. Kobayashi, K., Shirasaka, T., Yamaguchi, K., Sakamoto, S., Horn, E. and Furukawa, N. (2000) *Chem. Commun.*, 41–42.
95. (a) Shivanyuk, A., Paulus, E.F. and Böhmer, V. (1999) *Angew. Chem. Int. Ed.*, **111**, 3091–3094; (b) Shivanyuk, A., Paulus, E.F. and Böhmer, V. (1999) *Angew. Chem. Int. Ed.*, **38**, 2906–2909.
96. (a) Mathias, J.P., Simanek, E.E., Seto, C.T. and Whitesides, G.M. (1993) *Angew. Chem. Int. Ed. Engl.*, **105**, 1848–1850; (b) Mathias, J.P., Simanek, E.E., Seto, C.T. and Whitesides, G.M. (1993) *Angew. Chem. Int. Ed. Engl.*, **32**, 1766–1769; (c) Mathias, J.P., Seto, C.T., Simanek, E.E. and Whitesides, G.M. (1994) *J. Am. Chem. Soc.*, **116**, 1725–1736.
97. (a) Reinhoudt, D.N. and Crego-Calama, M. (2002) *Science*, **295**, 2403–2407; (b) Prins, L.J., Neuteboom, E.E., Paraschiv, V., Crego-Calama, M., Timmerman, P. and Reinhoudt, D.N. (2002) *J. Org. Chem.*, **67**, 4808–4820;

(c) Kerckhoffs, J.M.C.A., van Leeuwen, F.W.B., Spek, A.L., Kooijman, H., Crego-Calama, M. and Reinhoudt, D.N. (2003) *Angew. Chem. Int. Ed.*, **115**, 5895–5900; (d) Kerckhoffs, J.M.C.A., van Leeuwen, F.W.B., Spek, A.L., Kooijman, H., Crego-Calama, M. and Reinhoudt, D.N. (2003) *Angew. Chem. Int. Ed.*, **42**, 5717–5722; (e) ten Cate, M.G.J., Huskens, J., Crego-Calama, M. and Reinhoudt, D.N. (2004) *Chem. Eur. J.*, **10**, 3632–3639; (f) Kerckhoffs, J.M.C.A., ten Cate, M.G.J., Mateos-Timoneda, M.A., van Leeuwen, F.W.B., Snellink-Ruël, B., Spek, A.L., Kooijman, H., Crego-Calama, M. and Reinhoudt, D.N. (2005) *J. Am. Chem. Soc.*, **127**, 12697–12708.
98. (a) Fochi, F., Jacopozzi, P., Wegelius, E., Rissanen, K., Cozzini, P., Marastoni, E., Fisicaro, E., Manini, P., Fokkens, R. and Dalcanale, E. (2001) *J. Am. Chem. Soc.*, **123**, 7539–7552; (b) Zuccaccia, D., Pirondini, L., Pinalli, R., Dalcanale, E. and Macchioni, A. (2005) *J. Am. Chem. Soc.*, **127**, 7025–7032.
99. (a) Fox, O.D., Dalley, N.K. and Harrison, R.G. (1998) *J. Am. Chem. Soc.*, **120**, 7111–7112; (b) Fox, O.D., Dalley, N.K. and Harrison, R.G. (1999) *Inorg. Chem.*, **38**, 5860–5863.
100. (a) Davis, A.V. and Raymond, K.N. (2005) *J. Am. Chem. Soc.*, **127**, 7912–7919; (b) Davis, A.V., Fiedler, D., Seeber, G., Zahl, A., van Eldik, R. and Raymond, K.N. (2006) *J. Am. Chem. Soc.*, **128**, 1324–1333.
101. (a) Mugridge, J.S., Bergman, R.G. and Raymond, K.N. (2010) *Angew. Chem. Int. Ed.*, **122**, 3717–3719; (b) Mugridge, J.S., Bergman, R.G. and Raymond, K.N. (2010) *Angew. Chem. Int. Ed.*, **49**, 3635–3637.

9
Cyclodextrin-Based Switches

He Tian and Qiao-Chun Wang

9.1
Introduction

Cyclodextrins (CyDs) are a series of cyclic glucose oligomers with α-1,4 linkages and their structures consist of an inner hydrophobic cavity and a hydrophilic hydroxyl exterior. The most commonly known α-, β- and γ-CyD have six to eight glucose units and they are water soluble, nontoxic and commercial available. CyDs can bind to a wide variety of guest molecules due to the hydrophobic interactions between the cavity and the guests, and as a result, they continue to be the attractive components in constructing molecular recognition and self-organization systems in supramolecular chemistry [1–5].

The most obvious foundation for constructing a CyD-based switch is to change the interaction strength between the CyD host and the hydrophobic guest with an external stimulus, such as chemical, electrochemical, photochemical or even an environmental change, and the guest is thus included in or excluded from the CyD cavity. According to the motions of the host and guest during the switching process, three types of switching systems might be defined:

1) In and out switching. This is the simplest system that is composed of one host and one guest and the guest can be triggered to enter and leave the CyD cavity (Figure 9.1a).
2) Displacement switching. In this case, the system becomes more complex, as it may consist of either one CyD host and two different guests, or one guest and two different hosts (Figure 9.1b). In the former system, the CyD host can be switched to form inclusion complex between the two guests by altering their binding ability to the host; while in the latter, the guest is included alternatively between the CyD and the other host in response to the external stimuli.
3) Back and forth switching. Once the two guests are joined together through a linear linker between the two guests and one host displacement switching system, the CyD ring would shuttle along the linear rod between the two guests when the binding-ability sequence is changed (Figure 9.1c) and such a

Molecular Switches, Second Edition. Edited by Ben L. Feringa and Wesley R. Browne.
© 2011 Wiley-VCH Verlag GmbH & Co. KGaA. Published 2011 by Wiley-VCH Verlag GmbH & Co. KGaA.

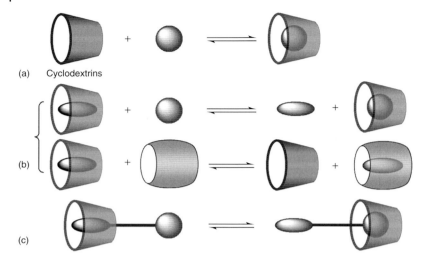

Figure 9.1 The in and out switch (a), displacement switch (b) and back and forth switch (c).

system can be vividly deemed as a back and forth switching system. It would be more interesting when we consider the function of a back and forth switch as the system can transform an external energy (used to trigger the switching function) to a mechanical energy (the shuttling of the CyD), so it is also known as a *molecular shuttle* [6–9]. Besides the switching, such a molecular shuttle can perform other beautiful functions, such as logical operations, information storage, and so on, which have been well depicted in many reviews and books [10–15] and will not be discussed in this chapter.

Apart from the above three switching patterns, coordination switching and rearrangement switching, where the relative host–guest migrations may not be involved, are two other important switching modes in CyD-based switches. In the case of coordination switching, a functional side arm is introduced to the CyD to catch ions through coordination. These ions exert physical or chemical effects on the chromophore on the side arm or in near the CyD cavity and thus, activate the switching function. As for rearrangement switching, the CyD units may first occupy a stable conformation, and then the CyD units reorder when an external stimulus is applied and switch to a metastable conformation. As the CyDs are chiral molecules, such a switching process always brings about an optical activity output.

Besides the switching mode, the type of signal to be controlled in a CyD-based switching system is another key element. Like an electric switch that can control electrical current, and a hydrant that can adjust water flow, a CyD-based switch can tune the absorption, fluorescence, optical activity or even the morphology of the system:

1) Absorption. The modification of the absorption of a CyD-base switching system can be achieved through the solvent effects on absorption spectra

of a chromophore – that is the positions, intensities and shapes of the absorption bands are usually modified when the absorption spectra are measured in solvents of different polarity [16–18]. According to the rule of like dissolves like, when the unit size of a hydrophobic guest or the hydrophobic unit of a guest is suitable to enter the hydrophobic CyD cavity in an aquatic environment, the microenvironment of the guest will be changed (from a high polarity hydrophilic environment to a low polarity hydrophobic cavity). So if the guest contains a chromophore, the colour of the system may change during the switching motions.

2) Fluorescence. There are many parameters that can affect the fluorescent quantum yields and lifetimes of a molecule: temperature, pH, polarity, viscosity, hydrogen bonding, presence of quenchers, and so on [19]. So there are many opportunities to obtain a fluorescent variation output in a CyD switch when a guest fluorophore comes in and out from the CyD cavity. For instance, pH value, polarity, viscosity and hydrogen bonding may change because of the microenvironmental variation; the rigidity of the CyD ring to the guest fluorophore may increase the fluorescence quantum yield of the guest fluorophore due to the reduced internal rotations that often provide additional channels for nonradiative relaxation; the quenching effects by a quencher in the aqueous solution may be shielded when a guest fluorophore enters the CyD cavity; and the capturing of a quencher by a CyD side-arm may also quench the fluoresence of the guest molecule in the CyD cavity.

3) Induced circular dichroism. When an achiral guest chromophore is located in a chiral host, the guest becomes optically active and induces circular dichroism (ICD) signals. General rules have been derived for the ICD of chromophore/CyD systems [20, 21]:

   a. An electronic dipole transition moment of the chromophore parallel to the axis of CyD brings about a positive ICD while, perpendicular to the axis it gives a negative ICD (Figure 9.2a).

   b. The sign of ICD becomes reversed when the chromophore stands outside the CyD cavity, as shown in Figure 9.2b.

   From the above rules, it can be deduced that when a chromophore is induced to move in and out from a CyD cavity, or rotate relative to the CyD axis, the intensity or even the sign of the ICD signals may change. A CyD-based switching system with ICD readout could be thus obtained by the careful integration of the chromophore, CyD and a trigger.

4) Morphology. In this case, a complex forms by the combination of a guest molecule and the CyD host on a solid waffle. The physicochemical properties of the new complex might be vastly different from the individual properties and thus leads to the morphology variation of the waffle surface.

The following sections will review CyD-based molecular switches in a sequence of switching patterns.

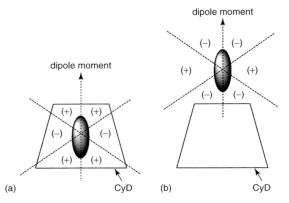

Figure 9.2 (a, b) The sign of ICD effect as a function of the location and the direction of the electric dipole transition moment with respect to the CyD cavity.

9.2
In and Out Switching

CyDs can include a wide range of hydrophobic guest molecules in aqueous solution and exclude these molecules again in many cases when an organic solvent is added. Switches can thus be fabricated by taking advantage of this solvent effect. One of the most interesting guest molecules is phenolphthalein, as shown in Figure 9.3 [22]. It is common knowledge that the colour of the phenolphthalein in aqueous solution will change from the colourless phenolic form to the red quinonoid one when the

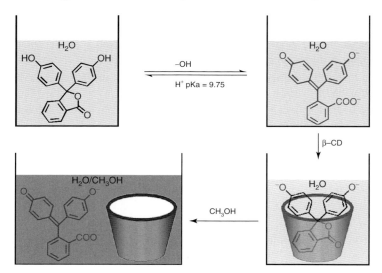

Figure 9.3 The switching mechanism of phenolphthalein in response to pH, β-CyD and solvent changes.

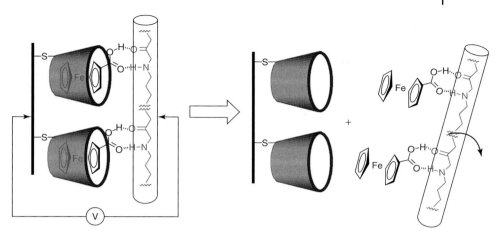

Figure 9.4 The electrocatching and detaching of ferrocene nanotube.

pH of the solution is greater than 9.75. When β-CyD is added at this time, phenolphthalein will be included in the cavity of β-CyD and changes to its lactonoid dianionic form with an accompanying colour change to colourless. More interestingly, the lactonoid dianion is pulled out from the β-CyD cavity when methanol is added to the aqueous solution, and the colour of the solution will change to red again.

Matsui and coworkers [23] constructed an electrochemically driven in and out switch, as shown in Figure 9.4. A peptide was firstly immobilized on a carbon nanotube and the subsequent anchoring of ferrocenecarboxylic acid on the resulting peptide nanotube gave the ferrocene nanotube building block. At the same time, self-assembled thiolated β-CyD monolayers on Au surfaces were also fabricated. The ferrocene nanotubes could be easily absorbed onto the β-CyD self-assembled monolayer (SAMs) by mixing them in aqueous solution, which can be clearly seen in scanning electron microscopy (SEM) images. However, after an electric field was applied (5×10^2 V/m, 1 kHz) in the stirred solution for 5 min, detachment of ferrocene nanotubes took place. In this way, the morphology of the SAM surface can be tuned by applying electric fields.

Liu et al. [24] also reported an interesting pH-controlled inclusion switching in the following compound – pyridine-2,6-dicarboxamide-bridged bis(β-CyD), as shown in Figure 9.5. In a pH 7.2 buffer, the pyridine group locates externally between the two CyD hosts, where the transition moment of the ^1L_a band around 213 nm of the pyridine is nearly perpendicular to the CyD axis and thus induces a positive circular dichroism (CD) signal at 213 nm. However, in a pH 2.0 buffer, the pyridine chromophore is shallowly included in the CyD cavity, where both the ^1L_a and ^1L_b transition moments around 220 and 270 nm are nearly perpendicular to the CyD axis, resulting in two negative Cotton effect peaks. The bis(β-CyD) compound also emits stronger fluorescence in a pH 2.0 buffer than in a pH 7.2 buffer.

Figure 9.6 shows another novel photodriven in and out switch that was set up in Reinhoudt's group [25–27]. Two β-CyD hosts were introduced as side arms

Figure 9.5 The acid/base triggered self-inclusion of the pyridine bridged bis(β-CyD).

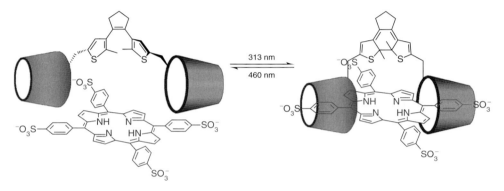

Figure 9.6 The mechanism of photocontrolled release and uptake of a porphyrin guest.

into the dithienylethene. Irradiation with light results in the conversion of the dithienylethene between an open and a closed form, and consequently the coconformation of the two β-CyDs was switched between the staggered and face-to-face form. The face-to-face dimer has strong binding ability to tetrakis-sulfonatophenyl porphyrin (TSPP) in aqueous solution, while the staggered one binds weakly to TSPP, and as a result, the catching and releasing of the porphyrin driven by light can be achieved. Such a photoswitchable in and out system leads to strong absorbance changes at 413 nm.

9.3
Back and Forth Switching

In most cases, the back and forth switching systems are those rotaxanes system where the CyD ring can shuttle between two different stations in response to external stimuli, such as pH, light, solvent, temperature, and so on. Among these stimuli, light is especially of priority because it brings about fast response, remote sensitivity and cleanliness. Reversible photoisomerization of photoactive groups can be exploited to modulate the physical and chemical properties, and thus may generate driving forces for the shuttling motions among molecular shuttles.

If a light-driven molecular shuttle can be combined with a fluorescent output, the photons can thus be explored for both causing the translational changes (input) and monitoring the different coconformational states (output). Such a system would have great advantages in constructing optical molecular devices because of its fast response, low cost and remote sensitivity. Tian and coworkers [28] have been focusing their interests on the construction of this kind of CyD-based switches. Figure 9.7 shows a 'lockable' light-driven rotaxane-based molecular switch set up in Tian's group utilizing stilbene as the photoactive group. In an acidic environment, the hydrogen bonds between the isophthalic acid and the hydroxy groups on the CyD hold the CyD ring tightly. The CyD ring consequently resides over the stilbene unit and is far away from the 4-amino-1,8-naphthalimide-sulfonate (ANS) fluorophore. However, when the pH of the solution is adjusted to 9.6 by Na_2CO_3, the hydrogen bonding is disrupted and the CyD ring becomes 'free'. Once the resulting alkaline aqueous solution is irradiated with 335-nm UV light, the photoisomerization of the stilbene unit is induced and the CyD ring is forced to shuttle from the stilbene station to the biphenyl unit and closer to ANS. The rigidity of the ring to the methylene and phenyl groups results in an increase in the fluorescence intensity of the fluorophore by 63%. The back isomerization can be reversibly obtained by the irradiation at 280 nm and the fluorescence decreases to the original level.

There are many other reports on switchable CyD-based rotaxanes with fluorescence responsiveness that have been reported from the Tian group, for example the report on a light-driven rotaxane with dual fluorescence addressing [29]; on a thermodriven rotaxane [30]; on the synthesis of isomer-free CyD rotaxanes [31]; on an abacus-like [3]rotaxane [32]; rotaxanes that can serve as logic gates because of their complex switching functions [33, 34]; and on switchable rotaxane sol-gel systems [35]. Apart from these fluorescent systems, Tian and coworkers also set up switchable CyD rotaxanes with another kind of optical output signal, that is the ICD. Figure 9.8 is a pseudo[4]rotaxane example of this kind [36]. The inclusion between the α-CyD and the viologen-azobenzene-viologen rod shape molecule in aqueous solution gives a pseudo[2]rotaxane with the CyD ring residing on the azobenzene station. The consequent mixing of pseudo[2]rotaxane with 2 equiv. of CB[7] in aqueous solution affords pseudo[4]rotaxane. The photoinduced E/Z isomerization of the azobenzene unit forces the CyD ring to shuttle back and forth between the azo and the phenyl ether station with an obvious variation in the intensity of ICD signal at around 365 nm. It should be noted that the two bulky CB[7] terminals of the pseudo[4]rotaxane exploit enough dimensional space for the shuttling motions of the CyD along the rod unit and as a result, the light-driven switching processes continue in sol-gel or even in solid-state thin film, which provides a useful site for the construction of solid-state molecular devices.

Tian and coworkers [37, 38] also reported another azobenzene-based [1]rotaxane switching system, as shown in Figure 9.9. The fold-followed-by-self-inclusion of the azobenzene-modified β-CyD in aqueous solution gives the self-included intramolecular complex with the phenyl halide unit sticking out of the CyD cavity, and the resulting Pd-catalysed Suzuki coupling with naphthalimide-N-benzyl boronic acid in alkaline solution brings about the formation of the [1]rotaxane. The

308 | *9 Cyclodextrin-Based Switches*

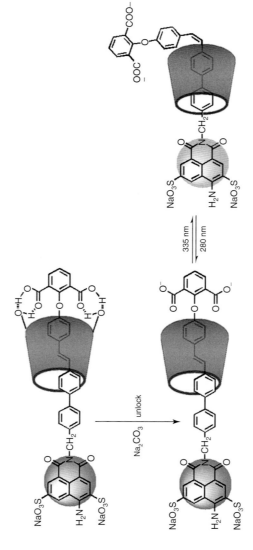

Figure 9.7 The 'lockable' molecular shuttle powered by light with a fluorescent output.

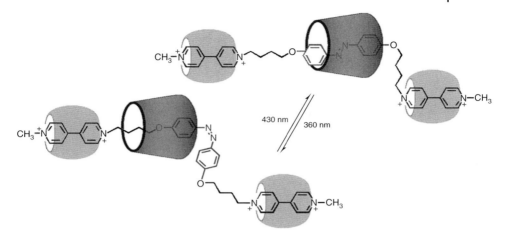

Figure 9.8 The reversible light-driven shuttling motions of the pseudo[4]rotaxane cooperating with an azobenzene group.

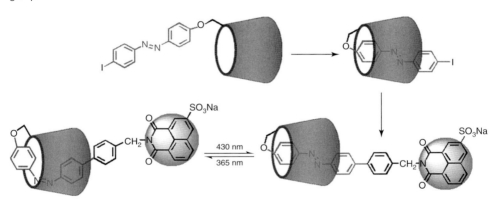

Figure 9.9 The construction of the [1]rotaxane and its light-driven switching processes.

azobenzene can also be induced to perform $Z \rightarrow E$ photoisomerization and the CyD ring shuttles slightly forward along the azo unit with an obvious increase in ICD intensities at the negative 322 nm band and the positive 433 nm band.

More recently, Harada reported the synthesis of polyethylene glycol (PEG) substituted β-CyD with an azobenzene group at the terminal of the PEG chain Azo-HC-PEG-CD, as shown in Figure 9.10 [39]. At low concentration at 80 °C, Azo-HC-PEG-CD exists as an extended form with the long side chain sticking out of the CyD cavity. However, when the temperature is lower to 60 °C, the azobenzene part is included in the cavity and Azo-HC-PEG-CD adopts as a self-threading conformation. Once the temperature is decreased to 1 °C, the cinnamyl aromatic moiety exhibits higher affinity with the CyD than the *trans*-azo group and the CyD ring shuttles to the aniline station. More interestingly, the CyD ring can be

310 | 9 Cyclodextrin-Based Switches

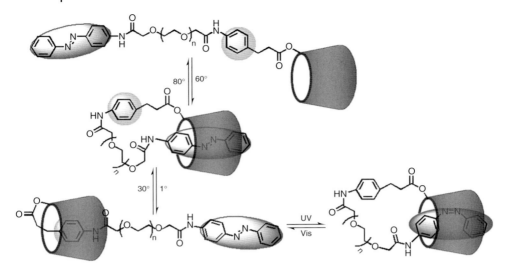

Figure 9.10 The themo- and light-driven switching processes of the [1]pseudorotaxane.

switched again to the azo station with UV-light to induce the $E \rightarrow Z$ isomerization of the azobenzene group.

9.4
Displacement Switching

In a host–guest system, the addition of another more appropriate guest would change the initial balance of the molecular recognition and destabilize the original complexation. As a result, the former guest is kicked out from the host and the conformation of the system is switched. Figure 9.11 shows a molecular switch of this kind [40]. The system comprises a cinnamide substituent attached to β-CyD. In aqueous solution, the *trans*-cinnamide part enters into the CyD annulus to form the intramolecular conformation. If 1-adamantanol is present in the solution, the larger complexation constant of the adamantanol than the aryl part results in replacement with the adamantanol entering the CyD cavity and the cinnamide unit being pushed out. The removal of the adamantanol by extraction with hexane leads to the conformational recovery to the original intramolecular complexation. Such a molecular switch can also act as a nanosized internal combustion engine in which the CyD ring serves as the cylinder, the aryl group as the piston and the adamantanol as fuel. Moreover, the photoisomerization of the cinnamide between *trans* and *cis* conformations would pull or push the cinnamide in and out of the CyD cavity and thus turns the switching function on and off.

Molecular recognition in biological systems plays important roles during vital process. However, these processes are very complicated and the binding strength and selectivity of the interaction may be greatly impacted by architectural and

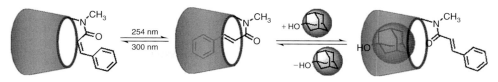

Figure 9.11 The chemical-fuelled molecular machine with a photocontrollable on/off switch.

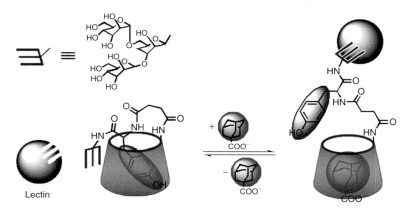

Figure 9.12 The activation/deactivation of the carbohydrate–protein recognition through host–guest interactions.

orientation factors. Figure 9.12 shows a biological switching system utilizing the lectin–carbohydrate interactions [41]. This system comprises of a β-CyD platform, a succinylamido, a tyrosinyl and a trisaccharide segments linked to each other. The phenol part in the tyrosinyl segment could self-include into the CyD cavity in water and the succinylamido chain folded, so that the trisaccharide recognition motif becomes closed to the CyD scaffold and is not accessible to the protein binding site. Once adamantine carboxylate (**AC**) is added to the mixture, the stronger binding ability of the adamantane pushes the phenol unit out of the cavity, the oligosaccharide thus stretches and is fully exposed to interact with the lectin, causing a dramatic increase in lectin–saccharide binding affinity. The removal of **AC** with an appropriate scavenger switches the system to the initial state where the lectin is released.

The structural character of CyDs bearing OH groups present around the hydrophilic cavity offers CyDs the opportunity to serve as an artificial enzyme. The most commonly known catalytic activity of CyDs is to accelerate ester hydrolysis [42]. Lee and Ueno [43] reported an interesting system where the catalytic activity can be switched on and off by light, as shown in Figure 9.13. A histidine residue as the catalytic functional unit and azobenzene as the light-responsive unit were introduced to the β-CyD rim. In the *trans* conformation, the azobenzene unit is self-included into the CyD cavity, the ester substrate resides outside and

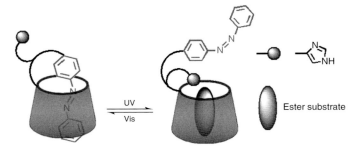

Figure 9.13 The photoswitching on and off of the catalytic activity in ester hydrolysis.

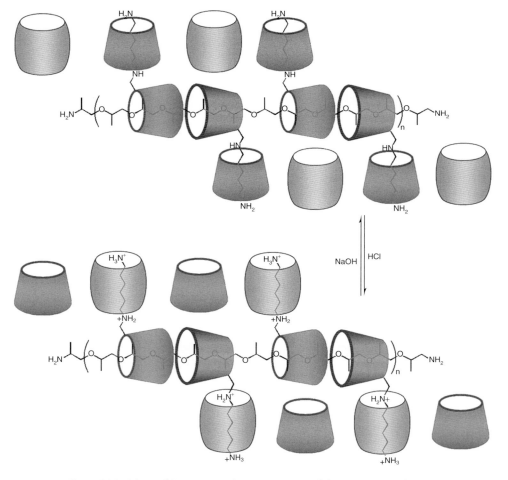

Figure 9.14 The acid/base-triggered interconversion of the two 2D pseudorotaxanes.

no obvious catalytic activity is found. However, when the *trans* conformation is changed photochemically to a *cis* one, the *cis*-azobenzene moiety is drawn out and the system provides an appropriate cavity to accommodate the ester substrate, as a result, an obvious acceleration of the hydrolysis is detected.

Liu *et al.* [44] reported a simple strategy to obtain a 2D pseudorotaxane where cyclic molecules were threaded onto both the polymeric main chain and its side chains. A hexane-1,6-diamine was firstly introduced to β-CyD and the consequent threading with cyclic cucurbit[6]uril (CB[6]) gave a CyD-stopped pseudorotaxane. The CyD-stopped pseudorotaxane was finally stringed together through the CyD cavities with a linear polypropylene gycol (PPG) polymer chain to give a 2D pseudorotaxane with CyD ring on the main chain and CB[6] on side chains. This novel 2D pseudopolyrotaxane can be changed to a main-chain pseudopolyrotaxane by dethreading the side CB[6] rings in the presence of base, and the consequent threading α-CyD rings onto the side chains gives another 2D pseudopolyrotaxane, as illustrated in Figure 9.14. The latter 2D pseudopolyrotaxanes can be reversibly converted to the original one by an acid stimulus.

9.5
Coordination Switching

In general, the coordination of a chromophore with ions would change the energy levels of the chromophore and thus vary its emitting properties including the wavelength or intensity. Many chemosensors and biosensors have been constructed based on this strategy. The sensing of ions can also be deemed as an ion-triggered switching process. CyDs are water soluble, nontoxic, commercially available and they are readily modified or bind to a wide variety of ligand molecules therefore, they are an excellent platform for the construction of ion sensors or switches. Figure 9.15 is a zinc-triggered molecular switch [45]. The supramolecular system consists of an 8-aminoquinoline modified β-CyD (1) and the included 1-adamantaneacetate (ADA). The nitrogen atoms on the quinoline and the carboxylate anion on the adamantane are prone to catch Zn^{2+} to form a ternary complex with a decreasing fluorescent intensity at 416 nm and a sharp increasing intensity at 490 nm

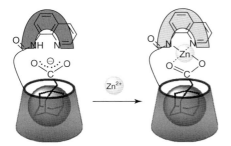

Figure 9.15 The formation of the ternary complex and the switching of the fluorescence wavelength.

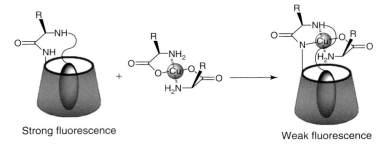

Figure 9.16 The ion-induced fluorescence switching processes that can sense the chirality of amino acids.

($I/I_0 = 6.8$). It is notable that this system shows excellent selectivity to Zn^{2+} over possible competing cations in aqueous solution, which meets the requirements of biochemosensors and shows potential application in physiological fields for Zn^{2+} assay.

In many cases, the fluorescence of a fluorophore would be quenched by some specific cations due to the nonradiative energy transition in the process of the electron or energy transfer between the ions and the fluorophore. A CyD-based molecular switch can thus be constructed in which a fluorophore is arranged in the CyD cavity and an ion-catching arm is fixed on the CyD rim. The catching of ions would result in the quenching of the fluorescence and the fluorescence switching can thus be expected. Figure 9.16 demonstrates a switching system of this kind [46]. A side arm consisting of a copper(II) binding site and a dansyl end-group was introduced to β-CyD rim. In aqueous solution, the dansyl group adopts self-inclusion within the CyD cavity and shows strong emission. The addition of a Cu(II) complex of an amino acid to the solution gives rise to the formation of the ternary complex of higher stability. As a result, the quenching effect from ligand-to-metal causes the fluorescence to be switched off. It is interesting that the copper(II) binding side arm is chiral and forms ternary complexes of different stability with the L- or R-amino acid, consequently, such a system shows enantioselectivity for amino acid sensing.

9.6
Rearrangement Switching

Rearrangement switching occurs overwhelmingly in those macromolecular systems, where the CyD units act coincidently or cooperatively to keep the macromolecule in a stable conformation. Once a stimulus such as environmental changes in temperature, pH or solvent occurs, or the appearance of a strong guest molecule, the harmonious actions of the CyDs are destroyed and the original conformation of the macromolecule collapses to give rise to a new conformation.

Yashima and coworkers [47] reported a series of CyD-appending polymers that are synthesized from the polymerization of ethynylbenzamide-β-CyD with a rhodium catalyst. These polymers have a predominantly one-handed helical conformation induced by the cooperative interaction between the adjacent CyD pendants so that they exhibit an intense ICD in the long absorption region of the conjugated polyacetylene backbones at 25 °C. However, at high temperatures, the helicity of the polymer backbones inverts with a different twist angle of the conjugated double bonds, as shown in Figure 9.17. The inversion of the helicity causes the sign in the ICD patterns to be inverted; and the change in the helical pitch of conjugated double bonds bring about a colour change from red to yellow. Beside the thermochromism, these polymers also exhibit solvatochromism. It should be noted that the entering of small guest molecules such as 1-adamantanol, as well as the chiral solvation with optical compounds such as 1-phenylethylamine, can also cause helicity inversion and colour change. So it can be seen that these polymer switches can also act as sensors for small molecules and environmental changes.

Figure 9.18 shows another rearrangement switching system based on methylated-α-CyD polyrotaxane [48]. A polyrotaxane was firstly prepared by the threading of α-CyD onto a PEG chain with coverage of 28%, and the resulting

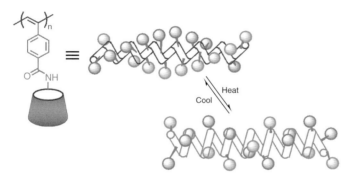

Figure 9.17 The synthesis of the polymer with β-CyD pendants and its thermoinduced helicity inversion.

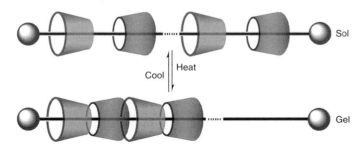

Figure 9.18 The thermocontrollable sol-gel conversion of the CyD-based polyrotaxane.

methylation of α-CDs in the polyrotaxane with iodomethane and sodium hydride in dimethyl sulfoxide (DMSO) gives the target polyrotaxane. At low temperature, the polyrotaxane aqueous solution adopts a separating conformation with the CyD rings scattering around the PEG chain. However, at high temperature, the hydrophobic interactions between the CyD rings gather the rings together and two segments – the hydrophilic naked PEG and the hydrophobic methylated CyD-tube – appear in the polyrotaxane, as a result, the viscosity of the aqueous solution increases with increasing temperature and the solutions eventually form an elastic gel. Interestingly, the methylation ratio to the OH groups of the CyD, which can be controlled by the mole equivalent of the iodomethane added to the naked CyD polyrotaxane solution, is the key element that determines the sol-gel transition temperature.

9.7
Conclusion and Perspective

This chapter focuses mainly on introducing CyD-based molecular switches, where the CyD ring plays an important role during the switching processes. Most examples are relatively typical ones and of course there are also many other examples. These include in and out switching systems [49–53]; switchable CyD-based rotaxanes [54, 55]; coordination switching [56]; rearrangement switching [57] and displacement switching [58], which are not dealt with in this chapter. It should be said that there are also molecular switches with CyDs, in which the CyD ring does not devote direct functions to the switching process, for example the CyD ring may only acts as a platform to support the main body of the switch [59–61], or a crosslinking group to hold the switches together [62], or a bulky group to increase the inter-switch distance [63], which are also not reviewed in this chapter.

CyDs can include a wide range of hydrophobic guest molecules and exclude these molecules again when appropriate stimuli such as temperature, pH value, and so on, are applied. Such reversible host–guest interactions will give scientists further inspiration, beside those interesting ideas mentioned above, to construct switchable materials. For example, tunable materials that can switch between monomer and the self-assembling macromolecule, or between monochain and polychain in response to temperature can thus be carefully designed, as shown in Figure 9.19. At a relatively low temperature (for example ambient temperature), the inclusion initiates the polymerization, while at a relatively high temperature, the exclusion occurs and the formed macrosystem is dissociated. These intelligent materials may find potential applications in materials science and medicine.

Moreover, it is encouraging that CyD-based switching materials are of increasing interest and an increasing number of scientists are engaged in this beautiful area. It can be foreseen that more CyD-based materials of amazing innovations and attractive function will come forth in the near future.

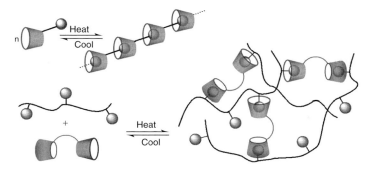

Figure 9.19 The cartoon illustration for the thermocontrolled self-polymerization and crosslinking polymerization.

Acknowledgement

The authors thank the funding of the NSFC/China (50673025, 90401026, 20603009), National Basic Research 973 Program (2006CB806200), the Key Project of Chinese Ministry of Education (107044), the Foundation for the Author of National Excellent Doctoral Dissertation of PR China (200758), Shanghai Science and Technology Development Funds (07QA14012) and the Scientific Committee of Shanghai.

References

1. Feringa, B.L. (2001) *Molecular Switches*, Wiley-VCH Verlag GmbH, Weinheim.
2. Szejtli, J. (1982) *Cyclodextrins and their Inclusion Complexes*, Budapest, Akadémiai Kiadó.
3. Szejtli, J. and Osa, T. Volume Eds. (1996) Cyclodextrins, in *Comprehensive Supramolecular Chemistry*, vol. 3 (eds. J.L. Atwood, J.E.D. Davies, D.D. MacNicol and F. Vögtle), Pergamon, Oxford.
4. Hapiot, F., Tilloy, S. and Monflier, E. (2006) *Chem. Rev.*, **106**, 767.
5. Villalonga, R., Cao, R. and Fragoso, A. (2007) *Chem. Rev.*, **107**, 3088.
6. Balzani, V., Credi, A., Raymo, F.M. and Stoddart, J.F. (2000) *Angew. Chem. Int. Ed.*, **39**, 3348.
7. Balzani, V., Credi, A. and Venturi, M. (2003) *Molecular Devices and Machines: A Journey into the Nano World*, Wiley-VCH Verlag GmbH, Weinheim.
8. Flood, A.H., Ramirez, R.J.A., Deng, W.-Q., Müller, R.P., Goddard III, W.A. and Stoddart, J.F. (2004) *Aust. J. Chem.*, **57**, 301.
9. Kay, E.R., Leigh, D.A. and Zerbetto, F. (2007) *Angew. Chem. Int. Ed.*, **46**, 72.
10. Nepogodiev, S.A. and Stoddart, J.F. (1998) *Chem. Rev.*, **98**, 1959.
11. Harada, A. (2001) *Acc. Chem. Res.*, **34**, 456.
12. Wenz, G., Han, B.-H. and Müller, A. (2006) *Chem. Rev.*, **106**, 782.
13. Tian, H. and Wang, Q.-C. (2006) Rotaxanes with Cyclodextrin, in *Cyclodextrin Materials Photochemistry, Photophysics and Photobiology*, Chapter 13 (ed. A. Douhal), Elsevier, Amsterdam, pp. 285–302.
14. Tian, H. and Wang, Q.-C. (2006) *Chem. Soc. Rev.*, **35**, 361.
15. Credi, A. and Tian, H. (Guest editorial) (2007) *Adv. Funct. Mater.*, **17**, 679 (Special issue on Molecular Switches and Molecular Machines).
16. Gough, T.E., Irish, D.E. and Lantzke, I.R. (1973) Spectroscopic Measurements

(electron absorption, infrared and Raman, ESR and NMR spectroscopy), in *Physical Chemistry of Organic Solvent Systems* (eds. A.K. Covington and T. Dickinson), Plenum Press, London, New York.
17. Jauquet, M. and Laszlo, P. (1975) Influence of solvents on spectroscopy, in *Solutions and Solubility, Techniques of Chemistry* (eds. M.R.J. Dack and A. Weissberger), Wiley-Interscience, New York.
18. Reichardt, C. (2003) *Solvents and Solvent Effects in Organic Chemistry*, Wiley-VCH Verlag GmbH, Weinheim.
19. Valeur, B. (2001) *Molecular Fluorescence: Principles and Applications*, Wiley-VCH Verlag GmbH, Weinheim.
20. Harata, K. and Uedaira, H. (1975) *Bull. Chem. Soc. Jpn.*, **48**, 375.
21. Kodaka, M. (1993) *J. Am. Chem. Soc.*, **115**, 3702.
22. Taguchi, K. (1986) *J. Am. Chem. Soc.*, **108**, 2705.
23. Chen, Y., Banerjee, I.A., Yu, L., Djalali, R. and Matsui, H. (2004) *Langmuir*, **20**, 8409.
24. Liu, Y., Chen, G.-S., Chen, Y., Ding, F., Liu, T. and Zhao, Y.-L. (2004) *Bioconjug. Chem.*, **15**, 300.
25. Mulder, A., Jukovi, A., Lucas, L.N., van Esch, J., Feringa, B.L., Huskens, J. and Reinhoudt, D.N. (2002) *Chem. Commun.*, 2734.
26. Mulder, A., Juković, A., Huskens, J. and Reinhoudt, D.N. (2004) *Org. Biomol. Chem.*, **2**, 1748.
27. Mulder, A., Jukovi, A., van Leeuwen, F.W.B., Kooijman, H., Spek, A.L., Huskens, J. and Reinhoudt, D.N. (2004) *Chem. Eur. J.*, **10**, 1114.
28. Wang, Q.-C., Qu, D.H., Ren, J., Chen, K. and Tian, H. (2004) *Angew. Chem. Int. Ed.*, **43**, 2661.
29. Qu, D.-H., Wang, Q.-C., Ren, J. and Tian, H. (2004) *Org. Lett.*, **6**, 2085.
30. Qu, D.-H., Wang, Q.-C. and Tian, H. (2005) *Mol. Cryst. Liq. Cryst.*, **430**, 59.
31. Qu, D.-H., Wang, Q.-C., Ma, X. and Tian, H. (2005) *Chem. Eur. J.*, **11**, 5929.
32. Wang, Q.-C., Ma, X., Qu, D.-H. and Tian, H. (2006) *Chem. Eur. J.*, **12**, 1088.
33. Qu, D.-H., Wang, Q.-C. and Tian, H. (2005) *Angew. Chem. Int. Ed.*, **44**, 5296.
34. Qu, D.-H., Ji, F.-Y., Wang, Q.-C. and Tian, H. (2006) *Adv. Mater.*, **18**, 2035.
35. Zhu, L., Ma, X., Ji, F., Wang, Q. and Tian, H. (2007) *Chem. Eur. J.*, **13**, 9216.
36. Ma, X., Wang, Q.-C., Qu, D.-H., Xu, Y., Ji, F.-Y. and Tian, H. (2007) *Adv. Funct. Mater.*, **17**, 829.
37. Ma, X., Qu, D.-H., Ji, F.-Y., Wang, Q.-C., Zhu, L.-L., Xu, Y. and Tian, H. (2007) *Chem. Commun.*, 1409.
38. Ma, X., Wang, Q.-C. and Tian, H. (2007) *Tetrahedron Lett.*, **48**, 7112.
39. Inoue, Y., Kuad, P., Okumura, Y., Takashima, Y., Yamaguchi, H. and Harada, A. (2007) *J. Am. Chem. Soc.*, **129**, 6396.
40. Coulston, R.J., Onagi, H., Lincoln, S.F. and Easton, C.J. (2006) *J. Am. Chem. Soc.*, **128**, 14750.
41. Smiljanic, N., Moreau, V., Yockot, D., Benito, J.M., Fernández, J.M.G. and Djedaïni-Pilard, F. (2006) *Angew. Chem. Int. Ed.*, **45**, 5465.
42. Bender, M.L. and Komiyama, M. (1978) *Cyclodextrin Chemistry*, Springer-Verlag, Berlin.
43. Lee, W.-S. and Ueno, A. (2001) *Macromol. Rapid Commun.*, **22**, 448.
44. Liu, Y., Ke, C.-F., Zhang, H.-Y., Wu, W.-J. and Shi, J. (2007) *J. Org. Chem.*, **72**, 280.
45. Chen, Y., Han, K.-Y. and Liu, Y. (2007) *Bioorg. Med. Chem.*, **15**, 4537.
46. Corradini, R., Paganuzzi, C., Marchelli, R., Pagliari, S., Sforza, S., Dossena, A., Galaverna, G. and Duchateau, A. (2003) *Chirality*, **15**, S30.
47. Maeda, K., Mochizuki, H., Watanabe, M. and Yashima, E. (2006) *J. Am. Chem. Soc.*, **128**, 7639.
48. Kidowaki, M., Zhao, C., Kataoka, T. and Ito, K. (2006) *Chem. Commun.*, 4102.
49. Castro, R., Godínez, L.A., Criss, C.M. and Kaifer, A.E. (1997) *J. Org. Chem.*, **62**, 4928.
50. Mirzoian, A. and Kaifer, A.E. (1999) *Chem. Commun.*, 1603.
51. Retna Raj, C. and Ramaraj, R. (1999) *J. Photochem. Photobiol. A: Chem.*, **122**, 39.
52. Bergamini, J.-F., Belabbas, M., Jouini, M., Aeiyach, S., Lacroix, J.-C., Chane-Ching, K.I. and Lacaze, P.-C. (2000) *J. Electroanal. Chem.*, **482**, 156.

53. Lock, J.S., May, B.L., Clements, P., Lincoln, S.F. and Easton, C.J. (2004) *Org. Biomol. Chem.*, **2**, 337.
54. Willner, I., Pardo-Yissar, V., Katz, E. and Ranjit, K.T. (2001) *J. Electroanal. Chem.*, **497**, 172.
55. Tsuda, S., Aso, Y. and Kaneda, T. (2006) *Chem. Commun.*, 3072.
56. Santra, S., Zhang, P. and Tan, W. (1999) *Chem. Commun.*, 1301.
57. Cheng, X.L., Wu, A.-H., Shen, X.-H. and He, Y.-K. (2006) *Acta Phys. Chim. Sin.*, **22**, 1466.
58. Nelissen, H.F.M., Schut, A.F.J., Venema, F., Feiters, M.C. and Nolte, R.J.M. (2000) *Chem. Commun.*, 577.
59. Nozaki, T., Maeda, M., Maeda, Y. and Kitano, H. (1997) *J. Chem. Soc., Perkin Trans. 2*, 1217.
60. Aime, S., Botta, M., Gianolio, E. and Terreno, E. (2000) *Angew. Chem. Int. Ed.*, **39**, 747.
61. Heck, R., Dumarcay, F. and Marsura, A. (2002) *Chem. Eur. J.*, **8**, 2438.
62. Kretschmann, O., Choi, S.W., Miyauchi, M., Tomatsu, I., Harada, A. and Ritter, H. (2006) *Angew. Chem. Int. Ed.*, **45**, 4361.
63. Callari, F., Petralia, S. and Sortino, S. (2006) *Chem. Commun.*, 1009.

10
Photoswitchable Polypeptides
Francesco Ciardelli, Simona Bronco, Osvaldo Pieroni, and Andrea Pucci

10.1
Photoresponsive Polypeptides

The photoresponsive polypeptides discussed in this chapter are obtained by introducing photochromic units, such as azobenzene or spiropyran groups, into the macromolecules derived from α-aminoacids. As described in Chapter 1 of this book, photochromic compounds can exist in two different states, such as two isomeric structures, that can be interconverted by means of a light stimulus, and whose relative concentration depends on the wavelength of the incident light.

The occurrence of two different structures that can be interconverted by means of an external light stimulus can be the basis of a molecular switch. Moreover, when photochromic molecules are incorporated into polymeric compounds, their photoisomerization can affect the structure and the physical properties of the attached macromolecules.

Some fine review articles dealing with various aspects of photochromic polymers are reported in the literature [1–10] and several photoresponse effects have been described. They include light-induced conformational changes, photostimulated variations of viscosity and solubility, photocontrol of membrane functions, and photomechanical effects. Here, we provide an overview of the photoresponse effects observed in the field of photochromic polypeptides.

Polypeptides and poly(α-aminoacid)s have a unique position amongst synthetic polymers. The reason is that most common synthetic polymers have very little long-range order in solution and their properties are the result of statistical random coil conformations. Polypeptides, by contrast, can adopt well-defined ordered structures typical of those existing in proteins, such as α-helix and β-structures. Moreover, the ordered structures can undergo conformational changes to the random coil state as cooperative transitions, analogous to the denaturation of proteins.

The most widely known ordered structure of polypeptides is the α-helix. When L-configuration aminoacid residues are used as the repeating units, a right-handed α-helix is obtained, that is the backbone of the chain follows the thread of a right-hand screw. Each helix turn is composed of 3.6 residues, and hence the helix

Molecular Switches, Second Edition. Edited by Ben L. Feringa and Wesley R. Browne.
© 2011 Wiley-VCH Verlag GmbH & Co. KGaA. Published 2011 by Wiley-VCH Verlag GmbH & Co. KGaA.

is nonintegral. However, every 18 monomeric units sees an exact repeat of the structure, which corresponds to five turns of the helix and a linear translation of 27 Å along the axis and a 5.4 Å pitch. The diameter of the helix is approximately 6 Å, neglecting side chains.

The pattern of the hydrogen bonding is a key feature of the α-helix. All of the carbonyl groups point up and all of the N–H groups point down when the α-helix is viewed down from the C-terminus to the N-terminus. The spatial positioning of the constituent atoms is such that all of the C=O groups form a hydrogen bond parallel to the helix axis with the N–H group, that is four residues distant, which is parallel to the helix axis. A single polypeptide chain can therefore form a well-defined, ordered structure, that is stabilized from within the chain and does not require intermolecular interactions.

All of the side chains of the L-aminoacid residues point away from the axis, and hence the structure accommodates essentially any type of flexible side chain, and it is not necessary for the macromolecules to incorporate chemically identical repeat units. Units bearing differing side groups can be accommodated therefore within the same structure, as the backbone of the chain is the same for all the polyamino acids.

The β-structure is yet another frequently encountered ordered structure formed by polypeptides. In β-sheets the macromolecules form an almost planar zig-zag geometry. C=O and N–H groups form hydrogen bonds; however, in this case the bonding is *interstrand* instead of *intrastrand*. This arrangement results in the formation of sheets through the linking together of two polypeptide chains. These sheets pucker and the side chains are found above and below the sheet's plane; the side chains in adjacent macromolecules run parallel and are in sufficient proximity to stabilize the structure by hydrophobic interactions. Alternate chains align either parallel or antiparallel to each another. For the parallel β-structure, the chains are all aligned along the same direction, while for the antiparallel β-structure the chains alternate in opposite directions. The antiparallel case is encountered more frequently in artificial polypeptides. Finally, it goes without saying perhaps that polypeptide chains can also form essentially completely disordered conformations (i.e. random coils).

The most sensitive and widely employed technique in the determination of ordering in the structure of proteins is circular dichroism (CD) spectroscopy. Right-handed α-helices that are formed by polypeptides of amino acids in the L-configuration are by and large the best characterized of the structures. The α-helix structure exhibits a very distinctive CD spectrum (Figure 10.1) [11, 12]. Two negative bands at 222 nm ($[\Theta] = -35\,000$) and 208 nm ($[\Theta] = -33\,000$), respectively, and a positive band at 190 nm ($[\Theta] \approx +70\,000$) are the most prominent spectral features. The negative band at 222 nm is assigned to the peptide n–π* transition, whereas the negative band at 208 nm and the positive band at 190 nm arises from the exciton splitting of the π–π* transition of the peptide [12]. Naturally, there are minor variations in the specific intensity of the maxima/minima in the CD spectra of α-helices, however, the spectrum in Figure 10.1 is typical of simple α-helical polypeptides.

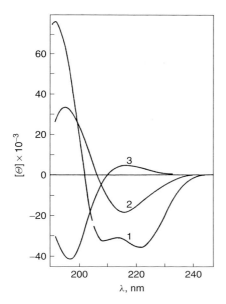

Figure 10.1 Standard circular dichroism (CD) spectra of the most common polypeptide structures: (1) α-helix; (2) β-structure and (3) random coil. Reproduced with permission from Ref. [15]. Copyright Wiley (1972).

The calculated CD spectrum of α-helical poly(L-alanine) [13] was found to agree closely with the experimental spectrum and although variation in solvent and the specific amino acid residues present do not usually affect the CD spectrum of the α-helix CD spectrum qualitatively, subtle variations in the relative amplitude of the 222- and 208-nm bands are observed [12, 14].

For polypeptides with a β-structure the CD spectra exhibit a negative band at approximately 216 nm ($[\Theta] \approx -18\,000$) and a positive band of more or less equal amplitude close to 195 nm (spectrum 2 in Figure 10.1). The CD spectra of β-structures are generally much more sensitive to variations in solvent and the specific amino acid residues present than α-helices are, in terms of signal amplitude and band position [12].

This variability may be a consequence of the more limited CD data available for β-structures, which has been found to be more challenging than for α-helices. Furthermore, theoretical studies have suggested that parallel and antiparallel β-structures cannot be distinguished readily on the basis of CD spectra alone [15].

The CD spectral features associated with random coil structures is typically characterized by an intense negative band at approximately 200 nm and a weak positive band at approximately 220 nm (spectrum 3 in Figure 10.1). The CD spectrum, however, can show considerable variation both between different polypeptides and depending on solvent composition and temperature [16–18].

When the side chains bear chromophoric groups near the backbone, for example in polymers with aromatic amino acids, the CD spectrum differs completely compared with the standard spectra of polypeptides. This is as a consequence of the interactions between the amide and aromatic transitions, in addition to the contribution of the chromophoric side chains to the CD spectrum in the peptide region. In general though, aromatic substituents at carbon atoms further than the γ-position do not, or only negligibly, contribute to the CD spectrum. For example, both poly(γ-benzyl-L-glutamate) and poly(N$^\varepsilon$-carbobenzoxy-L-lysine) show the expected CD spectra for α-helices, which indicates that in these polypeptides the aromatic groups are sufficiently removed from the backbone so as not to perturb the spectra to a significant extent [12].

10.2
Light-Induced Conformational Transitions

10.2.1
Azobenzene-Containing Polypeptides

Seminal studies in 1966–1967 reported a series of polypeptides derived from p-phenylazo-L-phenylalanine and γ-benzyl-L-glutamate prepared by polymerization of the corresponding N-carboxy-anhydrides [19]. These polypeptides displayed photochromic behaviour due to the photoisomerization of azobenzene moieties. The changes to the chiroptical properties between 300 and 500 nm could be assigned to changes accompanying the *trans–cis* isomerization of the azobenzene units that occurred without changes to the macromolecular conformations themselves.

These changes were observed for poly(L-glutamic acid) containing azobenzene units in the side chains (Figure 10.2, Structure I) up to 80 mol% also [20, 21].

The photochromic reaction is shown in Figure 10.2. All of the azo groups are in the *trans* configuration at room temperature in the dark. In this configuration the azobenzenes are planar and fully conjugated. Irradiation at 350 nm results in isomerization to the *cis* isomer, which is nonplanar and less-well conjugated. The reverse reaction to the *trans* form is achieved by irradiation at 450 nm or thermally in the dark.

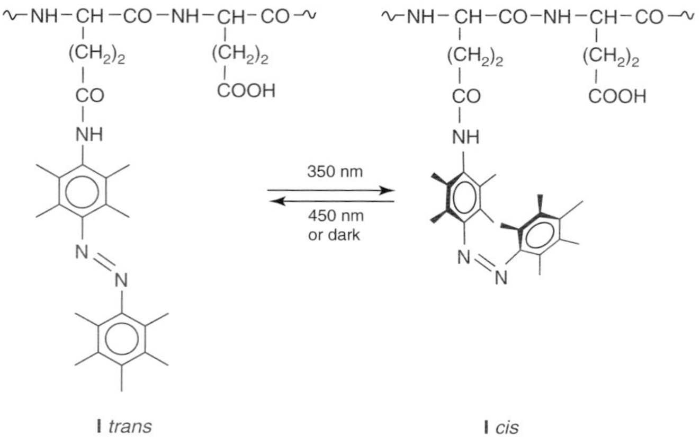

Figure 10.2 Photochromic reactions of poly(L-glutamic acid) containing azobenzene units in the side chains (I).

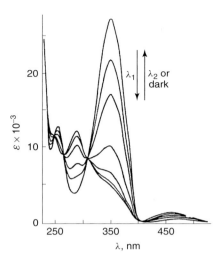

Figure 10.3 Reversible light-induced variations of the absorption spectra in azo-modified poly(L-glutamic acid) (I). Reproduced with permission from ref [21]. Copyright Wiley (1984).

The photoisomerization is accompanied by large changes in the absorption spectra (Figure 10.3). In particular, the *trans* → *cis* isomerization results in a large decrease in the intensity of the band at circa 350 nm, assigned to a $\pi-\pi^*$ transition, and a concomitant increase in the intensity of the band at 450 nm assigned to the $n-\pi^*$ transition of the azo chromophore [22].

Maximum photoconversion to the *cis* isomer (85%) is obtained by irradiating between 350 and 370 nm, while the maximum conversion in the reverse from the *cis* to the *trans* isomer (80%) is obtained by irradiating at 450 nm. With a 200-W lamp, irradiation for 1–2 min is usually enough to achieve the photostationary state in optically dilute solutions. The thermal reversion (in the dark) is much slower at room temperature, taking more than 200 h to revert fully to the fully *trans* isomeric composition. The photochromic cycles are fully reversible and can be repeated multiple times without evident fatigue [21, 22].

In organic solvents, for example trimethylphosphate (TMP) or trifluoroethanol, the CD spectrum of the azo-modified polymers are typical of the α-helix structure, with the two expected minima at 208 and 222 nm (Figure 10.4).

In TMP, the thermally reverted samples exhibit an intense couplet of bands centred at 350 nm also, assigned to the $\pi-\pi^*$ transition of the azo chromophore in the *trans* configuration. This couplet is associated with dipole–dipole interactions between azo side chains within the α-helix structure and is absent in trifluoroethanol. Irradiation, which leads to *trans* → *cis* photoisomerization results in a complete disappearance of the CD signals of the side chains in the 250–450 nm region, but does not result in changes to the CD spectra in the peptide region (below 250 nm). This indicates that, in these solvents, light drives the *trans–cis* isomerization of the azo side chains but this does not result in a change to the main chain structure of the polypeptide [22]. The typical pattern of the α-helix and the absence of variations below 250 nm upon *trans–cis* isomerization are strong

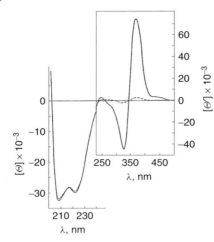

Figure 10.4 Poly(L-glutamic acid) containing 56 mol% azobenzene units in the side chains (I). CD spectra in trimethylphosphate before (continuous line) and after (dashed line) irradiation at 360 nm. Below 250 nm, molar ellipticity is based on the mean residue weight; above 250 nm, the ellipticity is referred to one azo-Glu residue. Reproduced with permission from ref [22]. Copyright Elsevier (1983).

evidence that the azo chromophores do not contribute to the CD spectra in the peptide region.

Similar findings have been reported for a number of polypeptides containing p-phenylazo-L-phenylalanine residues (Scheme 10.1, residues **II**) [23, 24]. Upon irradiation, the polypeptides showed reversible changes in optical rotation at 589 nm in agreement with the *trans–cis* photoisomerization of the azo units, despite the fact that the photoisomerization of the side chains does not change the α-helical conformation of the main chain. The photoreversible change in optical rotation is a potentially useful tool in achieving chiroptical photorecording, in which a digital record written by photoisomerization (*trans–cis*), can be read out *via* the change in optical rotation at wavelengths longer than those used for the recording. This avoids destruction of the stored information by the read-out process [23, 24].

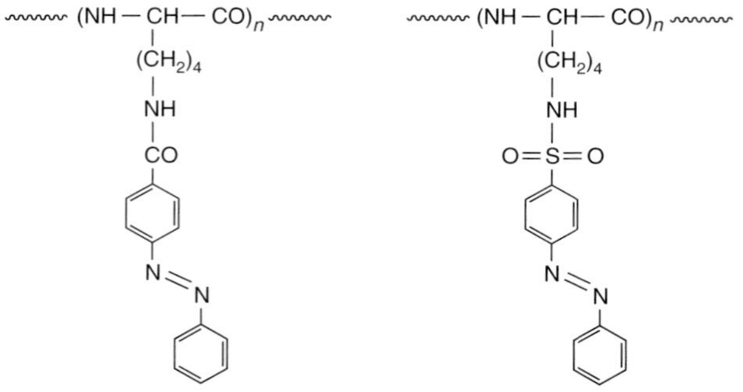

Scheme 10.1 Chemical structures of poly(N^ε-p-phenylazobenzoyl-L-lysine) (**II**), and poly(N^ε-p-phenylazobenzenesulfonyl-L-lysine) (**III**).

The conformational behaviour can be very different in aqueous solutions. Below pH 5, a sample of poly(L-glutamic acid), which contains circa 30 mol% azobenzene units, takes a β-structure that is unaffected by irradiation. Above pH 7, the polypeptide is present in a random coil conformation, which again is unaffected by the photoswitching of the azo side chains. However, in the pH range 5–7 (close to the pK of the conformational transition), irradiation results in a substantial decrease in the order of the structure; a change that is reversed fully over time in the dark [20, 21].

The mechanism of the photoresponse was rationalized as follows. the *trans–cis* photoisomerization of the azo side chains is the primary event; although a simple variation in the geometry of the azo units does not appear to be enough to drive substantial changes in the conformational state of the macromolecules (indeed in organic solvents no effect was observed). The primary factor responsible for the photoinduced conformational transition is most probably variation in the polarity of the immediate environment of the macromolecules, which is a result of the differences in polarity and hydrophobicity between the *trans* and *cis* azo isomers. In fact, the apparent pK_a of the unmodified COOH groups is decreased by switching the neighbouring azo units from the more hydrophobic *trans* configuration ($pK_a = 6.8$) to the more polar *cis* configuration ($pK_a = 6.3$). As a consequence, the *trans* → *cis* photoisomerization of the azo units proceeds with a concomitant increase in the degree of deprotonation of the neighbouring COOH side chains, thereby increasing ionic interactions among side chains and resulting in unfolding of the polypeptide.

Analogous polypeptides that were prepared by reacting poly(L-glutamic acid) with 4-amino-azobenzene-4-sulfonic acid sodium salt exhibited pH and azo content dependent conformational and photoresponsive behaviour [25]. A polypeptide, which contained a very low number (1.9%) of azobenzene sulfonate units, exhibited a pH-dependent α-helix/coil transition, whereas a polypeptide with 46 mol% azo units was, at all pH values, a random coil. Nevertheless, irradiation with UV light and the resulting *trans* → *cis* isomerization of the azo units failed to induce conformational changes for either of the polypeptides at any pH value. By contrast, for a polypeptide containing 9.3 mol% azobenzene sulfonate units substantial conformational changes upon UV irradiation were observed at certain pH values. At pH 4.3, irradiation resulted in changes to the CD spectrum that corresponded to a decrease in the α-helical structure content from 96 to 45%.

The effect was rationalized as due to a change in the geometry of the azo moieties, as a consequence of the *trans–cis* photoisomerization, which may result in an increase in local charge density around the helical backbone, thereby destabilizing the ordered structure [25]. However, the conformational change induced by irradiation was irreversible and neither irradiation at $\lambda > 390$ nm nor thermal reversion resulted in restoration of the original structure.

Irradiating of a 20% azo-modified poly(L-glutamic acid) (**I**) in the presence of the surfactant dodecyl ammonium chloride (DAC) allowed for amplification of the photoresponse to be observed. CD spectroscopy indicated that at pH 7.6, when surfactant was not present, the polymer is fully in a random coil state and unaffected by irradiation. At the same pH, where DAC was present but below the critical micellar concentration, both the thermally reverted and the irradiated samples are found to be α-helical. With surfactant at its critical micellar concentration, a coil-to-helix transition is induced by irradiation at 350 nm (*trans* → *cis* isomerization). The change is reversed fully by thermal reversion in the dark the sample or by irradiation at 450 nm (*cis* → *trans* isomerization). Hence, in the presence of DAC micelles, the conformation of the polypeptide can be photomodulated by alternating between light and dark conditions or by irradiation at the two appropriate wavelengths [26].

The mechanism by which the photoresponse occurs was rationalized tentatively as follows. When the azo units are planar and relatively apolar, that is the *trans* configuration, they dissolve within the hydrophobic core of the micelles. This forces the polypeptide chains to assume a coil conformation. Isomerization of the azo units to the bent and polar *cis* configuration reduces hydrophobic interactions and results in the azo units leaving the micelles, thereby allowing the polypeptide chains to adopt the α-helix structure that is favoured in the absence of micelles.

Furthermore, poly(*L*-glutamate) containing 13 mol% azobenzene units in the side chains [27] and that is esterified partially yields a similar response when incorporated into the bilayer membrane of vesicles that are comprised of distearyl dimethyl ammonium chloride. Irradiation of the vesicles with UV light results in a transfer of the polypeptide molecules from the hydrophobic interior to the hydrophilic surface of the bilayer membrane. This synthetic system mimics certain biological photoreceptors, including photopigments found in the retinal membranes of frogs, which change their localization between the aqueous interface and the hydrocarbon core of the membrane, depending on whether the photopigment is under irradiation or is in darkness [28].

Poly(*L*-lysine) containing azobenzene units tethered to the side chains by an amide moiety are shown in Scheme 10.1, structure **II**) [29–31]. Polymers containing various amounts (20–90 mol%) of azobenzoyl-*L*-lysine residues (**II**) are soluble in hexafluoro-2-propanol (HFP) and exhibit very similar CD spectra, independent of the concentration of the azobenzene unit. The thermally reverted samples (azo groups in *trans* configuration) show a CD pattern below 250 nm typical of an α-helix. Weak bands are also observed in the wavelength range 250 and 500 nm, which arise from the dissymmetric perturbation of the azo chromophores by the polypeptide chains. Irradiation at 340 nm, and alternately at 450 nm, produced reversible changes to the CD spectrum at >250 nm, however it did not result in modification of the CD spectra in the peptide region (<250 nm) [29]. Furthermore, for poly(*L*-lysine) where the content of azo moieties is 97 mol%, a decrease in the CD signal was observed only after extended irradiation (>4 h), albeit irreversibly [30]. It could be concluded that, in HFP, light causes the *trans–cis* isomerization of the azo side chains, however, this is not accompanied by changes to the conformation of the macromolecular backbone.

Poly(*L*-lysine) in which azobenzene units are linked to the side chains through a sulfonamide moiety (Scheme 10.1, Structure **III**), were soluble in HFP, in which it exhibited the intense photochromism of the *trans–cis* photoisomerization of the azobenzene units, as observed for related sulfonated azobenzene compounds [32].

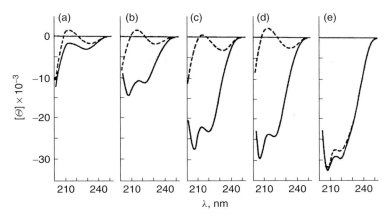

Figure 10.5 CD spectra of poly(N$^\varepsilon$-p-phenyl-azobenzensulfonyl-L-lysine) (**III**) in various HFP/MeOH solvent mixtures (v/v): (a) 0%; (b) 2%; (c) 5%; (d) 8%; (e) 15%. Continuous line, kept in the dark or irradiated at 417 nm; dashed line, irradiated at 340 nm. Reproduced with permission from ref [29]. Copyright Wiley (1987).

This azobenzensulfonyl-based poly(L-lysine) (**III**) exists in a random coil conformation in pure HFP, with *trans–cis* photoisomerization of the azo side chains having no effect on the disordered conformation. Where sufficient amounts of cosolvents, that is 1,2-dichloroethane (DCE) or methanol (MeOH), were added to the HFP containing **III**, the system was found to become response to irradiation with reversible changes to the conformation of the polypeptide (Figure 10.5). In HFP alone, the CD spectra are typical of random coil polypeptides both when the sample is irradiated at 340 nm (where the azo units are in *cis* configuration) or at 417 nm (where the azo units are in *trans* configuration). When the methanol content is higher than 15%, however, under both conditions the CD pattern typical of an α-helix is observed. The intensity of the CD band at 222 nm corresponds to that measured in HFP solution for poly(N$^\varepsilon$-carbobenzoxy-L-lysine) ([Θ]$_{222}$ = −28 900), which is consistent with formation of 100% of the α-helix [12]. When the methanol concentration is between 2 and 15%, irradiation at 340 and 417 nm, alternately, results in photoinduced changes to the helical content, with the degree of the photoresponse observed being dependent on the exact solvent composition. Photoinduced changes in the helical structure up to about 80% are obtained with concentrations of methanol between 8 and 10%.

If the intensity of the CD band at 222 nm, which can be considered as a measure of the α-helix content, is plotted as a function of the concentration of methanol, then it is apparent that addition of methanol results in a change in the macromolecular conformation from random coil to α-helix transition (Figure 10.6). However, the relative amount of methanol required to induce this conformational transition is dependent on whether the sample is irradiated at 340 nm (*cis* azo units) or at 417 nm (*trans* azo units). As a result, two distinct curves are obtained for the two samples. At solvent compositions intermediate between the two curves, irradiation

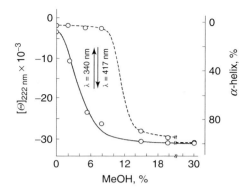

Figure 10.6 Poly(N^ε-p-phenylazobenzensulfonyl-L-lysine) (III) in HFP/MeOH solvent mixtures: ellipticity at 222 nm and α-helix content per cent, as a function of methanol concentration, for the samples irradiated at 417 (continuous line) and 340 nm (dashed line). Reproduced with permission from ref [29]. Copyright Wiley (1987).

at 340 and 417 nm results in folding or unfolding of the macromolecular chains. The photoresponse can be obtained only in a narrow range of solvent mixture compositions and hence it is an example of a *gated photoresponse* [33, 34].

The differences in the conformational behaviour of the azobenzoyl- and azobenzensulfonyl-L-lysine polymers was rationalized on the basis that the monomeric units **III** can interact with HFP differently than the units of **II** (Scheme 10.1). In fact, the relatively acidic solvent HFP ($pK_a = 9.30$) [35] is known to form electrostatic complexes with a wide range of organic compounds, for example amines and dimethylsulfoxide [36]; furthermore, sulfonamides are readily protonated in acid media [37]; so it can be presumed that protonation and formation of electrostatic complexes will occur for azobenzensulfonyl-L-lysine residues, also. Therefore in HFP, polypeptides of structure **II** can adopt an ordered α-helix structure, whereas polypeptides of structure **III** are forced as a consequence of the electrostatic interactions, due to complexation with HFP, to remain in a disordered conformation.

Naturally the stability and the formation of 'HFP·azosulfonyl-Lys' complexes is disfavoured upon addition of methanol and more so DCE to the HFP solutions. Therefore the *trans–cis* photoswitching of the azo units, the primary photochemical event, appears to be insufficient to induce appreciable changes to the conformation of the backbone. Indeed, photoinduced conformational changes are not observed in pure HFP or at high concentrations of MeOH. Under specific solvent conditions, the *trans–cis* photoisomerization of the azosulfonyl units **III** should result in a change in the protonation state of the sulfonamide moiety, which should be the key parameter in the photoregulation of the conformation of the polypeptide [38].

Poly(α-amino acid)s homologues of poly(L-lysine), such as poly(L-ornithine) [39, 40], poly(L-α, γ-diamino-butanoic acid) [41] and poly(L-α, β-diamino-propanoic acid) [42], provide considerable information with respect to the photoresponse of azo-modified polypeptides as they allow for the photochromic units to be attached to the macromolecules through different lengths of spacers.

Poly(L-ornithine)s, with a range of contents of azobenzene groups, from 20% to nearly 100%, were found to adopt an essentially α-helical conformation in HFP before irradiation. The CD spectra exhibited a couplet of bands at circa 320 nm also, which is assigned to be due to the electronic interactions between the azo side chains in the *trans* configuration. Photoisomerization from *trans* to *cis* leads to a loss in the CD bands due to the side chain and reduced helix content to about half of the original value. Irradiation at 460 nm led to only a partial reverse photoconversion [39, 40].

A sample of poly(L-ornithine), that had a content of azo units of 48 mol%, was found to form an α-helix structure in HFP/water = 1/1. For this solvent mixture, irradiation at 360 nm and subsequently at 460 nm resulted in the expected *trans–cis* photoisomerization of the azo moieties, however, no change in conformation of the backbone was observed. Addition of the surfactant sodium dodecyl sulfate to the HFP/water solvent mixture resulted in the CD spectrum displaying an intense CD couplet from the side chain and a negative band at circa 225 nm that was assigned to the formation of a β-structure. The CD bands underwent almost complete cancellation upon *trans* → *cis* photoisomerization.

A polymer of L-α,γ-diamino-butanoic acid that was substituted, essentially quantitatively, with azobenzene units in the side chains was not fully soluble in HFP when the sample was kept in the dark. The initially slightly turbid solution clarified upon irradiation at 360 nm and underwent photoconversion of the azo moieties from the *trans* to the *cis* configuration (for photosolubility effects see below). The '*cis*' polymer was found to exist essentially in a random coil conformation. Irradiation at 460 nm and the resulting reverse isomerization of the azo units to a isomeric composition of 70/30 *trans–cis* results in a reversible photoinduced transition from a random coil to an α-helical structure (helix content, circa 60%) [41].

The analogous polymer prepared from L-α,β-diamino-propanoic acid exhibited photochromic behaviour similar to that observed for the other homologues. However, irradiation at 360 nm resulted in changes to the CD spectra in the peptide region, and structural changes to the macromolecules, which were irreversible [42].

Another photochromic polymer containing azobenzene units has been prepared by modifying the naturally occurring microbial poly(ε-L-lysine) and studied by UV–vis absorption and CD spectroscopy [43]. The structure of the polymer, however, does not correspond to the structure of poly(amide)s of α-amino acids and hence the results cannot be interpreted by direct comparison with the typical polypeptide structures (α-helix, β-structure, random coil) and their CD spectra.

A distinct photochromic behaviour was found for poly(L-aspartate)s, which adopt helical structures with both left-handed and right-handed screw senses; the relative stability of two helices is dependent on the chemical structure of the ester group of the side chains [44]. Furthermore, for poly(β-benzyl-L-aspartate), the inclusion of substituents, for example methyl, chloro or nitro groups on the benzyl ring leads to helical polypeptides that form left-handed or right-handed helices depending on the

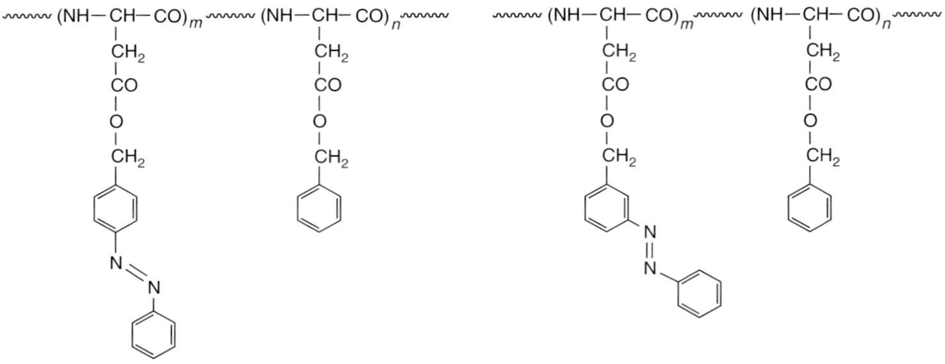

Scheme 10.2 Chemical structure of poly(β-L-aspartate)s having various contents of *para*- (**IV**) and *meta*-phenylazobenzyl (**V**) units in the side chains.

position of the substituent [45]. A series of poly(L-aspartate)s bearing azobenzene units on the side chains (Scheme 10.2), have been studied [46–49].

In 1,2-DCE, polypeptides incorporating *para*-phenylazo-L-aspartyl residues (Scheme 10.2, Structure **IV**) show CD spectra with a positive CD band at circa 220 nm, which indicates the presence of a left-handed helical structure. The band was unaffected by irradiation between 320 and 390 nm, where the azo content was less than 50 mol%. For two copolymers, which contain 59 and 81 mol%, respectively, of *para*-phenylazobenzyl-aspartyl residues, the behaviour was found to be very different: prior to irradiation a positive band at 220 nm is observed, which switched to a negative band upon irradiation. The change in sign was taken as evidence for the reversal of the sense of the helix driven by the *trans–cis* photoisomerization of the azo units. For analogous polypeptides bearing *meta*-phenylazobenzyl-L-aspartyl residues (Scheme 10.2, Structure **V**) a reversal of the 220-nm band was not observed upon irradiation, rather only a decrease of the intensity of the band. This difference is indicative of the formation of significant amounts of the random coil structure [46–48].

Where the photoswitching was carried out at appropriate solvent compositions, large responses were observed. A copolypeptide comprising of 33 mol% β-benzyl-L-aspartate and 67 mol% *para*-phenylazo-L-aspartate (**IV**) was found to give different responses to irradiation depending on the composition of solvent employed. In DCE/HFP = 95/5, irradiation at 320–390 nm resulted in an increase in the content of the right-handed helix; in DCE/HFP = 74/26, a light-induced change in conformation from a left-handed helix to a random coil conformation was observed; and finally in DCE/HFP = 65/35, irradiation resulted in the reversal of the helix sense [50, 51].

Two polymers with a 8 and 10 mol% content of *meta*-phenylazo-L-aspartyl residues (**V**), respectively, were found to form left-handed helices in pure DCE, and right-handed helices in pure TMP. The inversion of the helix was observed when the solvent composition had a 20–50% TMP concentration; however, the dependence of the helix sense on the concentration of TMP was different for the samples kept in the dark (azo units in *trans* configuration) and the irradiated samples (azo units in *cis* configuration). Large changes in the CD spectra were observed when irradiation was carried out in solvents with a concentration of TMP between 20 and 50%. Especially when the solvent contained 25–30% TMP, irradiation resulted in an inversion of the CD band at 222 nm, indicating a substantial change in conformation from a left-handed to a right-handed helix, even for polypeptides containing a relatively small number of photochromic units [52–54].

Copolymers containing *p*-phenylazobenzyl-*L*-aspartate and *n*-octadecyl-*L*-aspartate residues [55, 56] with a 50 mol% azo residue content, display CD spectra, at 25 °C, that are consistent with the presence of right-handed helical conformations, and are unaffected by irradiation at 320 nm. In contrast, for copolymers with a content of azobenzene groups between 68 and 89 mol%, irradiation resulted in the reversal of helix sense from the left to right handed.

The conformations of these polypeptides were highly dependent on temperature, so the variation in photoresponse could be obtained provided that the irradiation was carried out under appropriate conditions of azo-content and temperature. A copolymer containing 47 mol% azo units, which was unaffected by irradiation at 25 °C, was observed to undergo a photoinduced helix reversal upon irradiation at 60–70 °C. The octadecyl side chains most probably change the orientation of their array simultaneously due to the photoinduced structural changes of the main chains, and hence the system is an example of an environmental change induced by irradiation.

The study of the light-induced conformational changes was extended to solid films of azobenzene-containing poly(*L*-aspartate)s also, however, conformational change was not induced by the photoisomerization of the azobenzene units. This was ascribed to the limited mobility of the polypeptide chains within the films [57].

The azo-modified elastin-like polypeptide **VI** shown in Scheme 10.3 undergoes a so-called 'inverse temperature transition': that is the compound forms crosslinked gels that remain swollen in water at temperatures below 25 °C, yet deswell and

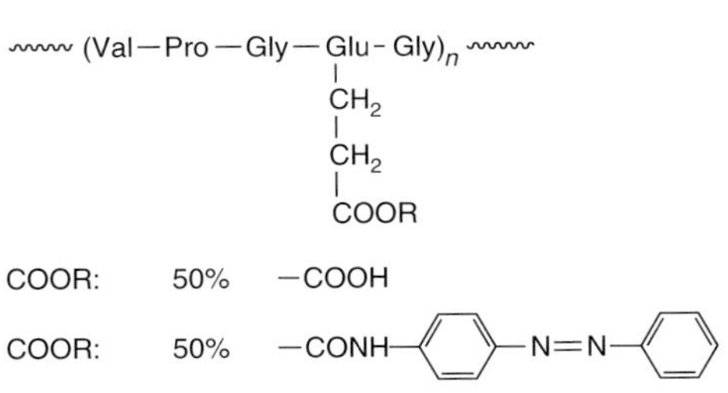

~~~~ (Val—Pro—Gly—Glu-Gly)$_n$ ~~~~
  |
  CH$_2$
  |
  CH$_2$
  |
  COOR

COOR: 50%   —COOH

COOR: 50%   —CONH—⟨ ⟩—N=N—⟨ ⟩

VI

Scheme 10.3 Chemical structure of the modified elastin-like poly(pentapeptide) **VI** which was found to give photomodulated inverse temperature transition [58].

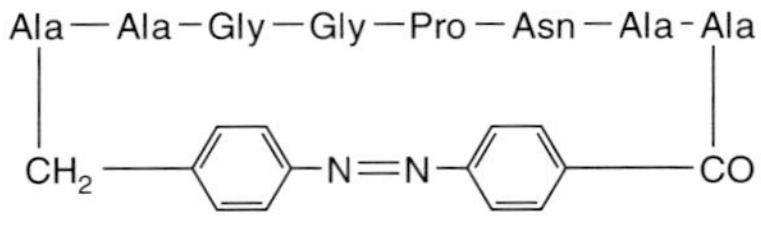

Ala—Ala—Gly—Gly—Pro—Asn—Ala-Ala
| |
CH₂—〈 〉—N=N—〈 〉—CO

VII

Scheme 10.4 Chemical structure of the photochromic cyclic peptide **VII** [59].

contract with an increase in temperature. The *trans–cis* photoisomerization of the azo units, which occurred upon irradiation at 350 and 450 nm, alternately, provides for the photomodulation of the inverse temperature transition [58]. The result indicates that modification of the polymer with a small number of azobenzene chromophores is enough to render the inverse temperature transition of elastin-like polypeptides photoresponsive and provides a route to protein-based polymeric materials that display photomechanical transduction.

Although it is perhaps obvious, it is nevertheless important to note that conformational changes occur when the azobenzene moiety is inserted in the polypeptides backbone rather then grafted to the side chains. This refers to both cyclic or open chain oligo- and polypeptides.

Thus, for the cyclic peptide **VII** (Scheme 10.4) [59] with azobenzene in *trans* configuration (samples kept in the dark), the peptide exhibited an extended, even if cyclic, configuration, whereas after the photoisomerization to the *cis* form (samples irradiated at 310–410 nm), the peptide adopted a 'β-turn' having a decreased cycle area.

In the case of the bicyclic peptide bcAMPB, obtained by connecting head to tail the octapeptide fragment H-Ala-Cys-Ala-Thr-Cys-Asp-Gly Phe-OH with (4-amino-methyl)-phenyl azobenzoic acid and the two cys residues through a disulfide bridge, a replica exchange molecular dynamics (REMDs) simulation in dimethyl sulfoxide solution [60] confirmed that the *trans* isomer of the azobenzene peptide exhibits a well-defined structure [61], while the *cis* isomer is a conformational heterogeneous system; that is the *trans* isomer occurs in two well-defined conformers, while the *cis* isomer represents an energetically frustrated system that leads to an ensemble of conformational isomers.

The calculation of time-dependent probability distributions along various global and local reaction coordinates reveals that the conformational rearrangement of the peptide is rather complex and occurs on at least four timescales: (i) after photoexcitation, the azobenzene unit of the molecule undergoes nonadiabatic photoisomerization within 0.2 ps, (ii) on the picosecond timescale, the cooling (13 ps) and the stretching (14 ps) of the photoexcited peptide is observed, (iii) most reaction coordinates exhibit a 50–100 ps component, reflecting a fast conformational rearrangement; (iv) the 500–1000 ps component observed in the simulation accounts for the slow diffusion-controlled conformational equilibration of the system. This simulation is in remarkable agreement with time-resolved optical and infrared (IR) experiments, although the calculated cooling as well as the initial conformational rearrangements of the peptide appear to be somewhat too slow [62].

A hairpin, the smallest β-type structures in peptides and proteins, incorporating an azobenzene-based photoswitch, allowed study of time-resolved folding of β-structures with high time resolution. Light-induced isomerization to the *cis*-azo form led to a predominantly extended and parallel conformation of the two peptide parts, by the 3-(3-aminomethyl)-phenylazo]phenylacetic acid. By contrast, in the original sequence the dipeptide Asn-Gly forms a type I' β-turn that connects the two strands of the hairpin. The β-hairpin structure was determined and confirmed by NMR spectroscopy, but the folding process can be monitored by pronounced changes in the CD, IR and fluorescence spectra.

The photochemical investigation and conformational analysis of a stilbene-type β-hairpin mimetic was also reported [63]. It was shown that the incorporation of the photochromic group into a cyclic peptide like β-hairpin allows for light-triggered switching between different conformations.

An original approach to photocontrol the peptide and protein conformation was based on the use of an azobenzene-based thiol reactive bifunctional derivative creating a crosslinking between $i, i+4$ or $i, i+11$ cysteine residues in the peptide sequence. Indeed the *trans*-to-*cis* photoisomerization increases the helix content in the $i, i+11$ case [64, 65].

The elastic behaviour of the photoswitchable polyazobenzene peptide composed of (Lys-Azo-Gly) azotripeptide units was simulated by molecular dynamics. The maximal calculated extension of 1.7 Å/per unit from the *trans* to *cis* isomer was experimentally found to be only 0.6 Å. This result was explained on the basis of the length limitation by the interlinking lysine residues. An increase of the work output was then obtained by replacing lysine with stiffer proline units [66].

10.2.2
Spiropyran-Containing Polypeptides

Photochromism of spiropyran compounds involves two photoisomers, the neutral spiro form and the zwitterionic merocyanine form, which are characterized by large differences in geometry and polarity (Figure 10.7).

The interconversion of these two forms upon exposure, even only to sunlight, can cause large structural changes in the attached macromolecules.

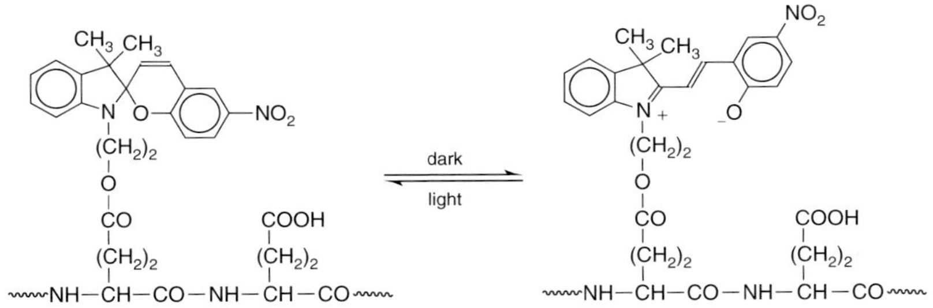

Figure 10.7 Structure and reverse photochromic reactions in hexafluoro-2-propanol of poly(L-glutamic acid) containing spiropyran units in the side chains.

Poly(L-glutamate)s containing various molar percentages of spiropyran units in HFP, in which they display a reverse photochromism, that is photochromic behaviour that is opposite to that normally observed in most common organic solvents. At room temperature in the dark, they yield coloured solutions because of the presence of the merocyanine form; upon irradiation with visible light or exposure to sunlight complete bleaching of the solutions occurs because of the formation of the colourless spiro form. The reverse reaction occurs in the dark and the original coloured state can be fully recovered. The reversed photochromism observed in the highly polar solvent HFP is ascribed to stabilization of the charged merocyanine form more than the apolar spiro form [67–69].

The coloured solution of a poly Glu with 85% spiropyran kept in the dark shows two intense bands at 500 and 370 nm in the UV–vis absorption spectrum as a result of the merocyanine species. Irradiation with visible light (500–550 nm) or exposure to sunlight causes a disappearance of the intense band in the visible region and results in a spectrum corresponding to the spiro form, with absorption maxima at 355 and 272 nm. In the dark, the original spectrum gradually recovered, with an isosbestic point in the spectra monitored over time at 295 nm (Figure 10.8).

The photochemical reaction proceeds rapidly with exposure to sunlight even for a few seconds effecting full conversion of the merocyanine to the spiro form. The reverse reaction in the dark proceeded at a much slower rate: it takes between 150 and 250 min for the various polymers to recover half of their original absorbance [67, 68]. The photochromic cycles appear to be fully reversible. It is likely that irradiation with low-energy visible light (reverse photochromism) instead of high-energy UV light (normal photochromism) can reduce undesirable photochemical side reactions and associated *fatigue* phenomena.

Before irradiation, the coloured solutions show the CD spectrum of a random coil conformation (Figure 10.9, 1). After exposure to sunlight, the colourless solutions display the typical CD pattern of the α-helix (Figure 10.9, 2), thus indicating that the isomerization of the side chains causes the coil-to-helix transition of the polypeptide chains.

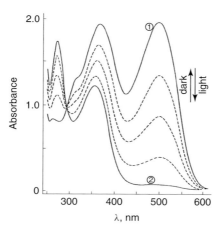

Figure 10.8 Absorption spectra of poly(L-glutamic acid) containing 85% spiropyran units, in HFP: (1) sample kept in the dark and (2) exposed to sunlight; dashed lines: intermediate spectra during decay in the dark. Reproduced with permission from ref [68]. Copyright Wiley (1993).

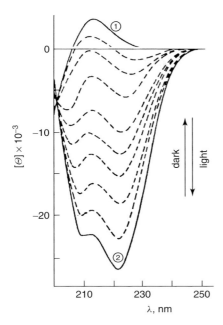

Figure 10.9 Effect of irradiation and dark adaptation on CD spectra of poly(L-glutamic acid) containing 85 mol% spiropyran units, in HFP: (1) kept in the dark and (2) exposed to sunlight; dashed lines: intermediate spectra during decay in the dark over 8 h. Reproduced with permission from ref [68]. Copyright Wiley (1993).

The photoinduced conformational variations are fully reversible: on dark adaptation, the helix content progressively decreases and the original disordered conformation is restored.

On the basis of fluorescence measurements, the driving force for the photoinduced conformational change was attributed to the interactions between the photochromic side chains, which are different when they are in the zwitterionic merocyanine form or in the apolar spiro form. In the dark, the merocyanine units have a strong tendency to give dimeric species; as a result the macromolecules are forced to adopt a disordered structure. When the side chains are photoisomerized to the spiro form, such dimers are destroyed, and the macromolecules assume the helical structure [70].

The kinetics of the helix-to-coil reaction in the dark for a polypeptide containing 33 mol% spiropyran units were investigated by means of CD and Fourier transform infrared (FTIR), as well as molecular-dynamics simulations [69, 71]. The polypeptide was found to undergo a slow transition according to the mechanism 'helix/solvated-helix/coil'. During the 'helix/solvated-helix' step, approximately 25% of the α-helix hydrogen bonding broke and formed hydrogen bonds between the unmodified carboxylic and merocyanine groups. No changes in carboxylate hydrogen bonding were observed during the 'solvated-helix/coil' step and the breakup of the helix [69].

When spiropyran-modified poly(L-glutamate) was dissolved in HFP and a small amount of trifluoroacetic acid (TFA, $c = 5 \times 10^{-4}$ g/ml) added to the solution [68] the photoisomerization did not result in any conformational variation of the macromolecular main chains, and CD spectra showed that the macromolecules

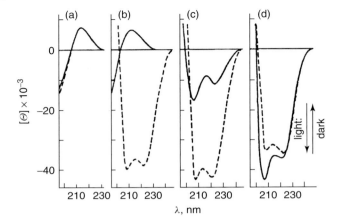

Figure 10.10 Poly(L-glutamic acid) containing 85 mol% spiropyran units in the side chains. Effect of irradiation on CD spectra in various HFP/MeOH solvent mixtures in the presence of trifluoroacetic acid (TFA, $c = 5 \times 10^{-4}$ g/ml). MeOH: (a) 0–5%; (b) 10%; (c) 20% and (d) 40%. Continuous line, dark adapted; dashed line, irradiated samples. Reproduced with permission from ref [68]. Copyright Wiley (1993).

were random coils both in the dark and after light exposure. However, when appropriate amounts of methanol were added as a cosolvent, the system again responded to light giving random coil → α-helix transitions (Figure 10.10).

A possible interpretation of this behaviour is that in HFP acidified upon addition of TFA, spiropyran compounds are present as protonated merocyanine MeH^+. Exposure to light converts the species MeH^+ into the ring-closed spiro species SpH^+. In the presence of acid, therefore, the photochromic side chains are present as cationic species in the dark as well as in light conditions. In both cases the repulsive electrostatic interactions among the charged side chains force the macromolecules to adopt an extended random coil structure and no photoinduced conformational change is observed as a result of photoisomerization.

When appropriate amounts of methanol are added to the HFP solution, the protonated species MeH^+ present in the dark is not altered, but the equilibrium between protonated and nonprotonated spiro units present in the irradiated solution is shifted towards the neutral form. Under these conditions, the photochromic species in the side chains are charged in the dark and are neutral under light, so irradiation induces formation of α-helix as it does in HFP without acid. Formation of α-helix even in the dark-adapted samples at high methanol concentration may be due to the same effect observed for other poly(α-aminoacid)s with ionic side chains such as poly(sodium L-glutamate) [72a] and poly(L-lysine hydrochloride) [72b] that are random coils in water but become helical upon addition of methanol in excess. Such an effect seems to be due to the ability of methanol to favour 'contact ion pairs' between polymer charges and counterions, thus providing a shielding effect among the charged side chains and stabilizing the helical structure [72].

Similar photochromic behaviour in HFP was observed for the polymers of
L-lysine containing spiropyran units in the side chains, whereas conformational
and photoresponsive behaviour was quite different [73, 74].

In fact, while spiropyran-modified polymers of L-glutamic acid undergo coil →
α-helix transitions upon exposure to light, the analogous polymers of L-lysine do
not give light-induced conformational changes in pure HFP, and their structure
is always random coil, either when the samples are kept in the dark or when they
are exposed to light. This different conformational behaviour is likely to be due
to the unmodified lysine side chains that are most likely protonated by the acid
solvent HFP. As a result, the macromolecules are essentially polycations that adopt
extended coil conformations, which are not affected by the photoisomerization
of the photochromic units. However, when appropriate amounts of triethylamine
(NEt_3) are added to the HFP solutions, the system again responds to light, giving
coil → α-helix conformational changes [73, 74].

Two separate curves are hence observed: exposure to light and dark conditions
at solvent compositions in the range between the two curves produces reversible
photoinduced conformational changes. The described system is an example of
a photoresponsive system having a *gated photoresponse* [33, 34], in the sense
that the photoisomerization of the side chains is able to trigger the coil → helix
transition of the macromolecular chains only in a narrow 'window' of environmental
conditions.

The role of triethylamine is not clear. A possible effect could be the removal
of protons from the unmodified amino side chains. Under these conditions the
macromolecular conformation could be controlled by the isomerization of the
photochromic groups as occurs in poly(spiropyran-L-glutamate). Alternatively, the
system might behave like other polypeptides that are random coils in pure solvents
such as dimethyl sulfoxide or dichloroacetic acid, but become helical in a mixture
of the two solvents [75]. The effect was attributed to the formation of a complex
derivative between the solvent components, which decreases their ability to solvate
the polypeptide chain and therefore favours the coil/α-helix transition. For the
present system, indeed, mixing of HFP and triethylamine was found to be strongly
exothermic, and definite evidence for formation of a HFP·NEt_3 salt complex is
reported in the literature [36]. In any case the concentration of acid/base complex,
and therefore the amount of triethylamine needed to allow the formation of
the α-helix structure should be different for the dark-adapted sample and for the
irradiated one, thus explaining the occurrence of two separate curves (Figure 10.11).

10.2.3
Thioxopeptide Chromophore

The incorporation of a thioamide linkage both between the residues of a β-turn
and within a helical peptide resulted only in minor changes to the native hairpin
and α-helical structures [76]. The thioxopeptide bond –CS–NR– (R=H, alkyl)
represents an isosteric replacement of the normal peptide bond with only a
slight change in the electron distribution in the ground state. This single-atom

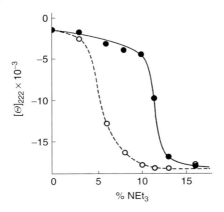

Figure 10.11 Poly(L-lysine) containing 46 mol% spiropyran units in HFP/NEt$_3$. Variation of ellipticity at 222 nm as a function of triethylamine concentration for the sample kept in the dark (continuous line) and the irradiated one (dashed line). Reproduced with permission from ref [75]. Copyright Wiley (1975).

Scheme 10.5 Schematic representation of *trans–cis* conformers of a secondary thioxo amide peptide bond (A and B are amino acid residues).

O/S substitution in a biologically active oligopeptide is of considerable interest because of the effect on conformational restriction, enhanced proteolytic stability and modulated activity and selectivity. In addition, the replacement of oxygen with a sulfur in a single peptide unit may provide a method to study the dynamics of photoinduced conformational changes in practically unperturbed peptides and proteins since thioamide isomerization takes place on a sufficiently fast timescale (Scheme 10.5). In thioxopeptides, the absorption spectrum of the peptide group is red-shifted from a range <230 nm to about 260 nm for the π–π* transition and light absorption at this wavelength induces *trans* to *cis* photoisomerization without decomposition.

The photoswitching mechanism [76] occurs with secondary thioxopeptide bonds. For example, the UV/Vis spectrum of F-ψ[CSNH]-A peptide changed significantly upon irradiation (Figure 10.12), indicating that the amount of *cis* conformer increased with irradiation. The well-anchored isosbestic point at 274 nm and the reversibility of the UV/Vis spectral changes showed that the photoswitching is reversible and only the *cis* and *trans* conformers were involved.

Significant photoswitching ability was found for secondary thioxo amide peptide bonds, and was used to characterize their *cis–trans* isomerization. In particular, the large increase of the *cis* isomers at the photostationary state (under experimental conditions, an apparent first-order rate constant of $k^*_{270} = (4.05 \pm 0.02) \times 10^{-3}$ s^{-1} was found), slow thermal re-equilibration rates (first-order rate constant of $kc/t = (9.78 \pm 0.02) \times 10^{-4}$ s^{-1}) and the site specificity of the photoswitching effect of

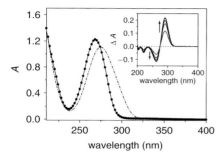

Figure 10.12 UV/Vis absorption spectra of peptide F-ψ[CSNH]-A before and after irradiation. 1.4×10^{-4} mol dm^{-3} F-ψ[CSNH]-A in 5.0×10^{-2} mol dm^{-3} sodium phosphate buffer (pH = 7.0), 16 °C. Before irradiation (solid line), after 3 min of irradiation at 254 nm (dot-dash line), re-equilibrated peptide after three cycles of irradiation-re-equilibration (solid circles). Inset: evaluation time course of the difference UV spectrum during irradiation (cis isomer has a stronger absorbance in the 275–325 nm region, which was a maximum at 290 nm). Reproduced with permission from ref [76]. Copyright RSC (2003).

the secondary thioxopeptide bonds should offer an unprecedented opportunity for studying the conformation–activity correlations of peptides or proteins.

The thioxo prolyl bond acts [77] as a probe of the isomeric state and on the ability to photoswitch between *cis* and *trans* isomers when this bond is present in the backbone of endomorphins. The thioxylated endomorphins Tyr-ψ[CS-N]-Pro-Trp-Phe-NH$_2$ and Tyr-ψ[CS-N]-Pro-Phe-Phe-NH$_2$ are effective μ-receptor agonists whose affinity for the μ-receptor is comparable to that of the natural endomorphins. UV/Vis and CD spectroscopy of the thioxylated peptides in aqueous solution showed the characteristic absorption band for the $^1\pi-^1\pi^*$ transition of thioxo peptide bonds and a negative Cotton effect (Figure 10.13).

Moreover, the authors found a positive Cotton effect for the $^1n-^3\pi^*$ transition, which is indicative of an asymmetric centre in close proximity to the chromophore. The time dependence of the light-induced *cis*–*trans* photoisomerization of the Tyr-ψ[CS-N]-Pro-bond (by excitation of the $^1n-^3\pi^*$ transition with a N$_2$ laser at 337 nm) demonstrated the existence of an isosbestic point at 251 nm, as could be expected for a uniform transition between two conformers. In particular, the dramatic isomer specificity of the Cotton effect for the $^1\pi-^1\pi^*$ transition was used for kinetic investigations of the *cis*–*trans* isomerization of thioxo peptides providing rate constants of $(5.38 \pm 0.02) \times 10^{-4}$ s^{-1} at 313 K for the sum of the isomerization in the forward and reverse directions. Thioxo prolyl bonds represented therefore sensitive probes for detecting isomer specificity in biological signalling, allowing the induction of well-defined changes of the backbone conformation by *cis*–*trans* photoisomerization.

Very recently investigations were extended to the study of ψ[CS-NH]^4-RNase S, a site specific modified version of RNase S obtained by thioxylation (O/S exchange) at the Ala^4–Ala^5–peptide bond, in order to evaluate the impact of protein backbone photoswitching on bioactivity [78]. A reversible photoisomerization with a highly

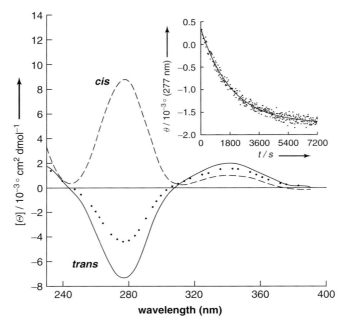

Figure 10.13 CD spectra of the pure conformers of peptide Tyr-ψ[CS-N]-Pro-Phe-Phe-NH$_2$ and the equilibrated solution in 0.01 M sodium phosphate buffer (pH = 7.4), 2 × 10^{-4} M, 313 K; equilibrated peptide (•••), trans (—) and cis (----) forms. Inset: time dependence of the cis–trans isomerization of the peptide: 0.01 M sodium phosphate buffer (pH = 7.4), 2 × 10^{-4} M, 313 K. Reproduced with permission from ref [77]. Copyright Wiley (2000).

increased *cis–trans* isomer ratio of the thioxopeptide bond of ψ[CS-NH]^4-RNase S in the photostationary state was observed under UV irradiation conditions (254 nm). The slow thermal reisomerization ($t^{1/2} = 180$ s) permitted determination of the enzymatic activity of *cis* ψ[CS-NH]^4-RNase S by measurement of cytidine 2′, 3′-cyclic monophosphate (cCMP) hydrolysis. Comparison with RNase S revealed similar thermodynamic stability of the complex an unperturbed enzymatic activity towards cCMP. Despite the thermodynamic stability of *cis* ψ[CS-NH]^4-RNase S, its enzymatic activity is completely lost but it is recovered after reverse isomerization. Photoswitching *trans* ψ[CS-NH]^4-RNase S thus provided the unique feature of giving an enzyme the opportunity to respond to a well-defined one-bond conformational change situated at a predefined peptide bond of the enzyme.

10.3
Photostimulated Aggregation–Disaggregation Effects

Azo-modified polypeptides have been reported to undergo reversible aggregation–disaggregation processes upon exposure to light or dark conditions [79]. Samples of azo-modified poly(L-glutamic acid) (Scheme 10.2, Structure **IV**) stored in the dark or irradiated at 450 nm (azo units in *trans* configuration) showed variations in their CD spectra on aging in TMP/water solution. The time dependence of the CD spectra was characterized by progressive distortions of the α-helix pattern, typical of those produced by formation of aggregates of polypeptide chains. Formation of aggregates was also accompanied by a progressive increase in light-scattering intensity. Irradiation at 360 nm (*trans* → *cis* isomerization) at any aging time, produced the erasure of light scattering and the full restoration of the initial CD spectra, thus indicating dissociation of the aggregates. The spectra reverted again to the distorted ones by irradiation at 450 nm or by dark adaptation of the samples, thus confirming the reversibility of the process.

The investigation of poly(L-glutamic acid) having a high content of azobenzene side chains (more than 80%) provided confirmation of photoinduced aggregation–disaggregation processes together with the occurrence of large photosolubility effects [80, 81]. The polypeptide stored in the dark was soluble in HFP where it assumed the α-helix structure; addition of a small amount of water (15% by volume) to the HFP solution caused formation of aggregates followed by the total and quantitative precipitation of the polymer as a yellow material. Irradiation of the suspension for a few seconds at 350 nm caused the complete dissolution of the polymer, while irradiation of the solution at 450 nm drove polymer precipitation. In this solvent mixture, therefore, the 'precipitation-dissolution' cycles could be controlled by irradiation at the two different wavelengths.

Irradiation experiments carried out with light at several wavelengths allowed measurement of the dependence of the polymer solubility on the *cis–trans* isomeric composition of the azobenzene side chains [81]. The results are illustrated in Figure 10.14. The solubility of the polymer, as a function of the *cis–trans* ratio of azobenzene side chains is described by a sharp sigmoidal curve: the polymer is fully insoluble when more than 60% of the azo groups are in *trans* configuration; by contrast, the maximum amount of photosolubilization is achieved when more than 60% azo group are in the *cis* configuration; the midpoint of the transition is located in correspondence of 50/50 *trans–cis* composition.

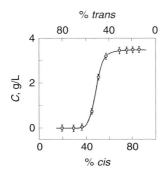

Figure 10.14 Poly(L-glutamic acid) containing 85 mol% azobenzene units. Variation of the solubility in HFP/water = 85/15 as a function of the *trans–cis* isomeric composition of the azo side chains. Reproduced with permission from ref [81]. Copyright ACS (1989).

Similar photosolubility effects have been observed for azo-modified poly(L-ornithine) [39] and poly(L-α, β-diamino-propanoic acid) [42], by monitoring the transmittance at 650 nm as a function of irradiation time. The initially turbid samples in HFP/water became clear upon irradiation at 360 nm as a consequence of the *trans* → *cis* isomerization. On subsequent irradiation at 460 nm, the clear solutions became turbid again as a consequence of the reverse *cis*-to-*trans* isomerization of the azo chromophores.

If the higher polarity of the *cis* isomer were the decisive factor causing the dissolution, polymer solubility should gradually increase with increasing *cis* content. However, the variation in solubility as a function of the *trans*–*cis* isomeric composition was described by a sharp sigmoidal curve typical of a phase transition [81]. Therefore, the photosolubility effect was interpreted in terms of supramolecular association, through hydrophobic interactions and stacking of azobenzene in the planar *trans* configuration. When the azo moieties are photoisomerized to the skewed *cis* configuration, interactions and stacking between azo-groups are inhibited, so disaggregation of the macromolecules takes place, and polymer dissolution occurs.

Photostimulated polymer precipitation and dissolution may find application in photoresist technology [82, 83] and be responsible for photoregulated processes in biology.

10.4
Photoeffects in Molecular and Thin Films

The reversible photoinduced changes either in surface pressure or in surface area of the monolayers have been observed. Therefore, these investigations are of increasing interest in the design of nanostructured systems, and may also be important as energy-conversion media from light to mechanical work.

Poly(L-lysine) **V**, containing about 40 mol% of *p*-phenylazobenzoyl units, was reported to form a stable monolayer at the water/air interface [84]. When the polypeptide monolayer was kept at constant area, irradiation at 365 nm produced a decrease of the surface pressure, which reversibly reverted to the original value upon irradiation at 450 nm. At constant pressure, irradiation with 365- and 450-nm light, alternately, produced reversible changes of the surface area of the monolayer.

IR spectra [85] were typical of oriented α-helices with parallel dichroism of the amide A (3300 cm^{-1}) and Amide I (1652 cm^{-1}) bands, and perpendicular dichroism in the Amide II band (1545 cm^{-1}). The result was independent of whether the specimen was prepared from monolayer illuminated with 450-nm light (azo units in *trans* configuration) or with 365-nm irradiation (azo units in *cis* configuration). This evidence suggests that the polymer does not undergo a conformational change upon irradiation in the monolayer state. The photomechanical effects seem to be simply due to the *trans*–*cis* isomerization of the azobenzene groups that occupy a different area in the interface when they are in the *trans* or in the *cis* configuration.

Monolayers of p-alkyl azobenzene-containing poly(L-glutamate)s showed photomechanical effects opposite to those described above for azo-modified poly(L-lysine). Indeed, they expanded when exposed to UV light (*trans* → *cis* isomerization), and shrank when exposed to visible light (*cis*-to-*trans* isomerization). The expansion was found to be smaller than that expected from the comparison of the monolayer isotherms obtained from irradiated and nonirradiated solutions due to the lower yields of photoisomerization in monolayers than in solution [86].

The photoresponsive polypeptide [87] consisting of two α-helical chains of poly(L-glutamate) of $M_w = 11\,000$, linked by an azobenzene moiety (Scheme 10.6, **XII**) formed monolayers at the water/air interface and the photoresponsive behaviour of the monolayer was investigated. The *trans* → *cis* photoisomerization produced a bending of the main chain of the molecule. As a result, a contraction of the area of the monolayer was observed. On the basis of the decrease in the limiting area per molecule, it was estimated that the bending angle between the two α-helical rods, produced by irradiation with UV light, was about 140°.

Photomechanical effects have been also observed in monolayers obtained from poly(L-glutamic acid) modified with carbocyanine [88] and spiropyran dyes [89]. In the latter case, irradiation at 254 nm produced changes in the molecular conformation which caused photomodulation of the surface pressure and surface area of the films. From all these examples, it appears that photoresponsive monolayers are quite fascinating systems; which can be eventually regarded 'as a machine to transform light into mechanical energy' [86].

Langmuir–Blodgett (LB) films and polymeric liquid crystals have been intensively investigated for application in optical data storage and the design of photoswitchable devices.

Poly(L-glutamate) [90–96] bearing azobenzene units in the side chains, with alkyl spacers as well as tails of different lengths have a so-called 'hairy rod' structure, that is a rigid rod-like helical backbone with flexible side chains [97] and liquid-crystalline behaviour. A water/air interface form monomolecular films that can be transferred to substrates using the LB technique. The resulting stable and homogeneous films are built up of macromolecules arranged in layers, with the main chains preferentially oriented in the dipping direction, and the photochromic azo side chains preferentially oriented normal to the surface [93, 96].

On the photoisomerization of the azobenzene chromophores the preferred orientation of the main chains is retained, but the layered structure of the

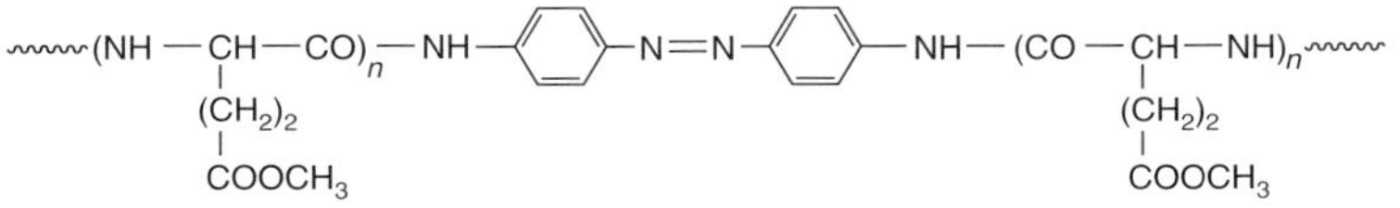

$$\sim\!\!\sim\!(NH\!-\!CH\!-\!CO)_n\!-\!NH\!-\!\langle\bigcirc\rangle\!-\!N\!=\!N\!-\!\langle\bigcirc\rangle\!-\!NH\!-\!(CO\!-\!CH\!-\!NH)_n\!\sim\!\!\sim$$
$$(CH_2)_2 \qquad\qquad\qquad\qquad\qquad\qquad (CH_2)_2$$
$$COOCH_3 \qquad\qquad\qquad\qquad\qquad\qquad COOCH_3$$

XII

Scheme 10.6 Chemical structure of the polypeptide **XII**, consisting of two α-helical chains of poly(γ-methyl-L-glutamate) ($M = 11\,000$) linked by an azobenzene units [87].

azobenzene moieties between the layers of the stiff poly(L-glutamate) rods is lost. The described LB films may be promising materials in optical switching and image-recording technologies [96].

The LB films of hairy-rod azo-poly(L-glutamate) have been used to prepare photoresponsive waveguides also [98]. High optically anisotropic structure is observed when the azo moieties are in the *trans* configuration, whereas they have an essentially optically isotropic structure when the azo molecules are in the *cis* configuration. So, it was shown that the refractive index could be reversibly switched by irradiating the films with light of appropriate wavelengths. A perfect 'on/off' switch of the optical anisotropy was observed by irradiating alternately at 360 nm (*trans* → *cis* isomerization) and 450 nm (*cis* → *trans* isomerization).

The same poly(L-glutamate)s with photochromic azobenzene side groups were found to be thermotropic [99]. The initial LB films have a well-defined bilayer structure in which the rod-like azobenzene moieties are tilted towards the α-helical backbones and form H-aggregates. Aggregation and orientational order of the azobenzene chromophores were found to change upon irradiation and annealing. The lamellar order and inplane anisotropy of the chromophores were irreversibly lost on UV irradiation. New ordered structures with a more symmetrical distribution of the side chains around the main chain helix, a modified spacing, changed aggregation and a different inplane anisotropy were established after subsequent visible irradiation or annealing.

The photobehaviour under polarized light of spin-coated films obtained from a hairy-rod poly(L-glutamate) with azobenzene in the side chains in the presence of a linearly polarized pump light beam align themselves perpendicular to the pump-beam polarization direction. The *cis* created state is aligned perpendicular to the polarization of the pump beam [100].

Thin films of photochromic polypeptides may have promise as possible nonlinear optical materials [101, 102]. In fact, the alignment of neighbouring macromolecules having helical conformation produces a greater opportunity for noncentrosymmetric side chain orientation, a requirement for nonlinear optical materials. Indeed the rod-like α-helical conformation of polypeptides is ideal to restrain the orientation of the dye side groups to a greater extent than in comparably modified synthetic polymers such as poly(methacrylate) [101]. Another possible application is as holographic materials from spiropyran-modified poly(L-glutamic acid) [103].

The use of photochromic compounds for optical data storage has been proposed. A photochromic dye consisting of an anthraquinone covalently linked to an azobenzene moiety was doped in cholesteric liquid-crystalline gels or thermotropic cholesteric films prepared from α-helical polypeptides. The dye showed large induced optical rotations and CD bands in the region of absorption bands of the anthraquinone moiety, the magnitude of which changed reversibly with the *trans*–*cis* photoisomerization of the azobenzene moiety. Therefore the photoisomeric state of the photochromic group could be detected by the large induced CD bands of the anthraquinone dye that occur at much longer wavelengths than the wavelengths used to produce the *trans*–*cis* photoisomerization of azobenzene moiety. This chiroptical system thus provides a tool for a nondestructive read-out technique [104–107].

10.5
Photoresponsive Polypeptide Membranes

Photochromic polymers have been used in order to develop artificial membranes whose physical properties and functions, such as permeability, conductivity, membrane potential, can be switched on/off or controlled in response to light [7, 108, 109].

Poly(L-glutamic acid) containing 12–14 mol% azobenzene units in the side chains was used to prepare membranes that were obtained by coating a porous Millipore filter with a 0.2% chloroform solution of I. Irradiation at 350 nm was found to increase the membrane potential and crossmembrane permeability. The photoinduced variations of the membrane functions were completely reversible and could be controlled by irradiation and dark adaptation.

The results [110, 111] were explained on the basis of the observation that the water content of the membrane is increased by irradiation at 350 nm, most likely as a consequence of the different polarity and hydrophobicity of the *trans* and the *cis* isomers. The increase in hydration of the membrane should be then probably accompanied by an increase of the dissociation degree of the unmodified COOH side chains, thus producing an increase in the negative charge of the membrane and enhancing the diffusion of ions.

Analogous membranes have been prepared from poly(L-glutamic acid) containing about 14 mol% azobenzene-sulfonate groups in the side chains [112].

Insoluble membranes were also prepared from crosslinked samples [113]. Irradiation of the membranes with UV light at pH values in the range 5–9 induced large variations in hydration and membrane potential. CD measurements showed that such variations were also accompanied by photoinduced conformational changes corresponding to a decrease of α-helix content from 76 to 46%. It was suggested that the *trans* → *cis* isomerization of the azo chromophores might result in a higher degree of dissociation of the unmodified COOH side chains, which should be the key factor responsible for the photoresponse effects. In fact, a shifting of the equilibrium from neutral COOH to ionic COO^- groups would increase the electrostatic interactions among side chains, thus causing the unfolding of the macromolecules and changing the charge distribution on the membrane. Unfortunately, the photoinduced variations in hydration and membrane potential, as well as in macromolecular structure were found to be irreversible [113].

Photoresponsive membranes were also prepared from polypeptides chemically modified with triphenylmethane dyes [114–117]. The photochromic behaviour of such compounds involves the ionization of the dye under UV irradiation to give the intensely coloured triphenylmethyl cation; the cation thermally recombines with the counteranion in the dark. Poly(L-glutamic acid) was reacted with pararosaniline to give a polymer containing about 10 mol% dye groups in the side chains

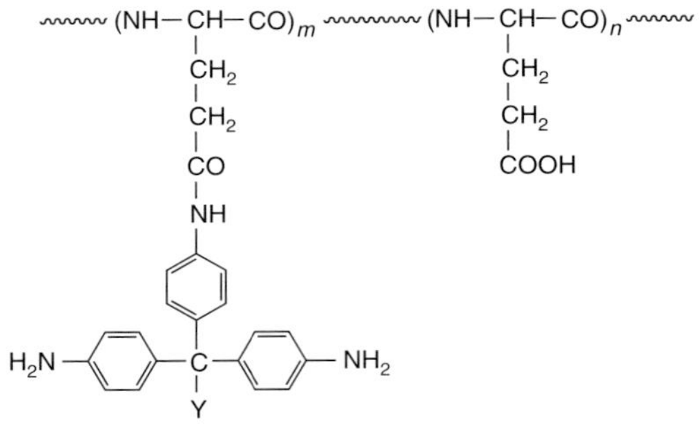

Scheme 10.7 Chemical structure of poly(L-glutamic acid) containing triphenylmethane dyes in the side chains (**XIII**, Y = −OH and −CN).

(Scheme 10.7, Structure **XIII** (Y = −OH)). The membrane obtained by casting a dimethyl formamide solution of the polymer was no longer soluble, indicating that a fraction of the dye molecules may act as a crosslinking agent during the casting process.

Irradiation of the membrane with light having 250–380 nm was found to produce photoinduced conformational changes only at critical pH values of the aqueous solution at which the irradiation was carried out [114].

Experiments of permeation of substrates across the membrane showed that irradiation with UV light at pH 8.6 induced an increase in permeability together with an increase in swelling of the membrane. Both permeability and swelling degree returned to their original values after 100 min in the dark [115, 116].

Large photoresponse effects have been observed in membranes prepared from poly(L-glutamic acid) containing leucocyanide (triphenylmethyl cyanide) groups in the side chains (Scheme 10.6, Structure **XIII** (Y = −CN)) [117]. For a membrane prepared from a polymer containing 38 mol% photochromic groups, at pH 5.3, exposure to UV light induced large variations in the degree of swelling, membrane potential and permeation coefficient of KCl through the membrane. All parameters and membrane function returned to their original value when light was removed and the membrane kept in the dark. The photoinduced variations of the membrane functions are consistent with the photodissociation of the dye molecules and formation of triphenylmethyl cations in the side chains of the macromolecules, and the consequent polarity change of the membrane.

The corresponding membranes prepared by casting DCE solutions of the polymers were stable in TMP [118, 119].

It was observed that the permeability of various substrates across the membranes from polyvinyl/polypeptide graft copolymers by attaching branches of p-phenylazobenzyl/β-benzyl-L-aspartate to poly(hydroxyethyl methacrylate) and poly(butyl methacrylate), was enhanced under irradiation with UV light and was suppressed on irradiation with visible light. The photoinduced permeability

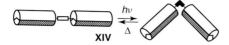

Figure 10.15 Schematic illustration of the preparation and photoresponsive behaviour of the polypeptide **XIV**, consisting of two amphiphilic helical rods linked by an azobenzene unit.

changes were correlated with the photoinduced and reversible conformational variations of the polypeptide branches grafted to the hydrocarbon backbone of the macromolecules. The photoregulation of permeability across the membrane is achieved by means of conformational changes of the polypeptide chains, without any changes in the electrostatic charges of the macromolecules [119].

A photoresponsive amphiphilic helical polypeptide, that is a helical polypeptide in which all polar residues are located on one side and all hydrophobic residues are located on the opposite side of the helical cylinder, was prepared at the water/air interface by selective saponification of the ester methyl groups only on the water side of the monolayer, so that the final polypeptide **XIV** consisted of helical rods having COOH side chains on one side (hydrophilic face) and COOCH$_3$ side chains on the other side (hydrophobic face) (Figure 10.15) [120–124].

The compound **XIV**, consisting of two amphiphilic helical rods linked by an azobenzene moiety, was found to form micelles and ordered aggregates in aqueous solution in the dark, where the azo moiety is in the *trans* configuration. Photoisomerization of the azo linkage to the *cis* configuration and the consequent bending in the structure of the molecules, induced disaggregation and disruption of the micelles [121, 122].

In nature, polypeptides having amphiphilic structure are known to form transmembrane channels formed by an assembly of several helices, so as to present their polar faces inward and their apolar faces outward. Considering such behaviour, the photochromic amphiphilic polypeptide was incorporated in a cationic bilayer membrane composed of dipalmitoyl phosphatidyl choline [123]. Fluorescence and microscopic measurements provided evidence that the polypeptide was able to form bundles of helical molecules analogous to the natural ones, which acted as transmembrane channels for K$^+$ ions. Irradiation and the consequent *trans* → *cis* isomerization of the azobenzene link, caused a bending of the molecular structure and a destabilization of the transmembrane bundles. Therefore, formation of ion permeable channels was favoured or inhibited depending on whether the azo moiety was in the *trans* or in the *cis* configuration, thus allowing photoregulation of membrane permeability [123].

The investigation has then been extended to a monolayer formed from dipalmitoyl phosphatidyl choline and the same amphiphilic photochromic polypeptide **XIV** [124]. When the monolayer was kept in the dark, the polypeptide molecules arranged perpendicularly to the membrane (the water/air interphase) and formed a bundle of helices that could be observed as a transmembranous particle of about 4 nm in diameter by atomic force microscopy. Irradiation with UV light and the consequent *trans* → *cis* isomerization of the azobenzene moiety caused a bending of the molecular main chain that, in turn, produced a destabilization and denaturation of the bundle of helices in the monolayer. After removal of the light, the polypeptide molecules returned to their original bundle structure [124].

10.6
Summary and Recent Developments

Photochromic compounds that can be reversibly switched by a light stimulus of appropriate wavelength, when incorporated into macromolecules may induce extended structural changes. Accordingly, they work as photochemical molecular switches, and the photochromic polymers may provide the basis for constructing light-driven switching systems.

The starting light signal associated with the photoisomerization of the chromophore is usually weak, and 'amplification' is necessary to construct photoswitchable devices. Substantial amplification can be achieved when the primary photochemical reaction is coupled with a subsequent event that occurs after absorption of light as in vision. Thus, in the case of polypeptides containing photochromic units the photoisomerization of their photochromic groups can produce 'order ⇌ disorder' conformational changes. These photostimulated structural variations, such as random coil ⇌ α-helix, take place as highly cooperative transitions; therefore photochromic polypeptides actually work as amplifiers and transducers of the primary photochemical events occurring in the photosensitive side chains. One of the most studied aspects using the above approach concerns protein folding that bridges the gap between the information held in the genetic sequence and protein structure. Its detailed understanding, that is the possibility to predict protein structure and function from genetic sequences, opens up therefore unprecedented applications in biotechnology and medicine. During the folding process, ultrafast rotations occur around single bonds on the picosecond timescale, whereas the formation of secondary structures and their rearrangement on the microsecond to second timescale [125]. The dynamics of tertiary and quaternary structure extends from milliseconds to seconds and even longer. Several experiments are performed to study structural dynamics occurring in small proteins or oligopeptides. Conformational transitions of peptides and proteins are most commonly triggered by changing the environment of the molecule, thereby shifting the equilibrium constant of the process under investigation. Among these methods, laser-induced, pH- and temperature-jump experiments have been shown to allow for subnanosecond time resolution [126–129]. Even faster triggering with a possible subpicosecond time resolution can only be achieved with a molecular switch incorporated in the peptide chain. Ultrafast-light-induced changes of the switching molecule initiate structural dynamics of the peptide on the picosecond timescale by changing a conformational restraint. The design of peptides with built-in chromophores that enable fast conformational changes by irradiation with monochromatic light is a particular powerful approach to investigate structural transitions during protein folding [130].

Typically these systems are bistable, that is sufficiently long lived in both isomeric forms and small enough to resolve changes in secondary structure using spectroscopic techniques including CD, NMR and ultrafast optical spectroscopy experiments.

For example, femtosecond spectroscopy in the visible and near-UV spectral range not only detects the dynamics of the molecular switch, but that the chromophoric moiety also acts as a built-in probe for conformational changes of the peptide. Nevertheless, ultrafast spectroscopy in the mid-IR range allowed observation of directly the motions of the peptide backbone [125, 131, 132].

The results were reported [133, 134] of ultrafast visible spectroscopy of two water-soluble azobenzene peptides with eight amino acids, a mono- and a bi-cyclic peptide (cAMPB and bcAMPB, respectively), this last with a disulfide bridge between two cysteine residues. The short peptide chain Ala–Cys–Ala–Thr–Cys–Asp–Gly–Phe with the active site motif Cys–Ala–Thr–Cys was used. According to the authors, the cyclic octapeptides showed similar dependences of the folding dynamics on solvent viscosity as much larger peptides or proteins. The steady-state spectroscopy showed that UV–vis absorption spectra of the two water-soluble azopeptides were almost identical with absorption features very similar to the absorption of the DMSO-soluble azopeptides: $\pi-\pi^*$ transition at 336 and 300 nm in the *trans* and *cis* state, respectively, and the $n-\pi^*$ transition at 430 nm in both states. In the transient absorption measurements, the long-wavelength side of the $n - \pi^*$ band of AMPB at 480 nm was excited and the signal was monitored with the UV–Vis white-light probe in the spectral range from 360 to 630 nm. The time constant associated to the isomerization mechanism was obtained by a three exponential algorithm fitting of the experimental data collected. The excitation at 480 nm led to a fast movement with a time constant of a few hundred fs out of the Frank–Condon region in the S_1 state following internal conversion to the ground state on the timescale of a few picoseconds. A large fraction of the molecules in the *cis* → *trans* isomerization and a smaller fraction in the *trans* → *cis* reaction were able to find a conical intersection within that initial motion.

Concerning the *trans* → *cis* isomerization, the movement out of the Frank–Condon region on the subpicosecond timescale contributed to the total absorption dynamics. The spectral shape of the associated amplitudes was similar for azopeptides in DMSO and water. The internal conversion to the ground state was found to be a biexponential process: a fast component that is assigned to relaxation and reorganization on the S_1 potential energy surface S_0 (1.3 ps for cAMPB, 1.6 ps for bcAMPB, water as solvent) was followed by the transition to the ground state S_0 (6.5 ps for cAMPB and 5.8 ps for bcAMPB, water as solvent). A separate vibrational cooling process in the S_0 state did not display an additional kinetic component. The *cis* → *trans* isomerization in azopeptides is generally dominated by an ultrafast pathway to the ground state with time constants of 0.1–0.2 ps. This strongly driven process with large amplitude led directly to the ground state and was not affected by the viscosity of the solvent. Absorption dynamics on the timescale of >10 ps reflected the motions of the peptide moiety. In fact over this time range all processes related to the azo-chromophore isomerization are terminated. The remaining changes in the AMPB-absorption spectrum were due to the strain exerted from the unrelaxed peptide backbone on the chromophore. In contrast to ultrafast relaxation, the slower reactions (5–10 ps) were twice as fast in water as in DMSO due to the fact that structural rearrangement of the peptide moiety was dominated by the friction imposed by the solvent, which counteracts the (chromophore induced) structural changes *via* its viscosity.

The more complex system in which the azobenzene AMPB chromophore was incorporated into an α-helical peptide composed of 30 amino acids [135] was compared to the neat AMPB azobenzene unit. The excitation pulses were centred at 475 nm and the photoinduced absorbance changes monitored in the spectral range of 390–680 nm for more than 30 individual wavelengths (Figure 10.16).

The 30-mer azopeptide was photoresponsive similar to that observed for the smaller azopeptides previously investigated. The photodynamics of the *trans* → *cis* reaction took place on the order of 200 fs, in agreement with results reported for other azopeptides, almost unaffected by the surrounding peptide. By contrast, the subsequent kinetics were changed for the azopeptide (2 and 12 ps, respectively) in comparison to the AMPB chromophore alone (1.7 and 6.6 ps, respectively), indicating that the specific properties of the entire peptide dominated the slower components of the device photoresponse.

Ultrafast IR spectroscopy, used to monitor the nonequilibrium backbone dynamics of a cyclic peptide in the amide I vibrational range, indicate [125, 136] that immediately after photoswitching of the azobenzene unit from *cis* to *trans*, a transient red shift of the amide I band of the peptide was observed, which decays after 4 ps and vanishes within 14 ps. A slower process led to the transient blue shift of the amide group observed after 20 ps. These frequency shifts were attributed to rearrangements of the individual carbonyl groups and/or from the opening and closing of intramolecular hydrogen bonds as a result of such rearrangements. In

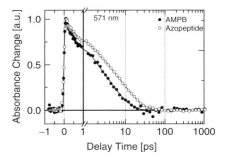

Figure 10.16 Comparison of transient absorbance changes of the azopeptide 3 with AMPB at a probing wavelength of 571 nm. Solid lines represent results of a multiexponential global fit analysis. Reproduced with permission from ref [135]. Copyright Wiley (1987).

particular, because the blue shift of the amide I band can be directly related to the change of backbone structure, the authors concluded that stretching of the peptide conformation and the subsequent relaxation of the peptide ensemble to equilibrium is governed by a discrete hierarchy of timescales, extending from 20 ps to 16 ns.

While the incorporation of photoswitchable azo chromophores in the peptide backbone is considered an important issue for the study of structural dynamics of small proteins, distinct changes to the local or global structure of the oligopeptide structure are univocally imposed. Recently, the incorporation of a thioamide linkage both between the residues of a β-turn and within a helical peptide resulted only in minor changes to the native hairpin and α-helical structures. N-Methylthioacetamide (NMTAA) has recently been studied by femtosecond transient absorption and IR spectroscopy [137]. In these experiments, the spectroscopic changes occurring upon π–π* excitation of *trans* NMTAA in water have been monitored. Transient IR spectra indicated that the formation of the *cis* isomer and the recovery of the ground-state *trans* form occurred on a dual timescale (biphasic kinetics) with a fast component of 8–9 ps and a slow time constant of about 250 ps. *Ab initio* theoretical calculations performed in vacuo on the photochemical reaction path indicate that the *trans* → *cis* isomerization event took place on the S_1 and/or T_1 triplet potential energy surfaces and it was controlled by very small energy barriers, in agreement with the experimentally observed picosecond timescale. In addition, the authors reported that only the 250-ps component is observed in the transient absorption experiment in the visible region probably due to a much lower time resolution of the IR technique with respect to the timescale of relaxation of the NMTAA unit. However, according to the results shown, the force field of the NMTAA chromophore might not be able to efficiently force and perturb a peptide chain out of its equilibrium conformation.

Analogously, femtosecond transient [138] absorption studies in the visible and near-UV region on linear peptides where one peptide bond was replaced by its thioxo unit showed that after excitation of the π–π* transition a strong visible absorption at around 550 nm emerged on the subpicosecond (~1 ps) and picosecond (~300 ps) timescale. The decay of visible absorption occurred in the range of 150–600 ps into an intermediate electronic state, most probably a triplet state (Scheme 10.8).

According to the scheme proposed, it is probable that the formation of the *cis* isomer terminates on the 100 ps timescale. This shows that the incorporation of

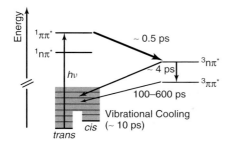

Scheme 10.8 Relative energetic ordering of the excited states of the thioxo compounds and possible deactivation paths following the excitation of the $^1\pi\pi^*$ state.

the thioxoamide group in oligopeptides of at least 20 amino acid residues may allow study of peptide dynamics on timescales of >100 ps. In addition, according to IR experiments, the *cis* → *trans* isomerization around the thioxopeptide bond terminated within less than 1 ns.

Additional recent developments include some innovative experimental approaches to making polypeptide or protein molecules photoswitchable. Bovine serum albumin (BSA) undergoes, in the presence of a cationic azobenzene surfactant, photomodulated chain unfolding to a greater degree with the visible-light (*trans*) form than with the UV-light (*cis*) form, the former being more hydrophobic than the latter. FT-IR was used to collect quantitative information on the secondary structure elements in the protein [139].

Photochemical control of cell adhesion on surfaces was observed by modifying the distance from polymethylmethacrylate through the light induced E-Z isomerization. The photoswitchable 4-[(4-aminophenyl)-azo]benzocarbonyl was used as a spacer between the acrylamide anchor and the cyclic RGD peptide [140].

Extension to proteins of the basic approach used for photoswitchable smaller and simple peptides is of basic importance for a better understanding of protein conversion from well-folded and soluble helices to β-sheets and related aggregates.

An interesting example concerns with the 36 residue avian pancreatic polypeptide for which a crystal structure is available. The β-turn segment between the helices, comprising residues Asp10-Asp11-Ala12 was replaced with [(p-methylamino-)-p-azobenzene phenyl] acetic acid. The photoirradiation of the modified protein allowed switching from one form to another of the possible tertiary structures [141].

Recently the discovery was reported of a novel photoswitchable green fluorescent protein called *Dronpa* from a Pectimidal coral [142]. This protein exhibits fast photoswitching between a bright and a dark state that can be observed by fluorescence correlation spectroscopy [143]. The excited states of Dronpa were characterized by time resolved spectroscopy. The ultrafast transient absorption spectroscopy investigation allowed identification of the rate of the first photoconversion step from the neutral dark form, to the anionic fluorescent form [144].

Studies of this last type are expected to bring more information about the conformational behaviour of the protein as well as to identify new approaches for developing advanced molecular switches.

10.7
Towards More Complex Biorelated Photoswitchable Polypetides

Photochromic compounds that can be reversibly switched by a light stimulus of appropriate wavelength, when incorporated into macromolecules may induce extended structural changes. Accordingly, they work as photochemical molecular switches, and the photochromic polymers may provide the basis for constructing light-driven switching systems.

The starting light signal associated with the photoisomerization of the chromophore is usually weak, and 'amplification' is necessary to construct photoswitchable devices. Substantial amplification can be achieved when the primary photochemical reaction is coupled with a subsequent event that occurs after absorption of light, as in vision. Thus, in the case of polypeptides containing photochromic units the photoisomerization of their photochromic groups can produce 'order $\rightleftharpoons$ disorder' conformational changes. This is dramatically shown by the developments towards improved biological model structures incorporating photoresponsive moieties. A method that enables spatiotemporal photoregulation of cell adhesion was recently developed by using a culture dish coated with a caged arginine-glycine-aspartate (RGD) peptide [145]. The method is based on involves an RGD peptide, which has been identified as a major integrin ligand motif in extracellular matrices (ECMs) such as fibronectin and laminin, and has been used to modify biomaterials to enhance cell adhesion. The prepared caged RGD peptide consists of a sequence (YAVTGRGDSPASS) that is the longest conserved sequence in vertebrates ranging from teleosts to mammals but containing a nitrobenzyl group as a cage. The photoresponsive culture dish was prepared by modifying a commercially available culture dish coated with poly-l-lysine by using a bifunctional crosslinked polyethylene glycol and the caged RGD peptide.

The synthesized amphiphile peptide (PA) that contains both the photocleavable 2-nitrobenzyl group as well as the bioactive epitope Arg-Gly-Asp-Ser (RGDS) [146] undergoes a sol-to-gel transition in response to light. This small structural change can induce a significant change in the supramolecular structure from nanospheres to nanofibres, affecting the response of cells surrounded by the nanostructures [147].

The azobenzene chromophore was further used to develop photoswitches for biological applications [148] through the preparation of a series of azobenzene derivatives in which longer switching wavelengths (up to 530 nm) are combined with good photochemical yields and stabilities of the *cis* isomers. These derivatives, based on 4,4'-diacetamido azobenzenes bearing amino substituents in the 2,2'-positions that enhance water solubility, can be used for directed photocontrol of biomolecular structures in intracellular environments. In this way, several RNA aptamers were obtained by *in vitro* capable of to reversible binding to a photoresponsive peptide (KRAzR; Lys-Arg-azobenzene-Arg) containing azobenzene chromophore [149]. Upon irradiation at 360 nm on the KRAzR-immobilized surface, the binding of each aptamer to the surface was significantly decreased. Subsequent photoirradiation of the same surface at 430 nm restored the aptamer binding to the surface. The first example of a light-responsive β-hairpin, model peptide of a biologically important protein domain, was reported to show considerably different binding affinities for the target protein that are dependent on the isomerization state of the embedded photoswitch [150]. The best example of internal ligand recognition is found in the extended PDZ domain (which is a common structural domain of 80-90 amino-acids) of neuronal nitric oxide synthase (nNOS) that interacts with the PDZ domain from α-1-syntrophin or the second PDZ domain from PSD95, which is a neuronal PDZ protein. In particular, a cyclic light-directed ligand was developed through the incorporation of an azobenzene-ω-amino acid (photoswitch) in a peptide ring of appropriate size. Its *trans* form shows no binding, while the *cis* form features overall binding comparable to the corresponding nonswitchable model peptide that adopts a structure similar to the β-hairpin in the native protein in aqueous solution. The interaction of a biologically important β-sheet with a protein domain has been modulated by a light-induced conformational change without destabilizing the system suggesting that the peptide may serve as a suitable model for a light-triggered β-sheet for use in cells.

References

1. Irie, M. (1990) *Adv. Polym. Sci.*, **94**, 27.
2. Kongrauz, V.A. (1990) in *Photochromism: Molecules and Systems*, Chapter 21 (eds. H. Durr and H.H. Bouas-Laurent) Elsevier, Amsterdam, p. 793.
3. McArdle, C.B. (1992) *Applied Photochromic Polymer Systems*, Blackie, Glasgow.
4. Sisido, M. (1992) *Prog. Polym. Sci.*, **17**, 699.
5. Cooper, T.M., Natarajan, L.V. and Crane, R.L. (1993) *Trends Polym. Sci.*, **1**, 400.
6. Pieroni, O. and Ciardelli, F. (1995) *Trends Polym. Sci.*, **3**, 282.
7. Kinoshita, T. (1995) *Prog. Polym. Sci.*, **20**, 527.
8. Willner, I. and Rubin, S. (1996) *Angew. Chem. Ed. Engl.*, **35**, 367.
9. Pieroni, O., Fissi, A. and Popova, G. (1998) *Progress Polym. Sci.*, **23**, 81.
10. Dugave, C. and Demange, L. (2003) *Chem. Rev.*, **103**, 2475.
11. Greenfield, N. and Fasman, G.D. (1969) *Biochemistry*, **8**, 4108.
12. Woody, R.W. (1977) *J. Polym. Sci. Macromol. Rev.*, **12**, 181.
13. Woody, R.W. (1968) *J. Chem. Phys.*, **49**, 4797.
14. Parrish, J.R. and Blout, E.R. (1971) *Biopolymers*, **10**, 1491.
15. Madison, V. and Schellman, J. (1972) *Biopolymers*, **11**, 1041.
16. Mattice, W.L., Lo, J.T. and Mandelkern, L. (1972) *Macromolecules*, **5**, 729.
17. Deaborn, D.G. and Wetlaufer, D.B. (1970) *Biochem. Biophys. Res. Commun.*, **39**, 314.
18. Woody, R.W. (1985) in *The Peptides*, vol. 7 (eds. J. Udenfriend and S. Meienhofer), Academic Press, Orlando, FL, p. 16.
19. (a) Goodman, M. and Kossoy, A. (1966) *J. Am. Chem. Soc.*, **88**, 5010; (b) Goodman, M. and Falxa, M.L. (1967) *J. Am. Chem. Soc.*, **89**, 3863.
20. Pieroni, O., Houben, J.L., Fissi, A., Costantino, P. and Ciardelli, F. (1980) *J. Am. Chem. Soc.*, **102**, 5913.
21. Ciardelli, F., Pieroni, O., Fissi, A. and Houben, J.L. (1984) *Biopolymers*, **23**, 1423.
22. Houben, J.L., Fissi, A., Bacciola, D., Rosato, N., Pieroni, O. and Ciardelli, F. (1983) *Int. J. Biol. Macromol.*, **5**, 94.
23. Sisido, M., Ishikawa, Y., Itoh, K. and Tazuke, S. (1991) *Macromolecules*, **24**, 3993.
24. Sisido, M., Ishikawa, Y., Harada, M. and Itoh, K. (1991) *Macromolecules*, **24**, 3999.
25. Sato, M., Kinoshita, T., Takizawa, A. and Tsujita, Y. (1988) *Macromolecules*, **21**, 1612.
26. Pieroni, O., Fabbri, D., Fissi, A. and Ciardelli, F., (1988) *Makromol. Chem.: Rapid. Commun.*, **9**, 637.
27. Higuchi, M., Takizawa, A., Kinoshita, T. and Tsujita, Y. (1987) *Macromolecules*, **20**, 2888.
28. Blasie, J.K. (1972) *Biophys. J.*, **12**, 191.
29. Fissi, A., Pieroni, O. and Ciardelli, F. (1987) *Biopolymers*, **26**, 1993.
30. Yamamoto, H. and Nishida, A. (1986) *Macromolecules*, **19**, 943.
31. Yamamoto, H. (1986) *Macromolecules*, **19**, 2472.

32. Inscoe, M.N., Gould, J.H. and Brode, W.R. (1959) *J. Am. Chem. Soc.*, **81**, 5634.
33. Irie, M., Miyatake, O. and Uchida, K. (1992) *J. Am. Chem. Soc.*, **114**, 8715.
34. Irie, M., Miyatake, O., Uchida, K. and Eriguchi, T. (1994) *J. Am. Chem. Soc.*, **116**, 9894.
35. Middletown, W.J. and Lindsey, R.V. Jr (1964) *J. Am. Chem. Soc.*, **86**, 4948.
36. Purcell, K.F., Stikeleather, J.A. and Brunk, S.D. (1969) *J. Am. Chem. Soc.*, **91**, 4019.
37. King, J.F. (1991) in *The Chemistry of Sulfonic Acids, Esters and their Derivatives* (eds. S. Patai and Z. Rappoport), John Wiley & Sons, Ltd, Chichester, p. 249.
38. Fissi, A., Pieroni, O., Balestreri, E. and Amato, C. (1996) *Macromolecules*, **29**, 4680.
39. Yamamoto, H., Nishida, A., Takimoto, T. and Nagai, A. (1990) *J. Polym. Sci.: Polym. Chem.*, **28**, 67.
40. Yamamoto, H., Ikeda, K. and Nishida, A. (1992) *Polym. Intern.*, **27**, 67.
41. Yamamoto, H. and Nishida, A. (1991) *Polym. Intern.*, **24**, 145.
42. Yamamoto, H., Nishida, A. and Kawaura, T. (1990) *Int. J. Biol. Macromol.*, **12**, 257.
43. Yamamoto, H., Miyagi, Y., Nishida, A., Takagishi, T. and Shima, S. (1987) *J. Photochem.*, **39**, 343.
44. Giancotti, V., Quadrifoglio, F. and Crescenzi, V. (1972) *J. Am. Chem. Soc.*, **94**, 297.
45. Erenrich, E.H., Andreatta, R.H. and Scheraga, H.A. (1970) *J. Am. Chem. Soc.*, **92**, 1116.
46. Ueno, A., Anzai, J., Osa, T. and Kadoma, Y. (1977) *J. Polym. Sci.: Polym. Lett.*, **15**, 407.
47. Ueno, A., Anzai, J., Osa, T. and Kadoma, Y. (1977) *Bull. Chem. Soc. Jpn.*, **50**, 2995.
48. Ueno, A., Anzai, J., Osa, T. and Kadoma, Y. (1979) *Bull. Chem. Soc. Jpn.*, **52**, 549.
49. Ueno, A., Anzai, J. and Osa, T. (1979) *J. Polym. Sci.: Polym. Lett.*, **17**, 149.
50. Ueno, A., Takahashi, K., Anzai, J. and Osa, T. (1980) *Macromolecules*, **13**, 459.
51. Ueno, A., Takahashi, K., Anzai, J. and Osa, T. (1980) *Bull. Chem. Soc. Jpn.*, **53**, 1988.
52. Ueno, A., Takahashi, K., Anzai, J. and Osa, T. (1981) *Chem. Lett.*, 113.
53. Ueno, A., Takahashi, K., Anzai, J. and Osa, T. (1981) *Makromol. Chem.*, **182**, 693.
54. Ueno, A., Takahashi, K., Anzai, J. and Osa, T. (1981) *J. Am. Chem. Soc.*, **103**, 6410.
55. Ueno, A., Nakamura, J., Adachi, K. and Osa, T. (1989) *Makromol. Chem., Rapid Commun.*, **10**, 683.
56. Ueno, A., Adachi, K., Nakamura, J. and Osa, T. (1990) *J. Polym. Sci.: Polym. Chem.*, **28**, 1161.
57. Ueno, A., Morikawa, Y., Anzai, J. and Osa, T. (1984) *Makromol Chem., Rapid Commun.*, **5**, 639.
58. Ulysse, L., Cubillos, J. and Chmielewski, J. (1995) *J. Am. Chem. Soc.*, **117**, 8466.
59. Strzegowski, L.A., Martinez, M.B., Gowda, D.C., Urry, D.W. and Tirrell, D.A. (1994) *J. Am. Chem. Soc.*, **116**, 813.
60. Nguyen, P.H., Mu, Y. and Stock, G. (2005) *Proteins: Struct. Funct. Bioinf.*, **60**, 485.
61. Nguyen, P.H., Gorbunov, R.D. and Stock, G. (2006) *Biophys. J.*, **91**, 1224.
62. Renner, C., Cramer, J., Behrendt, R. and Moroder, L. (2000) *Biopolymers*, **54**, 501.
63. Dong, S.-L., Löweneck, M., Figrader, T.E., Figreier, W.J., Zinth, W., Moroder, L. and Renner, C. (2006) *Chem. Eur. J.*, **12**, 1114.
64. Flint, D.G., Kumita, J.R., Smart, O.S. and Woolley, G.A. (2002) *Chem. Biol.*, **9**, 391.
65. Zhang, Z., Burns, D.C., Kumita, J.R., Smart, O.S. and Woolley, G.A. (2003) *Bioconjug. Chem.*, **14**, 824.
66. Schäfer, L.V., Müller, E.M., Gaub, H.E. and Grubmüller, H. (2007) *Angew. Chem. Int. Ed.*, **46**, 2232.
67. Ciardelli, F., Fabbri, D., Pieroni, O. and Fissi, A. (1989) *J. Am. Chem. Soc.*, **111**, 3470.
68. Fissi, A., Pieroni, O., Ciardelli, F., Fabbri, D., Ruggeri, G. and

Umezawa, K. (1993) *Biopolymers*, **33**, 1505.
69. Cooper, T.M., Obermeier, K.A., Natarajan, L.V. and Crane, R.L. (1992) *Photochem. Photobiol.*, **55**, 1.
70. Angelini, N., Corrias, B., Fissi, A., Pieroni, O. and Lenci, F. (1998) *Biophys. J.*, **74**, 2601.
71. Pachter, R., Cooper, T.M., Natarajan, L.V., Obermeier, K.A. and Crane, R.L. (1992) *Biopolymers*, **32**, 1129.
72. (a) Satoh, M., Fujii, Y., Kato, F. and Komiyama, J. (1991) *Biopolymers*, **31**, 1; (b) Satoh, M., Hirose, T., Komiyama, J. (1993) *Polymer*, **34**, 4762.
73. Pieroni, O., Fissi, A., Viegi, A., Fabbri, D. and Ciardelli, F. (1992) *J. Am. Chem. Soc.*, **114**, 2734.
74. Fissi, A., Pieroni, O., Ruggeri, G. and Ciardelli, F. (1995) *Macromolecules*, **28**, 302.
75. Wen, K.J. and Woody, R.W. (1975) *Biopolymers*, **14**, 1827.
76. Zhao, J., Wildemann, D., Jakob, M., Vargas, C. and Schiene-Fischer, C. (2003) *Chem. Commun.*, 2810–2811.
77. Frank, R., Jakob, M., Thunecke, F., Fischer, G. and Schutkowski, M. (2000) *Angew. Chem. Int. Ed.*, **39**, 1120–1122.
78. Wildemann, D., Schiene-Fischer, C., Aumüller, T., Bachmann, A., Kiefhaber, T., Lücke, C. and Fischer, G. (2007) *J. Am. Chem. Soc.*, **129**, 4910–4918.
79. Pieroni, O., Fissi, A., Houben, J.L. and Ciardelli, F. (1985) *J. Am. Chem. Soc.*, **107**, 2990.
80. Ciardelli, F., Pieroni, O. and Fissi, A. (1986) *J. Chem. Soc., Chem. Commun.*, 264.
81. Fissi, A. and Pieroni, O. (1989) *Macromolecules*, **22**, 1115.
82. Irie, M., Iwayanagi, T. and Taniguchi, Y. (1985) *Macromolecules*, **18**, 2418.
83. Ichimura, K. (1990) in *Photochromism, Molecular and Systems*, Chapter 26 (eds. H. Dürr and H.Bouas-Laurent), Elsevier, Amsterdam.
84. Malcolm, B.R. and Pieroni, O. (1990) *Biopolymers*, **29**, 1121.
85. Malcolm, B.R. (1989) *Thin Solid Films*, **178**, 17.
86. Menzel, H. (1994) *Macromol. Chem. Phys.*, **195**, 3747.
87. Higuchi, M., Minoura, N. and Kinoshita, T. (1995) *Colloid Polym. Sci*, **273**, 1022.
88. Menzel, H. and Popova, G.V. (1995) The 7th International Conference on Organized Molecular Films, Ancona, Italy, p. 99.
89. Munger, G., Popova, G.V., Fedorovsky, O.Y. and Salesse, C. (1995) The 7th International Conference on Organized Molecular Films, Ancona, Italy, p. 102.
90. Hallensleben, M.L. and Menzel, H. (1990) *Br. Polym. J.*, **23**, 199.
91. Menzel, H. and Hallensleben, M.L. (1992) *Polym. Bull.*, **27**, 89.
92. Menzel, H., Weichart, B. and Hallensleben, M.L. (1992) *Polym. Bull.*, **27**, 637.
93. Menzel, H., Weichart, B. and Hallensleben, M.L. (1993) *Thin Solid Films*, **223**, 181.
94. Menzel, H., Hallensleben, M.L., Schmidt, A., Knoll, W., Fischer, T. and Stumpe, J. (1993) *Macromolecules*, **26**, 3644.
95. Menzel, H. (1993) *Macromolecules*, **26**, 6226.
96. Menzel, H., Weichart, B., Schmidt, A., Paul, S., Knoll, W., Stumpe, J. and Fischer, T. (1994) *Langmuir*, **10**, 1926.
97. Wegner, G. (1992) *Thin Solid Films*, **216**, 105.
98. Büchel, M., Sekkat, Z., Paul, S., Weichart, B., Menzel, H. and Knoll, W. (1995) *Langmuir*, **11**, 4460.
99. Stumpe, J., Fischer, T. and Menzel, H. (1996) *Macromolecules*, **29**, 2831.
100. Sekkat, Z., Büchel, M., Orendi, H., Menzel, H. and Knoll, W. (1994) *Chem. Phys. Lett.*, **220**, 497.
101. Tokarski, Z., Natarajan, L.V., Epling, B.L., Cooper, T.M., Hussong, K.L., Grinstead, T.M. and Adams, W.W. (1994) *Chem. Mater.*, **6**, 2063.
102. Cooper, T.M., Campbell, A.L. and Crane, R.L. (1995) *Langmuir*, **11**, 2713.
103. Cooper, T.M., Tondiglia, V., Natarajan, L.V., Shapiro, M., Obermeier, K.A. and Crane, R.L. (1993) *Appl. Opt.*, **32**, 674.
104. Kishi, R. and Sisido, M. (1991) *Makromol. Chem.*, **192**, 2723.

105. Sisido, M., Narisawa, H., Kishi, R. and Watanabe, J. (1993) *Macromolecules*, **26**, 1424.
106. Sisido, M. (1994) in *Photo-Reactive Materials for Ultrahigh Density Optical Memory* (ed. M. Irie), Elsevier, Amsterdam, p. 13.
107. Narisawa, H., Kishi, R. and Sisido, M. (1995) *Macromol. Chem. Phys.*, **196**, 1419.
108. Anzai, J. and Osa, T. (1994) *Tetrahedron*, **50**, 4039.
109. Kinoshita, T. (1998) *J. Photochem. Photobiol. B: Biol.*, **42**, 12.
110. Kinoshita, T., Sato, M., Takizawa, A. and Tsujita, Y. (1984) *J. Chem. Soc., Chem. Commun.*, 929.
111. Kinoshita, T., Sato, M., Takizawa, A., and Tsujita, Y. (1986) *Macromolecules*, **19**, 51.
112. Sato, M., Kinoshita, T., Takizawa, A., Tsujita, Y. and Ito, R. (1988) *Polym. J.*, **20**, 761.
113. Sato, M., Kinoshita, T., Takizawa, A., Tsujita, Y. and Osada, T. (1989) *Polym. J.*, **21**, 533.
114. Kinoshita, T., Sato, M., Takizawa, A. and Tsujita, Y. (1986) *J. Am. Chem. Soc.*, **108**, 6399.
115. Sato, M., Kinoshita, T., Takizawa, A. and Tsujita, Y. (1988) *Poly. J.*, **20**, 729.
116. Sato, M., Kinoshita, T., Takizawa, A. and Tsujita, Y. (1988) *Macromolecules*, **21**, 3419.
117. Sato, M., Kinoshita, T., Takizawa, A. and Tsujita, Y. (1989) *Polym. J.*, **21**, 369.
118. Aoyama, M., Youda, A., Watanabe, J. and Inoue, S. (1990) *Macromolecules*, **23**, 1458.
119. Aoyama, M., Watanabe, J. and Inoue, S. (1990) *J. Am. Chem. Soc.*, **112**, 5542.
120. Higuchi, M., Takizawa, A., Kinoshita, T., Tsujita, Y. and Okochi, K. (1990) *Macromolecules*, **23**, 361.
121. Higuchi, M., Monoura, N. and Kinoshita, T. (1994) *Chem. Lett.*, 227.
122. Higuchi, M. and Kinoshita, T. (1998) *J. Photochem. Photobiol. B: Biol.*, **42**, 143.
123. Higuchi, M., Minoura, N. and Kinoshita, T. (1995) *Macromolecules*, **28**, 4981.
124. Higuchi, M., Minoura, N. and Kinoshita, T. (1997) *Langmuir*, **13**, 1616.
125. Bredenbeck, J., Helbing, J., Sieg, A., Figrader, T., Zinth, W., Renner, C., Behrendt, R., Moroder, L., Wachtveitl, J. and Hamm, P. (2003) *Proc. Natl. Acad. Sci. USA*, **100**, 6452.
126. Gilmanshin, R., Williams, S., Callender, R.H., Woodruff, W.H. and Dyer, R.B. (1997) *Proc. Natl. Acad. Sci. USA*, **94**, 3709.
127. Munoz, V., Thompson, P.A., Hofrichter, J. and Eaton, J. (1997) *Nature*, **390**, 196.
128. Duan, K.A. and Kollmann, P.A. (1998) *Science*, **282**, 740.
129. Bieri, O., Wirz, J., Hellrung, B., Schutkowski, M., Drewello, M. and Kiefhaber, T. (1999) *Proc. Natl. Acad. Sci. USA*, **96**, 9597.
130. Pieroni, O., Fissi, A., Angelini, N. and Lenci, F. (2001) *Acc. Chem. Res.*, 34, 9.
131. Bredenbeck, J., Helbing, J., Behrendt, R., Renner, C., Moroder, L., Wachtveitl, J. and Hamm, P. (2003) *J. Phys. Chem. B*, **107**, 8654.
132. Koller, F.O., Reho, R., Figrader, T.E., Moroder, L., Wachtveitl, J. and Zinth, W. (2007) *J. Phys. Chem. B*, **111**, 10481.
133. Satzger, H., Root, C., Renner, C., Behrendt, R., Moroder, L., Wachtveitl, J. and Zinth, W. (2004) *Chem. Phys. Lett.*, **396**, 191.
134. Wachtveitl, J., Spörlein, S., Satzger, H., Fonrobert, B., Renner, C., Behrendt, R., Oesterhelt, D., Moroder, L. and Zinth, W. (2004) *Biophys. J.*, **86**, 2350.
135. Rehm, S., Lenz, M.O., Mensch, S., Schwalbe, H. and Wachtveitl, J. (2006) *Chem. Phys.*, **323**, 28.
136. Bredenbeck, J., Helbing, J., Kumita, J.R., Woolley, G.A. and Hamm, P. (2005) *Proc. Natl. Acad. Sci. USA*, **102**, 2379.
137. Helbing, J., Bregy, H., Bredenbeck, J., Pfister, R., Hamm, P., Huber, R., Wachtveitl, J., De Vico, L. and Olivucci, M. (2004) *J. Am. Chem. Soc.*, **126**, 8823.
138. Satzger, H., Root, C., Gilch, P., Zinth, W., Wildemann, D. and

Fischer, G. (2005) *J. Phys. Chem. B*, **109**, 4770.
139. Wang, S.-C. and Lee, C.T. Jr (2006) *J. Phys. Chem. B*, **110**, 16117.
140. Auernheimer, J., Dahmen, C., Hersel, U., Bausch, A. and Kessler, H. (2005) *J. Am. Chem. Soc.*, **127**, 16107.
141. Jurt, S., Aemissegger, A., Güntert, P., Zerbe, O. and Hilvert, D. (2006) *Angew. Chem. Int. Ed.*, **45**, 6297.
142. Ando, R., Mizuno, H. and Miyawaki, A. (2004) *Science*, **306**, 1370.
143. Dedecker, P., Hotta, J., Ando, R., Miyawaki, A., Engelborghs, Y. and Hofkens, J. (2006) *Biophys. J: Biophys. Lett.*, **91**, L45.
144. Fron, E., Flors, C., Schweitzer, G., Habuchi, S., Mizuno, H., Ando, R., De Figryver, F.C., Miyawaki, A. and Hofkens, J. (2007) *J. Am. Chem. Soc.*, **129**, 4870.
145. Ohmuro-Matsuyama, Y. and Tatsu, Y. (2008) *Angew. Chem. Int. Ed.*, **47**, 7527.
146. Muraoka, T., Cui, H. and Stupp, S.I. (2008) *J. Am. Chem. Soc.*, **130**, 2946.
147. Muraoka, T., Koh, C.-Y., Cui, H. and Stupp, S.I. (2009) *Angew. Chem. Int. Ed.*, **48**, 5946.
148. Sadovski, O., Beharry, A.A., Zhang, F. and Woolley, G.A. (2009) *Angew. Chem. Int. Ed.*, **48**, 1484.
149. Hayashi, G., Hagihara, M. and Nakatani, K. (2009) *Chem Eur. J.*, **15**, 424.
150. Hoppmann, C., Seedorff, S., Richter, A., Fabian, H., Schmieder, P., Rück-Braun, K. and Beyermann, M. (2009) *Angew. Chem. Int. Ed.*, **48**, 6636.

11
Ion Translocation within Multisite Receptors
Valeria Amendola, Marco Bonizzoni, and Luigi Fabbrizzi

11.1
Introduction

Molecular switching is in most cases related to the reversible spatial movement of a portion of a molecule, of a supramolecule or of a polymer, between two definite positions, induced by an external stimulus of either a chemical or physical nature [1]. Classical examples include photoresponsive crown ethers [2], rotaxanes [3], catenanes [4] and polypeptides [5]. However, movement at the molecular level can be achieved also by translocating a particle, for example an ion, between two definite positions within a properly designed molecular system [6].

The first example of reversible ion translocation within a ditopic receptor was reported in 1974 by Lehn and Stubbs [7], who designed the cylindrical macrotricycle **1**, in which two 12-membered N_2O_2 crowns are linked together through the nitrogen atoms by diethyleneoxa spacers. On addition of 1 equiv. of an alkaline-earth metal ion (Ca^{2+}, Sr^{2+}, Ba^{2+}), the cation is observed to interact with one N_2O_2 ring, forming a 1:1 complex. Even a large excess of the alkaline-earth metal salt does not induce the inclusion of a second cation, due to repulsive electrostatic effects. The ^{13}C NMR spectrum of the $[Ca(\mathbf{1})]^{2+}$ complex in D_2O, at 4 °C, shows two sets of four resonances. When the temperature is raised, the signals of the same intensity within each set coalesce and a four-line spectrum is obtained. Such a behaviour indicates the occurrence of a fast intramolecular exchange of the Ca^{2+} cation between the two N_2O_2 binding compartments, as pictorially illustrated in Figure 11.1. The activation free energy for the process, $\Delta G^{\ddagger}$, is 64.4 kJ mol^{-1}, to which a residence time τ of the cation in each compartment of 30 ms at 25 °C corresponds. Moving down in the second group, both $\Delta G^{\ddagger}$ values for the cation exchange process (Sr^{2+}: 60.7 kJ mol^{-1}; Ba^{2+}: <56 kJ mol^{-1}), and residence times (Sr^{2+}: 7 ms; Ba^{2+}: <1 ms, at 25 °C) decrease. The observed process was thus described as a *degenerate* intramolecular motion, which anticipated a similar behaviour described later by Stoddart, with the reversible movement of a π-acceptor wheel between 2 equiv. π-donor stations of the rotaxane **2** (about 1800 traverses per second, back and forth, at 25 °C) (Figure 11.2) [8].

Molecular Switches, Second Edition. Edited by Ben L. Feringa and Wesley R. Browne.
© 2011 Wiley-VCH Verlag GmbH & Co. KGaA. Published 2011 by Wiley-VCH Verlag GmbH & Co. KGaA.

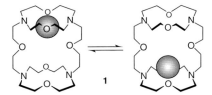

Figure 11.1 The reversible translocation of an alkaline-earth cation (Ca^{2+}, Sr^{2+}, Ba^{2+}) between the 2 equiv. binding compartments of a cylindrical macrotricycle. The frequency of the 'jumps' increases with temperature. Metal translocation is a process the system uses for dissipating the energy acquired from the environment. At 25 °C the Ca^{2+} ion makes 33 'jumps' per second, Sr^{2+} 150 and Ba^{2+} more than 1100 [7].

Noticeably, the activation free energy for the wheel's translocation process in deuterated acetone, $\Delta G^{\ddagger} = 54.4$ kJ mol^{-1}, is lower than that associated to the alkaline-earth metal translocation within **1**. This may reflect the flexibility of rotaxane **2**, which undergoes folding of the $O(CH_2CH_2O)_4$ chain bridging the two donor stations, allowing the occasional contact of the wheel with either phenoxy subunit. In this connection, one should note that in system **1** the metal ion cannot really 'jump' from one ring to the other, because jumping would involve an energetically unfavoured state in which the metal is 'suspended in the void', without any chemical interaction with the surroundings. A more realistic view of the process should consider an intermediate state in which the cation has weakened or broken some of the bonds with one ring and is interacting with the oxygen atom of one of the $O(CH_2CH_2O)_4$ spacers. This view could explain the fact that the translocation process becomes faster on increasing the radius of the cation ($Ca^{2+} < Sr^{2+} < Ba^{2+}$): the larger the ion, the easier the increase of its coordination number and the formation of a bond with an oxygen atom of the bridge.

11.2
Metal-Ion Translocation: Changing Metal's Oxidation State

Spontaneous ion translocation between 2 equiv. compartments is not a valuable process. In the final analysis, it is only a way the system uses for dissipating the thermal energy acquired from the environment. An external operator could only slow/accelerate the process, by modifying the temperature of the solution. In contrast, the principle of controlled molecular motion requires that the movement is determined from the outside, through a stimulus of chemical, electrochemical or photochemical nature. In other words, the operator should provide the energy necessary for the movement through an auxiliary reaction which takes place in solution. In this context, Lehn hypothesized the design of a heteroditopic receptor, suitable for electrochemically driven metal translocation [9]. The cylindrical macrotricycle **3**, shown in Figure 11.3, contains two inequivalent coordinating compartments: a N_2S_2 12-membered crown and a N_2O_3 15-membered crown.

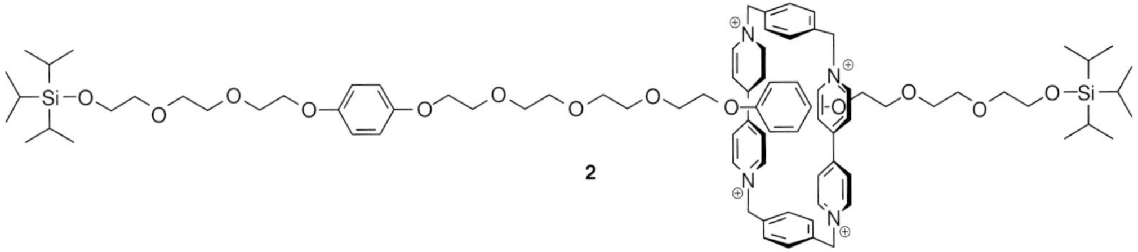

Figure 11.2 The degenerate shuttle of Stoddart. The π-acceptor bis-bipyridinium wheel moves back and forth from one diphenoxy π-donor station to the other. At 25 °C, in an acetone solution, the wheel makes 1800 traverses per second [8].

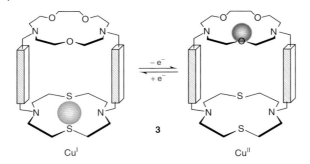

Figure 11.3 A hypothesized system suitable for the redox-driven translocation of a copper centre. When in the Cu^I oxidation state (d^{10}), the metal will chose the soft N_2S_2 donor set; on the other hand, the transition-metal ion Cu^{II} (d^9) will prefer the quinquedentate macrocycle N_2O_3, thus profiting from a higher coordination number. One-electron oxidation-reduction is expected to induce the reversible 'jumping' of the metal from one compartment to the other [9].

The N_2S_2 compartment is suitable for coordinating a Cu^I (d^{10}) centre, which accepts σ electrons from nitrogen and sulfur donor atoms, but can also donate electronic density from its filled $d\pi$ orbitals to the empty $d\pi$ orbitals of the sulfur atoms. On the other hand, the N_2O_3 ring would probably allow the binding of a Cu^{II} (d^9) ion, which, as a genuine transition metal, prefers five-coordination to four-coordination. Thus, in a solution containing an equivalent amount of **3** and of a copper salt, in either oxidation state, consecutive one-electron oxidation-reduction should induce the reversible 'jumping', up and down, of the metal ion. However, a ditopic receptor of type **3** has not been synthesized yet and Lehn's hypothesis on the translocation of a copper centre based on the Cu^{II}/Cu^I redox change cannot be verified.

The first example of redox-driven translocation of a metal centre was reported in 1993 by Shanzer and coworkers [10], who synthesized the triple-stranded ditopic receptor **4**. System **4** can act as a ligand for transition metal ions by offering two distinct coordinating compartments: one made by three hydroxamate subunits, in the lower part of the structural formula shown in Figure 11.4, the other consisting of three 2,2′-bipyridine fragments, in the upper part of the formula.

If 1 equiv. of an Fe^{III} salt is added to a water-methanol alkaline solution of **4**, the metal goes to occupy the tris-hydroxamate compartment: the solution takes a pale brown colour and an intense band develops ($\lambda_{max} = 420$ nm, $\varepsilon = 2400$ M^{-1} cm^{-1}). Tris-hydroxamate complexes are among the most stable complexes of iron(III) with a formation constant β_3 on the order of 10^{30}. On treating the solution with a reducing agent (e.g. ascorbic acid), the colour turns deep purple ($\lambda_{max} = 540$ nm, $\varepsilon = 4700$ M^{-1} cm^{-1}), a spectral feature typical of the $[Fe(bpy)_3]^{2+}$ chromophore. Thus, the Fe^{III}-to-Fe^{II} reduction induces the translocation of the metal centre from the lower to the upper compartment, as is pictorially illustrated in Figure 11.4. The lifetime τ for the translocation is 22 s, much higher than that observed for alkaline-earth metal ions within the degenerate ditopic

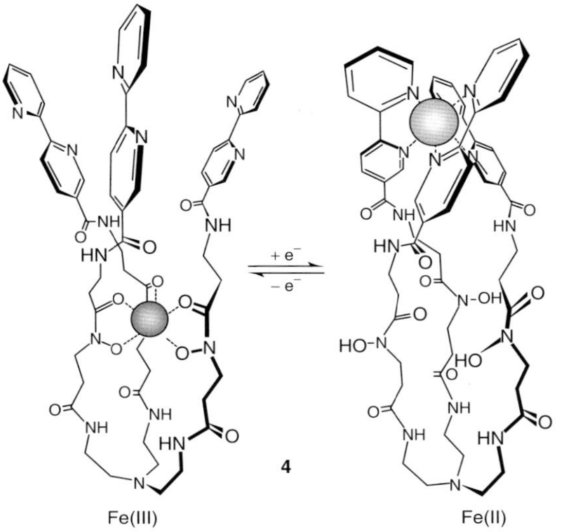

Figure 11.4 Iron translocation between two inequivalent compartments of a tripodal ditopic receptor, based on the Fe^{III}/Fe^{II} redox change. The Fe^{III} ion stays in the tris-hydroxamate compartment. Following chemical reduction to Fe^{II}, the metal moves to the tris-2,2′-bipyridine compartment. Further chemical oxidation brings back the Fe^{III} cation to the lower compartment, resetting the system [10].

receptor **1**. This points towards a complicated mechanism, which may involve severe conformational changes of the triple-stranded receptor. It is possible that the oxygen atoms of the amide groups on the spacers linking the hydroxamate subunits and bpy fragments help the translocation process through temporary coordination of the travelling metal ion. The iron centre can be translocated back to the starting compartment through the oxidation of Fe^{II} to Fe^{III} with peroxydisulfate. This process is especially slow and takes several minutes at 70 °C. The increased sluggishness of the translocation is probably related to the higher substitutional inertness of the Fe^{II} centre in the low-spin state (when coordinated by the three bpy fragments). In an analogous tripodal ditopic receptor, the iron centre was observed to travel from an inner compartment consisting of three deprotonated salicylamide fragment (providing six coordinating oxygen atoms and suitable for coordination of high-spin Fe^{III}) to an outer compartment made by three bpy subunits (accommodating a low-spin Fe^{II} cation), following chemical reduction (with ascorbic acid) and oxidation (with hydrogen peroxide) [11]. The higher rate of the process ($\tau \approx 10$ s in either direction) may reflect the more flexible nature of the receptor.

A further metal translocation process based on the Fe^{III}/Fe^{II} couple is illustrated in Figure 11.5 [12]. In the ditopic receptor **5**, two terdentate coordinating functionalities have been appended in 2 and 6 positions of a 4-methylphenol platform. One functionality contains a tertiary amine nitrogen atom and two phenolate oxygen atoms and, with the phenolate oxygen atom of the platform and two molecules

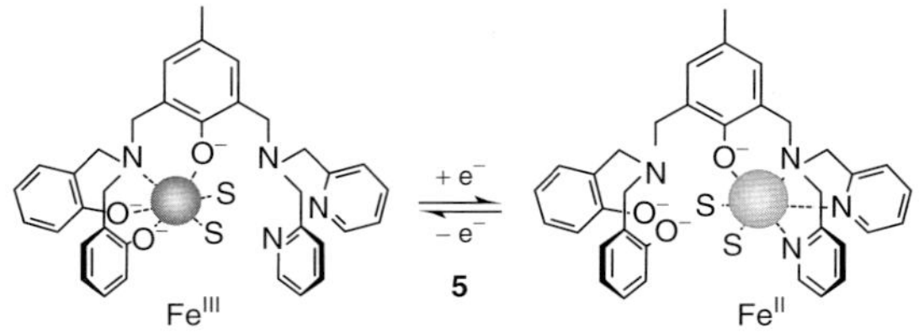

Figure 11.5 Electrochemically driven iron translocation between two inequivalent compartments of a tripodal ditopic receptor involving the Fe^{III}/Fe^{II} couple. Fe^{III} chooses the NO$_3$ + 2 S compartment (where S is a solvent molecule, e.g. MeCN); Fe^{II} chooses the N$_3$O + 2 S compartment. The phenolate oxygen atom of the central platform remains bound to the metal during redox and translocation processes, which may account for the high rate (τ < 0.5 s) [12].

of solvent (MeCN) provides a six-coordinating environment suitable for high-spin Fe^{III} binding. The other functionality contains a tertiary amine nitrogen atom and two pyridine nitrogen atoms, which, along with the phenolate oxygen atom of the platform and two molecules of solvent, provides a comfortable six-coordinating donor set for low-spin Fe^{II}. Indeed, the iron centre can be electrochemically translocated, back and forth, from one compartment to the other, according to a pendular motion. Noticeably, in the course of the translocation, the iron centre remains bound to the central phenolate oxygen atom, which acts as a hook. Due to the beneficial assistance of this oxygen atom and also to the short distance, the translocation process is very fast, the lifetime for both direct and reverse oscillation being lower than 0.5 s.

11.3
Metal-Ion Translocation: Changing through a pH Variation the Coordinating Properties of One Receptor's Compartment

In the experiment described in the previous section, the moving object, a metal M, decided to set out on the intramolecular journey following a change of its oxidation state. In particular, the M$^{(n+1)+}$–to–M^{n+} redox change was the driving force of the process. General aspects of the mechanism are pictorially illustrated in Figure 11.6.

To a first approximation, one could consider the cation in its higher oxidation state, M$^{(n+1)+}$, as a *hard* centre, thus prone to interact with the *hard* moiety of the ditopic receptor, and the cation in its lower oxidation state, M^{n+}, as a *soft* centre, thus showing affinity towards the *soft* compartment. This model fits quite well the behaviour of the iron centre in the ditopic receptor 4: the d^5 high-spin Fe^{III} ion is bound to the [O$_6$]$^{3-}$ tris-hydroxamate donor set, through predominantly electrostatic interactions (hard–hard); the d^6 low-spin Fe^{II} ion is bound to the [N$_6$] tris-2,2'-bipyridine donor set, through definitely covalent interactions, which involve ligand-to-metal σ donation and metal-to-ligand π donation (soft–soft).

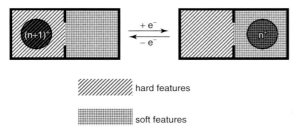

Figure 11.6 The translocation of a metal ion M induced by the $M^{(n+1)+}/M^{n+}$ redox change. In an idealized situation, the metal ion in its higher oxidation state, $M^{(n+1)+}$, displays *hard* features and prefers to be located in the *hard* compartment of the receptor; the metal in the lower oxidation state, M^{n+}, exhibits a *soft* nature and wants to stay in the *soft* compartment.

However, it may be also desirable to carry out a process in which the movable object (the cation) remains unchanged during the translocation. In such a case, the transfer must be induced by an externally controlled modification of the coordinating properties of one compartment. This situation can be achieved, for instance, in the presence of a negatively charged compartment A^{n-}, which is a good ligand for the movable cation M, but it is also a Brønsted base that, on addition of acid, gives AH_n and loses any coordinating tendency. The hypothesized system and the translocation process are schematically outlined in Figure 11.7. Further requirements for the occurrence of the translocation process are: compartment B does not show any Brønsted acid–base property and the affinity sequence towards the metal is: $A^{n-} > B > AH_n$. Under these conditions, the translocation of M between the two compartments can be induced through controlled addition of acid and base.

In particular, at low pH values, where AH_n is present, the metal ion chooses compartment B. Then, a standard base is added, which neutralizes AH_n and A^{n-} forms. At this point, the metal moves to the more coordinating compartment A^{n-}. The movement can be reversed by adding acid, so that the metal can be translocated

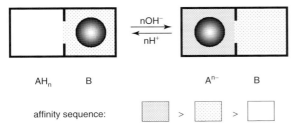

Figure 11.7 The translocation of a metal ion driven by a pH change. One of the compartments consists of a conjugate acid–base pair AH_n/A^{n-}, the other, B, does not show any acid-base properties. In more acidic conditions, where the AH_n form is present, the metal stays in compartment B after base addition.

back and forth at will between the two compartments through consecutive additions of calculated amounts of standard base and acid. The first deliberately designed system allowing pH-driven translocation of a metal ion is shown below (**6a**) [13].

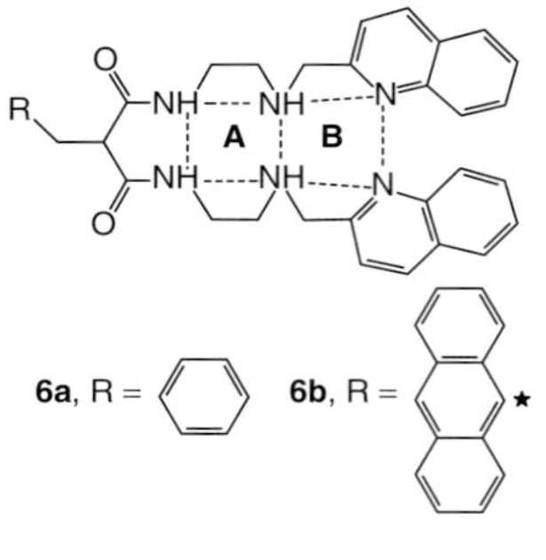

The compartment in the right side of formula **6a** consists of two pyridine nitrogen atoms and two secondary amine nitrogen atoms. In the range of pH suitable for metal translocation, all these nitrogen atoms will not undergo protonation and this compartment will play the role of B. The other compartment, on the left hand of the formula, is made of the same two secondary nitrogen atoms and of two amide nitrogen atoms. The nitrogen atom of the amide group shows negligible coordinating tendencies towards transition metals. However, when deprotonated, it becomes a strong donor for 3d block cations (e.g. Cu^{II}, Ni^{II}) [14]. The pH-controlled complexation of a divalent metal by a quadridentate ligand containing two secondary amide and two secondary amine groups is illustrated in Figure 11.8.

On base addition, the two amide groups deprotonate, while simultaneous chelation of the metal ion by the quadridentate ligand takes place, to give a neutral complex of square planar geometry. The negative charge resulting from deprotonation is not completely retained on each nitrogen atom, but it is partly delocalized over each NCO amide moiety. Notice that the deprotonation of the N–H fragment of an amide group is a very endergonic process, which makes amides very poor Brønsted acids ($pK_A = 14.5$). However, coordination may compensate such an endergonic effect and, in the presence of suitable metal ions, amide deprotonation

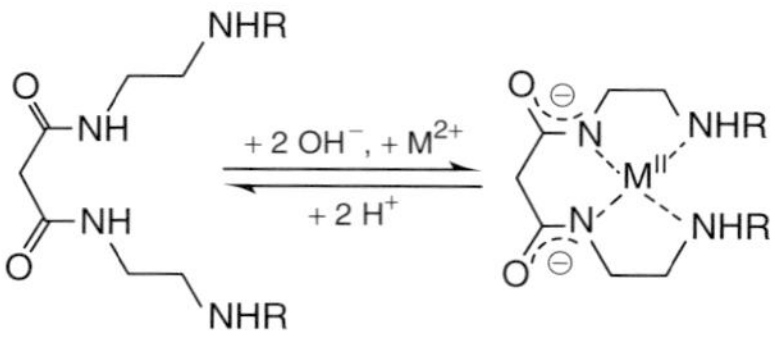

Figure 11.8 The pH-controlled complexation of a divalent metal by a quadridentate ligand containing two secondary amide and two secondary amine groups. On addition of 2 equiv. of OH^-, the N–H fragments of the amide groups deprotonate and neutral metal complex of square planar geometry forms.

can occur under neutral or moderately basic conditions. Noticeably, only divalent metal ions late in the 3d series (Ni^{II}, Cu^{II}) are able to promote amide deprotonation according to the equilibrium depicted in Figure 11.8. Ions early in the series (Mn^{II}, Fe^{II}, Co^{II}) do not establish coordinative interactions strong enough to compensate amide deprotonation and do not form complexes with ligands of the type shown in Figure 11.8, even at high pH values.

Related to this, the formation of 1:1 complexes between Ni^{II} and receptor **6a** (indicated in the following as LH_2) in an $MeCN/H_2O$ mixture (4:1, v/v), at varying pH, was investigated by carrying out titration experiments. Analysis of pH titration data indicated the formation of several complex species along the 2–12 pH interval, whose concentration profiles are shown in Figure 11.9.

Quite interestingly for translocation purposes, two major metal complex species are present over the 7–10 pH range: the complex of the neutral ligand, $[Ni^{II}(LH_2)]^{2+}$, which is present at 90% at pH 7.5, and the complex of the doubly deprotonated ligand, $[Ni^{II}(L)]$, which is present at 100% at pH $\geq$ 9. It is suggested that in the dicationic complex $[Ni^{II}(LH_2)]^{2+}$ the metal is located in compartment B, whereas in the neutral complex $[Ni^{II}(L)]$ the metal stays in the A^{2-} section. Such an assessment is based on the spectral features of the two complex species. The $[Ni^{II}(LH_2)]^{2+}$ complex (pale blue solution, adjusted to pH = 7.5) exhibits two weak metal-centred absorption bands, at 606 nm ($\varepsilon = 11\ M^{-1}\ cm^{-1}$) and at 820 nm ($\varepsilon = 5\ M^{-1}\ cm^{-1}$). These band are typically observed with a high-spin Ni^{II} ion in an octahedral

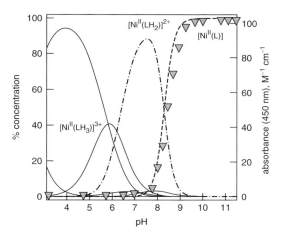

Figure 11.9 The concentration profiles of the species present at the equilibrium in a $MeOH/H_2O$ solution (4:1 v/v) containing equimolar amounts of Ni^{II} and of the heteroditopic receptor **6a** (per cent concentration in the left vertical axis). Species involved in the translocation process are: $[Ni^{II}(LH_2)]^{2+}$, 90% at pH = 7.5, in which the Ni^{II} centre occupies compartment B, and $[Ni^{II}(L)]$, 100% at pH $\geq$ 9.5, in which Ni^{II} has moved to the doubly deprotonated A^{2-} compartment. When in the A^{2-} compartment, the Ni^{II} centre shows a square stereochemistry (low-spin, yellow, d-d absorption band at 450 nm, $\varepsilon = 103\ M^{-1}\ cm^{-1}$): full triangles give the intensity of such an absorption band (molar absorbance on the right vertical axis). Reproduced from reference [13]. Copyright RSC (2000).

coordinative environment. It is therefore suggested that the donor set involves the two amine nitrogen atoms, the two quinoline nitrogen atoms and the oxygen atoms of two water molecules. On the other hand, the solution of the [Ni^{II}(L)] neutral complex is yellow and its absorption spectrum shows a relatively intense d-d band ($\varepsilon = 103$ M^{-1} cm^{-1}) centred at 450 nm. Such a band is typically observed with Ni^{II} complexes of square geometry and is formed with quadridentate ligands exerting strong in-plane interactions. This seems the case of the tetra-aza donor set consisting of two deprotonated amide groups and two amine groups. The extremely strong metal–ligand interactions induce pairing of the Ni^{II} ion, which becomes diamagnetic. The assignment of these geometrical and electronic features to the complex species of stoichiometry [Ni^{II}(L)] is corroborated by the finding that the absorbance at 450 nm (triangles in the diagram of Figure 11.9) superimposes well on the concentration profile of the neutral complex (dotted line). It shows that a pH variation from 7.5 to 9 makes the Ni^{II} ion move from compartment B to compartment A: the translocation process is signalled by the pale blue-to-yellow colour change and by the development of the absorption band of the [Ni^{II}(L)] complex, centred at 450 nm.

The possible occurrence of the translocation process has been demonstrated by steady-state investigations (i.e. through titration experiments, both potentiometric and spectrophotometric). However, the process had to be characterized in its temporal development through dynamic studies. In particular, the B-to-A translocation of the Ni^{II} ion was followed through stopped-flow spectrophotometry by monitoring the growth of the band at 450 nm, when a solution containing equimolar amounts of Ni^{II} and **6a**, adjusted to pH 7.5 (containing 90% of the [Ni^{II}(LH$_2$)]$^{2+}$ complex – syringe 1), was mixed with a solution buffered at pH 9.5 (CHES buffer – syringe 2). The family of spectra recorded over the course of the experiment is shown in Figure 11.10. The band at 450 nm reaches its limiting value within 1 s. In particular, the process shows first-order kinetics, with a lifetime $\tau = 0.25 \pm 0.01$ s.

On the other hand, when acid is added to the yellow solution adjusted to pH $\geq$ 9.5, containing 100% of the [Ni^{II}(L)] complex, and the pH is brought back to 7.5, the pale blue colour is quickly restored, indicating that the metal ion has been relocated in the B compartment. Again, the dynamics of the process was investigated through stopped-flow spectrophotometric studies. In particular, syringe 1 contained a solution of [Ni^{II}(L)], adjusted to pH 9.5, syringe 2 contained a solution buffered to pH = 7.5 with HEPES and the decay of the band at 450 nm was monitored. Quite surprisingly, the reverse translocation process (A-to-B) was found to be significantly slower than the direct process (B-to-A), being characterized by a lifetime $\tau = 2.2 \pm 0.1$ s. The difference of translocation rates has to be associated to the different kinetic properties of the Ni^{II} ion, whether in the high-spin state (labile with respect to substitution) or low-spin state (substitutionally more inert). Actually, the initial step of each translocation process, either direct or inverse, must involve the preliminary dissociation of the coordinative bonds. Thus, B-to-AH$_2$ translocation is faster because it involves the dissociation of the more-labile high-spin Ni^{II} ion, whereas the A^{2-}-to-B process is slower because it implicates the dissociation of the more inert low-spin Ni^{II} ion (Figure 11.11).

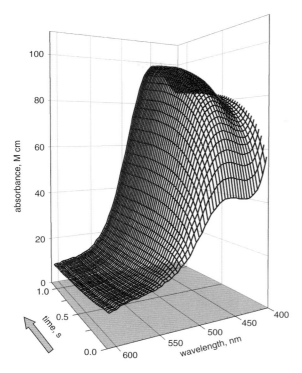

Figure 11.10 Kinetic aspects of the translocation of Ni^{II} from compartment B to A of the heteroditopic ligand **6a**, induced by a pH change from 7.5 to 9.5 and followed by stopped-flow spectrophotometry. Spectra were recorded with a diode array spectrophotometer. The band at 450 nm pertains to the square complex of Ni^{II} coordinated by two amine nitrogen atoms and two deprotonated amide groups (A^{2-}) and its growth indicates the occurrence of the B-to-A translocation. The process shows a first-order behaviour and is characterized by a lifetime $\tau = 0.25 \pm 0.01$ s. Reproduced from reference [13]. Copyright RSC (2000).

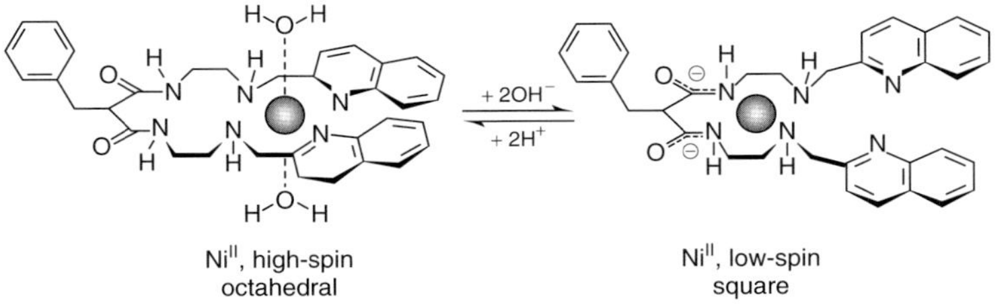

Ni^{II}, high-spin octahedral

Ni^{II}, low-spin square

Figure 11.11 Geometrical aspects of the translocation of Ni^{II} from compartment B to A of the heteroditopic ligand **6a**, induced by a pH change from 7.5 to 9.5 [13].

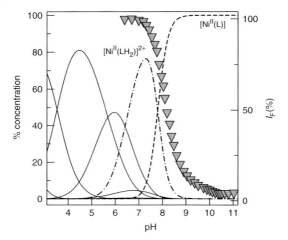

Figure 11.12 The concentration profiles of the species present at equilibrium in a dioxane/water solution (4:1 v/v) containing equimolar amounts of Ni^{II} and of the heteroditopic receptor **6b** (per cent concentration in the left vertical axis). Species involved in the translocation process are: $[Ni^{II}(LH_2)]^{2+}$, 80% at pH = 7.2, in which the Ni^{II} centre occupies compartment B, and $[Ni^{II}(L)]$, 100% at pH $\geq$ 9.0, in which Ni^{II} has moved to the doubly deprotonated A^{2-} compartment. Triangles give the fluorescence intensity of the anthracene fragment covalently linked to the framework of **6b**. When the Ni^{II} centre is in compartment B, the anthracene subunit displays its full fluorescent emission. When the metal moves to compartment A^{2-}, an electron-transfer process takes places from Ni^{II} to the excited fluorophore that induces complete quenching of the fluorescence. Reproduced from reference [13]. Copyright RSC (2000).

Colour change is a convenient means for monitoring the occurrence of a process in solution, but fluorescence provides a more powerful signal. A large variety of receptors have been equipped with fluorogenic subunits in order to sense all kind of analytes in solution (metal ions, anions, amino acids, etc.), thus generating complete assortments of fluorescent sensors [15]. In this context, the fluorogenic receptor **6b** was considered, in which an anthracenyl subunit is covalently linked to the carbon atom between the amide groups, with the aim to monitor the translocation process through a change in the fluorescence spectrum [16].

From potentiometric studies on a dioxane/water solution (4:1 v/v) containing equimolar amounts of **6b** and Ni^{II} the distribution diagram shown in Figure 11.12 could be calculated. It can be observed that the complex species $[Ni^{II}(LH_2)]^{2+}$ (metal in compartment B) reaches its maximum concentration (80%) at pH = 7.2 and that $[Ni^{II}(L)]$ attains 100% at pH $\geq$ 9.0. Thus, pH changes from 7 to 9 and *vice versa* should make the metal travel back and forth between the two compartments. Very interestingly, the two complexes exhibit different emission properties: the solution adjusted to pH = 7.2, in which the major species is $[Ni^{II}(LH_2)]^{2+}$, displays typical anthracene emission, whereas a solution at pH $\geq$ 9 (100% of $[Ni^{II}(L)]$) is nonfluorescent. Noticeably, the intensity of the fluorescence (triangles in the

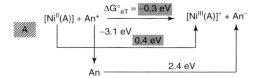

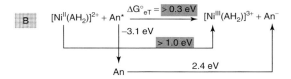

Figure 11.13 Thermodynamic cycles for calculating the free-energy changes $\Delta G°_{eT}$, associated to the intramolecular photoinduced electron transfer (eT) from Ni^{II} to the anthracene subunit of **6b**. When Ni^{II} is in compartment A ([Ni^{II}(A)] complex), the eT process is thermodynamically favoured ($\Delta G°_{eT} < 0$); when Ni^{II} is in compartment B ([Ni^{II}(AH$_2$)]$^{2+}$ complex), the process is disfavoured ($\Delta G°_{eT} > 0$).

diagram in Figure 11.12) decreases until complete quenching in correspondence with the formation of the [Ni^{II}(L)] complex species.

The ability of the Ni^{II} centre to quench, or not, the excited anthracene subunit An* cannot be explained in terms of the distance between the metal and the anthracene subunit (when close, Ni^{II} quenches An*, when more distant it does not), but can be fully accounted for on considering the thermodynamic aspects of the Ni^{II}-to-An* electron-transfer (eT) process.

In particular, as illustrated in Figure 11.13, the free-energy change, $\Delta G°_{eT}$ associated to the eT from Ni^{II} to An*, which can be calculated through the algebraic sum of pertinent photophysical and electrochemical quantities, is negative (favoured process) when the metal is located in compartment A, but it is positive (disfavoured process) when the metal stays in compartment B. This results from the different attitude of each compartment to favour the occurrence of the Ni^{II}-to-Ni^{III} oxidation process. Compartment A, in the A^{2-} form, exerts strong inplane coordination, thus raising the energy of the metal centred level (d_{xy}) from which the electron is abstracted on oxidation: the electrode potential associated to the Ni^{III}/Ni^{II} couple is unusually low (0.4 V). On the other hand, compartment B exerts rather poor coordinative interactions, which makes the attainment of the +3 oxidation state rather difficult and the electrode potential of the Ni^{III}/Ni^{II} couple much more positive (>1.0 V). In conclusion, the different metal interference with the photoexcited fluorophore is related to the more or less pronounced coordinating tendencies of each compartment, rather than to the Ni^{II}–An* distance.

Thus, the occurrence of the pH-controlled Ni^{II} translocation back and forth within receptor **6b** is signalled through quenching-revival of anthracene fluorescence. In particular, both direct and reverse processes were investigated in their kinetic aspects through spectrofluorimetric stopped-flow experiments and by monitoring the first-order decay or development of the anthracene emission band. The B-to-AH$_2$ translocation ($\tau = 12 \pm 1$ s) is faster than A^{2-}-to-B back translocation

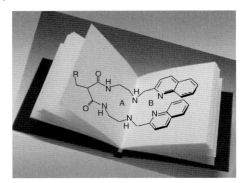

Figure 11.14 The 'facing pages' mechanism. Metal translocation takes place through the occasional folding of the two halves of the heteroditopic receptor (either **6a** or **6b**). Compartments A and B are laid on two 'facing pages of a book', whose spine passes through the two secondary amine nitrogen atoms. In this situation, it is expected that a bulky substituent (e.g. an anthracene fragment), due to steric repulsive effects, raises the energy of the transition state and slows down the 'passage' of the metal from one 'page' to the other.

($\tau = 66 \pm 12$ s), a pattern already observed in the case of system **6a** and ascribed to the different lability of the Ni^{II} ion, whether in the high- or low-spin state. However, the rates of both direct and reverse processes are markedly lower for **6b** than for **6a**. In this context, it is suggested that metal translocation is associated to the occasional folding of the ditopic receptor (**6a** or **6b**) around the ideal line passing through the two amine nitrogen atoms, which acts as a hinge. This mechanism is pictorially illustrated in Figure 11.14, in which compartments A and B have been laid down on two adjacent pages of a book. According to this model, the transition state for the translocation process corresponds to a situation of maximum folding, in which the two halves of the receptor (the two pages of the book) are brought one face to the other at the closest possible distance, an event that precedes metal transfer. In these circumstances, the steric repulsions exerted by the bulky anthracene substituent raise the energy of the transition state, thus reducing the rate of both direct and reverse translocation processes.

The Cu^{II} ion is expected to undergo a similar pH-driven translocation experiment within type-**6** ditopic receptors [17]. In particular, the occurrence of the translocation could be conveniently followed by looking at colour changes. In fact, when in compartment B, the Cu^{II} cation is bound by two amine nitrogen atoms, by two pyridine nitrogen atoms and probably also by a water molecule, to give a five-coordinate species of blue colour. Indeed, a slightly acidic solution containing equimolar amounts of **6a** and Cu^{II} shows a blue colour (absorption band centred at 615 nm). On the other hand, when in compartment B, the Cu^{II} ion should be bound by two amine nitrogen atoms and two deprotonated amide groups according to a square coordination geometry. Copper(II) complexes of this type, like that shown in Figure 11.8, show a pink-violet colour (the colour of the classical bis-biuretate complex), with a band centred at $\approx$500 nm. However, when the solution containing

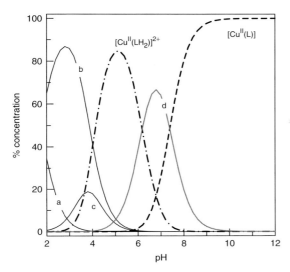

Figure 11.15 The concentration profiles of the species present at the equilibrium in a dioxane/water solution (4:1 v/v) containing equimolar amounts of Cu^{II} and of the heteroditopic receptor **6a** (per cent concentration in the left vertical axis). Species involved in the translocation process are: $[Cu^{II}(LH_2)]^{2+}$, about 90% at pH = 5, in which the Cu^{II} centre stays in compartment B, and $[Cu^{II}(L)]$, 100% at pH ≥ 9.5, in which Cu^{II} occupies the doubly deprotonated A^{2-} compartment, but is also coordinated by a quinoline nitrogen atom, to give a five-coordinate complex. Profiles a, b and c refer to metal-free forms of the receptors with different degrees of protonation (LH_5^{3+}, LH_4^{2+} and LH_3^+, respectively. Profile d corresponds to the $[Cu^{II}(LH_2)(OH)]^+$ complex, in which the metal is coordinated by the four nitrogen atoms of compartment B and by a hydroxide ion, to give a five-coordinate species. Reproduced from reference [17]. Copyright RSC (2001).

equimolar amounts of **6a** and Cu^{II} is brought to a distinct alkaline pH, the solution does not turn pink, but simply takes a more intense blue colour (band centred at 550 nm). pH titration experiments in a dioxane-water solution (4:1 v/v) containing equimolar amounts of **6a** and $Cu^{II}(CF_3SO_3)_2$ allowed for the determination of the complex species present at the equilibrium along the 2–12 pH interval, whose concentration profiles are shown in Figure 11.15.

At pH = 5 the doubly positively charged species $[Cu^{II}(LH_2)]^{2+}$ dominates (about 90%), which indicates that the metal is coordinated by the two amine nitrogen atoms and by the two quinoline nitrogen atoms of compartment B. In view of the pronounced preference of copper(II) towards five-coordination, it is suggested that a water molecule is also bound to the metal. On increasing pH, a species of stoichiometry $[Cu^{II}(LH_2)(OH)]^+$ forms, which reaches its maximum concentration, about 70%, at pH = 7 and should originate from the deprotonation of the metal-bound water molecule of the $[Cu^{II}(LH_2)]^{2+}$ complex. Most importantly for translocation purposes, on further pH increase, the neutral species $[Cu^{II}(L)]$ forms, which reaches 100% at pH ≥ 9. In this complex, Cu^{II} must be coordinated by two amine nitrogen atoms and by the two deprotonated amide groups of the

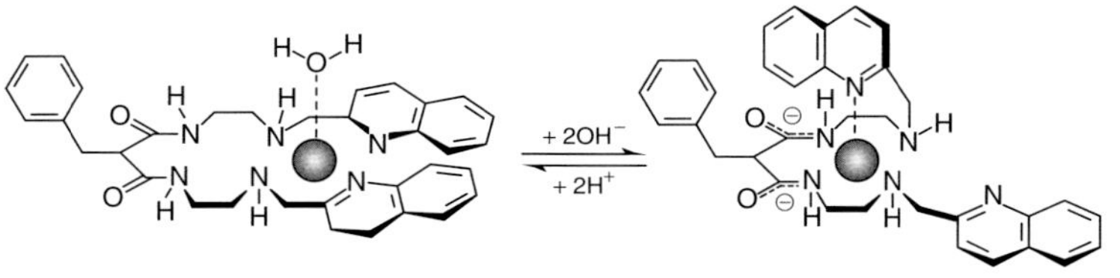

Figure 11.16 Geometrical features associated to the translocation of Cu^{II} from compartment B to A of the heteroditopic ligand **6a**, induced by a pH change from 5 to 9. Metal translocation is accompanied by a rearrangement of the receptor, which brings the quinoline pendant arm to coordinate the Cu^{II} ion in the A^{2-} compartment. This allows the Cu^{II} centre to keep the preferred five-coordination [17].

A compartment (in its A^{2-} version). However, the blue colour and the absorption band centred at 550 nm strongly suggests that a fifth nitrogen atom, from a quinoline subunit of the receptor, is bound to the metal. In fact, it is well documented, since the earliest studies in coordination chemistry [18], that binding of a further nitrogen-containing ligand to a Cu^{II} centre in a square N_4 donor set induces a distinct red-shift of the d-d band. On this basis, a translocation experiment can be carried out by adjusting the pH of a solution containing equimolar amounts of Cu^{II} and **6a** to 5, then adding standard base to bring the pH to 9 or more. The hypothesized geometrical aspects of the translocation process are tentatively sketched in Figure 11.16.

Notice that the pH-controlled process illustrated in Figure 11.16 involves both metal translocation and receptor rearrangement, a behaviour that derives from the pronounced tendency of the Cu^{II} ion to reach five-coordination. In order to realize a neat Cu^{II} translocation process, that is not complicated by drastic conformational rearrangements of the ligating frameworks, one should design a receptor in which compartments A and B are distinctly separated and the coordination of a fifth nitrogen atom from the other compartment is sterically prevented. Such an opportunity has been provided by the ditopic receptor **7**, which contains two discrete N_4 compartments, one consisting of the well-known diamine-diamide quadridentate subunit (compartment A, displaying Brønsted acid activity through the conjugate couple AH_2/A^{2-}), and the other, compartment B, made of two facing bpy fragments [19].

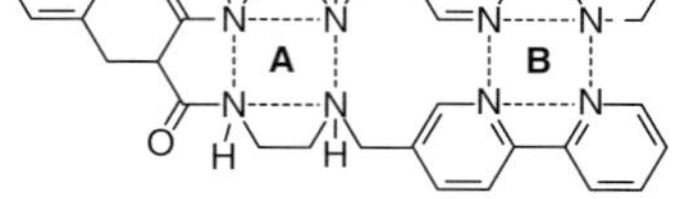

As usual, preliminary pH titration experiments on a dioxane–water solution (4 : 1 v/v) containing equimolar amounts of **7** and $Cu^{II}(CF_3SO_3)_2$ were carried out to define the complex species present at the equilibrium along the 2–12 pH

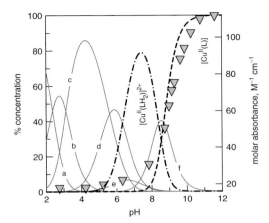

Figure 11.17 Percent concentration of the species present at the equilibrium in a solution containing equimolar amounts of **7** (= LH$_2$) and Cu^{2+} (left vertical axis); profiles a–d refer to the variously protonated forms of LH$_2$ (from LH$_6^{4+}$ to LH$_3^+$). Profile e refers to the metal containing protonated species [Cu^{II}(LH$_3$)]$^{3+}$. Profile f refers to the metal-containing species [Cu^{II}(LH$_2$)(OH)]$^+$. Triangles give the molar absorbance of the d-d band (λ_{max} = 502 nm) of the [Cu^{II}(L)] complex (right vertical axis). Reproduced from reference [19]. Copyright Wiley (2002).

interval (dioxane–water 4 : 1 v/v). Pertinent concentration profiles are shown in Figure 11.17.

Two major metal-containing species, relevant to the translocation process, are present beyond neutrality: (i) a species of stoichiometry [Cu^{II}(LH$_2$)]$^{2+}$, which reaches its maximum abundance (80%) at pH = 7.4 and (ii) a species of stoichiometry [Cu^{II}(L)], which is present at 100% at pH ≥ 11. Noticeably, the solution adjusted to pH = 7.4 is blue and its absorption spectrum is similar to that of a solution of the [Cu^{II}(bpy)$_2$]$^{2+}$ model complex (see spectrum **b** in Figure 11.18).

This indicates that, in the [Cu^{II}(LH$_2$)]$^{2+}$ form, the metal stays in compartment B. Also, in the present case, it is suggested that a water molecule completes the coordination polyhedron, which, as observed in a variety of [Cu^{II}(bpy)$_2$(H$_2$O)]$^+$ complexes crystallographically characterized in the solid state [20], should exhibit a slightly distorted trigonal-bipyramidal geometry. On increasing pH, a species of stoichiometry [Cu^{II}(LH$_2$)(OH)]$^+$ forms; this species, which reaches its maximum concentration of about 80%, at pH = 7, is thought to originate from the deprotonation of the metal-bound water molecule of the [Cu^{II}(LH$_2$)]$^{2+}$ complex. Very interestingly, when further base is added and the pH is adjusted to 12, with the formation of 100% of the [Cu^{II}(L)] species, the solution takes on a pink-violet colour (absorption band centred at 500 nm). This unambiguously indicates that the Cu^{II} ion has moved to compartment B in order to profit from the intense inplane interactions exerted by the two amine groups and by the two deprotonated amide groups, in a square planar geometry. This is confirmed by the evidence that the intensity of the band at 500 nm superimposes well on the concentration profile of the [Cu^{II}(L)] complex, as shown in the diagram in Figure 11.17. Thus,

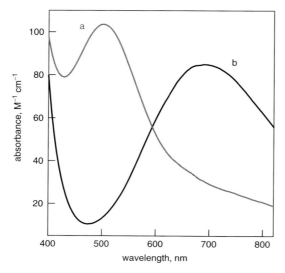

Figure 11.18 d-d absorption bands in the visible region recorded in a dioxane/water solution (4:1 v/v) containing equimolar amounts of **7** and $Cu^{II}(CF_3SO_3)_2$: a, pH = 12, major species $[Cu^{II}(L)]$, 100%, the metal is in compartment A (A^{2-} form), pink colour; b, pH = 7.4, major species $[Cu^{II}(LH_2)]^{2+}$, 80%, the metal is in compartment B, blue colour. Reproduced from reference [19]. Copyright Wiley (2002).

changing the pH from 7.4 to 12 induces Cu^{II} translocation from B to A^{2-}. The occurrence of the process is signalled by a readily-detected colour change: from blue to pink-violet. On the other hand, on addition of standard acid back to pH = 7.4, the solution turns blue again, indicating that the reverse translocation has taken place. The direct and reverse translocation processes can be repeated at will, in principle indefinitely. The detection limit is given by the progressive dilution of the solution, due to the consecutive addition of the standard solution of acid and base.

Dynamic aspects of the metal translocation, both direct and reverse, were investigated by stopped-flow spectrophotometric experiments. Figure 11.19 shows the family of spectra obtained with a diode array spectrophotometer when a solution containing equimolar amounts of **7** and $Cu^{II}(CF_3SO_3)_2$ and adjusted to pH 7.4 ($[Cu^{II}(LH_2)]^{2+}$ complex present at 80%) was mixed with a solution 2×10^{-2} M of NaOH. The absorbance profiles at 500 nm (increasing, monitoring the formation of the pink $[Cu^{II}(L)]$ complex) and at 685 nm (decreasing, monitoring the disappearance of the $[Cu^{II}(LH_2)]^{2+}$ complex) showed a first order behaviour, with a lifetime $\tau = 0.54 \pm 0.05$ s. A first-order behaviour was observed also on mixing a solution of the complex adjusted to pH 12 with a solution buffered to pH 7.4 and a lifetime $\tau = 0.58 \pm 0.05$ s was determined. The high rate of the process in either direction is related both to the substitutional lability of the Cu^{II} ion and to the flexible nature of the ditopic receptor. In any case, it is probable that the

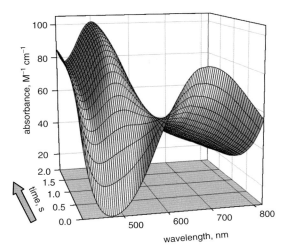

Figure 11.19 Spectra taken over the course of a stopped-flow spectrophotometric experiment for monitoring the B-to-A translocation of Cu^{II} within system **7**. One syringe contained a solution 2.5×10^{-3} M both in **7** and in Cu^{II}(CF$_3$SO$_3$)$_2$; the other contained a solution 2×10^{-2} M in NaOH. Medium: dioxane/water (4:1 v/v). The lifetime τ associated to the first order process is 0.54 ± 0.05 s. Stationary spectra are shown in Figure 11.18.

minor species [Cu^{II}(L)OH]$^+$ that forms at intermediate pH values (7.5–9.5) plays some role in the metal-transfer mechanism and makes the translocation smoother.

It has been previously observed with system **6b**/Ni^{II} that covalent linking of a fluorogenic fragment to the receptor provides a powerful tool for monitoring metal translocation. However, this approach may be time consuming due to the multistep synthesis of the receptor and, in any case, the occurrence of the correct signal-transduction mechanism (the metal quenches the fluorescent emission when in one compartment and does not when in the other) cannot be certainly predicted, but only verified *a posteriori*. However, it is possible to monitor the occurrence of the Cu^{II} translocation within receptor **7** by taking take advantage of an auxiliary fluorescent indicator. In particular, we considered coumarin-343, FlH (**8**). This molecule is a protic acid, as it contains a carboxylic group, whose pK_A, in the 4:1 dioxane-water solution, is 7.30 ± 0.02. The undissociated form FlH is strongly fluorescent with an emission band centred at $\lambda_{\max} = 490$ nm;

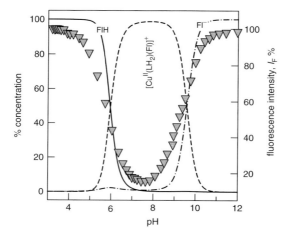

Figure 11.20 pH-controlled translocation of Cu^{II} within receptor **7**, signalled by the fluorescence of an auxiliary indicator. Dioxane/water solution (4:1 v/v), 4×10^{-4} M both in Cu^{II} and in **7** and 2×10^{-6} M in coumarine-343. Triangles: normalized fluorescence intensity at 490 nm (undissociated form of coumarine-343 FlH) and at 471 nm (dissociated form Fl$^-$), on the right vertical axis. Lines: concentration profiles of the species involving coumarine-343 present at the equilibrium over the 3–12 pH range: the undissociated indicator FlH (solid line); the dissociated form coordinated to the metal centre $[Cu^{II}(LH_2)(Fl)]^+$ (dashed line) the dissociated form Fl$^-$ (dash-and-dot).

the emission band of the dissociated form, Fl$^-$, is less intense (fluorescence intensity, I_F, is 75% of that of FlH) and is blue-shifted to a $\lambda_{max} = 471$. It was found that a solution of coumarin-343 is in any case fluorescent over the entire 2–12 pH range, even if a change of the fluorescence intensity I_F and λ_{max} is observed in the 6.5–8.5 pH interval. Thus, the translocation experiment was investigated in a solution 4×10^{-4} M in the copper(II) complex of **7** and 2×10^{-6} M with the fluorescent indicator **8** (a concentration 220 times lower than that of the complex). Figure 11.20 shows how the fluorescence of the coumarine indicator, either in FlH or Fl$^-$ form, varies over the course of the pH titration experiment.

In the acidic region FlH is present and displays its fluorescence then, at pH = 5.5 the $[Cu^{II}(LH_2)(H_2O)]^{2+}$ species begins to form: here it happens that the $-COO^-$ group of the dissociated form of the indicator Fl$^-$ goes to replace the water molecule and binds Cu^{II}, to give the $[Cu^{II}(LH_2)(Fl)]^+$ five-coordinate complex. Binding of Fl$^-$ to the transition metal causes quenching of the fluorescence, through either an eT or an energy-transfer process. Thus, the absence of the fluorescent signal in the pH range 6.5–8.5 indicates that Cu^{II} is located in compartment B. On increasing pH, the metal is translocated to A^{2-}, to give a planar species, which, due to the very strong in-plane interaction, does not exhibit any affinity towards further ligands. As a consequence, Fl$^-$ is released to the solution, where it displays its full fluorescence. It emerges that B-to-A translocation is signalled by the switching ON of the fluorescence and that,

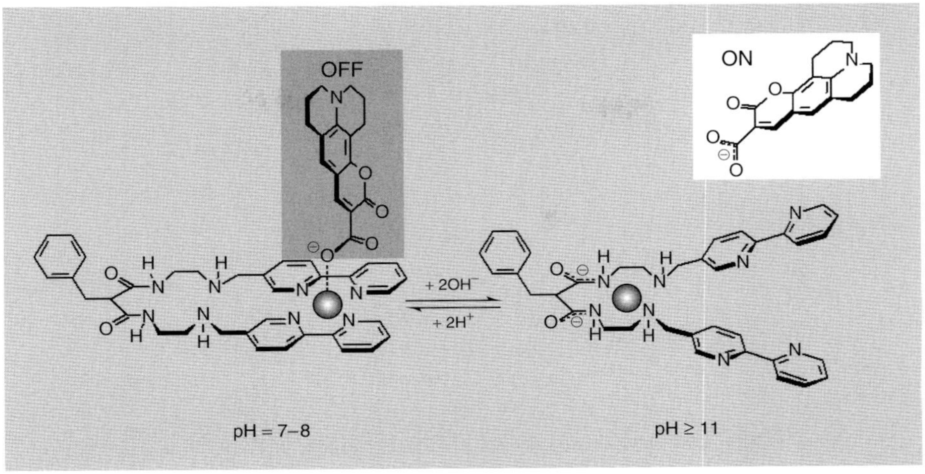

Figure 11.21 Geometrical aspects of the pH controlled translocation of Cu^{II} in the presence of the fluorescent indicator coumarine-343 [18].

conversely, A-to-B translocation, induced by bringing the pH back to 7–8, is signalled by the switching OFF of the fluorescence. Therefore, the operator can go back and forth with pH many times and, due to the efficiency of the fluorescent signal, can be sensed each time, both visually and instrumentally, the occurrence of the reversible translocation process. The geometrical aspects of metal translocation and of complexation/decomplexation of Fl$^-$ are illustrated in Figure 11.21.

11.4
The Simultaneous Translocation of Two Metal Ions

Moving a metal ion from one compartment to another can be compared to the act of passing a ball from one hand to the other, the simplest act of juggling. However, an experienced conjuror juggles with more than one ball. Doing the same at the molecular level, one should leave one metal translocation and try the experiment with two metal ions. Indeed, such a process was attempted and successfully accomplished with the macrocyclic receptor **9** [21].

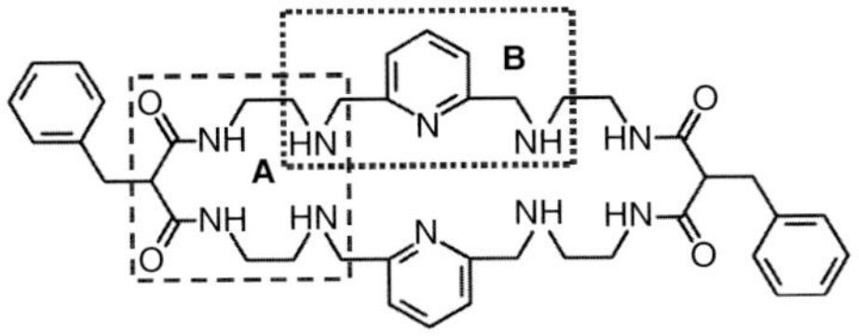

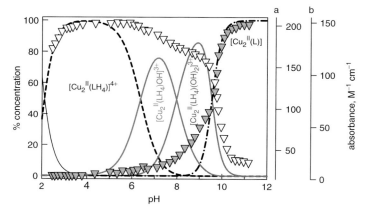

Figure 11.22 Lines, left vertical axis: per cent concentration of the species present at equilibrium in a solution containing **9**($=LH_4$) and 2 equiv. of Cu^{2+} (left vertical axis). Symbols, right vertical axis: filled triangles, molar absorbance of the d-d band with $\lambda_{max} = 660$ nm, pertaining to the complexes $[Cu_2^{II}(LH_4)(H_2O)_2]^{4+}$, $[Cu_2^{II}(LH_4)(H_2O)(OH)]^{3+}$ and $[Cu_2^{II}(LH_4)(OH)_2]^{2+}$, vertical axis (a) open triangles: molar absorbance of the d-d band with $\lambda_{max} = 515$ nm, pertaining to the complex $[Cu_2^{II}(L)]$, vertical axis (b). Reproduced from reference [22]. Copyright Wiley (2004).

The pH-sensitive compartment A is also in this case the well-known tetradentate diamine-diamide chelating agent (which in formula **9** has been inscribed within a dashed square), while compartment B is constituted by the terdentate 2,6-diamine-pyridine moiety (inscribed in a dotted square); notice that, when bound to the diaminopyridine moiety, the metal can reach square planar-coordination through the interaction with a solvent molecule or with an auxiliary ligand. Both A and B have their twin compartments on the other side of the macrocycle.

pH titration experiments were carried out in an $EtOH/H_2O$ solution (4 : 1, v/v) containing **9** (LH_4) and 2 equiv. of Cu^{II}. The equilibrium constants of all the species present at the equilibrium over the 2–12 pH interval were determined, which allowed us to draw the concentration diagram shown in Figure 11.22. A main dimetallic species of stoichiometry $[Cu_2^{II}(LH_4)]^{4+}$ is present in the acidic region between pH 3 and 5. It is suggested that in this species each Cu^{II} centre is coordinated by the diaminepyridine terdentate subunit and by a water molecule, as sketched in the left side of Figure 11.23.

On increasing pH, one of the coordinated water molecules of the $[Cu_2^{II}(LH_4)]^{4+}$ dinuclear complex (more correctly written $[Cu_2^{II}(LH_4)(H_2O)_2]^{4+}$) deprotonates, to give the $[Cu_2^{II}(LH_4)(H_2O)(OH)]^{3+}$ species; on further pH increase, the second water molecule also deprotonates and the $[Cu_2^{II}(LH_4)(OH)_2]^{2+}$ complex forms. The three species are present in the 3–9 pH interval and show a d-d band centred at 660 nm, to which a blue colour corresponds. On moving from $[Cu_2^{II}(LH_4))(H_2O)_2]^{4+}$ to $[Cu_2^{II}(LH_4)(OH)_2]^{2+}$, the band does not change its shape, but simply shows a slight decrease of intensity. After pH 9, the intensity of the

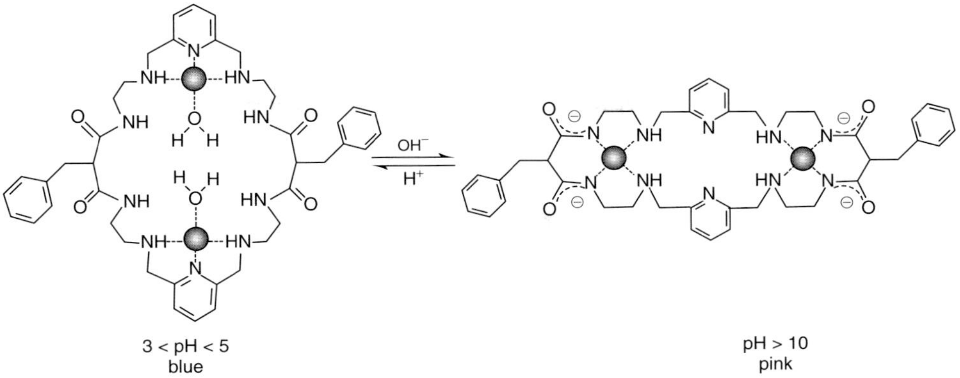

Figure 11.23 pH-controlled translocation of two Cu^{II} ions within receptor **9** [20].

band at 660 nm abruptly decreases, a new band centred at 515 nm forms and develops, while the blue solution turns pink. Colour change is associated to the formation of the neutral species $[Cu_2^{II}(L)]$, as is corroborated by the satisfactory superposition of the absorbance at 515 nm on the concentration profile. Thus, in a solution containing **9** and 2 equiv. of Cu^{II} and adjusted to pH 4, the two metal ions stay in compartments of type B. If standard base is added to bring the pH to 12, the two metals move to the compartments of type A, in their doubly deprotonated version, as is tentatively sketched in Figure 11.23. Quick reverse double metal translocation is obtained if the solution is acidified back to pH 4. Back and forth metal translocation is signalled by the blue-to-pink colour change and induces a severe conformational rearrangement. Things are different also from a coordinative point of view; in fact, in the pink form each Cu^{II} ion profits from strong inplane interactions in square geometry and does not interact with any further ligands. On the other hand, in the blue form, each Cu^{II} ion is bound by three nitrogen atoms and reaches four coordination by interacting with a water molecule. The two water molecules can be replaced by further ligands, for example anions. In particular, the blue form at pH 7, that is containing at 80% the $[Cu_2^{II}(LH_4)(H_2O)(OH)]^{3+}$ species, forms 1 : 1 adducts with dicarboxylates (oxalate and malonate), phosphates (phosphate and pyrophosphate), and azide, in which the anion bridges the two metal centres. Of special interest is the inclusion of imidazole (imH). imH is a protonic acid weaker than water ($pK_A = 14.5$). However, in the presence of two Cu^{II} ions, prepositioned in a given coordinating system, for example a macrocycle, imH deprotonates and simultaneously bridges the two metal centres acting as an ambidentate ligand (see the sketch below) [22].

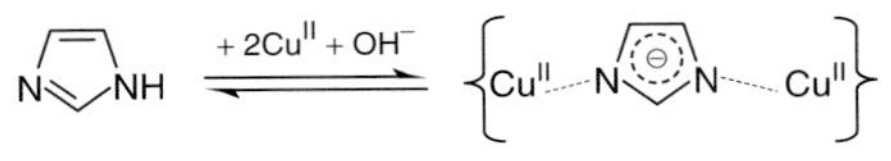

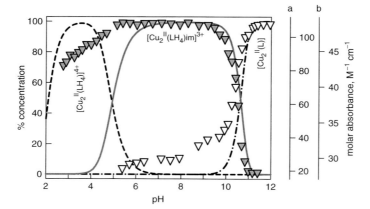

Figure 11.24 Lines, left vertical axis: per cent concentration of the species present at the equilibrium in a solution containing 1 equiv. of **9** (= LH$_4$) 1 equiv. of imidazole (imH) and 2 equiv. of Cu^{2+} (left vertical axis). Symbols, right vertical axis: filled triangles, molar absorbance of the d-d band with λ_{max} = 666 nm, pertaining to the complexes [Cu$_2^{II}$(LH$_4$)(im)]$^{3+}$, vertical axis (a) open triangles: molar absorbance of the d-d band with λ_{max} = 515 nm, pertaining to the complex [Cu$_2^{II}$(L)], vertical axis (b).

Figure 11.24 shows the concentration profiles of the species present at the equilibrium over the 2–12 pH range in a solution containing 1 equiv. of **9**, 1 equiv. of imH and 2 equiv. of Cu^{II}.

At low pH values (3–4) the blue complex [Cu$_2^{II}$(LH$_4$)(H$_2$O)$_2$]$^{4+}$ forms, in which a water molecule completes the four-coordination for each metal centre. On increasing pH, the two water molecules are replaced by an imidazolate ligand, which bridges the two Cu^{II} ions and the [Cu$_2^{II}$(LH$_4$)im]$^{3+}$ ternary complex is formed. This species shows a blue colour (λ_{max} = 666 nm) and is present at 100% over a large pH interval, from 5 to 9.5. On further pH increase, the following neutralization equilibrium takes place: [Cu$_2^{II}$(LH$_4$)im]$^{3+}$ + 3OH$^-$ $\rightleftarrows$ [Cu$_2^{II}$(L)] + imH + 3H$_2$O. The four amide groups deprotonate, one of the protons neutralizes im$^-$, which is released to the solution as imH, while the two Cu^{II} ions translocate to type-A compartments. Translocation is signalled by the blue-to-pink colour change. The decreasing and increasing profiles of the absorbance at 666 nm and at 515 nm ([Cu$_2^{II}$(L)] complex) superimposes well on the concentration profiles of the [Cu$_2^{II}$(LH$_4$)im]$^{3+}$ complex and of the [Cu$_2^{II}$(L)], respectively (see Figure 11.24). Quite interestingly, the presence in solution of the very stable [Cu$_2^{II}$(LH$_4$)im]$^{3+}$ imidazolate complex pushes the formation of the pink [Cu$_2^{II}$(L)] species to a distinctly higher pH. This generates an interesting situation for metal-translocation purposes. In fact, in a solution containing 1 equiv. of **9**, 2 equiv. of Cu^{II} and no imH and adjusted to pH 10.2, the pink neutral species [Cu$_2^{II}$(L)] is present at 95% (see Figure 11.22). But, if the solution also contains 1 equiv. of imH, the concentration of [Cu$_2^{II}$(L)] decreases to 12%, in favour of the blue imidazolate complex [Cu$_2^{II}$(LH$_4$)im]$^{3+}$, which is present at 87% (see Figure 11.24).

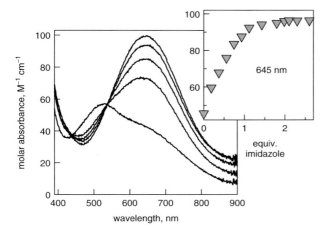

Figure 11.25 Selected visible spectra showing the d-d absorption bands taken over the course of the titration with imidazole of a solution containing **9** and 2 equiv. of Cu^{II}, buffered to pH 10.2. Inset: titration profile on the band centred at 645 nm, pertaining to the dicopper(II) imidazolate complex $[Cu_2^{II}(LH_4)im]^{3+}$.

This suggests that if imH is added to a solution containing 1 equiv. of **9** and 2 equiv. of Cu^{II}, adjusted to pH 10.2 (and thus of pink colour due to the presence of the major species $[Cu_2^{II}(L)]$), double metal translocation should take place, with the concluding formation of the imidazolate bridged complex. Indeed, upon imH addition, the pink solution buffered to pH 10.2 turns blue, while significant spectral modifications are observed (see Figure 11.25).

Thus, double Cu^{II} translocation can be induced, at a defined pH value, by addition of imH, providing a unique example of metal translocation induced by a molecule, rather than by a change of pH or by the variation of the redox potential. The process, pictorially illustrated in Figure 11.26, is quite slow ($\tau = 45$ min).

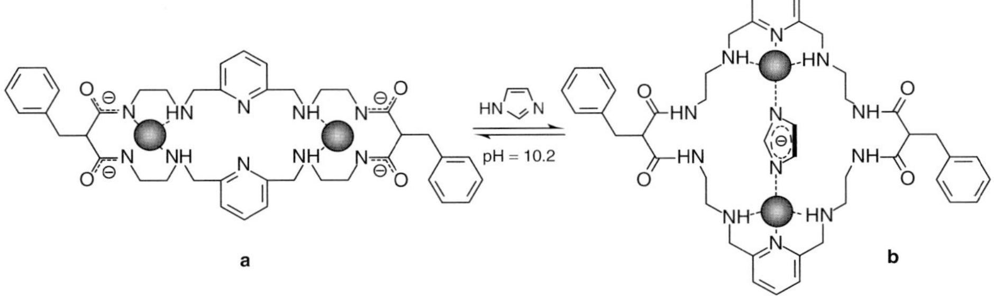

Figure 11.26 Translocation of two Cu^{II} ions within receptor **9** induced by imidazole addition at pH 10.2 [20].

The sluggishness of the process may be ascribed to the fact that, as a preliminary step of the translocation, the ambidentate ligand imH/imidazolate must enter into the complex **a** in order to bind the two Cu^{II} ions, thus facing severe steric repulsions by the macrocycle in its closed arrangement **a**.

The process illustrated in Figure 11.26 shows some relevant aspects from the point of view of molecular recognition. In fact, if a solution containing complex **a** and buffered to pH 10.2 is treated with a variety of ambidentate anions (N_3^-, PO_4^{3-}, $P_2O_7^{4-}$, $C_2O_4^{2-}$) in a fivefold excess, nothing occurs: the colour of the solution remains pink and no spectral modifications are observed. In the same way, nothing takes place on addition of an excess of representative amino acids and other biologically significant analytes: glycine, arginine, proline, glutamate, ADP and ATP. Only imH and imH-containing molecules (histidine, histamine) are able to displace the two Cu^{II} centres from the strongly coordinating square planar donor sets, each containing two amine groups and two deprotonated amide groups. This is due to the unique feature of imidazolate as a bridging ligand, which allows electronic communication between the two Cu^{II} centres and spin pairing. Thus, we are in the presence of a sleeping host (the complex in its **a** form), which can be awoken by only one guest: the imH fragment (like Sleeping Beauty, when awoken by the Prince, through a simple, but very selective interaction: a kiss, in the classic fairy tale by Charles Perrault) [23].

11.5
Redox-Driven Anion Translocation

Anions can be translocated within multisite receptors like metal cations. However, examples of anion translocation are rare, for a variety of reasons. First, anions establish much weaker interactions with receptors than metal ions: they can be electrostatic (very weak), hydrogen bonding (weak) and metal ligand (relatively strong) [24]. Indeed, the first reports of anion translocation refer to receptors containing coordinatively unsaturated metal centres [25].

As an example, the cyclam–tren conjugate **10** gives a heterobimetallic Ni^{II}-Cu^{II} complex suitable for a redox-driven chloride translocation [26]. The synthesis of the complex is not straightforward: first, **10** is made to react with 1 equiv. of Ni^{II} in refluxing MeCN. The Ni^{II} ion chooses to interact with the cyclam subunit in order to profit from the strong inplane interactions exerted by the 14-membered tetra-aza ring and from the thermodynamic macrocyclic effect. Thus, a low-spin square [Ni^{II}(cyclam)]$^{2+}$ complex is formed, yellow in colour. Then, 1 equiv. of Cu^{II} is added to the solution, at room temperature. The Cu^{II} ion would rightfully choose the more favourable macrocyclic cavity, but it cannot replace the Ni^{II} ion that coordinated first, which benefits from the inertness associated to the kinetic macrocyclic effect. Thus, Cu^{II} preferentially to forms with the tren subunit a five-coordinate complex in which the remaining axial position is occupied by a solvent molecule. The solution takes a green colour, which results from the combination of the yellow [Ni^{II}(cyclam)]$^{2+}$ and of the blue [Cu^{II}(tren)MeCN]$^{2+}$ complex. If chloride is added to the solution, it goes

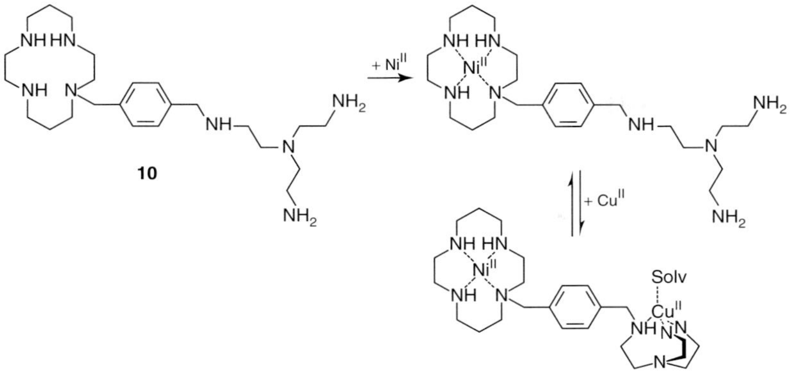

11.5 Redox-Driven Anion Translocation

Figure 11.27 The kinetically controlled synthesis of the heterodimetallic $Ni^{II} \sim Cu^{II}$ complex with the tren–cyclam conjugate **10**. The process has been designed in order to have Ni^{II} encircled by the cyclam subunit and Cu^{II} coordinated by the tren moiety. Notice that both Ni^{II} and Cu^{II} prefer coordination with cyclam rather than with tren and that Cu^{II} has a greater affinity for tetramines than Ni^{II}.

to bind the Cu^{II} centre, replacing the axially coordinated MeCN molecule. Ni^{II} does not compete for the anion because it preferentially maintains four-coordination and to profit from the greater ligand field stabilization energy experienced by a low-spin d^8 cation in a square-planer-coordinative environment. The $\log K$ associated with the chloride complexation equilibrium ($\log K = 5.66 \pm 0.09$) guarantees that, in a MeCN solution 10^{-3} M both in the Cu^{II}–Ni^{II} complex and in Cl^-, 95% of the anion is bound to the Cu^{II}–tren subunit. In an independent experiment, CV studies were carried out in a MeCN solution 10^{-3} M in the Cu^{II}–Ni^{II} complex and 0.1 M in [Bu_4N]ClO_4, using a platinum microsphere as a working electrode. A reversible CV wave was obtained with $E_{1/2} = 0.74$ V vs. Fc^+/Fc, which corresponded to the one-electron Ni^{II}-to-Ni^{III} redox change. On addition of Cl^-, a new peak developed at a much less positive potential (0.24 V), while the intensity of the peak at 0.74 V progressively decreased. The new peak reached its maximum intensity with the addition of 1 equiv. of Cl^-, while addition of further equivalents did not induce the appearance of any other peak. This behaviour points towards the formation of a [Ni^{III}(cyclam)Cl]$^{2+}$ five-coordinate subunit within the oxidized system Cu^{II}–Ni^{III}. The tendency of the Ni^{III}-cyclam complex (low-spin d^7) to bind halide ions is well documented. The steepness and lack of curvature of the current intensity vs. chloride equivalent profile indicate a binding constant higher than 10^7 (Figure 11.27).

Thus, in a solution containing the Cu^{II}–Ni^{II} complex and less than 1 equiv. of Cl^-, at the initial potential before scanning, for example at -0.10 V vs Fc^+/Fc, the chloride stays on the Cu^{II} centre, site **b** in Figure 11.28. Then, on increasing the potential, Ni^{II} is oxidized to Ni^{III} and Cl^- moves from site **b** to site **a**. On the reverse scan, Ni^{III} is reduced to Ni^{II}, the chloride anion is released and moves back to the Cu^{II} centre. The wave with $E_{1/2} = 0.24$ V appearing on Cl^- addition maintains a reversible profile, indicating that the translocation of the anion from one metal centre to the other is too fast a

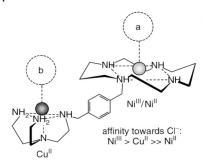

Figure 11.28 Electronic and geometrical rationalisation of the redox-driven chloride translocation in the Ni–Cu complex of the tren–cyclam conjugate system **10**. When the redox-active nickel centre is in the reduced form, Ni^{II}, the anion stays on the Cu^{II} centre (site **b**). Upon nickel oxidation, chloride moves to the Ni^{III} ion (site **a**). This process is fast and reversible [24].

process to be detected on the timescale of the CV experiment (highest potential scan rate explored 1 V s^{-1}). The ease of the process reflects an uncomplicated access to the transition state, in which the two metal centres should both be coordinated to the chloride ion, following the favourable thermal dangling of the cyclam and tren subunits appended to the 1,4-xylyl spacer. Thus, the high rate of the translocation process reflects the flexibility of the tren–cyclam conjugate system.

The question now is whether the translocation process is *intra*-molecular (in the sense that it is the Cl$^-$ anion staying on Cu^{II} that moves to Ni^{III}) or *inter*-molecular (X$^-$ goes from Cu^{II} into the solution, whereas a different X$^-$ anion from the solution comes to Ni^{III}). There is no doubt that, due to the substitutional lability of the involved metal centres (Cu^{II}, Ni^{II} and Ni^{III}), the metal-bound anion quickly exchanges with other X$^-$ anions present in the solution (either unbound or bound to a metal). However, we suggest that when the Ni^{II} centre is oxidized, the translocation of the Cl$^-$ anion bound to the proximate Cu^{II} centre is by far the most probable event, several orders of magnitude more probable than the transfer from a different $[Cu^{II}(Cl)-Ni^{II}]^{3+}$ molecular system dispersed in the solution. This statement is based on the following considerations: it is assumed that the Cu^{II}-X subunit moves within a sphere whose centre is the Ni^{III} ion and whose radius is given by the Cu–Ni distance (7.5 Å, as obtained from molecular modelling). As the volume of the sphere is 1766 Å^3 = 1.766×10^{-24} dm^3, it derives that the effective concentration of X$^-$ (bound to Cu^{II}) is 0.94 M. Such a concentration is much higher than that of Cl$^-$ dispersed in solution (either free or bound the Cu^{II} centre of a different $[Cu^{II}(Cl)-Ni^{II}]^{3+}$ complex, 10^{-3} to 10^{-4} M in a typical electrochemical process. This points towards a much higher probability for the occurrence of an intramolecular translocation process.

Not many anions can replace Cl$^-$ in a translocation experiment. They must be good ligands for transition metals, but they should also be resistant to the oxidation, two properties that may contrast each other. However, at least one other anion, cyanate, can do the same work as chloride. NCO⁻ gives a stable adduct with the [Cu^{II}(tren)]$^{2+}$ moiety of the Ni^{II}–Cu^{II} complex (log K = 4.4 ± 0.1, which, in a solution 10^{-3} M both in Ni^{II}–Cu^{II} and NCO⁻, guarantees the formation of 82% of the ternary complex). Moreover, in a CV titration experiment with [Bu$_4$N]NCO, a new wave develops with $E_{1/2}$ = 0.27 V vs. Fc⁺/Fc and reaches its limiting value after the addition of 1 equiv. of NCO⁻. This unambiguously indicates the occurrence of a fast and reversible redox-driven translocation process. NO$_3^-$ and HSO$_4^-$ are as resistant to oxidation as Cl⁻, but show only a moderate affinity towards both Cu^{II} and Ni^{III} centres. As a consequence, no new peaks appear in the CV profile in anion titration experiments and the translocation process, if any, involves only a small fraction of the Ni^{II}–Cu^{II} system.

Considering the well-known affinity of Zn^{II} tetramine complexes towards five-coordination, in particular, by carboxylates, Cu^{II} was replaced by Zn^{II} in the heterodimetallic complex of **10**. Indeed, in a MeCN/MeOH solution (1 : 1 v/v), the [Zn^{II}(tren)]$^{2+}$ moiety of the Ni^{II}–Zn^{II} system gives a stable adduct with the benzoate anion (log K = 5.6 ± 0.1). Unfortunately, for translocation purposes, this complex is more stable than that formed by benzoate with the [Ni^{III}(cyclam)]$^{3+}$ subunit. Thus, on benzoate addition, no modifications are observed in the CV profile, indicating that the benzoate ion first goes on the Zn^{II} centre and there it resists well the call of the close and freshly oxidized Ni^{III} ion.

In systems like the heterodimetallic Ni–Cu/(**10**) complex, both metal centres act as anion receptors, but the engine is represented by the Ni^{II}/Ni^{III} redox couple. New systems can be designed for redox-driven anion translocation: a metal-centred redox couple should be maintained as an engine, but the second anion-recognition site can be changed, for instance, replaced by a receptor of different nature (e.g. providing a hydrogen-bonding donor set).

Redox-driven translocation of the nitrate anion has been recently carried out in a system based on the molecular dication [**11**]$^{2+}$ (Figure 11.29) [27]. This

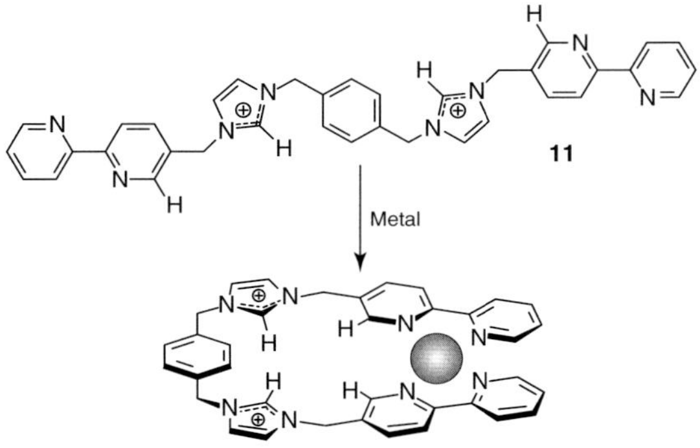

Figure 11.29 The interaction of the two bpy fragments of system [**11**]$^{2+}$ with a metal ion prone to four-coordination gives rise to a pseudomacrocycle possessing a H-bond-donating compartment, suitable for the interaction with an anion [26].

system possesses two bpy subunits, which can react with a metal centre to give a pseudomacrocycle affording two distinct sites for subsequent interaction with anions: (i) the metal centre itself, provided that it is coordinatively unsaturated and (ii) a cavity capable of donating four hydrogen bonds, two from the H_α atoms of the two imidazolium subunits and two from the H_β atoms of the bpy subunits, which have been activated by metal coordination. Copper was the metal of choice, in view of its capability to work as an engine through the Cu^{II}/Cu^I redox couple. In particular, both Cu^{II} and Cu^I form 1:1 complexes with $[\mathbf{11}]^{2+}$, in which the metal centre is coordinated by the two bpy fragments. However, an important difference exists: the $Cu^I(bpy)_2{}^+$ subunit, which exhibits a distorted tetrahedral geometry, is coordinatively saturated and is not prone to interact with anions. On the other hand, the $Cu^{II}(bpy)_2{}^{2+}$ moiety aspires to be five-coordinate and is strongly inclined to interact with anions. In particular, titration studies in a MeCN solution have shown that the $[Cu^{II}(\mathbf{11})]^{4+}$ complex interacts with a variety of anions according to two consecutive stepwise equilibria:

$$[Cu^{II}(\mathbf{11})]^{4+} + X^- \rightleftarrows [Cu^{II}(\mathbf{11})(\leftarrow X)]^{3+} \quad K(II)_1 \qquad (11.1)$$

$$[Cu^{II}(\mathbf{11})(\leftarrow X)]^{3+} + X^- \rightleftarrows [Cu^{II}(\mathbf{11})(\leftarrow X)(\cdots X)]^{2+} \quad K(II)_2 \qquad (11.2)$$

In the first equilibrium, the anion goes to interact with the metal centre, in the second it goes into the H-bond-donating compartment. The anion strongly prefers the metal–ligand interaction with respect to the H-bond interaction. In particular, for Cl^-, Br^- and NCS^- $\log K_1 > 6$, while $\log K_2$ values are: 5.0, 4.3 and 3.5, respectively. Only with the poorly coordinating anion NO_3^- could both $\log K(II)_1$ (5.3) and $\log K(II)_2$ (3.5) be determined. The behaviour of the $[Cu^I(\mathbf{11})]^{3+}$ system was much less favourable, as anion addition induced in most cases decomposition, with demetallation and formation of the more stable $[Cu^I X_4]^{3-}$ complex. Only in the case of NO_3^- was the formation of a stable H-bond complex ascertained, according to the equilibrium:

$$[Cu^I(\mathbf{11})]^{3+} + X^- \rightleftarrows [Cu^I(\mathbf{11})(\cdots X)]^{2+} \quad K(I) \qquad (11.3)$$

with $\log K(I) = 3.29 \pm 0.01$.

Thus, thermodynamics seem to favour the occurrence of nitrate anion translocation within the $Cu/[\mathbf{11}]^{2+}$ system, driven by the Cu^{II}/Cu^I redox change. In fact, in a MeCN solution containing equimolar amounts of $[Cu^{II}(\mathbf{11})]^{4+}$ and NO_3^- it should happen that the anion is coordinated to the Cu^{II} metal centre (in 5×10^{-3} M solution 96% of the total anion will be bound to Cu^{II}). On electrochemical reduction of Cu^{II} to Cu^I, NO_3^- has to be released from the metal centre and moves to the proximate H-bond-donor compartment (which, at the chosen 5×10^{-3} M concentration level should bind 73% of the total amount of the anion in solution). Consecutive oxidation–reduction processes at the metal centre should make the anion move back and forth between the two binding compartments, as illustrated in Figure 11.30.

Indeed, the occurrence of the redox-driven translocation process has been verified through CV studies. However, it is convenient to consider a preliminary

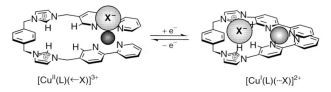

Figure 11.30 The electrochemically driven translocation of an anion (e.g. nitrate) within the Cu/[**11**]$^{2+}$ system (**11** = L). When the metal is in the +2 oxidation state, the anion is coordinatively bound to it; when Cu^{II} is reduced to Cu^I, the anion moves to the H-bond donor bis-imidazolium compartment [26].

experiment on a solution containing the model complex [Cu^{II}(Mebpy)$_2$]$^{2+}$ (Mebpy = 4-methyl-2,2'-bipyridine).

Figure 11.31a shows the CV profile obtained at a platinum working electrode on a solution 2.00×10^{-3} M of [Cu^{II}(Mebpy)$_2$]$^{2+}$ in MeCN (solid line). The process is quasi-reversible ($\Delta p = 150$ mV, at a scan rate of 500 mV s^{-1}), due to the kinetically complicated rearrangement of the coordination geometry of the metal centre: from distorted tetrahedral of Cu^I to distorted trigonal-bipyramidal of Cu^{II}. On nitrate addition, the CV profile becomes more reversible ($\Delta p = 120$ mV): this may reflect a beneficial involvement of the anion in stabilizing the activated complex that forms during the change of the coordination geometry. Then, it is observed that both reduction and oxidation peaks are shifted towards more negative potentials. In particular, the dashed line in Figure 11.31a refers to the CV profile taken in the presence of a large excess of [Bu$_4$N]NO$_3$, showing a $E_{1/2}$ value 55 mV

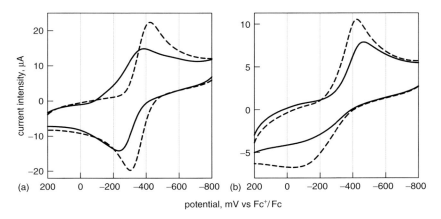

Figure 11.31 Cyclic voltammetry profiles obtained at the platinum working electrode (potential scan rate 500 mV s^{-1}) in a MeCN solution 0.1 M in [Bu$_4$N]ClO$_4$ (a) solid line, [Cu^{II}(Mebpy)$_2$]$^{2+}$ 2.00×10^{-3} M; dashed line, + excess [Bu$_4$N]NO$_3$; (b) solid line, 1.5×10^{-3} M in [**11**]$^{2+}$ and Cu^{II}(CF$_3$SO$_3$)$_2$; dashed line, +1.0 equiv. of [Bu$_4$N]NO$_3$.

more negative than that recorded in the absence of NO_3^-: such an effect results from the thermodynamic stabilization of the Cu^{II} complex, which takes advantage from the coordination of NO_3^- (which replaces a coordinated MeCN molecule). Furthermore the coordinatively saturated Cu^I complex cannot profit from nitrate binding.

Figure 11.31b shows the CV profile of a solution containing equimolar amounts of $[11]^{2+}$ and Cu^{II} (solid line). First, it has to be noted that the electrochemical reversibility is lower than that observed in the model system $[Cu^{II,I}(Mebpy)_2]^{2+,+}$ (Δp is now $\sim$400 mV, at 500 mV s^{-1}); this is an expected behaviour as the presence of the carbon chain linking the two bpy moieties makes the geometrical rearrangement associated to the Cu^I-to-Cu^{II} change much more difficult. Addition of 1.0 equiv. of NO_3^- renders the CV profile (dashed line) more reversible ($\Delta p = \sim$300 mV) and, most interestingly, makes the cathodic peak (associated to the Cu^{II}-to-Cu^I reduction) less negative. This behaviour is opposite to that observed with the $[Cu^{II,I}(Mebpy)_2]^{2+,+}$ system and ultimately demonstrates the occurrence of nitrate translocation. Actually, the stabilization of the $[Cu^I(11)]^{3+}$ form has to be ascribed to the fact that, on Cu^{II}/Cu^I reduction, the leaving NO_3^- ion does not move to the bulk solution, but goes to interact with the proximate bis-imidazolium compartment, thus profiting from a favourable energy term, not experienced by the reference system $[Cu^I(Mebpy)_2]^+$. It follows that, in the consecutive voltammetric cycles, NO_3^- moves back and forth, quickly and reversibly, between the metal centre and the H-bond-donating cavity.

11.6
Anion Swapping in a Heteroditopic Receptor, Driven by a Concentration Gradient

The tripodal molecule **12**, in which three nitrophenylurea subunits have been appended to a tren platform, can act as a receptor for two anions according to the cascade mechanism illustrated in Figure 11.32 [28].

First, a Cu^{II} ion is added, which interacts with the branched tetramine moiety, thus inducing the formation of a cavity suitable for anion inclusion. In particular, inside the cavity there is room for two anions X^- and Y^-: one, for example X^-, will go to interact with the Cu^{II} centre, occupying the vacant axial position of the trigonal bipyramidal coordination polyhedron; the other, for example Y^-, will occupy the lower portion of the cavity in order to receive up to six hydrogen bonds from the three facing urea subunits. Equilibrium studies in DMSO were carried out in order to determine the selective affinity of a given anion for the two binding sites. As an example, Figure 11.33a shows the family of spectra taken over the course of the titration of a solution of the $[Cu^{II}(12)]^{2+}$ complex with NaN_3.

Actually, UV-Vis spectra provide direct information on the nature of the interaction of the anion with the two available binding sites. In particular, in the 600–900 nm interval d-d bands pertinent to the $Cu^{II}(tren)^{2+}$ chromophore are

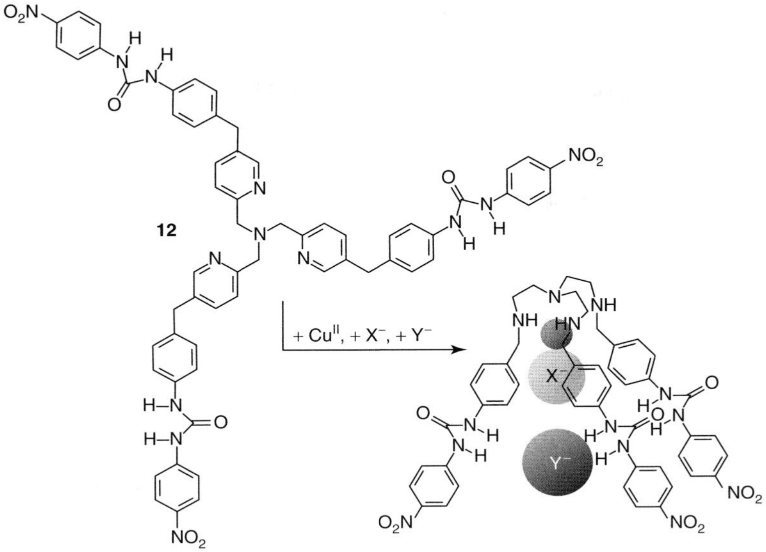

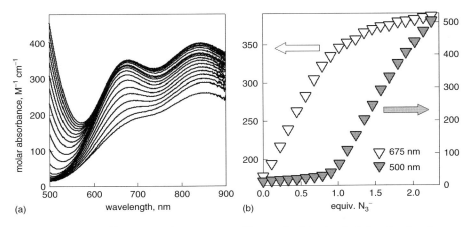

Figure 11.32 A ditopic receptor providing different interactions with anions: metal-ligand (from the $Cu^{II}(tren)^{2+}$ subunit) and hydrogen bonding (from the three facing nitrophenylurea fragments) [26].

Figure 11.33 (a) Visible spectra taken over the course of the titration of a 2.075×10^{-3} M solution of $[Cu^{II}(\mathbf{12})]^{2+}$ in DMSO with NaN_3. (b) Titration profiles at selected wavelengths: the d-d band at 675 nm gives evidence of the coordination of the first N_3^- ion at the metal centre; the absorbance at 500 nm, tail of the band of the nitrophenyl chromophore, monitors the entrance of the second N_3^- ion and its hydrogen-bonding interaction with the three urea fragments. Reproduced from reference [28]. Copyright Wiley (2007).

present (molar absorptivity $\varepsilon = 200$–$400\ M^{-1}\ cm^{-1}$ and their modification monitor anion interactions at the metal centre. On the other hand, the intense band of the nitrobenzene chromophore centred at about 350 nm ($\varepsilon > 50\,000\ M^{-1}\ cm^{-1}$) undergoes a distinct red shift on interaction of anions at the proximate urea fragments. The spectra reported in Figure 11.33a were taken on a solution 2×10^{-3} in the $[Cu^{II}(12)]^{2+}$ complex and allow a full monitoring of the d–d bands. At this concentration, the nitrophenyl band is completely out of scale; however, monitoring of the absorbance at 500 nm (tail of the intense absorption band at 350 nm) affords a direct investigation on what happens at the tris-urea compartment. Titration curves in Figure 11.33b show that on addition of the first equivalent of N_3^- the absorbance of the d–d band at 675 nm increases according to a saturation profile, while the absorbance at 500 nm remains constant. This indicates that the first azide ion goes to interact with the metal centre. The titration profile at 675 nm shows a nice curvature, to which a $\log K_1 = 3.86 \pm 0.03$ corresponds.

$$[Cu^{II}(12)]^{2+} + N_3^- \rightleftarrows [Cu^{II}(12) \leftarrow N_3]^+ \quad \log K_1 = 3.86 \pm 0.03 \quad (11.4)$$

$$[Cu^{II}(12)(\leftarrow N_3)]^+ + N_3^- \rightleftarrows [Cu^{II}(12)(\leftarrow N_3)\cdots N_3)] \quad \log K_2 = 2.2 \pm 0.1 \quad (11.5)$$

Notice that the symbol $\leftarrow$ in $[Cu^{II}(12) \leftarrow N_3]^+$ indicates interaction at the metal centre. On addition of the second equivalent of N_3^-, the absorbance at 500 nm starts increasing, which reveals anion interaction at the tris-urea compartment, as described by Equation 11.5. The $\cdots$ symbol in the formula $[(Cu^{II}(11) \leftarrow N_3)\cdots N_3]$ indicates a hydrogen-bonding interaction. The binding constant at the second site is 50-fold lower than that pertinent to the first site.

The $H_2PO_4^-$ anion displays a similar behaviour, as outlined by the titration profiles shown in Figure 11.34.

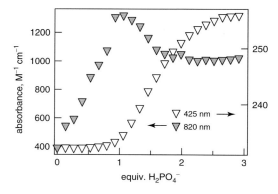

Figure 11.34 Titration profiles at selected wavelengths taken over the course of the titration of a 2.00×10^{-3} M solution of $[Cu^{II}(12)]^{2+}$ in DMSO with $[Bu_4N]H_2PO_4$: the absorbance at 650 nm provides evidence of the coordination of the first $H_2PO_4^-$ ion at the Cu^{II} centre ($\log K > 5$); the absorbance at 425 nm indicates the establishing of H-bond interactions of the tris-urea compartment of $[Cu^{II}(12)]^{2+}$ with the second $H_2PO_4^-$ ion ($\log K = 2.0 \pm 0.1$). Reproduced from reference [28]. Copyright Wiley (2007).

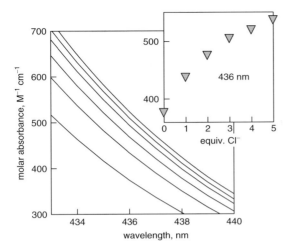

Figure 11.35 Spectra taken over the course of the titration with [Et$_3$Bn]Cl of a DMSO solution; 2.00 × 10^{-3} M in [Cu^{II}(**12**)]$^{2+}$ and [Bu$_4$N]H$_2$PO$_4$: the spectral region shows the tail of the intense absorption band of the nitrobenzene chromophore and monitors the interactions of the phosphate ion at the tris-urea compartment of the receptor. Inset: titration profile at 436 nm. Reproduced from reference [28]. Copyright Wiley (2007).

Also in the present case, the first H$_2$PO$_4^-$ seeks the Cu^{II} centre and forms an especially stable complex (log K_1 > 5, as indicated by the steep profile of the d-d band). On addition of the second equivalent of H$_2$PO$_4^-$, a smoother increase of the nitrophenyl band is observed, to which a log K_2 = 2.0 ± 0.1 corresponds. Halide ions interact with the metal centre (log K_1 = 4.7 ± 0.1 for Cl$^-$, 3.5 ± 0.1 for Br$^-$, 2.13 ± 0.02 for I$^-$), but, in view of their reduced tendency to receive H-bonds, do not interact with the tris-urea compartment.

Of special interest for anion-translocation purposes is the formation of heterodinuclear anion complexes, with a special reference to H$_2$PO$_4^-$ and Cl$^-$ anions. If to a DMSO solution (2.00 × 10^{-3} M in [Cu^{II}(**12**)]$^{2+}$) an equivalent amount of [Bu$_4$N]H$_2$PO$_4$ is added, then one hundred per cent of the [Cu^{II}(**12**) ← H$_2$PO$_4$]$^+$ complex is formed, in which the H$_2$PO$_4^-$ ion is bound to the metal centre. Then, to this solution, aliquots of a concentrated solution of [Et$_3$Bn]Cl in DMSO (0.620 M) are added. On chloride addition a distinct shift of the nitrobenzyl charge-transfer absorption band is observed, as shown in Figure 11.35.

As the Cl$^-$ anion is not able to interact with the tris-urea cavity of the [Cu^{II}(**12**)]$^{2+}$ receptor, even at high concentration, the spectral change has to be ascribed to the binding of the H$_2$PO$_4^-$ ion. Thus, it happens that chloride, due to a mass effect, displaces the phosphate anion from the metal centre, which translocates to the tris-urea compartment, according to equilibrium (Equation 11.6):

$$[Cu^{II}(\mathbf{12}) \leftarrow H_2PO_4]^+ + Cl^- \rightleftarrows [Cu^{II}(\mathbf{12})(\leftarrow Cl)\cdots H_2PO_4] \qquad (11.6)$$

The progress of the H$_2$PO$_4^-$ sliding from one compartment to another is illustrated by the titration profile in Figure 11.35, inset. We are thus in the presence of a novel

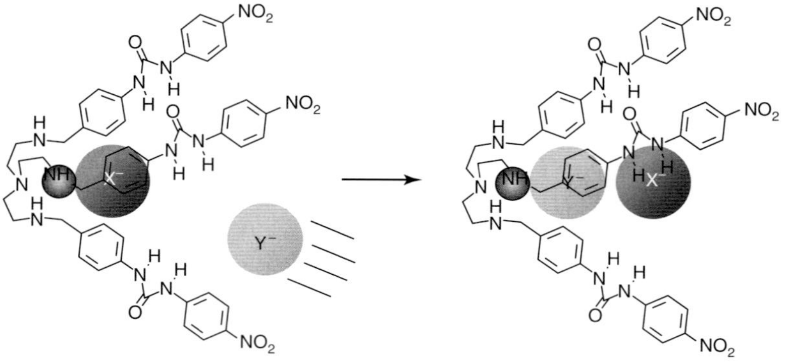

Figure 11.36 Playing billiards at the molecular level. Thanks to a concentration effect, the chloride anion (cue ball Y^-) displaces the phosphate anion (opponent's ball X^-) [26].

type of anion translocation (that of $H_2PO_4^-$) from one compartment to another of a heteroditopic receptor, which is promoted by a different anion (Cl^-). The development of the process is illustrated in Figure 11.36.

Using the imaginative and sensational language of supramolecular chemistry (and considering anions as balls, as they are often conventionally drawn, see Figure 11.36), the operator is playing Italian billiards at a molecular level: in particular, he has shot with his cue ball that of his opponent, which is being displaced to a desired position (in the metaphor, Cu^{II} is the red object).

11.7
Conclusions and Perspectives: Further Types of Molecular Machines?

A main question remains whether the systems described in this chapter can be technically considered (i) switches, (ii) mechanical switches and, more pretentiously, (iii) machines at the molecular level. Point (i): the systems described in the present chapter show two distinctive properties: (a) they exist in two states of comparable stability, each one displaying a definite optical property (light absorption/emission) and (b) each state can be addressed by an external stimulus (of a chemical or electrochemical nature). In this sense, the systems described here can be considered as molecular switches. Furthermore, as the switching mechanism is associated to a movement in a defined space, the investigated systems may be also correctly defined mechanical switches. Point (iii) requires a more circumstantiated answer. A molecular machine is typically constituted by two subunits, of which one is conventionally defined as stationary and the other movable. Chemical, electrochemical or photochemical modification of one subunit alters the equilibrium and causes the movable part to move in order to reach a new favourable equilibrium position. In this sense, multisite receptors including movable ions can be considered machines at the molecular level. However, in classical molecular machines, for example rotaxanes and catenanes, the stationary and the movable parts are mechanically linked (in particular, interlocked), as is usually observed in the machines of the macroscopic world. On the contrary, the movable parts of the systems considered in this chapter are not fixed and, after doing their mechanical work, they are quickly and continuously replaced by similar pieces coming from the solution, reflecting their kinetic lability. At this stage, a more practical reader could argue: 'I do not care about the more or less strict similarity, the important issue is that your systems, like macroscopic machines, do something useful'. With this statement, the question has become more general: what are multisite receptors and their included movable ions being designed for? A rather vague answer could be that these systems, like any type of molecular switch, are potentially applicable in the field of signal processing and data storage at the molecular level (as Stoddart's rotaxanes promise to do, hopefully in a not too distant future) [29]. Surely, ions, in particular, metal ions, used either as a movable piece or as an engine, exhibit unrivalled versatility, compared to the commonly used organic fragments. In particular, metal ions typically undergo one-electron oxidation–reduction processes that induce dramatic changes in their properties: colour, magnetism, geometrical preferences and reactivity. It may happen, for instance, that following a one-electron oxidation process (Cr^{II}/Cr^{III}, Co^{II}/Co^{III}, in the 3d series), a labile metal becomes inert, a feature that could be exploited to impart hysteresis in molecular motions, a valuable opportunity from the point of view of data storage at the molecular level. A change of the magnetic properties of a metal ion in a given oxidation state, when close to the spin crossover point (e.g. high- and low-spin Ni^{II} and Fe^{II}), can be induced through fine tuning of the donor properties of the donor set of one compartment, a useful feature for signalling purposes. Thus, it seems that there are no limits to the use of metal ions for the design of new molecular devices, performing unprecedented and long-awaited functions. They are waiting, in a variety of coloured salts on the shelf, anxious to demonstrate their many resources and qualities.

References

1. Lehn, J.M. (1995) *Supramolecular Chemistry. Concepts and perspectives*, Wiley-VCH Verlag GmbH, Weinheim, p. 124.
2. Shinkai, S., Nakaji, T., Nishida, Y., Ogawa, T. and Manabe, O. (1980) *J. Am. Chem. Soc.*, **102**, 5860.
3. Bissell, R.A., Córdova, E., Kaifer, A.E. and Stoddart, J.F. (1994) *Nature*, **369**, 133.
4. Livoreil, A., Dietrich-Buchecker, C.O. and Sauvage, J.P. (1994) *J. Am. Chem. Soc.*, **116**, 9399.
5. Irie, M. (1990) *Adv. Polym. Sci.*, **94**, 28.
6. Amendola, V., Fabbrizzi, L., Mangano, C. and Pallavicini, P. (2001) *Acc. Chem. Res.*, **34**, 488.
7. Lehn, J.M. and Stubbs, M.E. (1974) *J. Am. Chem. Soc.*, **96**, 4011.
8. Anelli, P.L., Spencer, N. and Stoddart, J.F. (1991) *J. Am. Chem. Soc.*, **113**, 5131.
9. Lehn, J.M. (1995) *Supramolecular Chemistry. Concepts and Perspectives*, Wiley-VCH Verlag GmbH, Weinheim, p. 134.
10. Zelikovich, L., Libman, J. and Shanzer, A. (1995) *Nature*, **374**, 790.

11. Ward, T.R., Lutz, A., Parel, S.P., Ensling, J., Gütlich, P., Buglyó, P. and Orvig, C. (1999) *Inorg. Chem.*, **38**, 5007.
12. Belle, P., Pierre, J.L. and Saint-Aman, E. (1998) *New. J. Chem.*, 1399.
13. Amendola, V., Fabbrizzi, L., Mangano, C., Pallavicini, P., Perotti, A. and Taglietti, A. (2000) *J. Chem. Soc. Dalton Trans.*, 185.
14. Fabbrizzi, L., Kaden, T.A., Perotti, A., Seghi, B. and Siegfried, L. (1986) *Inorg. Chem.*, **25**, 321.
15. (a) Martinez-Mañez, R. and Sancenón, F. (2003) *Chem. Rev.*, **103**, 4419–4476; (b) Suksai, C. and Tuntulani, T. (2003) *Chem. Soc. Rev.*, **32**, 192.
16. Amendola, V., Di Casa, M., Fabbrizzi, L., Licchelli, M., Mangano, C., Pallavicini, P. and Poggi, A. (2001) *J. Inclusion Phenom. Macrocyclic Chem.*, **41**, 13.
17. Amendola, V., Brusoni, C., Fabbrizzi, L., Mangano, C., Miller, H., Pallavicini, P., Perotti, A. and Taglietti, A. (2001) *J. Chem. Soc., Dalton Trans.*, 3528.
18. (a) Bjerrum, J. and Lamm, C.G. (1950) *Acta Chem. Scand.*, **4**, 997; (b) Romano, V. and Bjerrum, J. (1970) *Acta Chem. Scand.*, **24**, 1551.
19. Amendola, V., Fabbrizzi, L., Mangano, C., Miller, H., Pallavicini, P., Perotti, A. and Taglietti, A. (2002) *Angew. Chem. Int. Ed.*, **41**, 2553.
20. (a) Harrison, W.D., Hathaway, B.J. and Kennedy, D. (1979) *Acta Crystallogr., Sect. B: Struct. Crystallogr. Cryst. Chem.*, **35**, 2301; (b) Harrison, W.D. and Hathaway, B.J. (1979) *Acta Crystallogr., Sect. B: Struct. Crystallogr. Cryst. Chem.*, **35**, 2910–2913; (c) Potočňák, I., Burčák, M., Baran, P. and Jäger, L. (2005) *Transition Met. Chem.*, **30**, 889.
21. Fabbrizzi, L., Foti, F., Patroni, S., Pallavicini, P. and Taglietti, A. (2004) *Angew. Chem., Int. Ed.*, **43**, 5073.
22. Coughlin, P.K., Martin, A.E., Dewan, J.C., Watanabe, E.I., Bulkowski, J.E., Lehn, J.M. and Lippard, S.J. (1984) *Inorg. Chem.*, **23**, 1004.
23. Perrault, C., Johnson, A.E. and Doré, G. (1969) *Perrault's Fairy Tales*, Dover Publications, New York.
24. Sessler, J.L., Gale, P.A. and Cho, W.S. (2006) *Anion Receptor Chemistry*, Royal Society of Chemistry, Cambridge.
25. De Santis, G., Fabbrizzi, L., Iacopino, D., Pallavicini, P., Perotti, A. and Poggi, A. (1997) *Inorg. Chem.*, **36**, 827.
26. Fabbrizzi, L., Gatti, F., Pallavicini, P. and Zambarbieri, E. (1999) *Chem. Eur. J.*, **5**, 682.
27. Amendola, V., Colasson, B., Fabbrizzi, L. and Rodriguez-Douton, M.-J. (2007) *Chem. Eur. J.*, **13**, 4988.
28. Allevi, M., Bonizzoni, M. and Fabbrizzi, L. (2007) *Chem. Eur. J.*, **13**, 3787.
29. Flood, A.H., Stoddart, J.F., Steuerman, D.W. and Heath, J.R. (2004) *Science*, **306**, 2055.